Haying with Horses

Haying with Horses

Lynn R. Miller

Haying with Horses

Copyright © 2000 Lynn R. Miller

Publisher
Small Farmer's Journal Inc.
PO Box 1627
215 N. Cedar
Sisters, Oregon 97759
541-549-2064
800-876-2893

email: agrarian@smallfarmersjournal.com

Printed in the United States of America

Second Edition,
Fourth printing May 2016

ISBN 978-1-885210-17-3 Soft Cover

*On the cover: Front cover photos by Kristi Gilman-Miller. Top photo - Jess Ross, Lynn Miller
(buckrakes) Justin Miller (side delivery) at Singing Horse Ranch on the east side of the Cascade
mountains in Central Oregon. Bottom photo - Tony Miller, stackmaster extraordinaire on the ladder
with three different "experimental" haystacks at Singing Horse Ranch. The big white dog was Ziggy,
Tony's Great Pyrenees.*

*Back Cover: Top - Adriance Mower. Bottom - Flying Dutchman Hay Loader. Left - Justin
Miller with Red and Molly on number 9 McCormick Deering International High Gear at Singing Horse
Ranch in 1999. Right - Lynn Miller with Cali and Lana mowing same place in 1997. Photos by Kristi
Gilman-Miller.*

Inside Front Cover: John Deere Side Delivery Rake in factory colors.
Inside Back cover: International number 9 mower in factory colors

Dedication

*For my children who, whether they like it
or not, were me and are me into different
times and different places.*

*For my son Maxwell Henry
who left us in 1968*

*For my son Ian Lewis
who left us in 1985*

*For my son Justin Everett who continues
to demand humility*

*For my daughter Juliet Christene who
continues to demand poetry*

*For my daughter Scout Gabrielle who
continues to demand singing*

*Without you I would be a song without
music, a hollow heart,
a father without children.*

Other books by Lynn R. Miller

technical
The Work Horse Handbook
Training Workhorses / Training Teamsters
Horsedrawn Plows & Plowing
Horsedrawn Tillage Tools
The Horsedrawn Mower Book
Starting Your Farm
Ten Acres Enough: (out of print)
Why Farm: Essays
Farmer Pirates & Dancing Cows: Essays
Old Man Farming: Essays
Horses At Work (out of print)

fiction
The Glass Horse (a novel)

poetry
Thought Small

Also from Small Farmer's Journal
Horsedrawn Equipment Manual Reprints

Haying with Horses

PREFACE

This book is not an authorship by any one person, certainly least of all me. While there are chunks that no one else will want to take credit for, much of this work was an editing job. It is reasonable for anyone new to our publishing efforts to ask what credentials I, Lynn Miller, bring to this project. I have been depending on horses in harness, as motive power, for thirty years. Twenty five years ago we began a publishing venture the cornerstone of which is the international quarterly Small Farmer's Journal. I have travelled around North America doing work horse workshops and assorted presentations on animal power and small farm issues. I have authored twelve books including the *Work Horse Handbook* and *Training Workhorses/Training Teamsters*.

The success of Horsedrawn Plows and Plowing, the text which immediately precedes this one, guaranteed that this book also would be born. Separate from that, this book deserved to be done because the information, spread as scattered curiousities, was sure to gain an applicable vitality when collected and presented. The same is true

of the Plow Book. Underlying these publishing efforts is an idea that the parts, pieces, ideas and methods which are key to animal-powered agriculture MUST be preserved. We understand the risk of bottling up and shelving important information about methods. We don't want to do that. This book is less of a cataloging or recording effort and more of an instruction manual. The way we keep these ideas alive is to somehow encourage people to use them.

We hope to be able to publish a complete series which would include all aspects of animal power. But the challenges of such a project are being felt. And, though the future of the Plow Book is assured, we are less certain about the interest in this volume and subsequent ones. That said, it remains for us an important goal.

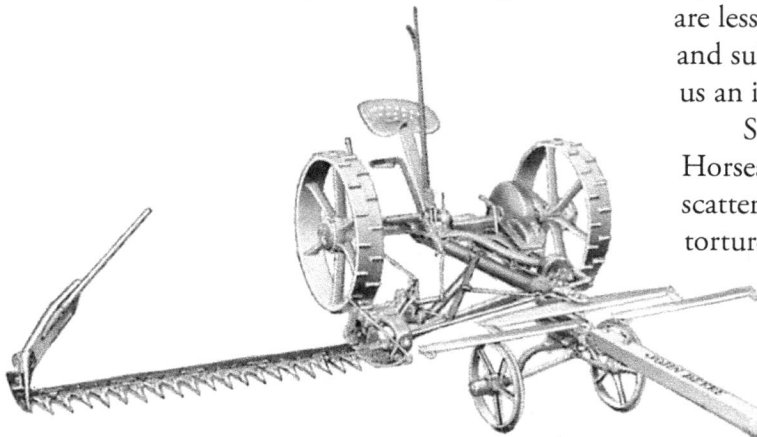

Selecting the subject of Haying with Horses and then putting into book form all the scattered, related information we hold in our tortured library and archives would be a difficult task in and of itself. Our insistence that this volume also serve as an accessible, all encompassing operator's manual has proven we are gluttons for punishment. But we couldn't help but try. In the right hands we happily suspect it may work. It may prove useful. But we are equally sure that careful examination will unearth embarrassing mistakes and awkward omissions. For these we apologize in advance. Knowing, with a project this size, that perfection would elude us we still felt compelled to push on.

Why, you might reasonably ask?

It is my thirty years personal experience using horses in harness which has repeatedly tweaked my missionary lobe when it comes to horse power. I am forever asking myself, whilst in the midst of some highly pleasureable and practical horse working routine, why aren't more people doing this?

Chapter One, the introduction to this book, delves into the subject of why anyone would choose, in the "virtual" mad twenty-first century, to make hay with horses. And it also talks about how to organize the great

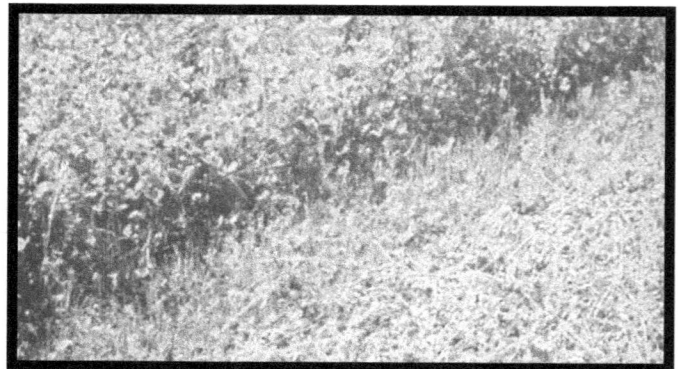

photo by Kristi Gilman-Miller

attraction of this way of working in order to use, to full advantage, the fact that so many people would like to help those of you crazy enough to go this way.

I have done it, this haying with horses business, should you want to know, because I loved it. I have known and continue to know what it is to be completely joyful in the working.

As was mentioned, I have attempted with some success and some failure to share this way of working with others in workshops and demonstrations around North America. That experience, and my desire to share through teaching while I continue to learn myself, gave shape to another wish for this book. I want to imagine this book could perform somewhat like a great big month-long haying workshop on some ranch, with every conceiveable equipment variable on hand to poke at and talk about, and poor and lovely crops to actually work with, and cantankerous and wonderfully willing and terribly green animals to put to the work, all with hundreds and perhaps thousands of interested folks. I was, and am, in that wish being purely selfish because I hunger for the changing look of folks faces in those settings. How they go from thirst to confusion to insight to consternation to exhiliration to fatigue to refreshed to wonder to determination to gratitude and finally to joyful exhaustion.

But with a book, we authors don't get to see those changing faces. We are stuck with our imagination of them. In my case I tend to fuel it mid-process by saying to myself 'wait til they read this or wait til they see these pictures!' That may explain for you why some things are in this book.

This book is a mixture of old and new imagery and information. We make no claim or point of this other than to say it is so and that we did it on purpose. To those who would argue that our goal should be for a progress which delivers us an ever more efficient system of haying, and indeed farming, we say phooey. Efficiency, if it remains a measure of how few man hours are required to get the haying done, is a misplaced concept of questionable value. This text, and indeed all of our publications, most notably the quarterly ***Small Farmer's Journal***, are laced with and driven by this philosophy. The dedicated independent devotee propelled by joy and purpose will always, ALWAYS out perform the efficient corporate citizen.

This is not a book about how to work horses. That complex subject requires more than one book in and of itself. That is why we published ***The Work Horse Handbook*** and ***Training Workhorses/Training Teamsters***. Someone new to the subject of haying with horses or indeed the practise of working horses or mules in harness will need more information than they will find here. They will need to know about the animals, their care, the harness, its function, the dynamics of hitch and pull, and how to drive the horses. Without this critically basic foundation the waterfall of implement and procedure information found in this big book will be a dangerous confusion.

But also, please don't let the complexity and weight of all this information discourage you. Read what you think applies and keep the rest as future reference. We want this book to encourage people, not discourage them. So it might surprise the reader to hear that we actually had to cut hundreds of pages of equipment technicalities and variables to get this down to the size that it is. Again, as with the Plow Book, space dictated editing cuts which notably included the evolutionary history of each of the implements presented here.

Haying with horses need not be daunting or frightening or too difficult. Especially if you are fortunate to have good help, good animals and good implements. Haying with horses can be its own reward even more than the finished, stored crop.

Mowing a field of fine hay gives the teamster the same sense of profound satisfaction that would come from plowing a rich soil. There is an easy and complete communion with process, partners and environment which fills the receptive working soul to over full. Putting that crop, after successful curing, in to its storage takes that farmer far up to the stratosphere of self satisfaction and completion.

As with any book of this scope, it is not a singular authorship, far from it. Many people here and now, and stretching far back in time, have contributed and slaved to make this what it is. I thank them as I know you do. I thank Doug Hammill, Bulldog Fraser, Tony Miller, Bud Evers, Jess Ross, Pete Lorenzo and scores of others. I thank Kristi Gilman-Miller for her incredible photos and editing help. I thank the Small Farmer's Journal staff, Suzanna Clarke, Amy Evers, Kathy Blann & Lisa Booher for hours of nail biting work on this project.

If we have succeeded in helping a handful of people experience the wonder of *Haying with Horses* our goal has been met. Because those people will surely infect others with the passion. And in this way, together we all keep a delicious way of working alive.

To that end I personally wish to thank each of you who have wished for and taken this book to heart. You made me do it, you made us do it. And we are pleased for the opportunity to give it our best.

- LRM Spring 2000

photo by Kristi Gilman-Miller

Young team on the scatter rake. Storm brewing. Mammoth Beaver Slide stacker.
Big Hole Montana. Hirshey Ranch 1983

Pretty Lady, Laurie Hammill. Clydes on the dump rake. Late afternoon.
Rocky Mountain front range, 1997.

You don't need to be born to this work in order to be able to do it. You can learn it, you can do it. And when you do your homework, and protect yourself with competent help, this whole process will be even more pleasurable than you imagined.

Introduction

Why Use Horses To Make Hay?

Politics & Agriculture
For the Crop
Satisfying Work
How Many Horses?
Organizing the Hay Crew

This is a book about making hay in a number of systems where horses and/or mules are the motive power source.

Making hay, if you're from a farm background, is no big leap for the imagination.

But putting horses in harness and using them to power the haymaking, that is a leap especially for the twenty-first century. It shouldn't be. For those who come to the subject of Haying with Horses with some measure of anxiety about public perception, it is necessary to discuss briefly the curious economics of social prejudice.

Politics and Agriculture

Interesting, and potentially useful, low-tech agricultural alternatives are by the establishment immediately considered suspect and/or dangerous. This should be seen as clear evidence that a political situation exists which holds a bias against social change in modern agriculture. For anyone considering the possibility of using animal power in <u>any</u> farming operation, it is important to prepare for a systemic bias from all government agencies and most banking institutions. The initial response from these quarters, and notably academia, is that animal power in agriculture is silly and archaic. They say only someone too poor or too stupid would attempt it. This is an unfortunate excuse and it conceals a far more important concern. Government and industry leaders have been busy on the propaganda front. Over forty years have been invested in selling to society the inevitability and supremacy of a chemically intensive high-tech industrialized agriculture. And the last fifteen years have been spent selling the notion that farming would be better done in the third world with cheaper labor and a cheaper land base, shipping food to the so-called 'developed nations', thereby 'saving' the North American land base for higher and better use as open space and population centers. Within this scenario any attractive ideas which return people, as craftsmen, to the North American landscape threatens the growing hold of the federal governments and big business.

The most prevalent argument against people on the land is the 'modern efficiency dictum': the fewer man hours the more efficient. The logical progression of this argument of course is that man's time would be better spent elsewhere. And that's where the theorists have terrible difficulties. Where is elsewhere?

Mankind belongs with fruitful, gainful, creative, and satisfying work! Why should that be difficult for bureaucrats, economists, and academics to understand? And NO ONE need apologize for choosing to work at something which contributes to the community, creates beneficial goods, is nonpolluting, builds heritage and legacy, keeps important knowledge alive, and gives satisfaction! But that said, be forewarned. Extension agents, bankers, and government bureaucrats will still view you as strange and deny your application or information request 9 times out of ten when they hear you choose to work horses or mules. So, heck with 'em! Let's just invite those who enjoy the same work and go get it done.

For The Crop & Satisfying Work

Let's divide the subject and see if it can be brought back together after a few paragraphs. Working horses or mules in harness; making hay; those are the two subjects.

First the hay segment. What follows this chapter is a great pile of theories, ideas, practises, issues, methods, and mechanics all tied to making a hay crop from various standing forage. The technicalities we hope, are covered there. Here we discuss why anyone would look to the job of making hay with any gleeful anticipation - because they do. And they are those people, in part, who have allowed themselves to choose their work rather than have their work choose them. That's not to say that the latter is always a negative but rather that the former may be a clear indication that within this person resides a heart open to the direct, magnificent pleasure that may come of satisfaction with work. Most any job or task can be made, or thought of as, unpleasant. And the reverse is also true.

Haying can be drudgery. Haying in a certain way can be more drudge for some people. But the putting up of cured forage can also be akin to a culinary art, to exterior decorating, to landscape sculpting, to process poetry, if the ingredients are understood and worthless notions of efficiency are thrown out.

For five of my early farming years I put up hay with tractors and balers. Horses weren't involved. I became a shade tree tractor mechanic and an 'expert' baler mechanic. If the neighbors couldn't get their baler to tie correctly, I was invariably called to help. After two seasons of driving the tractors, running the balers, and handling the bales and bale crew I developed an extreme dislike for the whole sheebang. For the last twenty years I have used the horses and put up loose hay in barns and in outside stacks. I have NEVER tired of the work or the working. I don't like picking up hay bales but I'll happily work hard on a pitchfork in 10 below

The author on a number 9 McD mower cutting a thick rich mixed grass/legume crop with Red Clover in full bloom.

The mowing is timed to take maximum advantage of root mass. The full bloom of clover indicates that the nitrogen fixation underground has reached its peak. Cutting at this stage releases that nitrogen as well as maximized soil-borne minerals, vitamins and chelating agencies. Also please note the pictorial evidence of just how hard the author is working while mowing.

zero or 102 degree weather. I don't like the smell of tractor exhaust or the dusty air of the slamming baler, but I love the smell of sweaty horses and loose hay. Matter of personal preference. But a matter, none the less, which is all important. If you don't like the work you'll find reasons. If you do like the work you'll find reasons.

It is possible to look upon haying as a craft. We suggest that to look at it in this way is to give yourself the best chance for satisfying work.

Haying can be seen as the simple direct process of cutting curing and storage (possibly selling) of the forage from your lands. Haying can also be in a larger context. In a few places in this book we propose that you think about the possible effect of that action on the land itself. It is this author's belief, born of experimentation and experience, that the act of cutting of the forage can be timed to have a dramatic effect on soil tilth and fertility. This theory is covered on future pages but here we wish to include the observation that the action of the horse and mule hooves on the soil while pulling hay implements is far gentler than the rolling pin action of tractor tires. Imagine that how we hay can improve the land in several ways. And all while we're enjoying ourselves.

How Many Horses?

"Just how big a deal is this haying with horses business? I mean how big a deal does it have to be? You see I've got forty acres of hay I put up with a tractor and baler now. Do I have to get several teams of horses or mules and a whole great big line of machinery to change over?"

It's not a big deal. In many cases it is possible for people to slowly dovetail into haying with horses by giving them pieces of the work. For example a team and a rake can be a big, cost efficient aid to most tractor haying operations. A team and wagon may entice neighbor folks to help get the bales picked up. A team and mower can mow some of the hay. There are thousands of outfits which are mixed-power (part tractor/part horse or mule). They start out that way because the people are looking for ways to work their animals. They want to work them because they enjoy them. So they work the animals in to the routine. Some of those people eventually move towards working the animals more and more. Some just stay with the mix. The hangup to increasing use of, or dependency on, the work animals seems to come, most often, from being unable or unwilling to allow the "suggested changes" to take place. Because the working animals WILL suggest changes. They suggest a more human pace to work. They suggest a closer connection to the crop and the land. They suggest spreading out the work. They suggest getting folks over to help. They suggest a complete rethinking of how and why the haying happens. They suggest craft and community over industry and corporate ethics.

Forty acres of hay can be put up with two animals (a third 'pinch hitting substitute' can save from unexpected

This photo shows the opening pass across that same hay field from the previous two photos. Note that the thick crop of mixed grass and legumes still stands even after mowing. This second cutting yielded in excess of 4 tons per acre. In the background are the stacks from the first cutting.

down time) one mower, one rake and a choice of buckrake, hayloader or baler. Not a big deal. Two teams can put up that same hay in half the time. Or twice as much in nearly the same time.

With larger operations, such as the Triangle Ranch and Hirshey Ranch examples illustrated in Chapter 12 and throughout the book, the number of animals and scope of the equipment can be substantial. It can become a big deal - if so chosen.

But even when it is a big deal, it works, and it works with and because of people. Although it doesn't have to be hard work, it is labor intensive (which as we've said is good - it's a good place for people.) You can be casual about haying crew concerns and make it a family and friends "work party" affair or you can be quite organized and formal. What follows is a set of examples of formal where scale dictates some formality:

Hay Crews
Crew Organization and Duty

The first thing to consider in planning the size of a haying crew and its organization is the acreage or tonnage of hay that has to be handled daily. This will depend to a considerable extent upon the total quantity of hay that is to be made and the number of good haying days that are likely to be available.

Alfalfa is a valuable and fragile crop. It should be cut and off the ground as quickly as is consistent with the equipment and help available in order to interfere as little as possible with the growth of the next cutting. With wild hays and legume/grass mixes timing is less of a factor. But with every hay crop there is a certain stage of growth at which cutting and stacking will produce the highest quality that can be obtained. The length of this stage varies greatly for different hay crops. With alfalfa it is usually only a few days, whereas with wild hays and mixes it may be considerably longer than this.

Methods of crew organization

There are two ways to approach crew organization. In the first method the cutting and stacking operations are carried on simultaneously with separate crews. In the second method the cutting and stacking is alternated and the same crew does both. Obviously with the former, different skills and abilities are required with each group, whereas with the latter, everyone needs to be capable to run all equipment and handle the animals.

With the first method, less equipment is required but a considerably larger crew is necessary to operate efficiently than with the second method. The larger crew is not necessarily objectionable in the larger haying operations but may be so on the small ones.

The second method has the greatest advantage in those localities where heavy dews or fog are common during the haying season. Such conditions make it impossible to start stacking until very late in the morning, but it need not necessarily retard mowing operations. By using this method it is possible to send the entire crew out to mowing during the mornings and to stacking in the afternoons when the hay is in the best condition to be put into the stacks. This method requires a greater investment in equipment, but the extra cost is usually more than made up in savings obtained through more efficient crew management and in producing a better quality of hay.

In situations (highly unlikely but possible for settings like living history museums) where it is customary to cure the hay in cocks, a somewhat different crew organization is necessary. In such sections, the hay is usually cut and then raked almost immediately, and after partially curing in the windrow it is made into cocks by a pitching crew which is separate from the cutting crew. After the hay is sufficiently cured to be stacked, the cutting and pitching crews are combined into a single crew for stacking.

The amount of necessary equipment will depend on the quantity of hay that is to be handled daily.

Mowing

The number of acres mowed daily with a given size of mower will vary considerably according to the kind and the yield of hay, the topography of the land, the size and number of horses, the weather conditions, and the length of the work day. On the average, a 5-foot horse-drawn mower will cut slightly over 10 acres in a 10-hour day; the larger mowers proportionally more. An easy way to figure the rate per day is to allow 2 acres daily per foot of width of the cutter bar. Where several mowers are kept going, the mowing operations can be speeded up considerably by having an extra man to sharpen the sickles and make repairs. (If the luxury is available the hay camp cook can often be employed for this purpose during his or her spare time but it necessitates a certain type of individual - one who can rebuild and replace a pitman stick in minutes, wash hands, and build a big pot of outstanding soup, all without any grumpiness. In some hay camps the cook gets paid more than the foreman - or should.)

Where the mowing and stacking are done simultaneously (i.e. those situations where the hay will cure quickly) the rate of mowing will depend entirely on the progress of the stacking crew. Just enough hay should be cut in advance to keep the crew provided with a supply of properly cured hay. Because of variations in the yield per acre on various parts of the hay land and in the rate at which the hay will cure on different days, the number of mowers that should be kept going will vary. This should be taken care of by shifting one or more of the rake or buckrake teams to an extra mower as necessity requires.

Raking

The rate of raking varies even more than that of mowing because of a greater variation in the speed the horses

Making hay with horsepower attracts people. Volunteer help for the few days that it is necessary to build hay stacks or fill a barn is often available to most such operations, especially if food, beverage, and camaraderie is planned for.

Amy Evers with Red and Molly and a #9 mower on Singing Horse Ranch. Evidence that this work can be a piece of heaven.

walk when pulling a rake. The area raked per day with a dump rake will vary from about 2 acres per foot in width when the horses walk slowly to 3 acres when they are made to walk fast. A side-delivery rake will average from 20 to 25 acres per day.

One rake for each two mowers will usually be sufficient, though an additional rake may be necessary at times for raking the scatterings.

Stacking

The size of the stacking crew may vary without materially affecting the cost per ton of stacking. The things to consider in organizing the stacking crew include providing enough people to put up the hay in a given time and to plan the duties of each person so there will be a minimum of lost time for each.

The number of people that should be provided for the different duties in stacking will vary, according to the size and location of the stacks, the kind and yield of hay, the type of stacker, and other equipment and methods used.

On ranches where the stock is fed in lots during the winter months, a great deal of time and labor in feeding is saved if a few or all the stacks are built in permanent stack yards adjacent to the barns or lots. Under such conditions, however, the length of haul in stacking is usually considerably greater than where the hay is stacked in the field.

The usual quantity of hay hauled with a buckrake at each load will vary according to the kind and dryness of the hay and the size of the horses and equipment used. On the average, a good team of horses will handle from 600 to 1,000 pounds of hay per load and for a 4-horse buckrake approximately a ton per load.

The tonnage handled daily by each sweep is governed by the size of load hauled, the kind of stacker used, the length of haul, and the interference of the buckrakes with each other, where several are used with the same stacker.

For the average length of haul, the quantity of hay handled with a 2-horse buckrake will vary between 20 and 30 tons per day when using slide stackers (also known as a Beaverslide) and between 15 and 25 tons when using other types of stackers.

Where a slide stacker is used, all that is necessary for the buckrake to do to deposit the hay at the stacker is to drive up in front of it and then back away, the plunger being able to pick up the hay and push it up on the stack from this position. With other types of stackers it is usually necessary to give the hay a second push after backing from it the first time. This requires some additional time and lowers the number of loads hauled by the buckrake.

If two or more sweeps are being used and the haul is short there may be some interference of the buckrakes when depositing the hay at the stacker. These delays will slow up the work of each buckrake and should be avoided as much as possible.

Where slips are used the quantity of hay handled daily by each slip is less than with a buckrake because of the longer time required to load and unload them. Where enough field or spike pitchers are provided to help with the loading this extra time will amount to very little. The average slip handles from 10 to 15 tons per day where there are extra pitchers to help in loading.

Where wagons are used for hauling, the labor required for loading and the length of the haul is usually greater and the quantity of hay hauled per day by a given size of crew is less than where buckrakes or slips are used.

Stacker Help

Usually one man, or sometimes a boy, is able to take care of the operation of the stacker and the work around the stack when not more than two buckrakes are used, but where three or more are used it frequently pays to have an extra man to help the stacker operator.

The number of men to use on the stack depends on the kind of hay being stacked, the size of the stack, the rate at which the hay is being brought in, and the type of stacker used.

Where alfalfa or mixed hay is being stacked and the hay is tangled, one man for each buckrake is the rule and in extreme cases two men per buckrake may be required. Where wild hay is stacked and frontboard and backboards are used with the stacker, one man can stack the hay from three buckrakes.

Changing Teams

Where sufficient horses are available a greater acreage of hay can be cut if the horses used on mowers are changed twice a day. Changing of teams on all implements except perhaps the sulky or dump rakes may be advantageous especially if long days are necessary.

Crew Organization and Management for Alfalfa Hay in 1925 & 1926

An efficient method of putting up alfalfa hay on an extensive scale was used on the Bruno Sheep Ranch, at Grandview, Idaho, where they had about 500 acres in alfalfa. The crop was cut three times, making a total of 1,500 acres of hay to be cut, cured, and stacked each season. Most of this hay was fed to sheep during the winter. The first cutting was made about June 1 and yielded around 2-1/2 tons per acre, the second cutting came about July 15 and yielded around 2 tons, and the third cutting about September 1 yielded around 1-½ tons, making a total of about 6 tons per acre for the season.

On this ranch the same haying crew was employed throughout the season giving better results than hiring new men for each cutting. In 1925 the crew consisted of 14 men who worked about eight hours per day. For cutting, five 5-foot mowers were used. Two 10-foot dump rakes followed directly behind the mowers. After the hay was raked into windrows, seven men put the green hay into cocks, where it cured about six days. This crew cut, raked, and cocked 40 acres of hay daily, all of which was later stacked in a single stack yard.

In 1926 certain changes were made in the raking and cocking crew which lowered the cost of haying and still maintained the same high quality. The same five mowers

(left to right) Harold Capps, Lynn Miller (on buckrake) Bob Oaster and Tony Miller. (Polly and Anna, three year old Belgians.) Part of one year's hay crew at Singing Horse Ranch.

are used. Two side-delivery rakes in place of the sulky rakes followed about one-half day behind the mowers, one goes around the field in the same direction as the mowers go and the other in the opposite direction, thus throwing one windrow on top of the other. The hay in the windrows is then bunched by one dump rake, thus avoiding the use of hand cockers. In other words, two men with side-delivery rakes and one man with a dump rake are now doing the work of seven hand cockers and two dump rakes, thus saving six men per day.

When the hay was stacked, seven slips built on runners are used to haul the hay to the stack. Three spike pitchers are in the field to help the drivers load their slips. Each slip made a round trip every 15 minutes, including about one and one-half minutes for unloading at the stack. A Mormon derrick was used for stacking. When the slip came to the stack the driver and trip man fastened the pulley hook to the two chains on the bed of the slip and the entire load of 1,000 to 1,500 pounds was pulled up on the stack. The derrick team driver and two men on the stack completed the crew. This crew stacked from 80 to 90 tons of hay daily depending on wind and weather.

Some ranchers had the idea that the Mormon derrick was cumbersome and hard to move, requiring at least two teams. This did not seem to be the case on this ranch, since about 10 minutes are required from the time the stakes are pulled and the stacker moved by one team to a new set until the first slip load of hay has been dropped for the new butt.

Some practices and devices used on this ranch could be of value on other ranches, since they tended to reduce the cost of haying. The old sections in the sickles were replaced with new at the beginning of the season and again when the season was half completed. The mowers were started out each morning with sharp sickles which were changed twice during the day. A power-driven emery wheel was used for sharpening, and when the mowers were in operation one man spent most of his time grinding the extra sickles. A one-half inch iron pipe substituted for the grass board prevented loss of time from breakage. A weight fastened to the under side of the mower seat relieved the horses of some of the weight of the mower. A truck was used to take the men to the ranch house for midday meal and back again, while the horses were fed and watered in the field.

Two men looked over the outfits at the end of each day and any weak bolts or parts were replaced to prevent their breakage and loss of time while the equipment was in use. No stops were made for small breaks during stacking unless absolutely necessary. An extra pair of sling chains were used at the stack so that the slip driver did not have to wait for the return of his chains before starting for another load.

Montana Crews

On some of the ranches in Montana, different sized crews were and are used for mowing and raking than for stacking. Depending on the elevation, two or three cuttings are made during the season. Except perhaps on some large ranches, the same men are not employed throughout the haying season.

For cutting, one 5-foot mower is used and one 10-foot sulky rake follows a few hours later. The hay is put into windrows, where it cures in one to four days depending on weather and wind. About 10 acres of hay are cut and raked daily.

For stacking, five men are used. The hay is brought into an overshot stacker by two sweep rakes. Two men stacking the hay and a man or boy to drive the stacker team complete the crew. From 30 to 40 tons of hay are stacked by this crew in one day.

Best practices call for stacking each cutting separately. On irrigated fields, the first crop makes the best hay for horses, the second crop is the poorest, as it is not so palatable, and the third crop is superior to the first crop.

Most ranches use sweep rakes and overshot stackers, as little attention is given to the quality of hay obtained by using different methods of stacking. Where the flood system of irrigation is used the alfalfa plants practically cover the ground, so that little dirt is picked up by hay on buckrakes as occurs in Idaho, where the corrugation system of irrigation was commonly used. The quantity of leaves shattered by using a buckrake is about the same wherever used. Using slips or wagons prevents most of this loss.

Colorado Crews

In 1926 most ranchers in and around Greeley, Colorado, put up alfalfa hay in the same way. The crop was cut three times. The first cutting, around June 20, yielded about 1-3/4 tons per acre; the second, from July 25 to August 1, yielded about 1 ton; and the third, around September 10, yielded about three-fourths ton. The entire crew for cutting and stacking was usually made up of 11 men. Two 6-foot mowers cut the hay, which lay in the swath one-half to two days, depending on the weather. The hay was put into windrows with a side-delivery or dump rake and then cocked with a dump rake. The hay remained in the cock from four to six days. This crew cuts and rakes from 20 to 24 acres of hay daily. For stacking, three slips are used and one spike pitcher helps the drivers in loading them. The stacker is of the Mormon type. Two men on the stack and a man to drive the stacker team complete the crew. About 30 tons of hay daily are stacked by this crew.

In some localities of the San Luis Valley there were large acreages of alfalfa hay. Usually only two crops are cut during the season, the first around July 10 and the second around August 15, each yielded about 1-½ tons per acre.

The usual crew was made up of 10 men. The hay was cut with two 6-foot mowers and was raked into windrows with a 12-foot dump rake, where it remained from four to five days. About 24 acres of hay were cut and raked daily.

Laurie Hammill mowing with two Clydesdales in Montana.

For stacking two sweep rakes were used together with an overshot stacker. Usually two men are on the stack, but some ranchers, by paying a little higher wages, employed only one man for this work. One on a dump rake to clean up the scatterings and one on the stacker team completed the crew. This crew stacked from 35 to 45 tons of hay daily.

The Big Horn Cattle Ranch at Walden, Colorado, in 1925 was a good example of the methods used in making wild hay on the ranches of North Park. About 7,000 acres of hay, averaging around 1 ton per acre, were put up yearly. During the growing season the hay was irrigated by flooding.

Six complete crews were employed during the haying season, each crew being made up of 14 men. For cutting, four 6-foot mowers were used in each crew and the hay cures for several days in the swath. After curing, the hay was raked into windrows with two 10-foot sulky rakes. When ready to stack, three homemade buckrakes were used to haul the hay to the foot of the slide stacker. A man driving the plunger team pushed the hay up the slide and delivered it at a point indicated by the two men on the stack. Two rakes for raking scatterings complete the crew. This crew puts up from five to seven benches (stacks), each of 10 to 12 tons, daily. Usually several of these benches are joined together to make one long stack.

In addition to the field crew a man is employed to grind and sharpen sickles. This man also carries water, drives the lunch wagon to the men in the field, and does other odd jobs. Each morning the mowers start out with sharp sickles

which are changed twice during the day. If the mowers get too far ahead of the stackers, one machine is taken off and a fourth buckrake put on. At noon fresh teams were put on all outfits with the exception of the dump rakes.

On many ranches in Routt County, Colorado, large acreages of irrigated timothy and alsike clover hay yielding from 3 to 5 tons per acre were grown when conditions were favorable. The most efficient haying crew in this locality was usually made up of nine men.

For cutting, two 6-foot mowers were used. The hay then lay in the swath for about three days, after which it was raked into windrows with a 10-foot dump rake. Three push rakes were used to haul the hay to the stack. Because of the heavy yield, buckrakes having four wheels were most frequently used. A slide stacker was commonly used to elevate the hay to the stack. A man to drive the plunger team and two men on the stack completed the crew. This crew put up about 60 tons of hay daily. Where only two buckrakes were used, the crew put up about 40 tons of hay in nine hours.

Nebraska for Wild Hay

The same system of putting up wild hay was used on most of the ranches in the sand hills of northwestern Nebraska. The size of crews varied slightly, but the tonnage of hay stacked daily varied between 4 and 5 tons per man irrespective of the size of crew.

A common-sized crew was made up of eight men. The hay was cut with two 6-foot mowers and lay in the swath one-half to two days. After curing it was raked into windrows with a 10 or 12 foot sulky rake. Two buckrakes were used to haul the hay to the overshot stackers from the windrows. In this area the backboard and frontboard were used for stacking so that only one man was required on the stack. A man or boy to drive the stacker team and a man to rake the scatterings completed the crew, which put up about 32 tons of hay daily. Increasing the crew to 13 men by adding two additional mowers, one sulky rake and two sweep rakes, about 60 tons of hay were put up daily. To handle this quantity of hay, the man on the stack had to be exceptionally good.

Whew! Lots of variables. And we shared them to give that top side picture to show Haying with Horses does work as an option for big operations as well as small. You might reasonably wonder about the cost of a 13 man crew to put up 60 tons of hay per day. In ten days time 600 tons of hay would be put up with, aside from what that crew might be paid, a total equipment and purchased input cost which would be negligble by any comparison to the tractors, swathers, haybines, balers, bale handlers, trucks etc. etc. which are traditionally employed on large haying operations today. The amortization of the cost of all that stuff has long puzzled ag economists because the hay it produces seldom ever covers the cost of the machinery and fuel.

But on the crew front it needs to be pointed out that with most of us our need for help in the haying is short term and may be filled more than adequately by friends, family and folks looking to share and learn. With a little planning and invitations, maybe even advertisements, a substantial contingent of help can be brought together from the ranks of folks who want to share the experience. With tents, lots of good food, a relaxed schedule and opportunity to learn, you'll be surprised at how many people will volunteer. And, if you have a shortage of horses and equipment think about making your hay, or a portion of it, part of a club or association weekend field project. Invite folks to bring their horses, mowers, and rakes and have fun helping you get your hay in.

But be careful. One year we made the mistake of sending out just such an invitation. Five teams and mowers went to the field and in two days a hundred acres of hay was down! Then everybody left and we were stuck with having to rake and stack that hundred acres alone! We lost quite a bit of it to the sun. But, live and learn. So if such an idea sounds good to you make sure that you have some hay down and ready to stack before the party starts. Divide folks into crews with some stacking hay, some raking hay and some mowing. Make it all work to your benefit and have fun.

Doubtless there are many folks reading this with no experience and they are scratching their heads. We apologize if it seems we jumped into the middle of a complex subject. Unfortunately it is difficult to know where to start in a book such as this. But take heart what follows is a good effort to explain all the stuff you've just read about.

As for your beginnings, we repeat; this is not the place to learn about how to work animals in harness. That is another assignment which we trust you will take seriously. We do hope your travels through this text will give you the correct sense that many people are doing this and that though it is complex you can learn how to do it.

You don't need to be born to this work in order to be able to do it. You can learn it, you can do it. And when you do your homework, and protect yourself with competent help, this whole process will be even more pleasurable than you imagined.

Lynn Miller with Cali and Lana after having fed the cattle on Singing Horse Ranch.

GRASSES & CLOVERS

Grasses in General

Legumes in General

Description of Varieties of Grasses & Clovers

Chapter One

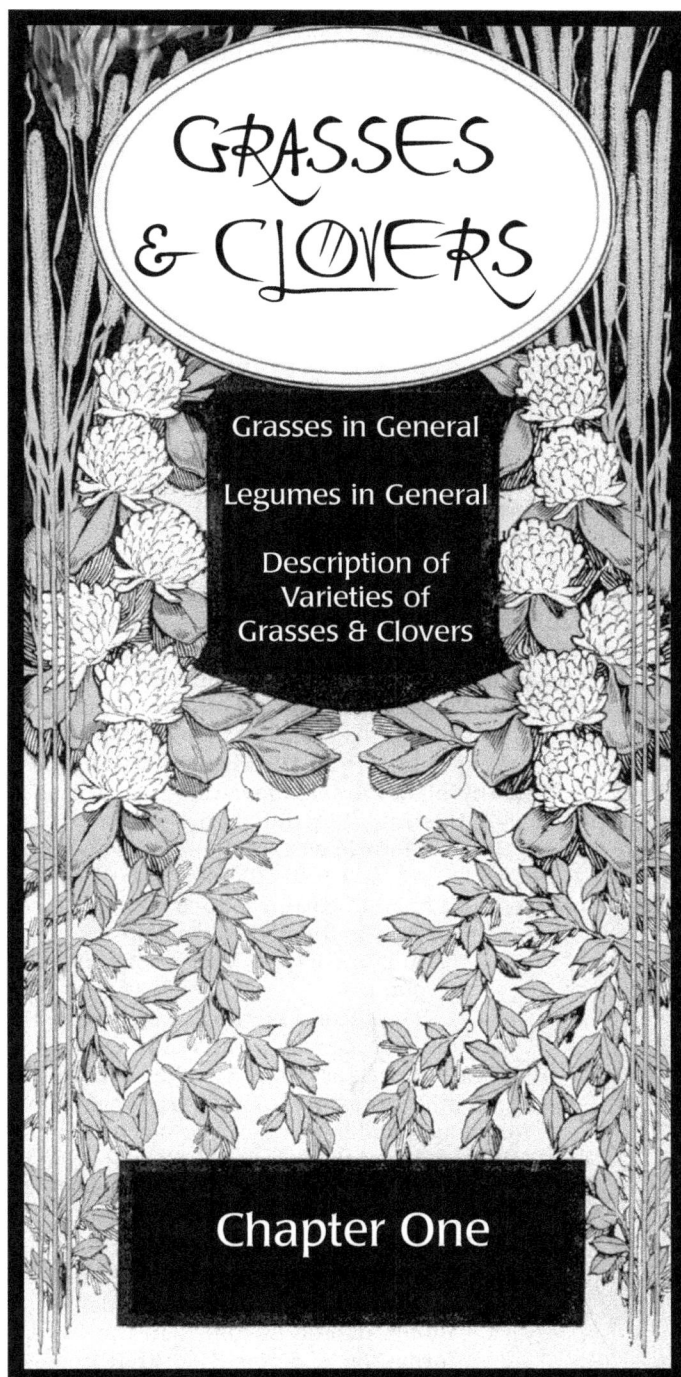

The bulk of the material in this chapter is an edited compilation of articles originally authored for John Deere Co. and The International Correspondence School agricultural course from earlier in this century.

Grasses in General

The term **grass,** as it is commonly used, is somewhat indefinite, and in order to avoid confusion we'll try to make our exact meaning clear. In some cases, the word grass is applied to all of the common plants that are fed to livestock; in other cases it is applied to a particular family of plants known to botanists as the **grass family**, all the members of which have certain common characteristics. The former use of the term is incorrect and the latter term confusing as it includes members of the family most North American farmers do not normally think of as grass (i.e. corn and wheat). The word grass will be used here to designate only certain members of the grass family.

The grass family is one of the largest families of plant life. Included in it are many of the commonest plants on the farm, such, for example, as timothy, oats, wheat, corn, lawn grass, etc. The grasses, together with a family of plants known as the legume family, which will be discussed later, comprise most of the plants that are commonly used for hay and pasture purposes.

The relationship between such apparently dissimilar plants as corn and wheat may not at first be apparent; nevertheless, they have certain common characteristics that distinguish them from plants belonging to other families and place them in the same general group. Some of these common characteristics are apparent from a casual examination of any member of the grass family. For example, the leaves of grasses are **parallel veined**; that is, the veins of the leaf lie in a parallel position, whereas in such common leaves as the oak, maple, potato, etc. they extend irregularly throughout the leaf. A leaf having this irregular arrangement of veins is called *netted veined.*

In grasses there is usually, in the central part of the leaf, a vein largely known as the **midrib**. The midrib in some grasses, corn, for example, is pronounced, as will be seen in Fig. 2 (a). The main upright part, *a,* of the plant above ground is called the **stem**, or **culm**, and this is divided into sections by enlarged rings, or knots, *b,* known as **nodes**. The sections of the culm between the nodes, and shown by *c,* are called **internodes**.

The leaves of grasses have three common parts, which are the sheath *d,* the blade *e,* and the ligule *f.* The sheath always starts from a node on the culm; in some cases, the edges of this sheath are completely grown together, forming a tube that encloses the stem; in other cases, the edges of the sheath merely overlap. The blade, which is the upper and most conspicuous part of the leaf, may be more or less erect or it may be drooping in habit. It may be 3 or 4 inches in width, as in some kinds of corn, or only a fractional part of an inch, as in some lawn grasses. The third part of the leaf, the *ligule*, which derives its name from a Latin word meaning tongue, is not so conspicuous as the other parts mentioned. It is a projection, usually thin and membranous, at the upper part of the sheath where the latter joins the blade. In some species of grasses, it is just a mere line; in others, it may be a fourth of an inch in length.

All grasses also possess what is known as a

Figure 1. Holstein heifers grazing a luxriant pasture.

fibrous root system. We mean by this that the roots are numerous and threadlike, as shown in Fig. 2 (b).

The *culm* of grass usually has a hard, smooth outer wall. In some varieties, the internode is hollow, as shown in Fig. 2 and 3, but the opening does not extend through the node. A bamboo fishing pole is an example of the hollow form of growth. In some varieties of grass, the internodes are filled with a soft, spongy material known as *pith*.

The grass blossom, although often somewhat conspicuous, is not so showy as that of many other kinds of plants. For this reason mainly, grasses are seldom used for ornamental purposes. An exception to this is grass for lawns and certain other grasses that have peculiarly colored leaves.

For the most part, grasses that are used for hay and pasture purposes are long lived; that is, when once established, they continue to grow for several years without reseeding. This character is of importance, as it greatly reduces the cost of seeding meadows and pasture, both in the matter of labor and in the cost of seed.

The leaves of grass form a considerable portion of the plant when growing and of the hay made from a crop of grass. The leaves are not easily broken off in making hay, which fact gives grasses a distinct advantage over legumes, such as clover, alfalfa, etc., for hay purposes, since the leaves of the latter plants are easily broken off and lost in the process of haymaking. Certain grasses, as, for example, blue grass, have a tendency to produce a profuse growth of leaves at the base of the plant. This character is doubtless of considerable economical importance when the grass is used for pasture, as a heavy growth of leaves is thought to aid considerably in preventing the evaporation of moisture from the ground during the growing season. Grasses having this habit, if otherwise satisfactory, make excellent pastures and lawns.

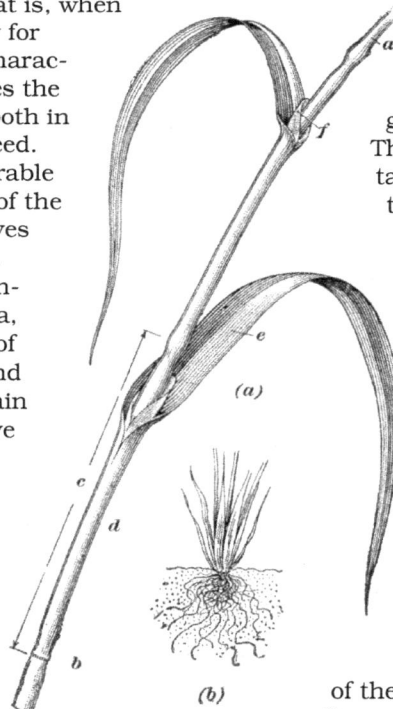

Fig. 2.

The value of grasses is greatly improved by a peculiar habit of growth of the leaf blade. The point at which growth of the leaf takes place is not at the tip, or extreme end, but at the base, as may be recognized, by examining a plant, from the lighter green color at this point. The extreme end, or tip, of the leaf blade may be removed without injuring the growing point; and, in fact, this is exactly what happens when grass is browsed in the pasture or the lawn is mown.

Other characteristics of grasses that make them of agricultural value are their palatability and healthfulness when fed to farm animals. Although there is probably little difference in grasses as to healthfulness, there may be considerable difference as to palatability and both will depend somewhat on the stage at which the grass is cut and the manner of curing it.

There are two general ways in which grasses reproduce or multiply. One of these is by the production of seeds, which is too familiar to require detailed description; the other method is by what is known as **vegetative multiplication**. It is by this process that certain grasses are able to grow on indefinitely and spread over large areas, without production of seed. Vegetative multiplication takes place in two ways, one of which is by what is knows as **tillering**, or **stooling**. Wheat and timothy plants are good examples of plants that reproduce by tillering. Starting with a single plant, bearing three or four leaves, offshoots are soon produced, as seen in Fig. 4. These offshoots start from the lower nodes, and may, in turn, produce other offshoots; the process continues until it is checked by the ripening of the plant.

The second method of vegetative multiplication is by means of underground stems, or runners, called **stolons**. The parent grass plant sends out horizontal shoots, which are usually just below the surface of the ground. These shoots, or runners, produce at regular intervals a new root system and a new plant above ground. Each of the plants so produced may send out stolons, and thus the process continues indefinitely, usually resulting in the formation of a close sod; that is, the soil near the surface becomes so filled with these stolons and roots that it is bound together into one compact mass. Kentucky Blue grass is an example of the grasses that reproduce by *stolons*. Fig. 5 shows a parent grass plant sending out stolons.

Importance of Grasses.--Grasses are of the greatest value to the human race, since all cereals, with the exception of buckwheat, are members of this family. These cereals furnish the staple foods for the vast majority of the world's population. But even if the cereals were omitted, grasses would still hold a prominent place in agriculture.

Aside from the direct use of grasses in agriculture, there is no doubt but that they are of great importance in the improvement of soils. When a good sod is plowed and the ground planted to other crops, it is observed that the soil has been improved by being left in grass a few years. By the decay of the grass roots much plant-food is undoubtedly added to the soil.[1] It has been noticed in the history of agriculture, in the plains and the prairie regions, that the newly broken soil generally produces excellent crops for a number of years without the addition of manure or fertilizer. The grass crops have decayed on these soils for ages, and there has doubtless been a constant addition of humus to the soils on account of this process.[2]

Grasses do not have, like the legumes, the peculiar power of fixing free nitrogen in the soil by the formation of nodules on their roots; it has been shown by experiment, however, that a considerable quantity of nitrogen accumulates in the soil even where only grasses are grown.

Uses of Grasses.--Grasses are the principal plants found in nearly all permanent pastures. This is probably due to the fact that grasses are less sensitive to soil conditions than are legumes. In many sections of the country, pastures are by far the cheapest means of furnishing feed for horses, cattle, and sheep.

There are sections in the western part of the United States where, owing to

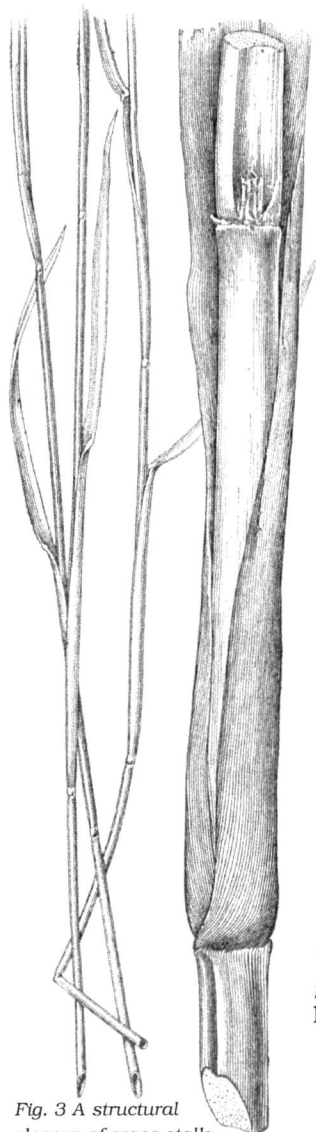

Fig. 3 A structural closeup of grass stalk.

peculiar climatic conditions, grasses go through a sort of natural curing process in the field after the growing season is over. This occurs in localities where there is practically no rain or snow after the close of the growing season. Animals may be pastured and kept in good condition on such dried grasses throughout the winter season. But grasses left exposed to the weather, in regions where rain and snow prevail, quickly lose their nutritive properties.

Grass Hay

Aside from their use in pastures, grasses play an important part in almost all agricultural systems by furnishing hay. The process of making hay from grasses consists in cutting the plants before they have reached maturity, allowing them to dry, or cure, in the open air until a certain part of the moisture has been given off, then storing them in barns, stacks, or sheds for use as feed for livestock.

Hay may be made from wild or from cultivated grasses. In the northeastern part of the United States, where timothy is grown, it is the great hay plant. In the north central section, just west of the timothy region, the wild prairie grasses become important. Continuing west, the grasses occupy a less important place as hay plants, their place being taken by the

1. As is noted further in Chapter Two, the phenomenon of root decay is a grossly under-rated, under-valued, biological truth which, in the hands of curious and ingenious individual farmers, may be directed tool-like to an increase and/or a decrease with amazing fertilizing advantage. LRM

2. Before the encroachment of 'civilized' man the great expanses of prairie grass had their growth ignited by the cyclical returns of massive herds of grazing animals (i.e. Buffalo) eating the forage to nubs and trampling in their own waste deposits. With each pass an immediate change took place underground as the nutritive storehouse of root mass began to sluff and decay setting off myriad beneficial chemical chain reactions. Civilized man, in his consumate arrogance, has ignored the clear history of such long term interconnected biological activity arguing, out of greedy destructive short-sightedness, that all life is controllable for corporate and/or social gain. In this regard the executives of the Sierra Club and the U.S. Forest Service are as much to blame as the executives of Pepsico and Monsanto for the destruction of the planet. Arrogance, the many shaded motives of control, and short-sightedness are the true villains. LRM

Fig. 4

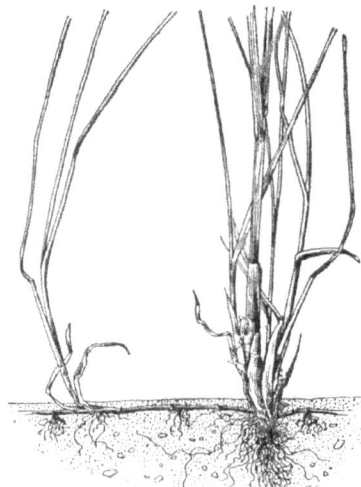

Fig. 5

legume alfalfa, which is better adapted for hay production in the western half of the country.

The fact that grasses may be cured into hay is of great importance. This method of preservation extends their use throughout the whole year, and even makes it possible for the farmer to lay in store a supply for a future season of scarcity. It is a common practice, in many localities, for the farmer to hold a part of his hay until he knows for a certainty that he will secure enough from his fields to carry his farm animals another year.

In some localities, grasses serve an important function as soil binders; that is, they prevent surface soil from being carried away by winds and water. This trouble is, of course, most common on steep embankments and terraces. In regions of sandy soils, the wind is likely to cause serious drifting unless the

soil is covered by vegetation. Permanent grasses are effective in checking or preventing these troubles. Grasses that multiply by underground stems after the manner of blue grass are the most satisfactory as soil binders, as they form a closer sod than those that do not multiply in this way.

Culture of Grasses

Soil Requirements for Grasses.--Briefly we look at how grasses are grown. In farm practice, it is generally considered that a fairly rich, heavy loam is best suited for grass crops. In the Timothy region, maximum yields of hay are grown on clay loams. The root system of the ordinary grasses consists of numerous rather fine roots, which are better suited for securing plant-food from finely divided soils than from the coarser sandy soils. There is no type of soil, however, on which some kind of grass cannot be grown if given opportunity. Grasses are not so sensitive to soil conditions as the legumes and other classes of farm crops. They will endure more acid conditions as well as poorer drainage than legumes.

Preparation of the Land for Grasses.--As a rule, when a meadow is seeded, it is expected that several crops will be harvested before the land is again plowed. The preparation of the soil before sowing the seed should be such as will fit the land for several years of cropping. If land is poorly fitted for tilled crops, it is

possible by a thorough system of cultivation afterwards to correct this error; this is not true in the case of meadows, for it is not possible to till the field after grass has begun to grow. For this reason it is of special importance to prepare the ground thoroughly before sowing grass for a meadow.

Deep and thorough plowing is essential for grasses. If the best results are expected, the land should be as carefully prepared as for the most exacting garden crops.

The addition of barnyard manure to the soil is an aid to the growing of large yields of hay. From 20 to 30 tons per acre is often used when applied immediately before sowing grass seed. It is a more usual custom to manure a previous crop, such as corn, in which case it is expected that the effect of the manure will extend to the grass crop.

Value of Mixed Plantings of Grasses.--In the culture of grasses, it is a common practice to sow a mixture of two or more different grass seeds, which is termed a mixed planting. There are certain advantages to be derived from this practice, one being that the varying soil conditions of a field can be better met by two or more grasses than by one; in other words, one grass will thrive on a particular portion of the field, perhaps crowding out all others, and another variety will be better adapted to some other portions, and hence excel there. Another advantage of a mixture, when it includes legume seed, is that the resulting leguminous plants add an appreciable quantity of nitrogen to the soil, and the grasses profit by the increased fertility. Still another advantage is the possibility, in the case of mixtures for pastures, of sowing a mixture that will yield a succession of forage growth, one variety of grass furnishing pasturage during the time that another is practically dormant.

If grass mixtures are sown for hay or pasture, each variety included in the mixture should be one that would succeed in the locality if it were sown alone. In the case of meadows, it is important that the varieties mature at about the same time, otherwise it will be necessary to cut some too green or others too ripe. *(See Chapter Two page 51 for a different attitude towards when crops should be cut for hay. LRM)*

Nurse Crops for Grasses.--It frequently happens, in seeding a certain grass, that it is desirable to sow at the same time a second crop with the idea that the latter will furnish protection to the former. The protective crop is known as a **nurse crop**, and is harvested as soon as it has served its purpose. Nurse crops commonly used with grasses are rye, wheat, oats, and barley.

Where a nurse crop is used with grasses, it is a common practice to sow the grasses in the fall with rye or wheat, although they may be sown in the spring with oats or barley. Fall sowing is generally preferable, as fall-sown grain is harvested earlier than spring-sown grain. Another reason why fall sowing of grasses is generally best is that the seeding is done at a time when there is sufficient moisture in the soil to start the young plants. Oats are not always desirable as a nurse crop for the reason that in dry seasons they take so much water from the soil that the young grass plants suffer.

Grasses do not grow rapidly while young. If sown alone, especially in the North, weeds are likely to appear in such numbers as to do considerable damage. When sown with grain, either in the fall or spring, the appearance of weeds is prevented, to a considerable extent, by the rapid growth of the grain. Although the nurse crop may retard the growth of grasses, they recover as soon as the nurse crop is harvested and are in better condition than if weeds had been allowed to occupy the ground. As grasses do not usually give a crop of hay the first season in northen sections, there is a decided advantage in sowing them with grain, for the grain crop is secured at little additional cost. In the South, it is thought by some to be better practice either to sow grasses alone or with a light seeding of barley or oats. The grain will prevent the growth of weeds and may be mown for hay before mature in order to prevent serious damage to the grasses.

Legumes in General

CHARACTERISTICS COMMON TO LEGUMES

Nearly all leguminous plants have certain common characteristics. The leaves are arranged around the stem in regular order--not two-rowed as in grasses. As shown in Fig. 6, the leaf consists of a stalk *a* and

leaflets *b*. At the base of the stalk is a pair of leaf-like outgrowths *c* called stipules. All legumes have a common form of blossom, examples being sweet peas, garden peas, and beans. As shown in Fig. 7, the root system of legumes comprises a large central root, called a tap root, from which numerous branches are sent out at varying distances. The roots of all legumes under favorable conditions bear tubercles, which are clearly shown in the figure. These tubercles, or nodules, as they are properly called, are caused by certain forms of bacteria that live in the soil. They have the power of assimilating the free nitrogen of the soil atmosphere and of transferring it to the root tubercles.

Although legumes have many common characteristics, they also vary greatly in some respects. Nearly all colors of blossom may be found among them. The color is usually consistent in any given species, as in Red clover, but on the other hand it may be of almost any shade of color, as in sweet peas, the blossoms of which vary from white to deep red. Leguminous plants vary extremely in size. Some are very small as White clover, and others are large trees, as, for instance, the locust.

While the leaves of different species of legumes resemble in general those of Red clover, they have certain distinguishing characteristics. The leaflets may be arranged along the sides of the leaf stalks, as in the locust, illustrated in Fig. 8, or they may radiate from the end of the leaf stalk, as in Red clover, Fig. 10. The leaf may be composed of an even number of leaflets or there may be an odd number, depending on the species. When there is an even number of leaflets, the midrib may end in a tendril, which serves the plant for

Fig. 6

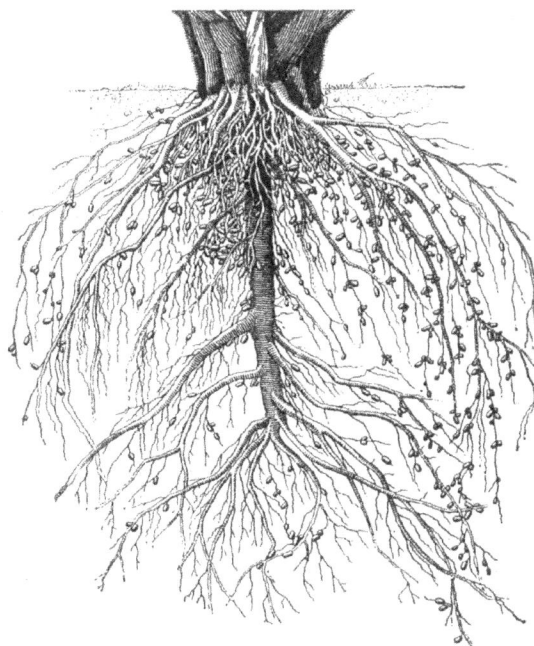

Fig. 7

climbing purposes. Vetch, Fig. 9, is an example of this kind of legume.

Uses of Legumes

Legumes for hay.--The two legumes most extensively used for hay in the United States are Red clover and alfalfa. The former occupies first place among legumi-nous plants for hay in the northeastern section, while west of the Mississippi River alfalfa is better adapted for the purpose. In southern states the cowpea is a plant of great value for hay, and Alsike clover is often used as a hay plant in northern states. Legumes are of value as hay plants primarily because of the high percentage of protein they contain. Alfalfa hay con-tains 11.4 per cent of digestible protein; Red-clover hay 7.1 per cent; cowpea hay, 13.1 per cent; and Alsike-clover hay 8.4 per cent. Besides containing a large percentage of protein, hay from legumes is rich in potash, richer, in fact, than hay from grasses in many instances.

Legumes for Pasture.--Most legumes have value as pasture plants. This is especially true of legumes that do not require reseeding each year. In many sections of the United States, alfalfa is an important pasture plant for the grazing of swine. Red and Alsike clovers are also very valuable pasture plants in those sections where they thrive. A

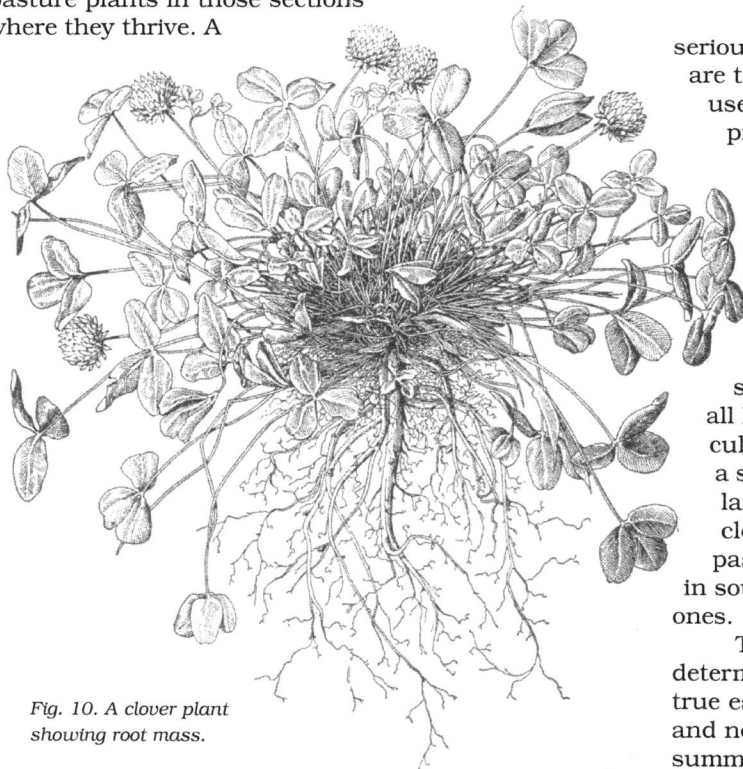

Fig. 8

Fig. 9

Fig. 10. A clover plant showing root mass.

serious objection, however, to these clovers is that they are too short-lived for permanent pastures. When used for pasture purposes, they are usually sown primarily for hay and then the aftermath is grazed; or, the meadow is turned into pasture land after cutting hay for a year or two. In such cases, grasses are sown with the clover. In a short time, the clover disappears and the grasses become the chief herbage.

White clover is by far the most important legume for pasturing, especially in northern sections of the United States. This clover is seldom sown in pastures, yet it appears in almost all kinds of soil shortly after the land ceases to be cultivated. Its habit of creeping, after the manner of a strawberry plant, makes it very valuable where land is closely grazed. In southern states, Japan clover is a legume of considerable importance as a pasture plant. It occupies somewhat the same place in southern pastures as White clover does in northern ones.

The value of legumes for pasture purposes is determined largely by their protein content. This is true especially where young stock or cows are pastured and no supplementary food is given to them during the summer. It has been shown, also, that by growing a mixture of plants in a pasture more forage can be secured under ordinary conditions than if only one kind of plant is sown; hence, a pasture of grasses and legumes mixed is better than one of either grasses or legumes alone.

Legumes in pastures also serve another important function--that of gathering nitrogen. Nitrogen is an exceedingly valuable plant-food for grasses. Therefore, when legumes and grasses are grown together, the decay of the tubercles on the legume roots furnishes the much-needed nitrogen for the grasses.

Legumes as Soil Renewers.-- All legumes are of importance in keeping up the fertility of soil. In many sections of the United States a profitable system of farming would be out of the question if it were not for clover and other legumes.[3] Besides being benefited by the addition of nitrogen, the soil in which legumes are grown is also directly benefited by the decay of the extensive root system of these plants.

Aside from the preceding valuable features, legumes are frequently sown for the purpose of plowing under as green manure. Crimson clover, in sections where it can be grown, is an especially valuable plant for this purpose. In the southern part of the United States, the cowpea is of great value in renovating unproductive soils. This plant is an excellent gatherer of nitrogen, and its root tubercles are large and numerous. In these states it is common to plow under a crop of cowpeas when preparing the land for other crops. The cowpeas are either turned under green or the plants are allowed to decay, after which they are worked into the soil by tillage implements. In many cases, however, a hay crop is cut from the field and only the stubble plowed under. In either case, a liberal quantity of plant-food, including nitrogen, phosphoric acid, and potash, is returned to the soil and a certain amount of nitrogen is added.

Fig. 11 Timothy

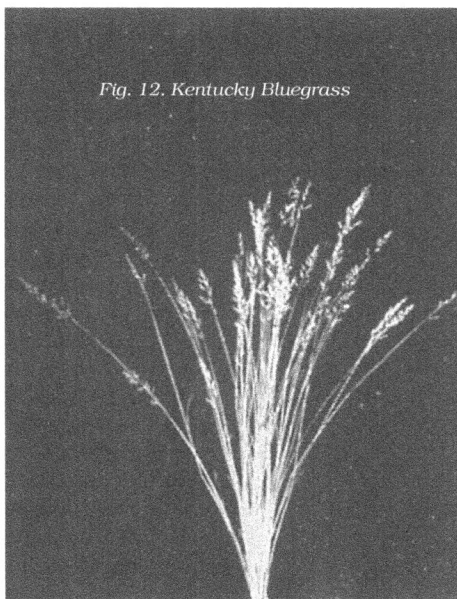
Fig. 12. Kentucky Bluegrass

Description of Grasses and Clovers

Timothy. Timothy is an important and widely adapted hay grass in the United States and should be used wherever suited as a part of pasture mixtures. It does best in a cool, humid climate and is excellent for clay and silt loam soils.

Timothy is a hardy perennial grass possessing a short rootstock. It reaches a height of three to four feet and propagates by seed and by new branches growing from corms at the base of the stems. New growth in the spring of the year usually comes from the corms or bulbs. It is one of the most palatable of pasture grasses and is relished by all classes of livestock. It yields well either alone or in combination with other grasses and legumes.

Timothy is easily sown. It may be seeded alone or in mixtures with bluegrass or legumes. When seeded with bluegrass, it furnishes pasturage until the slower growing bluegrass is large enough to graze. It is frequently used in a mixture with redtop on soils where bluegrass does not thrive. Large acreages of timothy are seeded with small grain, either in the spring or fall.

Timothy is used for pasture and for hay. Seeded alone for pasture, it does not withstand heavy grazing for long periods of time and weeds become troublesome if it is grazed too closely. Timothy is usually seeded with red and alsike clovers for hay.

Kentucky Bluegrass. Kentucky Bluegrass, sometimes called June grass, is one of the widely distributed and valuable grasses.

This grass is a long-lived perennial and grows from seed and from short, creeping rootstocks, although in rich soil the rootstocks may be long and extensive. Kentucky bluegrass becomes established comparatively slowly after seeding, but after establishment it is one of the earliest grasses in reaching maturity. It heads only once during the season, and after producing seed, the plant goes into a semi-

---

3. *Tragically this position, once held by farmers and scientists alike, has all but lost its hold. The tragedy lies in the fact that, outside of alternative agriculture circles, orthodox farming puts little or no credence in the biological regeneration of soils. By necessity the future must see a return to the notion of "growing" soils and value of all components to that end be they Legumes or practises.*

dormant stage unless good moisture conditions prevail throughout the summer. During the cool fall months it again develops a vigorous growth.

Kentucky bluegrass is adapted to fertile, sweet soils. It is winter hardy and withstands extremely low temperatures, but has the serious defect of becoming summer dormant during hot, dry weather and may be said to be drouth escaping in this respect.

The seeds are very small and often of low germination. In view of this fact, a good, firm seedbed is necessary. Seeded alone, the rate of seeding is 25 pounds to the acre. In the northeastern states, Kentucky bluegrass comes in naturally and once established it tends to maintain itself.

Canada Bluegrass. Canada bluegrass, sometimes called wire grass, is a plant closely related to Kentucky bluegrass. It is better suited than the former for drouthy soils. As a pasture grass, it is more palatable than Kentucky bluegrass but does not form as dense a sod. It is considerably later in maturing and does not become dormant as soon as Kentucky bluegrass. Sheep and cattle seem to prefer it to other pasture grasses in the early stages of growth.

The climatic adaptation of Canada bluegrass is similar to that of Kentucky bluegrass. It is a perennial and reproduces from seed and extensive underground rootstocks.

Methods of culture and seed harvest of Canada bluegrass are the same as for Kentucky bluegrass. Seed of both plants is harvested when the panicles have turned a deep yellow.

Redtop. Redtop is an important pasture grass in America. It is a perennial belonging to the bent grass group, and propagates by seeds and creeping rootstocks. It is later maturing than Kentucky bluegrass or orchard grass but under pasture conditions it rapidly forms a turf. Isolated plants will form a turf one to three feet in diameter.

Redtop has a very wide range of soil and climatic adaptation. No other pasture grass will grow under such a wide range of conditions. It thrives on moist soils and the moisture content of the soil is a

Fig. 13. Canada Bluegrass

Fig. 14. Redtop

greater factor in its successful growth than the type of soil; however, it is very drouth-resistant and has a high degree of cold resistance.

The seed is very small and requires a fine, compact seedbed. When seeded alone the average rate of seeding is 8 to 10 pounds per acre. It is seeded by broadcasting or with the drill with a grass seeder attachment. The time of seeding varies with seasonal conditions and locality, but it may be seeded either in the fall or early spring.

Redtop is used chiefly for pasture and is a common ingredient of pasture mixtures. It is palatable until it flowers, after which it seems to be less palatable. During the flush periods of growth--in spring and fall--it is relished by all classes of livestock. Where Kentucky bluegrass is well adapted it rapidly crowds redtop out. Redtop is used to some extent for hay, particularly in regions where seed is produced, and it is also commonly used as a turf grass.

Orchard Grass. Although orchard grass does not always withstand extremes of cold in the northern states, it is a valuable pasture and hay plant throughout the Corn Belt and the upper South. The plant is a long-lived bunch type perennial grass which grows to a height of 30 to 40 inches.

Orchard grass has a wide range of adaptation. It stands high temperatures very well, and it is recommended as a shade grass. It grows best on moist soils but also does well on sandy loams and on muck soils.

Because it is relatively light in weight, orchard grass is more difficult to seed than timothy--it is ordinarily seeded with small grain on a good, firm seedbed in the early fall or spring. Orchard grass is usually seeded in mixtures with other grasses. Fall seedings sometimes freeze out due to lack of snow cover. It is seeded by broadcasting or by a drill equipped with a grass seeder attachment.

While orchard grass is grown both for hay and pasture, its chief value lies in its use for pasture. Close pasturing does not seem to injure orchard grass--rather it lessens its habit of growing in bunches. The palatability of this grass is high, especially in the early stages of growth and in the fall. Its nutritive value is somewhat lower than that of Kentucky bluegrass. For hay it is usually cut in the bloom stage.

Orchard grass seed is ready to harvest when it is a bright straw color, usually about three to four weeks after it has flowered. Seed may be harvested by cutting with a binder and threshing, or by combining.

Bromegrass. No other forage grass has increased as rapidly in use and popularity within recent years as has smooth bromegrass, frequently sold by seedsmen under its Latin name, *Bromus inermis*. Where adapted it yields well, is palatable and nutritious to all classes of livestock, is persistent and aggressive as well as drouth and cold resistant. Its use is rapidly spreading in the north-central states for hay and pasture, either alone or in mixtures.

It is a perennial, sod-forming grass which for best growth requires a soil rich in organic matter. It grows to a height of 30 to 50 inches and retains its green color even after flowering and seed production.

Smooth bromegrass grows slowly the first year and does not reach its full growth until the second or third year after seeding. Adapted strains for regions south of the Corn Belt proper have not been produced, but its range southward is being gradually extended.

Smooth bromegrass, like other grasses, is best seeded on a good firm seedbed. The seed is large but very light and is difficult to seed except by broadcasting. It is usually seeded separately. It can be seeded with an endgate seeder by having the operator pour the seed through the hopper as the machine moves over the field.

The rate of seeding smooth bromegrass alone is 20 pounds per acre. In mixtures the quantity varies with the desires of the individual. It is seeded either in the fall or early spring.

Fig. 15. Orchard Grass

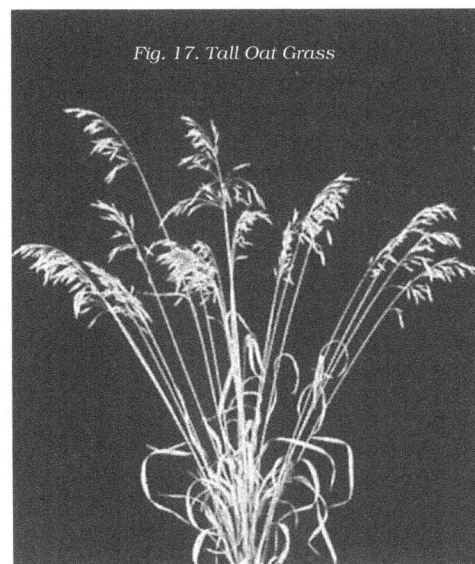

Fig. 16. Bromegrass

This plant is used chiefly as pasture and less frequently for hay. It makes an excellent pasture and hay of good quality. It is cut for hay when the plants are in bloom. Yearling cattle and sheep make good gains on bromegrass pasture. A mixture of bromegrass and alfalfa has found great favor in some localities in the north-central states.

Tall Oat Grass or meadow oat grass, as it is sometimes called, is a bunch type grass growing to a height of five or six feet. It is a deep-rooted, long-lived perennial adapted to about the same conditions of soil and moisture as orchard grass. It is very drouth-resistant and is excelled only by bromegrass and the wheat grasses.

Tall oat grass may be seeded either in the spring or fall, depending on the region. The seed is very light and difficult to sow. It has a twisted awn, which increases the difficulty, and because of this characteristic, the acreage of tall oat grass has been restricted. In order to obtain an even distribution of seed it is often mixed and sown with grain which is used as a nurse crop. The rate of seeding is high--25 to 40 pounds per acre.

This grass is used for hay or pasture. Continuous heavy pasturing reduces its productivity, but, because it is a top grass, it is of value in mixtures of alsike clover, orchard grass, lespedeza, and redtop. When grown for hay, it should be cut at the beginning of flowering; when it

Fig. 17. Tall Oat Grass

is in full bloom it becomes woody and unpalatable.

Rye Grass. The rye grasses, perennial or English rye grass and Italian rye grass, are being used more extensively with each passing year for early pasture, hay, pasture nurse crop, and for lawns. They are short-lived perennial grasses which form a sod rapidly, produce early palatable pasturage and abundant seed. English or perennial rye grass yields less than does the Italian rye grass in both pasturage and seed.

Much of the rye grass seeded in America is used for specific purposes; seeded in grass and clover mixtures it acts as a nurse crop and at the same time grows rapidly, forming a good temporary sod. It is adapted to the humid areas of the United States and like other adapted grasses thrives best on moist, cool soils. It may be seeded in the spring or fall. In the South, it is seeded in the fall and furnishes pasture in the late fall and the spring. It can be seeded in either fall or spring in the northern states, but spring seedings are recommended because of the danger of winter killing of fall seedings.

Meadow Fescue grows larger than most fescues, but like them is a densely tufted grass with numerous fine leaves. Even though it has no rootstocks it does not grow bunchy. It propagates only by seed. Red fescue and Chewing's fescue have short, creeping rootstocks, are adapted to shade and are used for forage and turf.

Meadow fescue, like timothy, grows best in a cool, humid climate. It also needs rich, moist soil. It stands shade about as well as orchard grass. Meadow fescue does not reach its highest state of productiveness as early as timothy, but usually persists much longer. It comes on early in the spring and remains late in the fall.

A good seedbed is

Fig. 18. Rye Grass

especially important for meadow fescue. Seeding is made at the rate of 25 pounds an acre without a nurse crop; in mixtures, at the rate of 10 to 15 pounds an acre.

Fescues are used in mixtures for grazing, and to some extent for hay. Meadow fescue, the only one that is very palatable, is especially valuable for fattening cattle.

The seed is produced in open panicles similar to Kentucky bluegrass, although it is much larger and more easily harvested.

Meadow Foxtail. Grown for many years in eastern United States and Canada, meadow foxtail is a long-lived perennial, producing loose tufts with many lower leaves, and resembling timothy to some extent. The stems frequently grow to a height of two feet or more. Growth begins very early in the spring, and the plant matures earlier than most grasses and recovers rapidly after cutting.

Meadow foxtail is one of the wet-land pasture crops. It is fairly cold resistant and is adapted to moist, cool regions. It is not particularly exacting as to soil, though it responds to rich soils well supplied with moisture.

Since the seeds of this plant are short, fine, soft, and somewhat fluffy, it is very difficult to get an even stand. A first-class seedbed is essential. Meadow foxtail is seldom sown alone,

Fig. 19. Meadow Fescue

Fig. 20 Meadow Foxtail

but usually with a nurse crop of small grain and at the rate of 20 to 25 pounds an acre.

Meadow foxtail is used in permanent pasture mixtures on wet land where early growth is desired.

Wheat Grass. This is a perennial bunch type grass and the most widely cultivated of the wheat grasses. It grows to a height of 30 to 50 inches and has attracted attention because of its ability to withstand drouth and cold. It is particularly useful in the northern and northwestern states.

Crested wheat grass is adapted to most well drained soils found in the northern states, but does not seem to do well south of the Corn Belt where high temperature and humidity are factors.

It is usually seeded on a prepared seedbed in the spring, although it can be fall seeded in north central states. The rate of seeding is 12 to 15 pounds per acre either broadcast and harrowed or seeded with the grass and clover drill.

Crested wheat grass has been very successfully used for regrassing land in the western plains states formerly plowed for cereal grains during the World War. It is very palatable to livestock and is extensively used for grazing as well as hay where well adapted.

Fig. 21. Wheat Grass

Slender wheat grass, blue bunch wheat grass, and western wheat grass are best adapted to the western plains states and have not been used to any extent east of the Mississippi River. All are good grasses for grazing in the less humid western areas where they are considered very desirable and relatively palatable grasses.

Reed Canary grass is a long-lived, coarse perennial with leafy stems and where adpated grows to a height of five to six feet. It spreads by creeping rootstocks and forms a dense, heavy sod, strong enough to support cattle on marshy land.

Reed canary grass is best adapted for wet, cool soils, but strains have been selected which do well on upland soils. Although it is a wet land grass, it shows marked drouth resistance. After it becomes established it is little affected by cold. It yields well and remains green and leafy during the hot, dry periods of the season.

Reed canary grass does best on a good seedbed. Some success has been attained by seeding without seedbed preparation. Because of its special adaptation to wet land it is best to prepare the seedbed during the dry part of the season, usually in July or August, following which it may be seeded. Where moisture is deficient, it should be spring seeded. The usual rate of seeding is 10 pounds to the acre. It is slow in establishing itself and it is frequently necessary to mow the weeds. The seed is either broadcast or sown with a grass and clover drill, or grain drill. Seed is smooth and "slick", and gray in color.

Reed canary grass is used for pasture or hay and has a large carrying capacity. It is coarse and of medium nutritive value; however, it makes good hay and a good stand will yield three to five tons of air-dry hay per acre. It is not as palatable as timothy or brome-grass.

Seed is difficult to harvest because it shatters easily. Until recent years much of the seed was harvested by hand. Where machine harvesters can be used, this crop has been successfully combined.

Fig. 22 Reed Canary

Fig. 23. Bent Grass

Bent Grass. The bent grasses are not generally used for pastures except in certain localities in the northeastern United States. They are generally adapted to poorer wet soil. Their root systems are shallow and an abundance of moisture is necessary for best yields. They form a close, fine turf, which withstands grazing well. Being closely related to redtop, they rank in palatability with it. Their principal use has been for lawns and golf greens, chiefly because of the high cost of the seed.

Legumes for Pasture and Pasture Improvement

The use of legumes in pastures needs no recommendation. The value of legumes not only as feed for livestock but also for their good effect upon soil fertility is well known. That they are nutritious feed is attested by the results obtained from feeding experiments and practical observation. Although there are many different legumes of value for pasture, the species considered of greatest general use are only briefly described here.

Alfalfa. This legume is a deep-rooted perennial requiring good, well-drained, sweet soil for best productivity. There are adapted strains for most parts of the country. Where wilt disease is not serious, stands last over a period of several years and occasionally a farmer reports a stand 20 years old.

Alfalfa requires a fine, firm, moist seedbed for best results. It may be seeded either in the spring or late summer, with or without a nurse crop. Where grasshoppers are a menace, spring seedings are recommended. As a rule northern grown seed is more resistant to cold and drouth. Seedings without a nurse crop are gaining favor, although on land foul with weeds a nurse crop probably should be used. The rate of seeding is 12 to 15 pounds per acre in the humid parts of the United States. In areas where drouth conditions are apt to occur the rate is proportionately less. It is either broadcast and covered lightly or seeded with a grass and clover drill or a grain drill with grass-seeder attachment.

The use of alfalfa for hay need not be enlarged upon. When the ground is wet, cattle on alfalfa pasture will damage the spring crop by trampling, and for that reason it is often

Fig. 24 Alfalfa

advisable to cut the first crop for hay and pasture the alfalfa when the fields are firmer. This is a double economy, for the farmer can pasture grass fields when they are at their best and follow with alfalfa when it is also best for pasture. Alfalfa and alfalfa mixtures may be pastured until killing frosts occur, but fall growth should be pastured lightly. Close pasturing increases the hazard of winter-killing.

In some sections of the country it is more or less general practice to cover alfalfa pastures with a good coat of coarse barnyard manure after the stock has been taken off in the fall. This helps to prevent winter killing and adds fertility to the soil.

Unless proper precautions are taken there is danger of bloat in pasturing alfalfa. Cattle and sheep are subject to bloat but experience in the past shows that when livestock is first turned into alfalfa it should also be allowed to pasture on grass. Salt and water should be available at all times and dry roughage should be kept conveniently placed. After livestock is started on alfalfa the stock may be kept on it for the balance of the season. Alfalfa in a pasture mixture adds very materially to gains made by sheep, cattle, and hogs.

Red Clover. Next to alfalfa, medium red clover is unsurpassed as a leguminous forage crop. It is widely adapted in the northwestern United States and Canada. It is treated as a biennial, although some plants frequently appear to be short-lived perennials. It is deep-rooted and for this reason, red clover, when well established, withstands drouth. It requires a rich, sweet, well-drained soil and it is useless to try to grow red clover on sour soils.

Red clover, being a biennial, is better adapted for rotation pastures than most other legumes. It is palatable to all classes of livestock and is high in feeding value.

Fig. 25 Red Clover

Fig. 26. White Clover

Fig. 27. Alsike Clover

White clover should be included in pasture mixtures, for it is one of the very few legumes which will persist in a good stand of bluegrass. It forms an important element in permanent pastures for it is not only palatable but furnishes nitrogen for grasses in the pasture, thus increasing productivity. On closely grazed, but not overgrazed, pasture, it is especially valuable.

It is seldom seeded alone, most farmers preferring to include two or more pounds per acre in pasturing mixtures. If seeded alone, the rate is six to eight pounds to the acre. High seed costs have restricted its extensive use.

Red clover is usually seeded in early spring with oats or barley or in winter grain at the rate of 8 to 12 pounds to the acre. Large acreages of red clover are seeded with timothy and in this mixture, as well as others, the rate of seeding red clover is proportionately reduced.

This legume is used chiefly for hay, but it is finding increased use in pasture mixtures in which it furnishes a large volume of pasturage during the first and second years. In some pastures, red clover appears without reseeding ten years after establishment. Red clover does not stand heavy grazing well and for this reason has a very definite place in short rotations and temporary pastures.

Seed of red clover is usually harvested by cutting and windrowing or cocking the hay and then threshing. Red clover seed is also harvested by combining, using the windrow-pick-up method. Depending on local conditions either the first or second crop is harvest for seed.

White Clover. White clover or white Dutch clover is widely distributed over the United States from Oregon to Maine and south to Louisiana. It is a perennial, spreading by creeping stems growing along the surface of the ground. It is not injured by cutting, but trampling by livestock on soft ground destroys it. It is adapted to well drained soil but needs an abundance of moisture to thrive. It appears spontaneously in permanent pastures presumably from seed in the soil or deposited by birds.

Alsike Clover. Alsike clover or Swedish clover is seldom used alone for pastures; however, it is extensively used in pasture mixtures. It has a wider climatic adaptation than red clover. It is more cold resistant but, having a shallower root system, does not have the same qualities of drouth resistance. It grows fairly well on most acid, poorly drained soils.

Seedbed preparation and time of seeding are about the same as for red clover. The rates of seeding are similar to those for white clover.

Alsike clover is primarily a pasture crop due to its ability to grow on a wider range of soil types. It is a short-lived perennial but by re-seeding itself persists longer in pasture mixtures than do other legumes. It is also an excellent erosion control crop, seeded alone or in mixture with grasses. Alsike clover can be harvested by the binder-thresher method, or windrowed and harvested with a combine equipped with pick-up attachment.

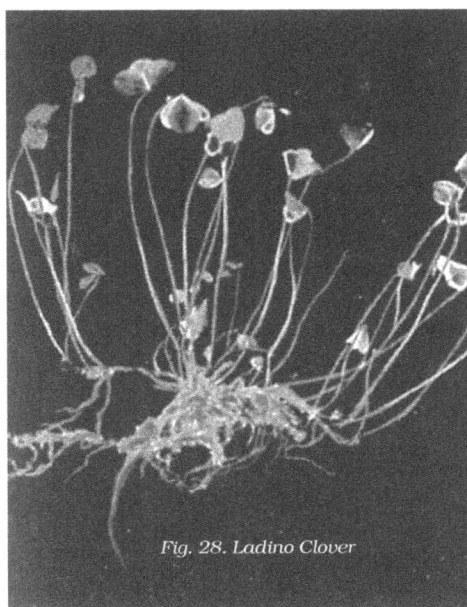

Fig. 28. Ladino Clover

Ladino Clover. One of the newer clovers gaining in popularity for pastures in some of the northeastern states is Ladino clover, a giant white clover. Large acreages of this clover are grown in the Pacific Coast states and increasing acreages are being seeded in the north Atlantic states.

It is a perennial clover with creeping stems similar to ordinary white clover. It is not as winter hardy as white clover and is subject to damage by leaf hoppers.

Fig. 29. Sweet Clover

Fig. 30. Lespedeza

It yields well and is palatable and nutritious for all livestock. Close, destructive grazing and excessive trampling destroy it. Sheep are especially fond of it and seek it out when it is used in pasture seeding mixtures. It is adapted to almost any well drained, well watered soil.

Ladino clover is seeded early in the spring, either in mixtures or alone. The rate of seeding is 6 to 10 pounds per acre. The high cost of the seed has been a retarding factor in the more extensive use of this clover. Ladino clover will cause bloat in cattle and sheep if precautions are not observed similar to those mentioned for alfalfa.

Sweet Clover. Sweet clover needs no introduction to most farmers. It is climatically the most widely adapted of all cultivated legumes. In early stages of growth it is difficult to distinguish from alfalfa. Most sweet clovers have a bitter taste, particularly when they have passed the early stage of growth. When bruised they have a characteristic pleasant odor from which the plant receives its name.

Sweet clover has a high lime requirement and will not grow satisfactorily on sour soil, if at all. On the other hand, it will grow well on raw subsoils where frequently nothing else seems to thrive, provided the soil is sweet.

The culture of sweet clover is similar to that of red clover. The rate of seeding is higher, ranging from 12 to 15 pounds an acre for hulled, scarified seed. Unhulled seed requires a heavier rate of seeding--20 to 25 pounds per acre.

When seeded for a rotation pasture or seed crop it is often seeded alone in small grain early in the spring. In pasture mixtures for new seeding or reseeding old pasture, sweet clover is being given an increasingly important place. Unhulled seed is recommended for fall or winter seeding on established pasture where it is desired to increase and maintain yields.

Lespedeza. The annual Korean and common lespedezas are small branched leguminous plants growing to a height of from five to 30 inches. The leaves are small and numerous and the stems small and slender. It blossoms in late summer or early fall.

A number of perennial lespedezas are being grown in some parts of the country experimentally and for soil conservation purposes, but their place in our agriculture has not been well defined.

The annual lespedezas rank with alfalfa and sweet clover as among the most drouth resistant legumes, but although they resist drouth their growth is retarded by lack of moisture.

The culture of lespedeza is similar to that of other small-seeded legumes. It is important that the variety best suited for the conditions at hand be chosen.

No special seedbed preparation is necessary where lespedeza is seeded on either winter or spring grain. But when seeded alone or in mixtures it requires a good seedbed. Difficulty is sometimes experienced in seeding lespedeza because of the light fluffy hull which prevents it from feeding through the seeder uniformly. Endgate and hand-horn seeders, however, have been used successfully.

On winter grain, lespedeza should be sown early in the spring, about the time for seeding red clover. On spring grain, or when seeded alone or in forage mixtures, it should be sown about the time spring grain is normally seeded. The rate of seeding varies from five pounds an acre in mixtures and in small grain where only a volunteer seeding is desired for the following year, to 25 pounds an acre where seeded alone.

Lespedeza fields should be clipped in order to control weeds, but should be cut high enough to avoid injuring the lespedeza plants.

Crimson Clover or incarnate clover is a winter annual belonging to the red clover group. It grows to a height of 18 to 30 inches, developing from crown branches. In a thick stand it grows upright producing beautiful crimson flower beads. The leaves and stems are hairy.

Crimson clover is best adapted to well drained soils, sandy or loamy in character. Sour soil decreases productivity. The crop is not winter hardy and cannot be successfully grown where temperatures fall below zero degrees Fahrenheit, although it grows under the influence of cool, moist conditions.

It requires a good firm seedbed and for this reason is often seeded in corn at the time of the last cultivation in the Cotton Belt. Farther north it is seeded in late August if sufficient moisture is present.

Crimson clover is grown chiefly for soil improvement but also affords grazing in late winter and early spring for all classes of live-stock. The hay can be used for all livestock except horses and mules. Hair from the leaves and stems has been known to cause death of horses and mules by formation of hair balls in the stomach. Crimson clover seed can be successfully harvested by straight-combining or by the windrow-pick-up method.

Fig. 31. Crimson Clover

Grass Mixes

Below are some mixtures recommended from the 1920's. Check with your local seed merchant for regional recommendations.

The following is suggested for temporary pasture (Seed for one acre.)

English Rye-grass.......................20 pounds
Timothy...5 pounds
Redtop..3 pounds
Red Clover..................................5 pounds
Alsike Clover...............................3 pounds
              Total   36 pounds

For soils that will grow blue-grass and timothy:

Timothy............................. 10 pounds
Red Clover............................5 pounds
Redtop...................................5 pounds
Orchard-grass.......................5 pounds
Kentucky blue-grass............10 pounds
White Clover.........................3 pounds
              Total   38 pounds

On poorer soils, too wet or dry hills low in lime:

Alsike Clover.......................5 pounds
Redtop....................................5 pounds
Orchard-grass.......................5 pounds
Canada blue-grass.........10-20 pounds
White Clover.........................3 pounds
              Total   28 pounds (to 38 pounds)

How To Make Grasses & Clovers Into Hay

Chapter Two

The process of making hay from grasses consists of cutting the plants before they have reached maturity, allowing them to dry, or cure, in the open air until a certain part of the moisture has been given off, then storing them in barns, stacks, or sheds for use as feed for livestock.

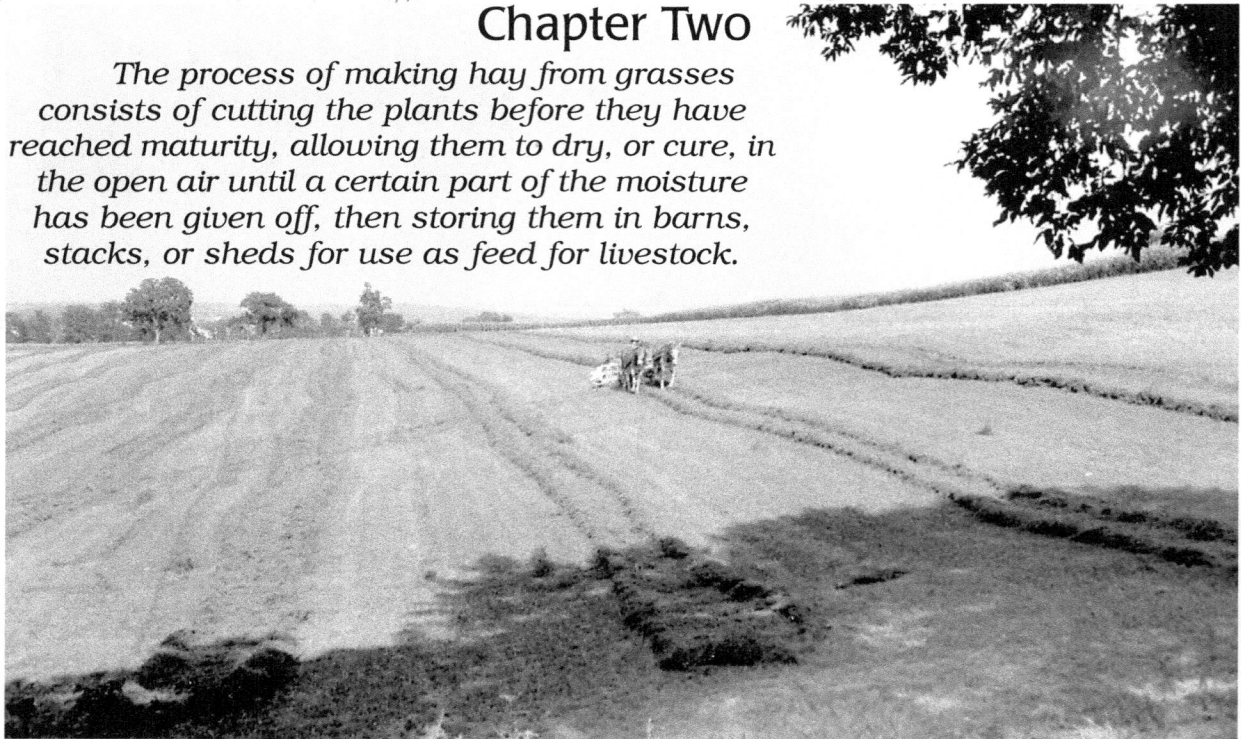

Figure 1. Two mules draw a modern side delivery rake across a Lancaster County Pennsylvania Amish hay field.

Why We Make Hay

This book will not escape some preaching on the subject of different "attitudes and approaches" to hay. We speak here first of the *Whys* of hay production and how they can, or should, influence the *how*.

There is a long standing institutionalized axiom (or argument, if you prefer) that the primary goal of any hay production should be for maximum efficiency and quality. "Efficiency" in this case translates to getting the forage put up with the fewest number of man hours, and "quality" translates to meeting or exceeding industrial standards for feed value and yield. It is our contention that this goal is actually counter-productive to the greater good of the small farmer and the soil.

We take this space to present the unusual notion that the production of hay should be a means to improving soil health and fertility. And to state that the adoption of this goal/approach will, paradoxically, result in superior forage even though it is seen within the operating design of the farm as a "conditional" waste product. We

Figure 2. The author, left, and Bud Dimick right, mowing on Singing Horse Ranch. This stand is being mowed with a goal of affecting the top soil. Gauging the near terminal stage of the forage growth and then cutting effectively returns to the soil the maximum bulk of root mass.

tilth.)

Best times in the northern hemisphere to mow forage for optimum soil effect:

June 21 thru July 22

August 22 thru October 23.

As you may recognize, in most northern climates these dates usually coincide with traditional forage harvest calendars so they pose little management challenge.

We are not so interested in the exact science of why ques-

say "conditional" because in a whole systems approach any forage should not be sold off the farm but rather should be fed to retained farm animals allowing all manures to return to the same soil. This is what might be called the condition of retention. If, however, any of the forage is sold off the farm the regenerative cycle is broken.

We do not mean to suggest by this that the retention of manure, in and of itself, is the sole best cause of increased fertility. Another and perhaps the most unique aspect of this goal/approach comes from the issue of timing with regard to cutting. And that timing can result in dramatically increased fertility.

We know that any grass or legume (or mix thereof) can be cut at a point in its cycle to minimize or maximize regrowth. (i.e. simplistically put - cut in early growth for rapid regrowth, cut late in growth for slow regrowth.) What may be less well known is that the stage of growth (and the relative phase of the moon) when the crop is cut, can have predictable effects on soil health and tilth. Improved tilth and fertility will result in increased plant growth. Over nearly two decades we have experimented with the effect of timing. Our conclusions are these:

1.) Mixed grass/legume stands are superior, in every regard, to any monocultural forage plantation.

2.) Spring pasturage of hay fields can often be most beneficial.

3.) If pastured, regrowth is improved if initial subsequent growth is mowed regardless of forage yield considerations.

4.) Mowing forage when legumes are at full bloom (and during the calendar periods listed) will cause the maximized subterranean root mass (including nitrogen root nodules) to be reabsorbed into the soil rapidly. (In this way the farmer orchestrates a "climax" growth management of root mass to improve fertility and

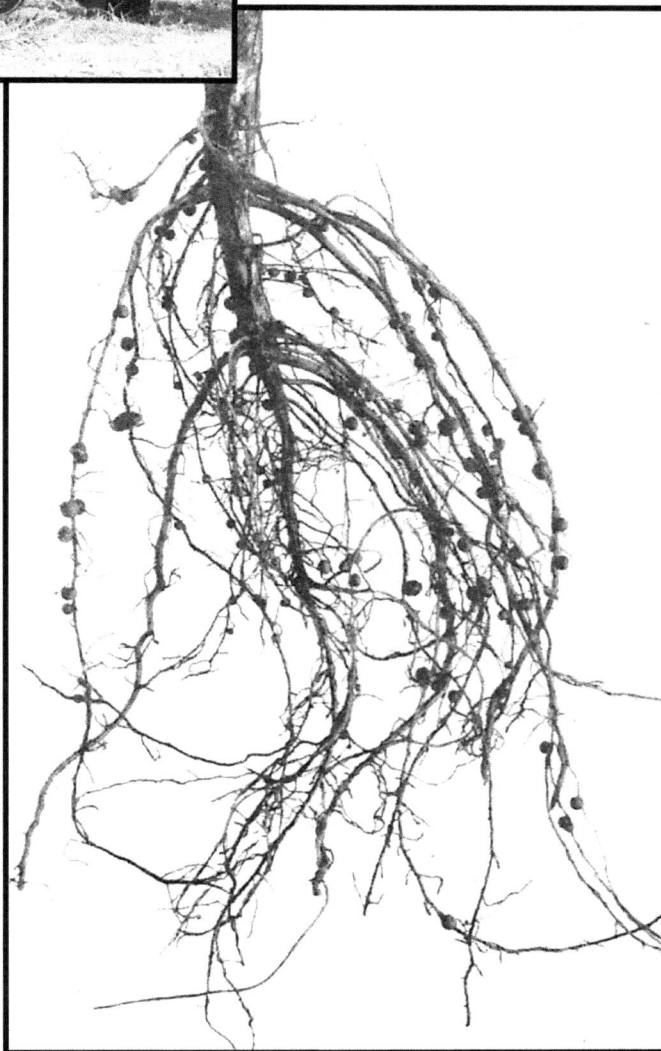

Figure 3. A soybean root mass clearly showing the nitrogen nodules in usual abundance. When this plant is cut this root mass sluffs back to a size commensurate with the remaining top growth. That sluffing process deposits into the soil decomposing root fibers and nodules. The fiber contributes to the tilth and water retention of the neighboring soil particles and the decomposition creates a catalytic environment of acids and heat.

tions of timing work. We leave those inquiries to others. We do, however, respect the predictability of these mysteries of nature and use them to tremendous benefit.

How this differs from the industrial model is this: instead of cutting at an early to mid bloom stage to achieve the highest TDN (Total Digestible Nutrient) ratings - we cut forage when it best benefits the soil. Rather than resulting with a lower quality feed, over time such an approach provides superior feed quality because the soil contains more nutrients (vitamins and minerals) in dynamic states ready to move up into the plant tissue. Also, in a regenerative organic biodynamic type approach the forage has additional superior properties which translate to healthier livestock and superior manure and urine for return to the soil. (Another radical notion: some manures and urine may be superior to others depending on the

Figure 4. This good stand of Timothy and Clover is being cut when the Timothy is in the early to full-bloom stage. The clover is in a very early bloom stage. Cutting at this stage returns less nitrogen to the soil and more grass root fiber (this may be a farmer's goal). Mowing at this time in the growth has, in this author's experience, resulted in a stronger subsequent growth for the legume portion of the mix. While mowing later, when the grass has begun to lose its seed and the legumes are near full bloom, usually results in a stronger return of the grass portion. These are clear indications of nutrient balances from phosphorus and calcium predominence to nitrogen

grazer's diet.)

The true challenge in such an approach is for the farmer to see his endeavors in the seasonal relationships of decades rather than seasonal relationships within a year or two.

Throughout the temperate regions of the world hay is the most important harvested roughage for livestock feeding. And this crop is subject to potentially heavy losses of value because of improper curing and/or storage (i.e. dangerous mold may result). In fact, improperly made hay, especially baled and loose hay stored in barn spaces, is subject to spontaneous combustion from unsafe amounts of water. Fires may result.

Most of these problems can be avoided with diligence and informed judgement.

Though we have made an introductory case for haying as a fertility aid we must recognize that many people who read this book will be tied, for various reasons, to traditional approaches.

What Is High Quality Hay?

If the goal is to maximize hay quality, conventional wisdom dictates that the plants must be cut at a sufficiently early stage of maturity. Next, modern industrial process would counter or challenge this outcome, high quality hay has been cured and handled so that it is leafy and green in color, the stems are soft and pliable, and it is free from mustiness or mold. Such hay will have an attractive fragrance adding to its palatability. And it should have little foreign matter, such as dust, weeds or stubble.

Hay thus described is most

Figure 5. A close-up of Birdsfoot trefoil, an important legume, showing flowers, seed pods and neighboring grasses. A view of such a bloom stage is clear indication that the root mass below the surface is reaching climax stage, and will not increase in volume perhaps even decreasing.

nutritious and palatable. And therefore has a higher feeding value per ton.

Some scientists believe there may be far more difference in actual value per ton between good and poor hay of ANY kind than there is between the different kinds of hay made from the various common hay crops. It is conventional wisdom to see legume hay, such as alfalfa or clover, as richer in protein, calcium and vitamins than grass hay of equal quality. This author would add qualifications: a balanced dynamic fertile organic soil which is being repeatedly "teased" and "flushed" by dramatic fluctuations in root mass decay and regrowth creates a morass of catalytic conditions which can and does make for increased vitamin and mineral uptake. (Something the scientific community may take issue with, as currently such a contention is difficult to quantify and test for.) It is this author's further insistence that a mixed stand of grasses and legumes, grown in such a manner, will always produce superior hay to any monocultural industrially produced legume stand. (This would include those crops demonstrating very high protein rates.) Modern agricultural science, in the mainstream, has not

Figure 6. Cowpeas and Johnson grass combined for hay. The creative and intelligent farmer has a wonderful smorgasbord of possible plant mixtures for hay and pasture, each with slightly different effect on the soil.

evolved far enough to give credence to the amazing evidence of plant and livestock health where these "organic" issues predominate. It is up to the individual farmer to decide the goal structure of his or her farm. We would encourage, in that decision making process, that health and fertility be given higher values than profitability and industrial efficiency. If the health and fertility are there the profitabilty will define itself and the rest be hanged.

Returning to the conventional hay model - In monocultures of say Timothy or Alfalfa or Fescue it would appear that the percentage of protein, the digestibility and the content of minerals and vitamins all decrease decidedly as hay crops advance in stage of growth. The rate of decrease in nutritive value would tend to be more rapid in most grasses than in legume hay crops. The textbook rule of thumb is *'cut it late and its worthless- cut it early and its worth more.'* This is simplistic, as most any seasoned farmer knows or suspects (we, all of us farmers, often are forced to cut when weather and time will allow - and not infrequently surprised by the outcome.) A better rule of thumb would be *'cut the forage when that action best benefits the soil'.*

As for issues of hay quality and procedure: how the hay is actually handled after cutting has potentially far more bearing on quality than when it is cut. This should be clearly demonstrated in the following text.

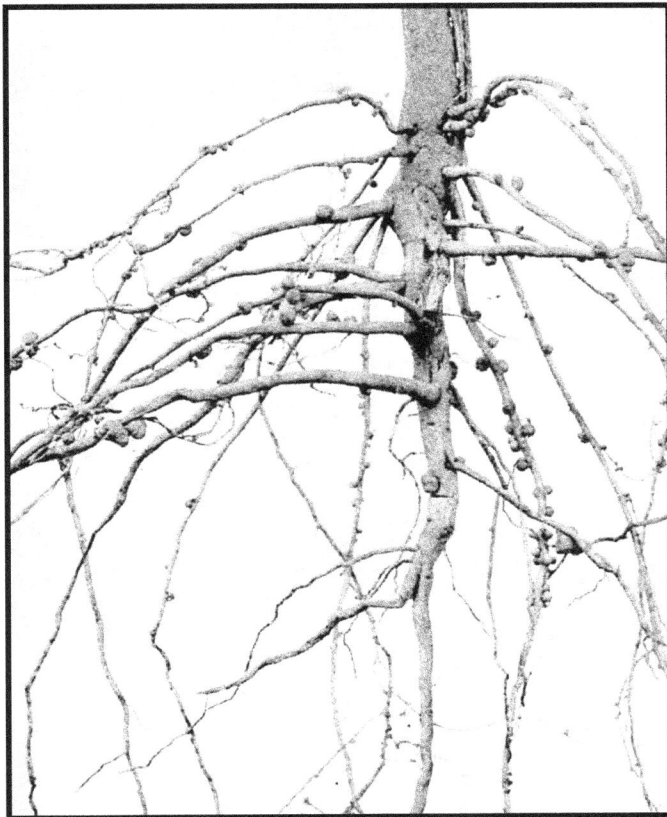

Figure 7. Roots of the Cowpea plant showing the development of nitrgen nodules.

How the cut hay is handled could be more important than when it is cut.

Haying of Grass & Legume Crops

A SIMPLE OVERVIEW

This entire book is about haying with true horsepower. What follows is a brief basic outline of the process by way of introduction. Each reader will, doubtless turn to the pages that apply to his or her equipment and system choice.

The first implement used in the operation of harvesting hay with horse powered machinery is the **mowing machine, or mower,** a common form of which is shown in Fig. 8. (Mowers are covered in great depth in two

Figure 8. The mowing machine at work

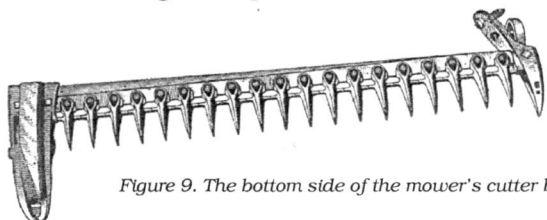

Figure 9. The bottom side of the mower's cutter bar.

Figure 10. (right) A hand crank sickle sharpener designed to clamp on to mower wheel.

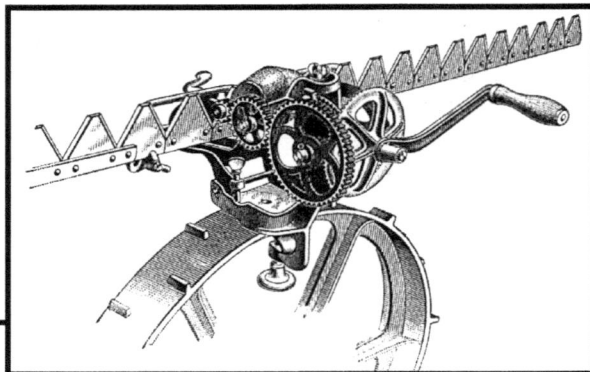

Figure 11. A hay tedder at work fluffing the hay. This implement was orginally designed for use before making windrows. Modern practises now have the tedder used solely to spread rained on hay to hasten redrying.

Figure 12. A single horse dump rake gathering hay and joining jags to form a long windrow.

Figure 13. Finished windrows that may have been made by either a good dump rake operator or a straight walking side delivery rake team.

Figure 14. An old wood frame style right-hand side delivery rake.

additional chapters of this book. Fig. 9 shows the cutter bar of such a machine. This cutter bar has, projecting from its front edge, a series of fingers known as *guards*.

The knife, which is simply a slender bar of steel to which is riveted a series of triangular plates having cutting edges on two sides, is made to move back and forth over the guards at a high rate of speed, thus cutting grass stems that come between the latter. The general form of the knife is shown in Fig. 10, which illustrates a knife in a grinding machine that will be described later. The rapid back-and-forth motion is given to the knife by a system of gears driven from the main wheels of the mower. The framework between the main wheels serves as a support for the mechanism and a seat for the driver of the team. Mowing machines of the ordinary size cut a strip, known as a **swath**, from 4 to 7 feet wide, and are drawn by two horses. Smaller machines that cut a swath of 3 or 3 ½ feet and are suitable for one horse were made and are a little more difficult to find.

As soon as hay is cut, the curing process begins. While lying in the swath, drying proceeds rapidly, particularly if the sun is shining, for a large part of the hay is exposed. Under some ordinary summer heat conditions, the hay is cured sufficiently within a few hours after cutting. When an extra heavy crop of hay has been cut or when an ordinary crop has been wet by rain, it is sometimes necessary to turn it over in the swath to prevent molding and get it properly cured. A machine called a **hay tedder** accomplishes this turning very rapidly and satisfactorily. Fig. 11 shows a hay tedder operation.

After the grass has cured in the swath for a sufficient length of time, it is raked into **windrows**, which are long rows of loose hay, as shown in Fig. 13. The raking is usually done with an implement known as a **hay rake**. Two types of these machines are common. One is known as a *dump or sulky rake*, which is made in both one-horse and two-horse sizes; the one-horse size is shown in Fig. 12. The second type is called the *side-delivery rake*; this is iillustrated in Fig. 14. The sulky rake is equipped with a row of teeth, which collect the hay from the swath. When a sufficient quantity has accumulated, the driver raises the teeth by means of a hand lever or a foot trip, which discharges the load, then the teeth return to their first position and begin collecting

Figure 15. A field of cocked hay with some of the cocks covered.

another load. On the next and subsequent trips around the field, each load is dumped in line with the one left on the previous round, thus forming long windrows, which extend at right angles to the direction in which the rake is moving. The side-delivery rake performs its work by a system of moving forks, which work the hay to one side of the machine and discharge it in a windrow parallel with the course of the rake, as shown in the illustration.

From the windrows, hay may be temporarily placed in cocks or be hauled directly to the stack or barn. (The cocking of hay is a very old method not much in use outside of historical museums or some old order Amish communities.) One reason to use hay cocks would be when the hay could not be quickly placed in a stack or under shelter. In this shape it is less likely to be injured by rain or bleached by the sun than when it is in windrows. Cocks are formed by gathering the hay into convenient piles with a pitch-fork, the piles being built in the form of a miniature stack so as to shed water. In certain regions or at times of heavy rainfall, it may be advisable to cap the cocks with canvas. Hay in cocks is illustrated in Fig. 15, which shows some of the cocks covered with canvas.

Stacking or mowing is the next step in haymaking. It consists in taking the hay from the field to a place of permanent storage, which may be a stack, a shed, or a hay mow. If the hay is to be placed in a stack in the meadow or in an adjoining field, a sweep rake may be employed for carrying the hay to the stack. This machine, which is illustrated in Fig. 16, will gather the hay from the swath, windrow, or cock and transport it to the stacker, a mechanism by means of which the stack is built up. Stackers are of various types; Fig. 17 shows a common form in operation. When ready to receive a load, the elevating fork of the stacker rests on the ground and the sweep or buck rake is drawn up to it in such a manner as to deposit its load on the fingers of the stacker fork. The sweep rake

is backed away and the horses operating the stacker are started, thus raising the load to a height that will permit its being swung over the stack. Fig. 17 shows the appearance of a stacker at this stage. The fork drops the load at the place desired and men with pitchforks smooth it out on the stack.

There is always some loss when hay is stored in stacks. At the top of the stack, if this is not protected, and at the sides a part of the hay will be of poor quality if not actually spoiled. However, it is probable that, in the drier sections, the loss from damaged hay is not equal to the cost of building hay barns. In stacking hay, the practice is followed of keeping the center of the stack well filled and tramped as well as possible. This prevents any water that may fall on the stack from running toward the center of the stack, which would be the case if the center were loose and poorly filled and the edges well filled and packed.

In stacking hay, it is good practice to place the stack on a slight elevation, so that water will not run under the stack and ruin the hay at the bottom. The soil on which the stack is built should be porous and well drained. Some farmers place a framework of poles on the ground and build the stack on these. This prevents the hay from coming in contact with damp earth.

If, instead of hay being stacked in or near the meadow, it is transported to a stack, barn, or shed at some distance from where made, it is necessary to load it on a wagon fitted with a **hay rack**. The hay rack is simply a framework displacing the regular wagon box. It extends the full length of the wagon and out over the wheels on each side, thus furnishing a platform from 14 to 18 feet long and from 7 to 8 feet wide. Loading the rack in the field may be done from the cocks or from the windrow by hand labor with pitchforks, or it may be done from the swath or the windrow by means of a machine known as a **hay loader**, which is illus-trated in Fig. 18. As will be seen, the loader is at-tached to the rear end of the hay wagon. As the wagon

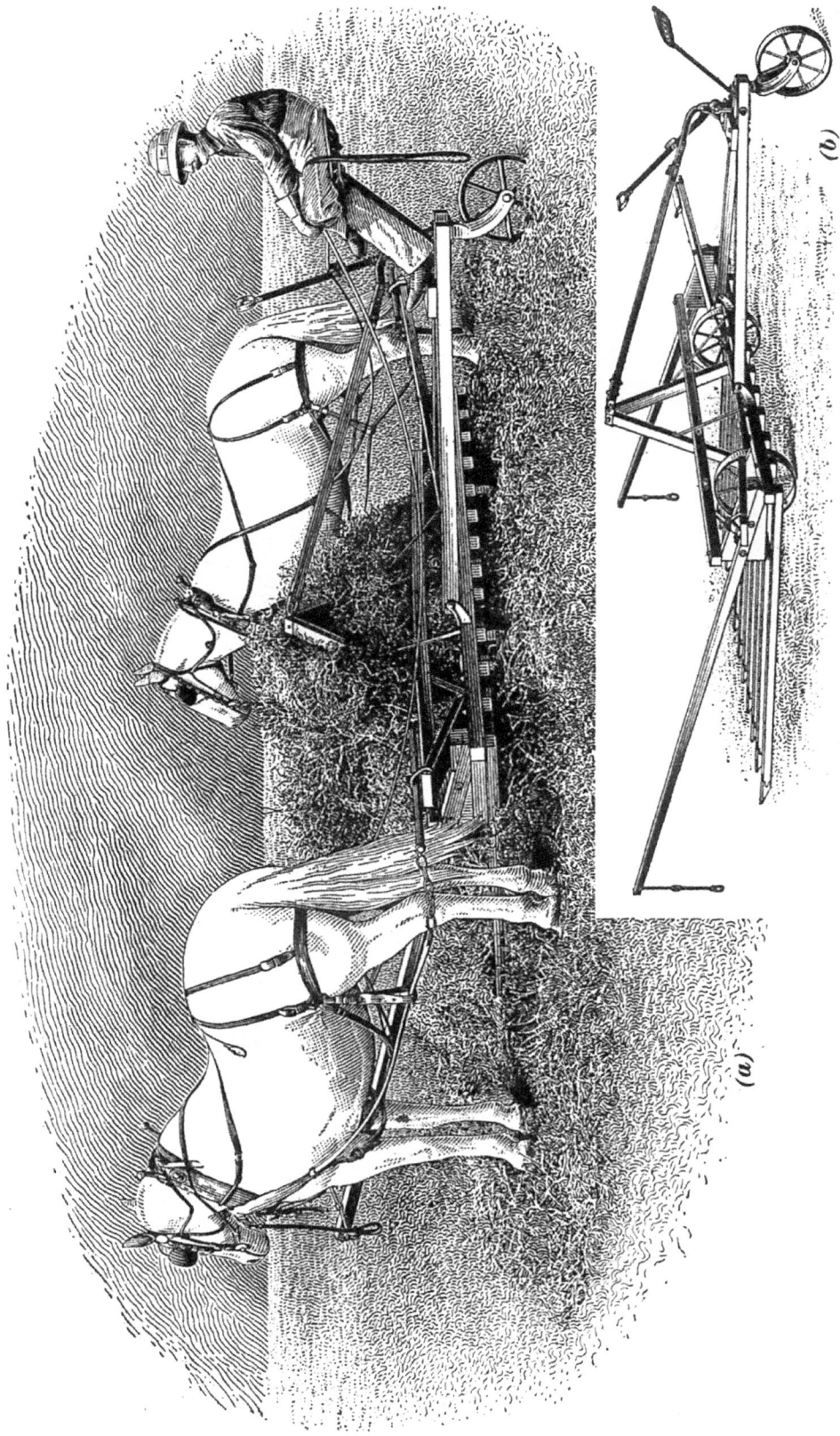

Figure 16. The buck or sweep rake is an excellent tool for gathering and pushing large amounts of loose hay to outside stacks as well as barns and shed. Pictured is an unusual buck rake with horses hitched up alongside the rake basket. It is this author's opinion, born of 12 years of buck rake experience, that this design might work well but it would be highly sensitive to the behavior of the horses. If either or both horses chose to pull sideways those side poles would be jeopardised. The handle near the teamster, when pulled back, would shove the A frame forward to start hay moving off as the horses back up. The pith helmet and moustache, though stylish, are not regular fare for buckrakers.

Figure 17. A form of swinging stacker. The buck rake pushes the jag of loose hay on to the stacker fork basket, which rests on the ground. As the team or power source pulls the stacker basket into the air the apparatus swings until the load is over the waiting stack. The load is then tripped and spread by the waiting stack crew. There are many different styles of stackers which have in common the fact that they make the building of large hay stacks an attainable goal for any farmer.

Fig. 18. A hay loader is attached to the rear of a horsedrawn wagon. As the wagon moves forward the hay is picked up and elevated on an endless belt (or in other styles pushed up a slide) until the hay falls off the top of the hayloader and on to the wagon where the material is distributed by waiting crew.

Into Hay

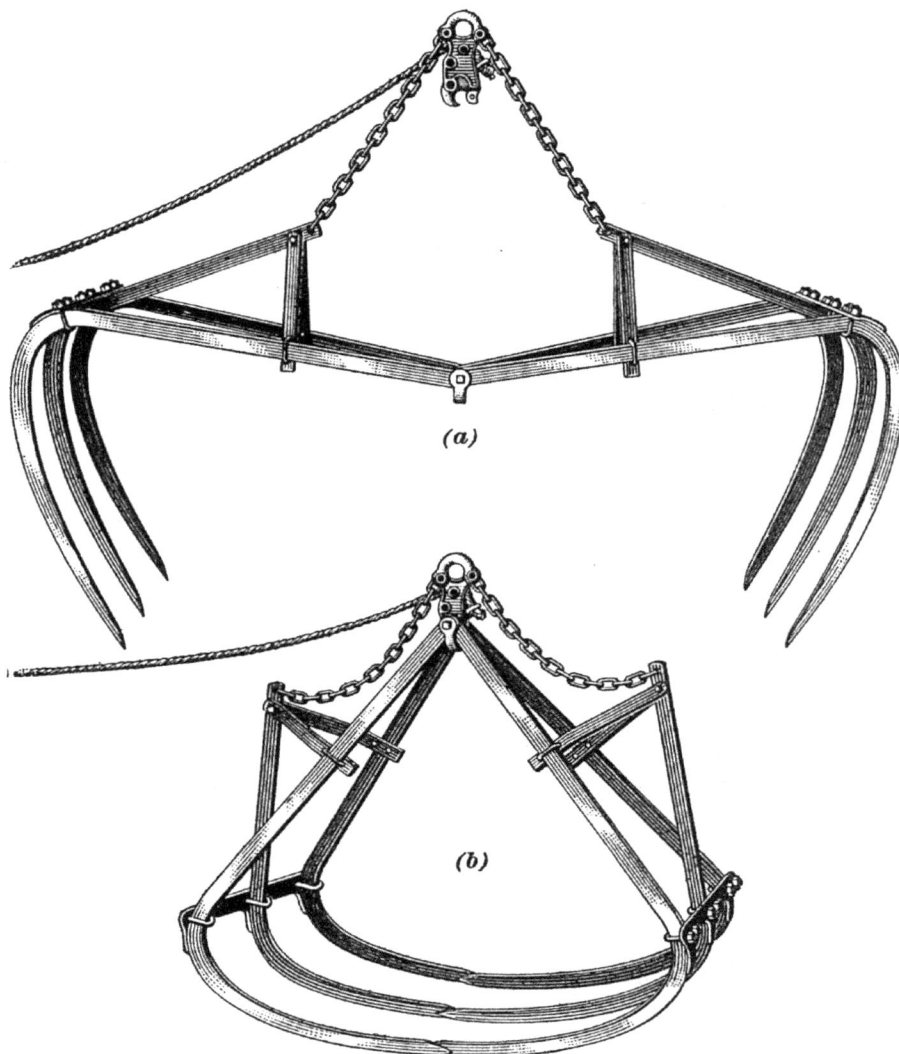

(a)

(b)

Figure 19. A grapple fork shown (a) open and ready to stick into a jag of loose hay (or having just released a load) and (b) closed and ready to lift hay.
Hay forks are designed first and foremost to use with track and trolleys used in hay barns. However these forks are also useful and practical with certain types of outside stackers (as will be shown).

moves forwards, the hay is picked up and placed on an endless belt, which elevates and drops it on the rear end of the rack, from which point it is distributed over the load by hand.

In humid sections, especially in the North, by far the largest part of the hay grown is stored in barns. The cost of buildings for storing hay is often considerable. It is essential that the framework of hay barns be strong and well built, but it is not so important that the walls be perfectly tight and weatherproof. The roof, however, must be water-tight. Some hay barns are made with a good roof supported by framework, but with the sides open.

Fig. 20. A view of one of many different setups for fork, track and trolley filling of barns.

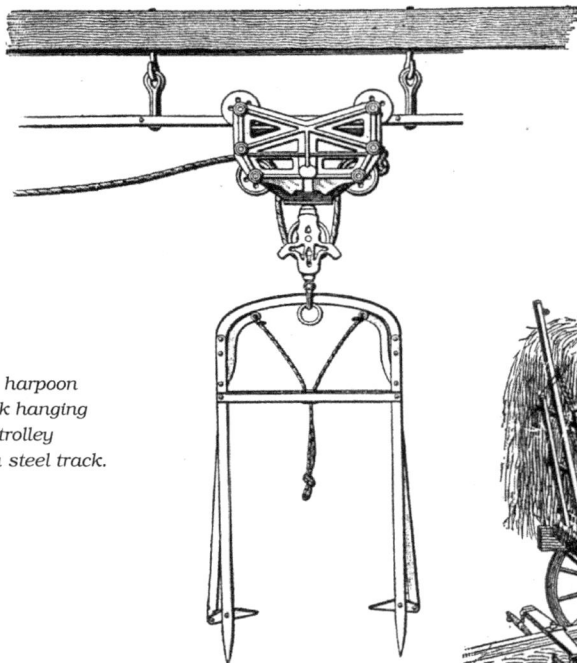

Figure 21. A harpoon style hay fork hanging from a barn trolley running on a steel track.

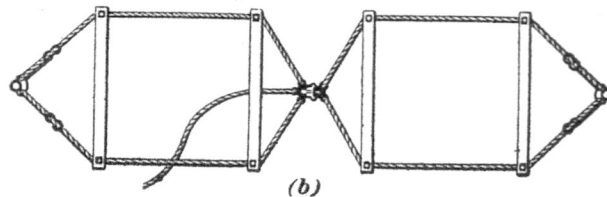

This is an economical and satisfactory way of building barns if they are intended only for storing hay or straw. Most barns are built for the double purpose of sheltering horses and cattle and of storing hay, grain, and other products. This is a great convenience, as much time is saved by keeping cattle and horses under the same roof where their winter supply of feed is stored.

Fig. 22. This wagon picture illustrates a hay sling system for offloading hay. This sling needs to be laid down before the hay is loaded on the wagon and care needs to be taken not to attempt too large a jag's lift.

When To Cut Hay?

But enough of this radical thinking for the time being. Many of you are expecting to have, in this book, conventional information on these issues. So the following is a slightly edited and annotated presentation of the prevalent approach to timing and quality issues.

The actual difference in feeding value per ton between hay cut in early bloom or before and that cut when the crop is in full bloom will depend on the class of stock to which it is to be fed. The difference is greatest where the hay is used as a vitamin supplement for swine or poultry. For this purpose, hay that is early-cut, leafy, and well-cured is essential, as the vitamin content is otherwise much lower. Early-cut hay should also always be used, if possible, for dairy calves and sheep. In storing the hay crop, some of the best hay should therefore be put where it will be accessible for these uses.

The difference in value between early-cut hay and that cut in full bloom is much less for dairy cows and beef cattle, especially when silage is fed in addition to the hay. For work horses or light horses hay cut when in bloom is preferred to that cut earlier, because hay cut extremely early is apt to be too laxative. Hay cut very late--in the seed stage or later--has a low value for all classes of stock.

Though hay cut when in early bloom or before had a slightly higher value per ton than that cut a little later, the difference was not sufficient to offset the lower yield. Hay cut when in seed was worth appreciably less than that cut when in bloom. The difference was greatest when the cows were fed only a small amount of concentrates in addition to the hay.

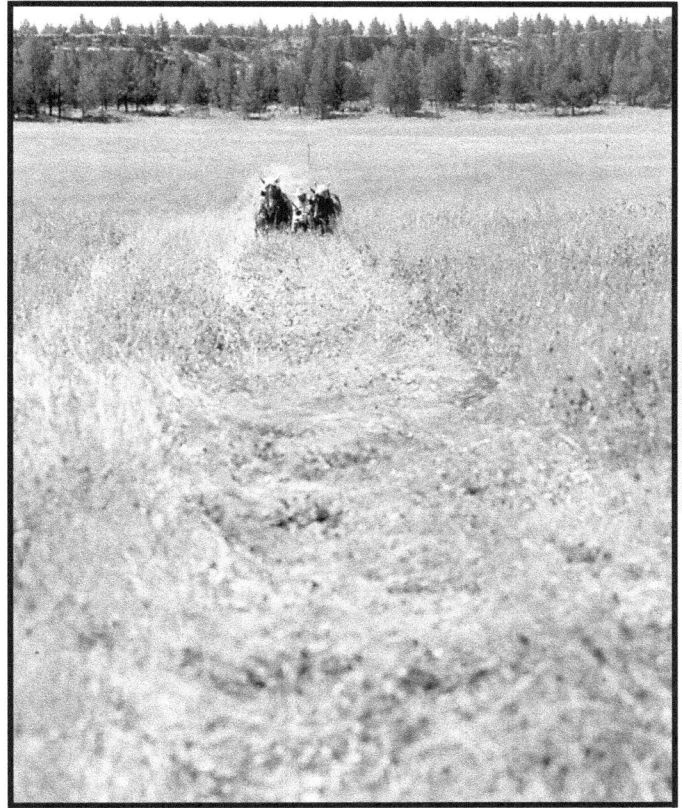

Figure 23. (above) Author mows grass, in arid Oregon high desert, at terminal growth to encourage legumes.

Figure 24. (below) Bud Evers mows the same field the following year. This is a second cutting showing phenomenal mixed grass/legume stand. No chemical fertilizers were used on this field for 10 years. The fertilizers were timing of grazing and mowing & top dressings of manures. These photos are a strong argument for a different approach to when to mow.

Figure 25. *Robert Clark of Whitehall, Montana baling with a team of drafters and a New Holland motorized unit.*

Moisture Content

The primary object in haymaking is to dry the green plants enough so that the hay can be safely stored without heating unduly or becoming moldy. For hay to keep safely in barn or stack, the water content must be reduced to not more than 25 per cent. If the hay is chopped or baled at time of storage the percentage of water should not be over about 22 per cent (16% is better). Hay containing too much moisture when stored will undergo pronounced fermentation and become very hot. The value may be greatly decreased because of mold or the losses of nutrients which occur in the extensive fermentation. Also, there is always danger of spontaneous combustion when hay is stored with too much moisture. The slight fermentation, or sweating, that occurs in properly-cured hay when it is stored does not cause any marked loss of green color or nutrients. It even seems to improve the aroma and palatability of the hay.

Some believe that it is safer to store hay with a high moisture content when this is due to the moisture content of the plants, than when the hay has been dampened by rain or dew. It is not wise to rely on this. The only safe way to avoid serious loss is not to store any hay when it is too high in moisture, no matter what the cause.

A practical method used by farmers to find out when hay is cured sufficiently for storage is to twist a wisp of it in the hands. If the stems are slightly brittle and there is no evidence of moisture when the stems are twisted, the hay can be stored safely. In excellent curing weather the hay may appear drier than it really is, especially if cured chiefly in the swath. This is because the leaves will be dry and brittle, while the stems are still too high in moisture.

Here is a simple method for determining whether or not hay is dry enough for safe storage. A large representative handful is carefully selected and then bent or twisted to break the stems somewhat. From the center of the sample a portion is cut, approximately as long as a quart glass jar and sufficient to fill it loosely. The sample is placed in the jar, a teaspoonful of fine-grained table salt is added, and the cover placed on the jar. The sample should not fit so snugly as to prevent free circulation of the salt.

The jar is shaken about 100 times to keep the salt and hay moving about. Then the jar is held upside down and the salt is shaken into the cover, where it can be examined. If the hay is dry enough for safe storage, the salt will still be in small grains. On the other hand, if the hay has more than about 25 per cent water, the salt will have taken up moisture and be gathered in clumps. Samples that are distinctly too wet will change the salt in about 30 seconds. In border-line cases, the sample should be shaken again and allowed to stand for a few minutes.

If it seems wise, because rain is threatening, to haul hay to the barn before it has reached the desired dryness, it should be spread out well in a mow to a depth not greater than about 3 to 5 feet. Special care should be taken not to leave any large, compact masses where the hay falls from the hay fork or sling. If possible, such hay should not be covered with other hay until it has cured out somewhat. Green hay, well spread over dry hay with room for air to flow, will dry nicely whether in the stack or in the barn.

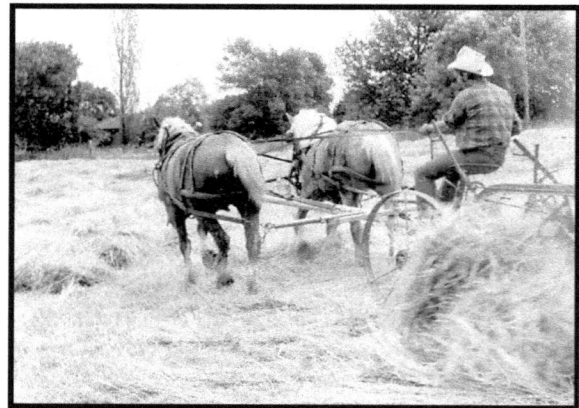

Figure 26. *The author raking hay with a pair of Belgians and a side delivery rake in Junction City, Oregon in 1974. Horsedrawn rakes travel at speeds far slower than tractors thereby turning the hay with greater gentleness. This means less leaf and bloom loss.*

Losses of Nutrients in Hay-making.

Some nutrients are always lost in the field-curing of hay, but under favorable conditions this loss is not large. However, the wastage may be great unless proper care is taken. These losses are not due to the mere drying out of the forage plants. Experiments have shown definitely that the drying of green grass or other green forage at ordinary temperatures does not reduce its digestibility. Also, when plants are dried without any bleaching or fermentation, they have a high content of the vitamins that are of importance in stock feeding.

Losses by shattering.--If legume hay becomes too dry, there will be heavy losses of leaves by shattering when the hay is raked or loaded. This loss is highly important, because the leaves contain 2 or 3 times as high a percentage of protein as do the stems. Also, the leaves are much richer than the stems in minerals and vitamins and are lower in fiber. The proportion of leaves is therefore one of the important factors in determining the feeding value of any particular lot of hay.

Unless great care is taken, the loss of leaves and fine stems is apt to be especially great in the making of alfalfa hay in dry climates. In Colorado tests, even under favorable conditions, the loss of leaves and stems amounted to 350 lbs. for each ton of hay taken off the field. When the hay was allowed to become too dry, one-half the total weight and even a much larger part of the feeding value was lost.

Losses by fermentations and bleaching.--Even under favorable conditions, the curing of hay in the field is not merely a simple process of drying. Fermentations take place in which some of the organic nutrients, especially the sugars and starch, are oxidized to carbon dioxide and water, thus being lost. Also,

Figure 27. A late model raker-tooth style hay loader common all across North America.

fermentation has a very destructive effect on the carotene in hay. If the weather is favorable and the hay is cured by proper methods, without undergoing pronounced heating, the losses by fermentation will be relatively small. On the other hand, if extensive fermentations take place, as in brown hay, a heavy loss of carbohydrates will occur and also the carotene and other vitamins will be largely destroyed.

If hay is badly bleached by long exposure to the sun, nearly all the carotene will be lost. In general, the amount of carotene in hay is proportional to the greenness in color. Green-colored hay is nearly always rich in carotene and straw-colored or brown hay very poor in it.

The losses by fermentation do not cease when hay is stored in the barn. If the hay is well dried and does not heat unduly in the mow, the loss of dry matter in barn storage for six months will usually not exceed five to seven per cent. When hay is under-cured and molding or severe heating occurs, the loss is much greater. A continual loss of carotene takes place in hay during storage.

Loss by leaching.--If hay that is already nearly cured is exposed to heavy and prolonged rain, especially when it is in the swath, severe losses may occur through leaching. At least 20 per cent of the protein and considerably more of the nitrogen-free extract may thus be lost by leaching.

However, unless the rain is so heavy that the hay is thoroughly soaked and washed by the rain, there will not be much loss by leaching. Also, severe leaching results only when the hay has dried out considerably before the rain comes. Hard rain soon after the crop is cut does not cause much loss.

Extent of total losses in hay.--The total loss of nutrients in hay during field curing and subsequent storage in the mow before feeding will vary widely. In fairly good hay-making weather and with proper meth-

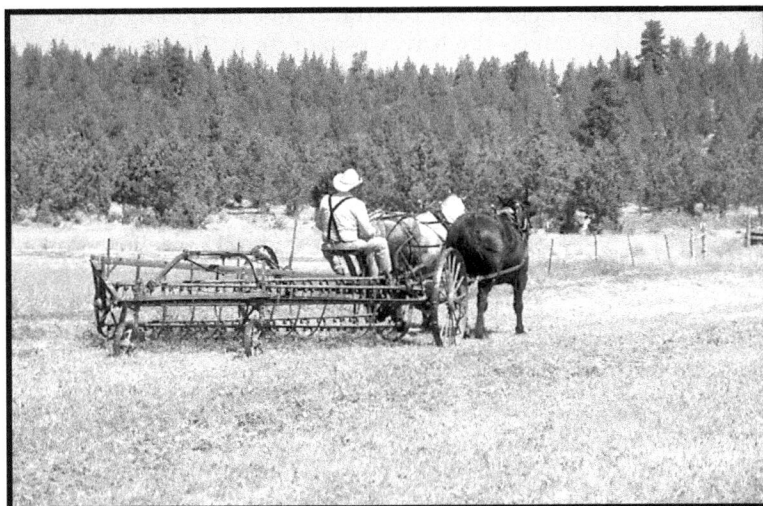
Figure 28. Bud Evers uses a New Idea side delivery rake to gather up "scatter" the short hay which fell through the buck rake teeth. This short hay would then be picked up clean by the hayloader and wagon.

Fig. 29. The grey team is waiting to pull up the stacker (hidden from view behind the stack) while two buck rake teams wait with additional jags. Such a crew will finish this mid-sized stack quite fast.

ods, the total loss of dry matter from the green crop to the manger should not exceed 20 to 30 per cent for legume hay and 10 to 15 per cent for grass hay. Under unfavorable conditions the loss will be considerably higher.

In extensive Vermont experiments back in the 1920's the average total loss of dry matter from the time the crop was cut until it was fed about 6 months later was 16.4 per cent for several lots of hay made in reasonably good weather. Most of this loss occurred before storage, the loss of dry matter in the mow averaging only 5.4 per cent. The total loss of dry matter was considerably greater for alfalfa or clover hay, averaging 23.4 per cent, than for timothy hay, in which the average total loss was only 9.5 per cent. This difference was undoubtedly due chiefly to shattering of some of the legume leaves.

Methods of curing hay.

Hay dries much more rapidly in the swath than in the windrow, even if the windrow is small and loose. However opinions vary and claims are often made to the contrary. The larger the windrow is, the slower will be the curing, and the rate is still slower in cocks.

Though hay cures rapidly in the swath, it is not advisable to cure it entirely there, because the leaves become dry and brittle long before the stems have dried out sufficiently. With legume hay, especially alfalfa, a heavy loss by shattering of the leaves will then take place when the hay is handled. Also, the prolonged exposure of most of the leaves to the sunlight will bleach the hay and destroy much of the carotene.

On the other hand, if the green forage is raked into a windrow, even a small, loose one, immediately after it is mowed, an unduly long time will be

needed for curing, except in a very dry climate. Fermentation and molding may occur, which will seriously damage the hay. Such slow curing also increases the risk of damage from rain.

To produce high-quality hay, leafy and green in color, it is usually best to let the crop lie in the swath until it is partly cured. Then it should be raked into small, loose windrows, preferably with a side-delivery rake. If good weather continues, the curing should be completed in the windrow and the hay hauled to the barn directly from the windrow. To avoid serious loss of leaves in very dry climates, it may be necessary to handle the hay only early in the morning before it has become too dry.

Opinions differ as to how much the crop should dry out in the swath before it is raked. Some advise windrowing just as soon as the plants are thoroughly wilted. Each farmer will have to judge his own crop.

When the weather is such that the hay cures rather slowly, it may be advisable in a few hours to turn the windrows partly over, in order to hasten the curing. Turning may also be necessary if the hay is wet by rain when in the windrow.

As long as the leaves remain alive, they may perhaps help to dry out the stems, by pulling water from them as the leaves evaporate it. This has been given as an added reason for windrowing the hay before the leaves have been killed by drying.

It is a common belief that the windrows made by a side-delivery rake cure more rapidly than those made with a dump rake. This is probably only because they are usually smaller. The hay tedder was formerly often used to hasten the curing of heavy cuttings of hay in the swath. This has been largely abandoned, because it tends to shatter the leaves of legume hay, unless the tedding is done early in the curing process.

A combination power hay mower and crusher, or haybine, causes the forage

Figure 30. A motorized forecart and four Belgians provide the power for the new John Deere mower/conditioner.

to pass between rollers after it is cut by the cutter bar of the mower, to crush the stems and hasten the drying the usual manner. Tests have shown that the curing time is usually reduced decidedly by crushing. But this author's experience in wetter Pacific Northwest climates has been that the crushed stems are more prone to reabsorb dew. These machines are large and intense and require three, four or more draft horses to pull them.

Time of Day to Mow Hay.

Opinions have differed as to whether hay will be ready for storage as soon if it is cut early in the morning when wet with dew, as it will be if cutting is delayed until the dew has dried. In New York tests (1920's) swaths of hay were cut on various days at 8 A.M. when wet with dew, and others were cut after the dew had dried, usually well towards noon. In most cases the hay that had been cut early was drier at 5 P.M. than that cut later. This indicates that there is no advantage in delaying cutting until the dew has dried.

In early experiments at Cornell University, it was found that because of the formation of carbohydrates during the day, hay crops contained appreciably more starch and sugar when harvested late in the afternoon than when cut in the morning. The total yield of dry matter was also significantly increased by cutting late in the day. A part of these readily available carbohydrates is used up in respiration during the night or is transferred to the roots. The content was therefore lower the following morning than the night before.

Other investigators from elsewhere have concluded from their studies that late-afternoon harvesting does not make enough difference in composition or yield of a hay crop to be of practical importance. If there is an appreciable difference in content of starch and sugar when the crop is harvested, considerable of the additional amount is used up in the respiration and fermentations that normally take place in hay making.

Curing hay in cocks.--Before the modern, labor-saving methods of making hay were developed, it was generally advised that to make hay of the best quality, regardless of expense, it should be put in well-made cocks as soon as it had dried enough so it would not heat or mold. This method preserves the color and carotene content, but it requires too much labor for use when there is any considerable acreage. Also, in a humid climate there is much risk of damage from rain before the cocks dry out, unless one goes to the further considerable expense of covering the cocks with hay caps. Therefore the cocking of hay is not now a common practice in most sections of this country.

However, it is probably wise to cock hay sometimes to lessen damage from rain. Hay is injured most by heavy rain when it is nearly dry enough to store. Therefore, if there is every prospect of a hard storm at that time, it is a good plan to hurry and put the hay into large, well-made cocks. Unless the rain is unusually heavy, these will not wet through, there will be practically no loss from leaching, and the color and vitamin A content will be largely preserved. When the weather clears and the outside of the cocks has dried, it may be necessary to open them somewhat to dry out the interior before the hay is hauled.

Curing devices for very rainy districts.--In some sections of northern Europe, especially in Scotland, Norway, Sweden, and Finland, where the summer weather is cool, the humidity high, and rain very frequent, it is nearly impossible to cure hay satisfactorily in the field by ordinary methods. In these regions, after the hay is partly cured, it is generally put on devices which allow the air to penetrate.

Similar methods are used to some extent in this country, especially in curing cowpea hay or peanuts in the southeastern states. All of these methods take much labor and are too expensive to be used where good hay can generally be made by ordinary methods.

A common method under such conditions is to cock the hay, after it is partly cured, on tripods, where it is left until dry enough for storage. The tripod consists of 3 poles, joined at the top, and generally with cross pieces at right angles to keep the hay off the ground. This permits making a large, well-ventilated cock that will contain as much as 500 lbs. or more of hay after curing. Sometimes single poles with crosspieces near the bottom are used instead of tripods. These are set in holes made in the ground. This method is often used in curing peanuts in the country. Another method used in some sections of northern Europe is to drape the hay in layers over the wires of temporary fences put up in the hay field.

In this country such crops as cowpeas are sometimes stacked before thorough curing, the stack being separated into layers with air spaces be-

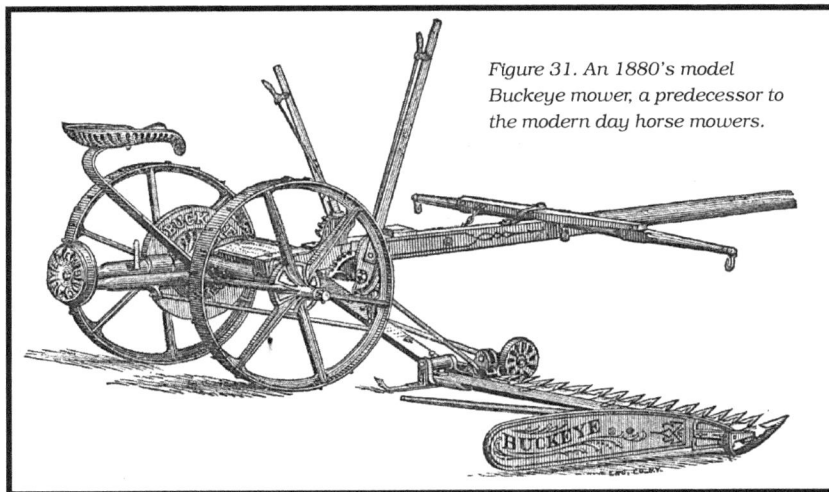

Figure 31. An 1880's model Buckeye mower, a predecessor to the modern day horse mowers.

tween, by means of rails supported at each end. Another method occasionally used in the southern states is to stack the hay on cheap sleds or two-wheeled trucks, when it is dry enough to cock. It is then covered with canvas or muslin covers and allowed to stand until thoroughly cured, when it is drawn to the barn on these sleds or trucks.

Storing various grades of hay.--No matter what method is used in handling and storing hay, good judgment should be used in putting the various kinds and qualities of hay in a place where they will be available when wanted, and not be covered with other hay. On a dairy farm, it is a mistake to put all the late-cut hay on top of the higher-quality, early-cut hay. If this is done, the cows may drop off severely in production in the fall when they get only the poorer hay.

The hay-loader method.--When the hay is loaded with a hay loader, it is generally unloaded at the barn with a hay fork or hay slings and mowed (or forked) away by hand.

When the hay is unloaded with a hay fork or slings, care should be taken to distribute it well in the mow and not leave it in large, compact masses where it falls from the fork or sling. Such masses are especially apt to heat badly, if the hay is not thoroughly dry.

The hay-loader method requires somewhat more man labor per ton than some of the newer methods, but the cost of equipment is much lower than where a windrow baler is used.

In the southeastern states, in certain of the New England States, and in the Pacific Coast States, most of the hay was still loaded by hand up until 1944. This is slower and requires more man labor per ton than when a hay loader is used. Pitching hay on a load by hand is also much harder work than loading it with a hay loader. Though there is a minimum expense for equipment, the total cost per ton of handling the hay is usually considerably higher than with a hay loader, except where the acreage of hay is very small.

The buck-rake, or sweep-rake, method.--In the Great Plains and the northern Mountain States most of the hay is hauled from the windrow with buck rakes, or sweep rakes. Commonly the hay is stacked in the field by hay stackers or other mechanical devices. Recently, many farmers in the Corn Belt and eastward, where hay is generally stored in barns or sheds, have adopted the buck-rake method of handling hay. The hay is gathered up from the windrow with the buck rake and hauled on it to the place where it is to be stored. Here it is put in the mow with hay sling, grapple fork, or blower. The hay is too loose for successful use of a harpoon hay fork.

Where the distance from the hayfield to the storage place is not long, the buck rake is the cheapest method of handling hay. The amount of labor per ton is small and the cost of equipment is relatively low.

Hay of good quality, free from mold and of good

Figure 32. Ten feet in the air Tony Miller and Pete Ridder do an artful job of spreading the hay and building strong walls for the haystack.

color, can be made with the windrow baler if certain precautions are taken.

The hay should preferably be a little drier than is safe for storage of loose hay in a large mass. The hay should certainly not contain more than 25 per cent water, and 20 to 22 per cent is better. If legume hay gets too dry before it is baled, the leaves will shatter badly. Otherwise, windrow baling saves the leaves well.

Barn-drying of hay.--When storing hay with a rather high percentage of moisture, some farmers sprinkle on 10 to 20 lbs. of salt per ton of hay in the belief that it aids in preventing mold or undue heating. Opinions differ widely concerning the value of this practice. In some of the experiments conducted to study the matter, salting has not improved the quality of under-cured hay. It may perhaps make poor hay somewhat more palatable.

It seems clear that salting hay is no insurance against spoilage, or even against spontaneous combustion if the hay is much too damp. In fact, too great reliance on salting may be dangerous. The only safe plan is not to store hay unless it is dry enough for safety.

Spontaneous combustion.--If hay or other dry forage containing too much moisture is put into a mow or stack, rapid fermentations take place in which a large amount of heat is produced. In a large mow or stack most of this heat is retained in the mass, causing a rapid rise of temperature. In these fermentations highly unstable organic compounds are apparently formed which are readily oxidized.

At temperatures of 150 degrees to 175 degrees F. all bacteria or molds are killed or made inactive, but the oxidations continue, and the mass may become extremely hot. Finally, the hay begins to char and spontaneous combustion may occur and the mass burst into flames. This generally happens a month or 6 weeks after the hay is stored, but it may occur sooner.

The only way to avoid such loss is never to store hay in a large mass unless it is thoroughly cured. Chopped hay must be drier than long hay for safe storage. For safety, mows should be inspected at least twice a week during the first 2 months after the hay is stored. If hay in a mow or stack heats badly within 2 or 3 days after storing, and pungent odors, with much vapor, are given off, it should be removed at once and spread out to dry. Removing the hay later may only hasten spontaneous combustion.

If danger threatens, it is wise to take the temperature down in the hot spots in the hay, by lowering, in a pipe driven into the hay, a thermometer that reads up to 200 degrees F. If the temperature goes above 160 degrees F., there is grave danger, and at 175 degrees to 185 degrees fire pockets may be expected. Before removing the hay then, a fire department should be called, if available, for the hay may burst into flame when the air reaches it. In some instances compressed carbon-dioxide gas from the containers used in soda fountains has been introduced into the mow through a pipe, to reduce the heating and prevent fire.

Stacks should never be built on old, rotten stack bottoms. Baled hay, grain in the sheaf, or other heavy material should not be put on top of hay in a mow which is going through the sweat, for it will prevent the escape of heat and gases. Crops upon which rain has fallen should receive extra care, and should not be housed until completely dry.

Measurement and shrinkage of hay.--The stack is measured and the volume and tonnage are then computed by one of several rules in common use. The density of the hay and therefore the number of cubic feet required per ton are affected by many factors. Hay that was stored when slightly damp will be more compact than that which was very dry. The coarseness of the hay also affects its density.

For hay 30 to 90 days in the stack, 485 cubic feet per ton for alfalfa; 640 for timothy and timothy mixed hay; and 600 for wild hay. For hay over 90 days in the stack, 470 cubic feet per ton for alfalfa; 625 for timothy and timothy mixed hay; and 450 for wild hay. The same approximate figures may be used for hay in a mow.

The volume of hay in a mow can readily be computed, but it is more difficult to determine the volume of a stack.

The volume of a rectangular or oblong stack in cubic feet is:

For low, round-topped stacks:
(0.52 x O) - (0.44 x W) x W x L
For high, round-topped stacks:
(0.52 x O) - (0.46 x W) x W x L
For square, flat-topped stacks:
(0.56 x O) - (0.55 x W) x W x L
The volume of a round stack is: (0.04 x O) - (0.012 x C) x C2

In these formulas O is the "over" (the distance from the base on one side of the stack, over the stack, and to the base of the other side); W is the width; L is the length; and C is the circumference of a round stack, or the distance around it. In measuring a round stack, the "over" should be an average of two measurements made at right angles to each other.

Hay stored in the mow will shrink in weight, due to drying out and also to fermentations taking place during the sweating process, in which nutrients are broken down into carbon dioxide and water. The shrinkage in weight will vary, depending on the water content of the hay when placed in the mow, and may reach 20 per cent or over. The losses of dry matter in hay during storage have been discussed previously.

When hay is stacked, the shrinkage in weight is greater than in a mow, since the outside of the stack is exposed to the weather. A stack 12 feet in diameter has about one-third of its contents in the surface foot.

The difference in the loss of dry matter in stacked hay and in that stored under cover will depend on how well the stack is made, on its size, and especially on the climatic conditions.

Figure 33. Justin Miller, the author's son, drives Molly and Red on a wide wheeled McCormick Deering #9 High gear mower at Singing Horse Ranch.

UNDERSTANDING THE PLANT WHICH WILL BECOME HAY

Making or curing hay is an important step which requires a knowledge of the plant and good management. Too often, after the crop is grown, a large per cent is damaged because of rains, heavy dews, by being exposed for a long time to the heat of the sun or to mismanagement in gathering. Rains and dews are especially detrimental to legumes. After alfalfa or clover has been wet, even slightly, the hay is very much like tea leaves after they have been steeped. Mild rains and heavy dews do not affect timothy and other grasses as much as the legumes. Hay exposed to a broiling sun for a day or more not only loses its appetizing flavor, but many of the leaves, on account of their being dried rapidly, break off and are wasted. A hot sun will dry and crinkle the leaves, preventing the escape of moisture from the stems, which should take place through the leaves. Moisture thus retained in the stems, when placed in the stack or mow, will mold. Hay should be cured and not sun-burned. In order to cure hay, it should not only be subjected to a reasonable amount of heat, but to the action of the air.

The old plan of curing hay was a very good one, but it entailed losses and unnecessary expense. After cutting, the hay would lie in the swath until it was dry. It was then raked into windrows and subsequently made into cocks where it would remain until cured.

The modern plan is to mow the grass and let it remain until it is wilted but not dry. It is then raked into windrows with a left hand side delivery rake. The left hand rake goes around the field in the same direction as the mower, inverting completely two swaths, forming the grass into a light, fluffy windrow. In order to understand the value of this method we should know something of nature's process of curing hay.

Grass in a green state contains a large per cent of moisture. As soon as the grass is cut nature attempts to throw the moisture off. A large amount of the moisture is in the stem, and the only way it can be gotten out is through the pores in the leaf. If the hay is left on the ground after it is cut until the leaves dry, the pores will be completely closed and the moisture will be retained in the stem, causing the hay to mold when placed in the mow or stack, although it may appear to be dry. By forming the grass into a fluffy windrow there is a free circulation of air which tends to hasten evaporation. If hay is raked with a dump rake it is placed in a compact bunch which prevents the free circulation of air and the curing process is materially lessened. If the hay is extremely heavy and the atmosphere is full of humidity, it may be necessary to use the rake the second time, rolling the mass over. In the event of rain after the hay has been raked and before it is gathered, it is always well to use the side delivery rake the second time. It has been demonstrated that it is perfectly feasible to use the left hand rake immediately behind the mower, especially in the second cutting of alfalfa and heavy clovers, but with the first cutting of heavy legumes it is advisable to permit it to lie in the swath until it has wilted.

It has been proven by many experiments that this process will put the hay in condition to be stored in one-half the time required by the old process, and the hay keeps in the stack or mow in perfect condition.

The time required to cure hay depends upon the locality, humidity, and the amount of moisture in the ground at the time of harvest. In semi-arid sections hay cures very rapidly, but in the more humid sections the process is necessarily slower. It should also be remembered that the first cutting of a legume is more difficult to cure than subsequent cuttings.

Figure 34. A three abreast of Belgians walk along pulling a forecart and mower/conditioner.

HAY Grades

In 1917 The National Hay Association came out with a little booklet with official grade standards for hay. Here they are for curiosity sake.

• *No. 1 Timothy Hay - Shall be timothy, with not more than one-eighth (1/8) mixed with clover or other tame grasses, may contain some brown blades, properly cured, good color, sound and well baled.*

• *Standard Timothy Hay - Shall be timothy, with not more than one-eighth (1/8) clover or other tame grasses, may contain brown heads and blades, otherwise good color, sound and well baled.*

• *No. 2 Timothy Hay - Shall be timothy, not good enough for No. 1, not over one-fourth (1/4) mixed with clover or other tame grasses, fair color, sound and well baled.*

• *No. 3 Timothy Hay - Shall include all timothy not good enough for other grades, sound and reasonably well baled.*

• *No. 1 Light Clover Mixed Hay - Shall be timothy, mixed with clover. The clover mixture not over one-third (1/3), properly cured, fair color, sound and well baled.*

• *No. 1 Clover Mixed Hay - Shall be timothy and clover mixed, with at least one-half (½) timothy, good color, sound and well baled.*

• *No. 2 Clover Mixed Hay - Shall be timothy and clover mixed, with at least one-fourth (1/4) timothy, reasonably sound and well baled.*

• *No. 1 Clover Hay - Shall be medium clover, containing not over fifteen (15%) percent timothy and five (5%) percent other tame grasses, properly cured, sound and well baled.*

• *No. 2 Clover Hay - Shall be clover, sound and reasonably well baled, not good enough for No. 1.*

• *Choice Prairie Hay - Shall be upland hay, of bright, natural color, well cured, sweet, sound, and may contain three (3%) percent weeds.*

• *No. 1 Prairie Hay - Shall be upland, and may contain fifteen (15%) percent midland, both of good color, well cured, sweet, sound, and may contain eight (8%) percent weeds.*

• *No. 2 Prairie Hay - Shall be upland, of fair color, and may contain twenty-five (25%) percent midland, both of good color, well cured, sweet, sound, and may contain twelve and a half (12 ½%) percent weeds.*

• *No. 3 Prairie Hay - Shall include hay not good enough for other grades and not caked.*

• *No. 1 Midland Hay - Shall be midland hay, of good color, well cured, sweet, sound, and may contain three (3%) percent weeds.*

• *No. 2 Midland Hay - Shall be of fair color, or slough hay of good color, and may contain twelve and a half (12 ½%) percent weeds.*

• *Packing Hay - Shall include all wild hay not good enough for other grades and not caked.*

• *Sample Prairie Hay - Shall include all hay not good enough for other grades.*

JOHNSON HAY

• *No. 1 Johnson Hay - Shall be Johnson hay with not more than one-tenth (1/10) mixed with clover or some other native grasses, good color, sound and well baled.*

• *No. 2 Johnson Hay - Shall be Johnson hay with not more than one-fifth (1/5) mixed with clover or other native grasses, sound, bright color and well baled.*

• *No. 3 Johnson Hay - Shall be Johnson hay with not more than one-fifth (1/5) mixed with clover or native grasses, sound, coarse, brown in color and well baled.*

• *Choice Alfalfa and Johnson Mixed Hay - Shall be three-quarters (3/4) alfalfa of good color, sound and well baled, mixed with one-quarter (1/4) of Johnson hay of good color.*

• *No. 1 Alfalfa and Johnson Mixed Hay - Shall be alfalfa and Johnson hay, one-half (½) of each, of good color, sound and well baled.*

• *No. 2 Alfalfa and Johnson Mixed Hay - Shall be alfalfa and Johnson hay not good enough for No. 1, and shall contain one-fourth (1/4) alfalfa and three-fourths (3/4) Johnson hay, with not more than ten (10%) percent of other native grasses. Same shall be sound and well baled.*

BERMUDA HAY

• *No. 1 Bermuda Hay - Shall be Bermuda grass mixed with not more than ten (10%) percent of native grasses, color of uniform greenish cast, sound, tender and well baled.*

• *No. 2 Bermuda Hay - Shall be Bermuda grass mixed with not more than one-fourth (1/4) native grasses, color greenish cast with not more than fifteen (15%) percent brownish blades, sound, tender, and well baled.*

• *No. 3 Bermuda Hay - Shall be Bermuda grass mixed with not more than one-fourth (1/4) native grasses, color of brownish cast, sound and well baled.*

LESPEDEZA HAY

• *No. 1 Lespedeza Hay - Shall be Lespedeza mixed with not more than fifteen (15%) percent native grasses, reasonably fine, well cured with leaves on, green color, sound and well baled.*

• *No. 2 Lespedeza Hay - Shall be Lespedeza mixed with not more than twenty (20%) percent native grasses, slightly brown color, well cured, sound and well baled.*

ALFALFA

• *Choice Alfalfa - Shall be reasonably fine leafy alfalfa of bright green color, properly cured, sound, sweet, and well baled.*

• *No. 1 Alfalfa - Shall be reasonably coarse alfalfa, of a bright green color, or reasonably fine leafy alfalfa of a good color and may contain two (2%) percent of foreign grasses, five (5%) percent of air-bleached hay on outside of bale allowed, but must be sound and well baled.*

• *Standard Alfalfa - May be of green color, of coarse or medium texture, and may contain five (5%) percent foreign matter; or it may be of green color, of coarse or medium texture, twenty (20%) percent bleached and two (2%) percent foreign matter; or it may be of greenish cast, of fine stem and clinging foliage, and may contain five (5%) percent foreign matter. All to be sound, sweet and well baled.*

• *No. 2 Alfalfa - Shall be any sound, sweet and well baled alfalfa, not good enough for Standard, and may contain ten (10%) percent foreign matter.*

• *No. 3 Alfalfa - May contain twenty-five (25%) percent stack spotted hay, but must be dry and not contain more than eight (8%) percent of foreign matter; or it may be of green color and may contain fifty (50%) percent of foreign matter; or it may be set alfalfa and may contain five (5%) percent foreign matter. All to be reasonably well baled.*

• *No Grade Alfalfa - Shall include all alfalfa not good enough for No. 3*

Hay System Variables

Chapter Three

Loose Hay in Stacks
versus Loose Hay in the Barn
versus Baled Hay
versus ?

While it is true that real horsepower, or the use of horses and/or mules as motive power, may dictate certain optimum hay system perimeters, the individual farmer's circumstances, goals and preferences will be shown to have far more bearing on the selection of an appropriate haying system.

It is a fact that some folks plow, till, mow and rake because it is something they enjoy doing with their horses. Certainly nothing wrong with that, but we do need to separate such motives from the so called practical concerns. Concerns like;

- *how many acres can I do with horses?*
- *can I get the haying work done with horses or mules as the power source?*
- *are there certain crops or crop circumstances which horses can't handle?*

and so on... These issues were addressed somewhat in the introduction but more does need to be said.

Figure 1. A young coyote works the swath looking for mice at Singing Horse Ranch.

First, when we speak of *hay systems* we're talking about the equipment, procedures, and layout employed to harvest, cure, pick up and store the hay crop. There are component parts/pieces/procedures/conditions which may be combined into a customized haying system. The component parts include but are not limited to:

- *number of acres to be hayed*
- *number of ready work animals available*
- *number of people available (& qualified) to help*
- *size[1] and number of mowers*
- *size and shape of fields and/or mowed lands*
- *nature of hay crop and goals of the farmer*
- *type and style of rake(s) employed*
- *the choice of loose hay or baled hay*
- *method of gathering and/or picking up loose hay*
- *method of baling hay and picking up bales*
- *stacking versus putting loose hay in barn(s) or shed(s)*
- *tripodding, polestacking or fence stacking*
- *quantity of, and class of, stock to be fed hay*
- *winter feeding procedures*

In all likelihood the individual farmer will need the more in-depth information from the following chapters, in addition to the climatological and cultural histories of the immediate region, in order to make an intelligent decision about the right hay system. But the introductory which follows may help organize thoughts on the subject.

What Do Horses in Harness Suggest About the System of Choice?

You have decided, or are leaning towards, the choice of horsepowered haying. To many of you this information will seem far too basic. To others this information may surprise you and sway your decision.

HOW MANY ACRES?

Two full-sized, well-conditioned draft horses, working at a brisk walk, can - with a five foot old-style mower that is well-timed, lubed and sharpened - cut ten acres of hay in one day. At the walking speed of horses any type of mower with a 6 foot cut (and without breakdowns) will cover 12 acres in 10 hours: a 7 foot mower - 14 acres, an 8 foot mower - 16 acres.

Those same two horses can rake 20 acres with a conventional side delivery rake

1. *'Size of mower' refers to cutter bar length but also to larger engine driven models, such as haybines, which may affect not only acres per day but drying time.*

Figure 2. A 4 abreast of Belgians pulls a motorized forecart and New Holland Round Baler at Horse Progress Days in Ohio.

Hay System Variables

in one day. A double width rake can be handled by two to four horses or mules doubling the acres raked.

Those same two horses can pick up 4 - 6 acres of hay (depending on the tonnage per acre and the distance to the stack or barn) with a hayloader and wagon.

Two horses on a buck or sweep rake can push 5 acres of raked hay to the barn or stack in one day.

Another way to look at these numbers is from the standpoint of the acreage or the tonnage totals.

One teamster/farmer with two good horses *(a third standby replacement horse is always a good idea)* could reasonably handle 40 acres of hay IF he or she

2. *By "lands" we mean to refer to that fractional portion of a designated field handled separately and deliberately. Also note: we are not saying that two horses cannot handle forty acres of hay in a single shot, just suggesting that breaking it into smaller units is a smarter move.*

Figure 3. The "Mammoth Swinging Stacker" delivers a last jag to Tony to cap the stack at Singing Horse Ranch.

Figure 4. Just how fancy or modern do you want to get? The technology exists. With motorized forecarts which feature power take off and hydraulics any tractor implement can trail along and do its work.

has good ground help (at the stack or barn) and IF the acreage is divided into smaller "lands"[2] or fields. Four to ten acre field or "land" unit sizes seem to be excellent for single team outfits. This translates to not having too much hay on the ground at any one time as inclement weather or delays might affect curing and handling time. Measured in tonnage; two horses and one teamster/farmer might expect to put up 80 to 160 tons of hay in outside stacks, loose in the barn or in bales.

One teamster/farmer with four to six good horses (and the help of a second teamster & a stack or barn crew) can easily handle 80 to 120 acres of hay crop. Again, much is to be gained by dividing the acreage into smaller units, in this case possibly 8 to 20 acre units. And the addition of a third teamster speeds things up greatly.

Measures taken to spread the risk of damage during curing may actually ease the work load and streamline the process.[3]

A hayloader, wagon, team and 3 people can put up 12 to 15 tons of hay a day with outside stacks built by a stacker (or into a barn loft by trolley).

Whereas a wagon loaded & unloaded by hand (with 3 men) might result in 9 tons in the stack per day.

(See the appropriate chapters for more examples specific to your chosen system.)

CAN I GET THE WORK DONE WITH HORSES?
YES.

Up front we have to admit (smiling a 'so what' smile) YES haying with horses is labor intensive. Yes, you can put up your hay with motorized modern implements and cut the actual man hours to very little. But the quality of hay suffers, most definitely. And what is your time for? Notice we didn't ask what is your time worth? If you want to be farming and

working your horses in the field then haying time is a pure and fulfilling joy! In fact, we've had people offer to pay us for the opportunity to do it!

If you're paying the mortgage on 50 to 100,000 dollars worth of haying equipment you don't enjoy operating, are you better off?

If you're independent, working horses that you love and trust, building up soil fertility, piling up gorgeous hay, and doing all this alongside of family and friends, are you better off?

What a supreme irony that so many people work a lifetime to be able to afford a few retirement years doing what they love? It is true that no amount of money will buy that complete satisfaction and happiness. But hard work will often get you there quick.

And the nature of the work is such that it does not require young athletes. Fact is knowing how to work, as in most facets of life, is worth more than store-bought biceps.

Haying with horses is labor-intensive and that is good.

Can I get the work done with horses? Yes, but good well-maintained animals, harness, and implements must be combined with good overall farming practises.

If you have no experience with working horses and 'hand-made hay' you cannot count on making a first hay season work without competent help, actually physical on-the-premises help. No amount of book learning will prepare you to deal patiently and fluidly with the little (and sometimes big) mishaps like loose mower guards, broken pitman sticks, snakes in the hay as it cascades up the hayloader, broken pitchfork handles, balky horses, jammed side delivery rakes, broken buck rake teeth, hormone-fueled show-offs (equine and/or human), gaps in the stack wall, flat tires on the loaded hay wagon, hay crew no-shows, old weak harness leather, sore shoulders on the animals, etc. etc.

To counter the above cautionary notes this author needs to reiterate statements from the book's introduction. You don't need to be born to this work in order to be able to do it. You can learn it, you can do it. And when you do your homework, and protect yourself with competent help, this whole process will be even more pleasurable than you imagined.

3. At Singing Horse Ranch we have regularly divided forty acre fields into four acre, quarter mile long, "lands" sometimes mowing several in one day. The lands are raked and handled individually and in order of cut. This allows half to two days difference in the relative curing stage of mown hay instead of finding ourselves with all forty acres ready to buck rake in the same few hours. Forty acres ready to put-up at the same moment often results in some hay being "over-cured".

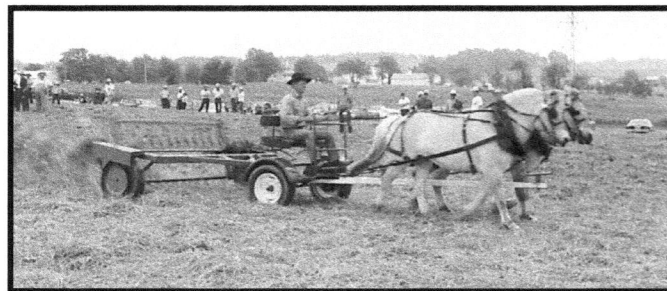

Figure 5. Two Fjords hooked to a forecart and pulling a tedder.

ARE THERE CERTAIN CROPS OR CROP CIRCUMSTANCES WHICH HORSES CAN'T HANDLE?

Maybe. Each time this author thought a crop condition was beyond horses in harness to handle we've been pleasantly surprised to discover otherwise. In our twenty-first century mentality, especially when it comes to farming, our cultural evolution has led us to assume more technological or motorized muscle is all that it takes to handle a procedural challenge. That's just not the case. Keep in mind that horses or mules in harness are but a power source to move implements through the field. If the implement is right for the task and properly adjusted the job can be handled. We have had monstrous hay crops with thick legume entanglements preventing mown forage from laying down. Our challenges with the crop were met not by adding muscle but by intelligently assessing the condition which needed focus. Adding curved 1/2 inch rod goosenecks to the swath board, with welded sharp edges where separational contact was made with the mown forage, pulled the forage sufficiently apart for us to find a cut line on the next round. And leaving the mown forage to stand somewhat hurt nothing and actually hastened the curing process.

The answer is that the capacity of true animal power is less of a concern when it comes to challenging crop circumstances than will be the farmer's ingenuity and the available implements.

Do Horses Dictate Certain Systems Over Others?

Not necessarily. The practical realities of walking speed and implement dimensions do pose some scale limits, but true to the other philosophies we advocate this is seen as less of a limitation and more of a form outline suggesting optimum scale and diversity.

If the choices are between 'loose hay in stacks' versus 'loose hay in barns' versus 'baled hay' the

Figure 6. Percherons pull motorized baler, bale thrower and bale wagon.

Figure 7. Two spotted ponies on forecart and modern finger rake.

harnessed animals, in and of themselves, do not dictate one option over another. There are other criteria which point to certain choices. Those criteria include;

existing facilities
experience
money
ease
preference

Each farmer will have to decide how to rate these criteria. For example if capital is not a problem, a farmer could simply elect to build a barn. However the struggling farmer may find that no barn for now dictates storing hay outside. A hay field three miles from where the hay will be stored tends to dictate options. A farmer with baling equipment who happens to want to put up loose hay with horses but who has no experience with the horses may elect to phase in slowly by mowing and raking a little with the horses the first year.

Following is a brief view of why someone might choose these different approaches. The chapters which immediately follow this one explain in a little more detail the advantages, challenges, and limitations of each system. In the equipment half of this book each aspect of the systems are covered in operational detail.

The Choices

LOOSE HAY IN STACKS

You mow the hay, rake it, pick it up or push it to a spot above standing water and you build a big old mound of it that you hope will shed water and last until winter with a minimum of waste from damage.

Within that description there are many variations. Big stacks or small ones. Hayloaders, pitchforks or buckrakes. Pitchforks or stackers. Within stackers there are pole systems, cable systems, derrick systems, overshots, beaverslides, swinging stackers and combination buckrake/stackers. This system can be made to work with a minimum investment (mower, rake, wagon, pitchforks) or lots of investment (i.e. many new imple-

Figure 8. Two Suffolks on forecart and tractor rake.

ments and many horses) with more or less people (1 to 10) with more or less acreage (5 to 100's of acres). And by covering the stacks this system can be made to work in almost any rainfall/climate zone. Properly cured and handled you have a superior forage with loose hay.

Although many people assume that this system is reserved for people with limited means or who don't know any better this author begs to differ. From years of expereince I can vouch for this being the most pleasurable system of haying I've ever been involved with. If you're inclined to appreciate craftsmanship in farming and enjoy being a part of a creative process, building ever more beautiful monolithic haystacks and thereby affecting the landscape for months, give this system a chance.

Figure 9. Any haying system begins with mowing.

Figure 10. With most systems the second stage is raking.

But on a more practical front. If you have hundreds of head of livestock to feed in the winter time you'll appreciate being able to get a hold of a lot of hay quickly. A barn stuffed with hay is not nearly so accessible as the outside haystack. Try pushing 4 or 5 tons of loose hay through a barn's hay mow door on a winter morning and you'll quickly appreciate the outside stacks. In the Big Hole of Montana where thousands of head of cattle are fed each morning, four-up teams of draft horses pull up to outside stacks, built by beaverslide, and quickly hydrafork tons of hay onto the feed sleds. This is the system of choice because it works.

LOOSE HAY IN THE BARN

You mow the hay, rake it, pick it up, and cram it into a barn or covered shed where it is kept, roof permitting, in pristine condition until winter feeding. Again, properly cured and handled you have a superior forage with loose hay.

Once again there are many variations within such a system. Big traditional barn with upstairs haymow, western style feed barn with floor to ceiling hay storage in the center, open shed, or what-have-you? Hayloader, pitchforks or buckrakes (yes, buckrakes can be used to push hay to the barn and in some cases if there is a big door to push it on in). And at the barn there are various ways to get the hay in. From homemade parbuckling setups, to pitchforking, to fork and trolley setups, the investment can be sizable for it takes a big barn to hold 100 tons of loose hay. But the equipment on the field and transport end need not be expensive. There are plenty of good reasons to put hay into a big old barn especially if that barn is winter home, in northern zones, to cows and horses. It's downright comfy to push down the morning's feeding for 2 to 5 dozen animals. And the hay in the barn will keep for several years if need be. This author

Figure 11. The hay loader is an important tool for any loose hay system. Even if buckrakes are the first choice there will still be room and value for the hay loader.

Figure 12. An overshot stacker.

Figure 13. A derrick stacker.

Figure 14. A cable stacker

Figure 15. An early horsedrawn baler.

Figure 16. A barn stuffed to the rafters with baled hay which horses had helped to make.

Hay System Variables

spent eight years putting loose hay into a 40 x 80 gambrel barn built in 1909. Our children grew up playing in that hay mow. Filling it chock full always gave us a sense of a circle completed. We love putting up loose hay stacks on our high desert ranch but we miss the wet coastal farm and the poetry of that big old barn.

BALED HAY

One of the more frequent questions we're asked in our work with Small Farmer's Journal is, *"Can I bale hay with a team of horses?"* The answer is yes. Keep in mind that, once again the animals in harness are merely a motive power source, in other words they pull the implement of choice through the field.

This is how the system works; you mow the hay, rake it with a side delivery rake and when it is well cured you pull a baler along and bale it. The bales are somehow picked up and hauled to stack or barn. As is covered in coming chapters, a baler equipped with a motor of its own can be pulled along windrows with a team of horses. Recent innovations have resulted in ground drive baler possiblities covered later. If the field is hilly and features severe gradients it may require additional animals to bale.

If the farmer is selling the hay off the place, or simply doesn't like loose hay, a baling system is certainly a bonafide option. But, and there are those who will argue this point, baled hay will ALWAYS be inferior to the same cured crop put up loose.

Recommendations: It is a good idea to keep options open. Be wary of a large initial investment that may lock you into a system that you will not be using in the near future. Regardless of the system you choose this author recommends that the prospective horsefarmer plan on having a ground-drive side delivery rake and, should the system be for loose hay, a hay loader even if buckrakes are the first choice for pickup. But more about that in the equipment section.

Why Loose Hay

in Outside Stacks?

Chapter Four

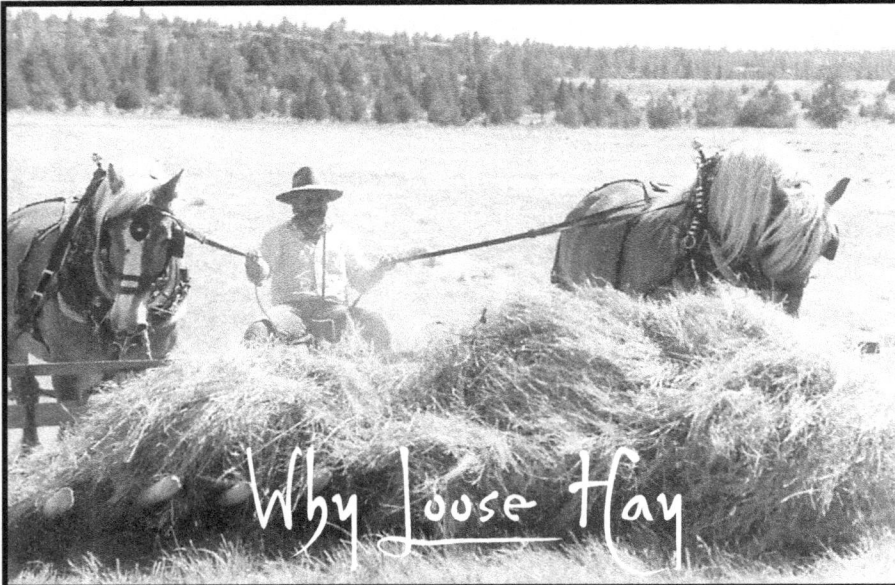

People might assume that anyone who puts up hay loose, or without baling it, does so because he or she has no choice. We're sure that is the case but only in a very few instances. Far and away the majority of people who put up loose hay do it because they have made the choice. They are attracted to this system and they enjoy it. This author is one of them. I can afford to bale my hay but have chosen not to. I much prefer dealing with loose hay than bales. I baled hay for my first ten years farming. I have put up hay loose, in barns and in hay stacks out of doors, for twenty plus years. It remains my preference. Whether or not it is, in any measure, a better way is up to the individual farmer. Obviously, if the farmer does not like loose hay and pitchforks the system won't work. Just as with horses or mules, if you don't like them they won't work for you.

Just imagine how direct, simple, cyclical, and lovely a system it can be: the hay is mowed and wilted in the swath before a side-delivery rake gently rolls it into windrows for further drying. Because it is going into a stack and not a bale there is less concern for the moisture content and the hay is a strong carotene-filled green color as the hay loader gently picks it up into the wagon - or, if preferred, as the hay is pushed by buckrake to the waiting stacker. The hay is then lifted onto the accumulated mound until a well-shaped stack, one which will shed rain, is formed. And, if horses have been used throughout this process, not a drop of gasoline or diesel has been consumed. The horses whose

Figure 1. Top, the author with Tuck and Barney on the buckrake.

Figure 2. (Right) Buckrake team and author watch as swinging stacker takes another load up. Singing Horse Ranch 1997.

Figure 3. In this example the buckrake pushes the hay to the ground crew which forks it up to the person waiting on the stack. Smallish stacks result with significant labor put in.

Figure 4. In this case the hay is pushed to beside a wagon. One man forks hay from the ground to the wagon and it is passed on up to the top of the stack. This simple method allows for slightly taller stacks but is still labor intensive.

manure helped to grow the crop have provided the power to put up feed that they will help to consume and thereby from which produce more manure and...

To answer the question posed by the title of this chapter:

for the economy of the system.
for the quality of the forage.
for the convenience of outside access to large amounts of feed in the winter time.
for the beauty of the system

NO baled hay, no matter the beauty of the standing forage and the attentiveness of the farmer, can compare with the same crop put up loose. Yes, there is likely damage to the loose hay stored in uncovered stacks. And for this reason barn stored loose hay will always be superior but a well-constructed stack will shed most rain and "grow" a "rind" which protects the inside of the stack for years.

The loose hay that is put in the stack may be stuck away with slightly more moisture than baled hay allowing that more color be retained. And hay that is turned at horse rake speed and gently lifted into a wagon by the hayloader loses very little leaf or blossom or aroma.

With a little ingenuity the independent farmer can set up an outside haystack as a self-feeder with its circumference girded by feeder panels.

In those areas where a great many animals need to be fed, and fed outside, these stacks provide many opportunities for ways to feed large quantities of hay quickly. In Montana and other western regions feed wagons and sleds are outfitted with a hydraulic fork (called a hydra-fork) run by a small gas engine. In other areas pitchforks, pull forks and cable mechanisms are used to pull hay on to vehicles. All this is covered in coming chapters.

There is no knocking the simple fact that putting hay up loose in hay stacks can be the cheapest way to make hay. Team(s), mower(s), rake(s), and hayloader(s) may all need to be purchased (or raised) but the other tools - buckrake(s) and stacker - can be jerry-rigged or homemade as the plans in this book will attest to.

We know folks who put up loose hay with one horse mower which has been in the family forever, one dump rake purchased for $15, and a homemade sweep or buck rake. They make smallish stacks pitching the hay up with pitchforks and for their ten acres it all works just fine. Hard to get much cheaper than that.

Figure 5. In this photo of days gone by, three teams are at work haying. Two are hitched to two factory made buckrakes and the third is pulling a side delivery rake. It is clear to see that a lot of hay can be moved this way. Far more than even two wagons and hayloaders may accomplish in the same time frame.

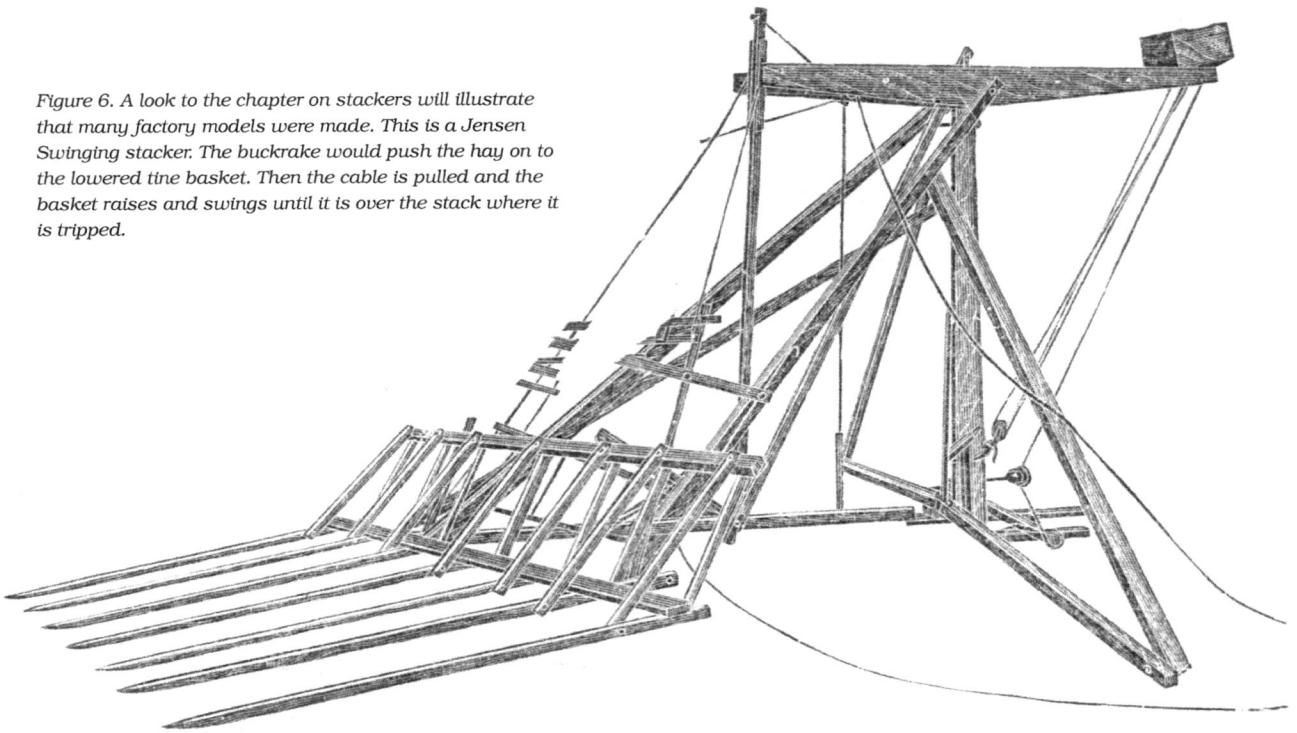

Figure 6. A look to the chapter on stackers will illustrate that many factory models were made. This is a Jensen Swinging stacker. The buckrake would push the hay on to the lowered tine basket. Then the cable is pulled and the basket raises and swings until it is over the stack where it is tripped.

Figure 7. A swinging stacker dumps its load on a respectable sized stack.

Figure 8. Capping a handsome stack with the last jags from the stacker.

Figure 9. Weights hang from cording holding down a cover on this stack. A basic affair such as this will take care of 80% of any weather related spoilage.

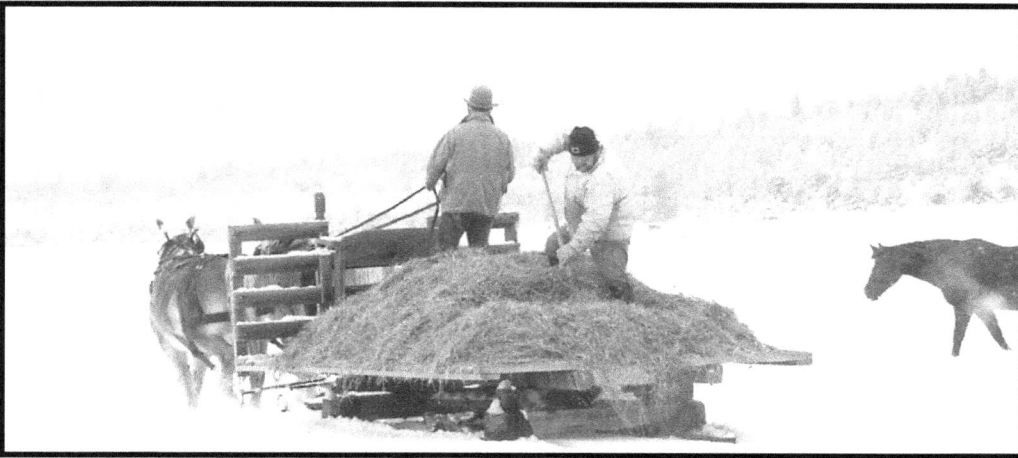

Figure 10. Finishing up the day's winter feeding on Singing Horse Ranch. A heavy bob sled gear with flatbed rack is pulled effortlessly through the snow by two Belgian mares.

Figure 11. (below) A John Deere overshot stacker at work.

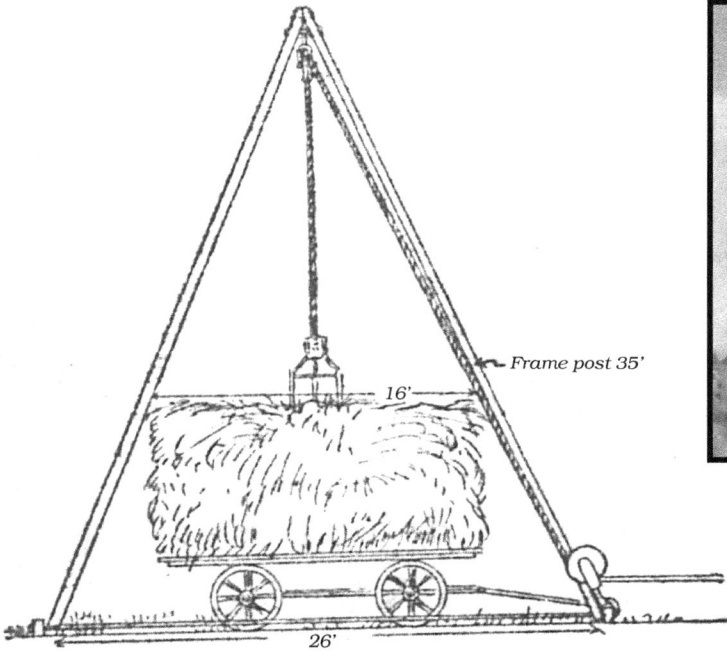

Frame post 35'

16'

26'

Fig. 12. One style of homemade stacker is illustrated in these two drawings. In this case the diagram should be self-explanatory. The A frame which hoists the hay from the wagon pivots on its legs, which are anchored to the 14' sill with end staked, to swing over the stack. As the hay jag swings out over the stack, the trip cord to the fork is yanked and the load deposited.

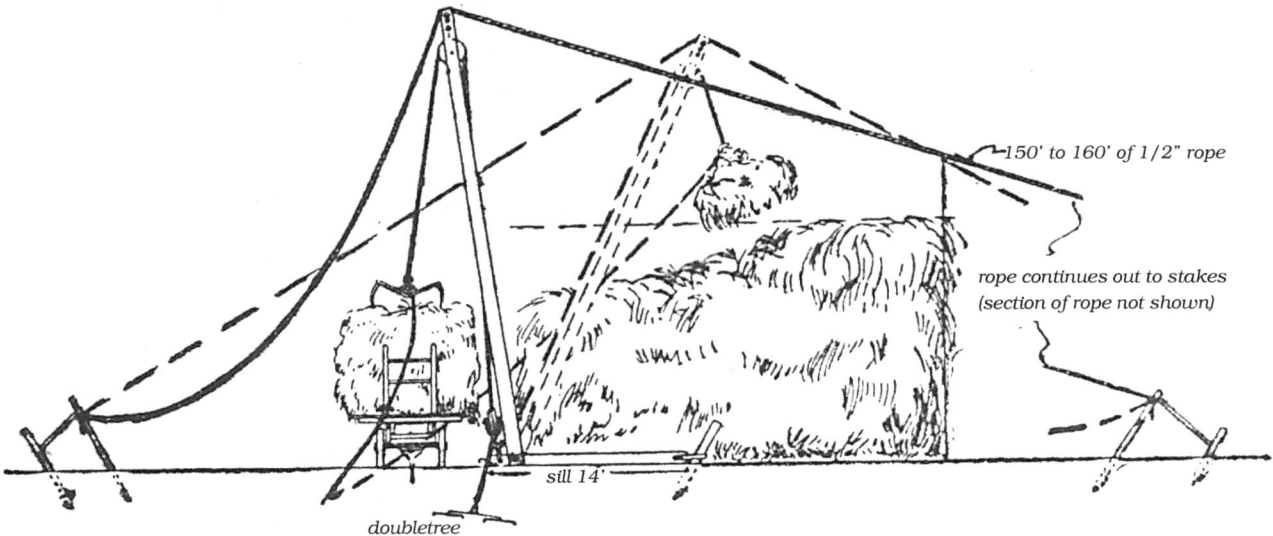

150' to 160' of 1/2" rope

rope continues out to stakes (section of rope not shown)

sill 14'

doubletree

Why Loose Hay in a Barn?

Chapter Five

Many agriculturalists see the period of 1914 to 1919 as the pinnacle of North American farming. And one inescapable icon of the time was the big classic multi-purpose loose hay barn. Triumphs of utilitarian architecture, these barns commonly featured a track system hung in the peak which carried a trolley or trolleys from which dangled forks or slings. They ran the length of the barn and out the big mow doors where the fork would be lowered to the waiting loaded wagon. The fork would be set into the loose hay and a signal given. Out back or on the side of the barn usually a single horse waited hooked to a long heavy rope which passed through a low mounted pulley. As the

Figure 1. The Myers Company was one of several outfits that sold plans for, and built, the great barns of the 1900's even going so far as to offer a complete line of barn hardware including the track and trolley systems. In this cutaway a two-way trolley is demonstrated. By changing from one side of the door to the other, and the other rope, the trolley could be pulled to either half of the barn. A barn of this design allowed that a feed wagon be drawn inside under cover in the winter to be loaded with feed.

Figure 2. The barn on the right had a huge hay mow capable of holding at least 125 tons.

Figure 3. Success with filling a barn in a timely manner really depends on the operation of the hay loader and the distance to the barn. It would be optimum if the teams and wagons did not have to travel much more than a quarter to three eights of a mile from the field to the barn with a load of hay. This is only because of the elapsed time. Road, wagon and grades considered, the horses themselves can handle far greater distances. With a considerable amount of hay it would be handy to have two teams and wagons. While one is being loaded in the field the other could be getting unloaded.

horse stepped ahead the ropes creaked through the length of the barn and a jag of hay was lifted up and through the waiting door.

For nearly ten years this author used just such a system to put up hay on a coastal Oregon farm.

To answer the question posed by the title of this chapter;

for the quality of the forage if for no other reason.

NO baled hay, no matter the beauty of the standing forage and the attentiveness of the farmer, can compare with the same crop put up loose into a barn. The loose hay that is put in the barn may be stuck away with slightly more moisture than baled hay allowing that more color be retained. And hay that is turned at horse rake speed and gently lifted into a wagon by the hayloader loses very little leaf or blossom or aroma. The real test is how the animals like it and this author can tell you that horses and cows will always prefer the barn-stored loose hay. Given the choice between a number one baled dairy-grade Alfalfa hay and a fork full of mixed grass and clover loose hay from the upstairs, our old Jersey milk cow would humiliate herself to get to the fork full.

Another all to obvious reason why loose

Figure 4. As will be shown in coming chapters there are a number of devices which work well to off load hay at the barn. One is the hay sling pictured below. Both ends can be hooked to the trolley for lift. The cord in the center is jerked to release the two halves at the middle and drop the hay.

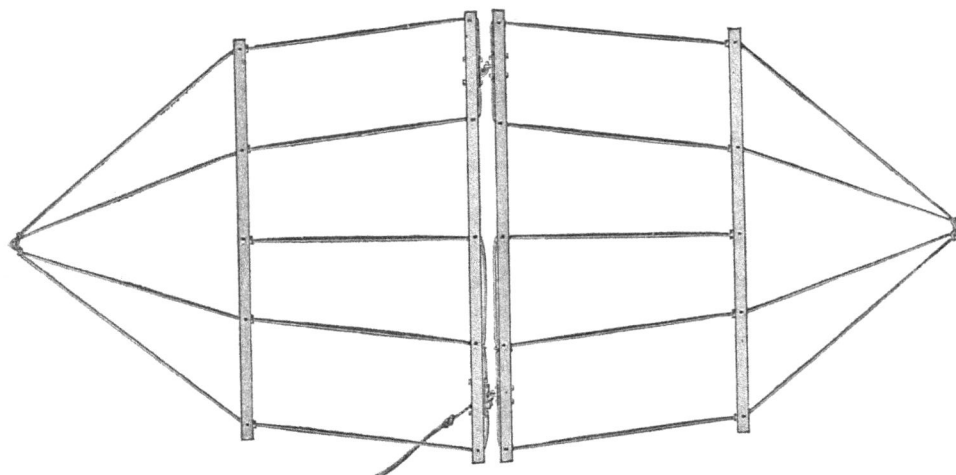

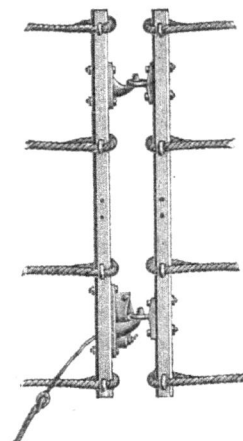

hay in the barn is chosen is weather damage. There isn't any, unless the barn roof leaks. All the hay you put in the barn stays fresh and sweet for up to three years before a significant loss of aroma and a brittleness starts to set in. Heard tell of some farmers who've kept hay in the barn for as much as five years and still had good feed to offer the animals.

On a visit to the Huevelton Amish community of upstate New York this author happened on a young horsefarmer who was filling barn number two. Barn number one had been half full at the beginning of this hay season, so he capped it off first. He explained that he divided the contents of the barns in half so that he could feed out the older portion first. As far as he was concerned it was his duty to make sure he had at least a year and a half's forage stored away. And to his way of thinking putting it up as baled hay would be too costly, too unwieldy, and it would have been short-changing his milk cows and horses.

Another reason for putting loose hay into a barn, depending on the barn's setup, might be for the convenience of its winter proximity to the

Figure 5. This photo is of the author's barn on the coastal farm. In this case the hay is being drawn up by jackson or harpoon fork. Our preferred hardware was the grapple fork as it allowed more leeway in what hay was actually lifted.

Figure 6. A fixed four-tine grapple fork. By fixed we mean that each pair of tines are rigid together.

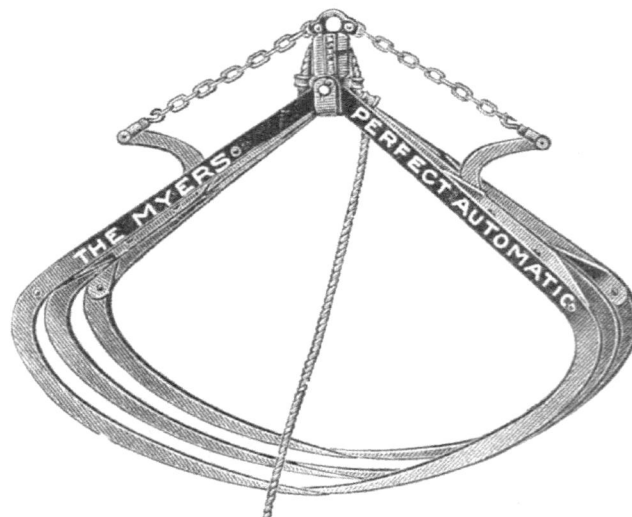

Figure. 7. A fixed six-tine grapple fork.

stabled livestock. In many parts of North America, winter is a limiting factor and prized stock are kept indoors, If loose hay is the choice or the only option, having it stored away in the same building as the stock, can be a great labor savings and comfort. And if the hay is overhead, chutes can be constructed so that the feed is dropped pretty near its final manger destination.

And a final reason may be aesthetic or sensual, it just feels good to fill a great big old barn plumb full of beautiful sweet smelling long stem hay.

Figure 8. Putting hay into the barn was and still is best done as a family affair.

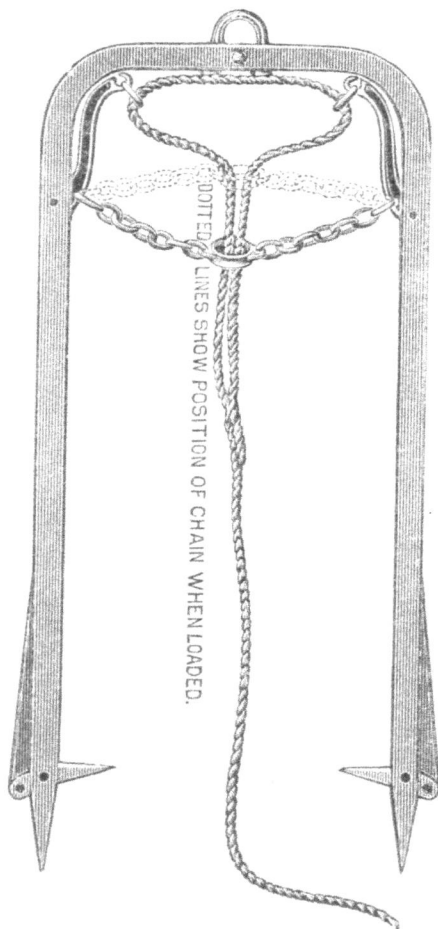

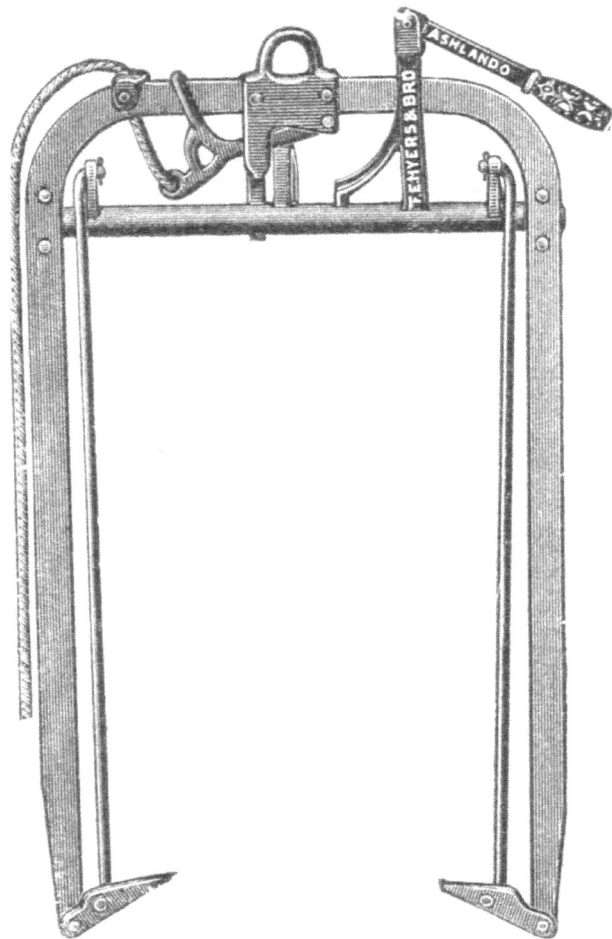

DOTTED LINES SHOW POSITION OF CHAIN WHEN LOADED.

Figure 9. Two of the many styles of double harpoon forks. These were jammed into the hay and as they were lifted, the tooth ends bent in to grab the hay.

Why Baled Hay?

Chapter Six

It's the twenty-first century and the evolution of forage handling has gone from loose hay into baled hay and beyond with haylage and hay pellets and who knows what else. Its an evolution that some people seem to think was inevitable and proper. Fact is many people in modern ag circles take the latest and greatest technology and methods quite seriously. They would be embarrassed to be seen doing something as backward as haying with horses in harness, they think it's silly. They can't see how things like baling hay, for instance, could be done with horses. Fortunately not all farmers feel that way.

Baling can be done with horses and in a variety of ways. We'll get to those varieties soon enough. Right now we need to discuss WHY someone would want to bale with horses or mules.

The name of the game hasn't changed. It is still to cut, cure, and store the forage for feeding to livestock (or for subsequent sale - this 'OR' is an important issue.)

Obviously with loose hay it is difficult or impossible to sell it off the farm. Baled hay, however, is highly transportable and more easily sold. Especially the smaller traditional bales. An argument has already been made in previous chapters for keeping all the forage on the farm as feed and selling livestock instead. But every farm has extenuating circumstances which can affect the best of plans. So, one reason to choose a baled hay system is for the option to sell hay.

A second reason, and perhaps the strongest and most prevalent, is the matter of personal preference. Some people would rather deal with bales. That is reason enough. This author enjoys feeding loose hay in the winter. Many people including folks who've been in the employ of Singing Horse Ranch prefer feeding baled hay in winter.

A third reason involves available qualified help. You can hook horses and bale the hay with little or no help. And, if the pick up is done by hand, it may be easier to find cheap

Figure 1. A four abreast of fine Haflinger draft ponies pull a baler while two Percherons on a bale wagon walk alongside to receive the goods.

Figure 2. The Haflinger four, hitched to a Pioneer forecart and New Holland baler wait their turn at the Indiana Horse Progress Days.

labor with bales. The loose hay systems require at least competent teamsters and good stack hands can be a real plus.

If the money is available there are bale elevators and throwers which coupled with bale wagons can reduce baling labor demands even further. So, if help is short, baling hay would be the way to go.

To most people attracted to working horses or mules in harness the picture of loose hay is enticing but truth be known there are far more people baling hay than putting it up loose. People would love to put up loose hay but instead they put it up in bales. One reason for this, we've found, is that the baling of hay is understood by most farm folks. Loose hay handling however has slipped back two or three generations and seems a laborious and difficult way to go. So the baling

just appears, or feels, like it would be easier.[1]

Another aspect of this is that many of the people who are considering changing to a work horse program, or adding it, are already set up to farm with tractors. They have a baler. So they try mowing with horses and like it. Then they try raking with horses and they like that. If they have the courage, the next seemingly obvious step is to hook the horses to the baler. That seems a direct shot, a logical move. Where as leaving the baler in the shed and trying a hayloader or buckrake seems a leap to the moon.

As for any appearance of difficulty with horsedrawn balers: once again we mention that horses

---

1. *This is an unfortunate assumption, as in truth baling hay is far more costly, complex, and limiting, than a well planned loose hay system.*

Figure 3. A rear view showing the same New Holland baler and following bale wagon.

Figure 4. The first balers were stationary affairs powered by 'horsepowers', a remote gearing apparatus which, when horses walked around and cranked, converted that motion to a spinning drive shaft. The precursor to the power take off. The baler chamber was filled by hand and the wires were tied by hand.

Figure 5. Modern balers with their own motors were originally designed to be pulled by small tractors. Turns out they are ideal for use with horses. In this photo Addie Funk drives a Hammill Clydesdale team up a Montana grade while they bale hay.

or mules in harness are but a motive power source. Provide an auxillary power source to run the baling process and any baler, large or small, may be pulled by the animals.

Look to chapter 13 for particular information on mechanical and power concerns with baling.

The Convenience of Bales

They can be stacked inside barns and sheds and outside under tarps. They can be loaded on wagons, sleds, car trunks, pickups and trucks to be fed out or moved to a new owner's facility. They offer some measure of precise uniformity that allows quick and simple measures of bulk (i.e. *it'll take twelve bales to feed that set of cows.*) The smaller conventional bales, like those Doug Hammill is standing on in this picture (figure 6), can, most of the time, be handled by any adult. If you like them there's no arguing with you. If you don't you ought to try loose hay.

Figure 6.

Figure 7. A big New Holland round baler powered by a Pioneer motorized forecart and pulled by four Belgian horses at the Ohio Horse Progress Days. The only thing stopping you from using this process is the decision to do it and the money to afford it.

The pictures below demonstrate Doc Hammill's ingenious way of getting round bales to his log barn and up in through the hay mow door. All with his Clydesdale horses. In figure 8 the bale is rolled and drug to the barn. In figure 9, at the barn, ropes are set up to lift and parbuckle the bale in through the opening. Figures 10 and 11 show the job accomplished.

Figure 8.

Figure 9.

Figure 10.

Figure 11.

Baled Hay

Miscellaneous Ways to Make Hay

Chapter Seven

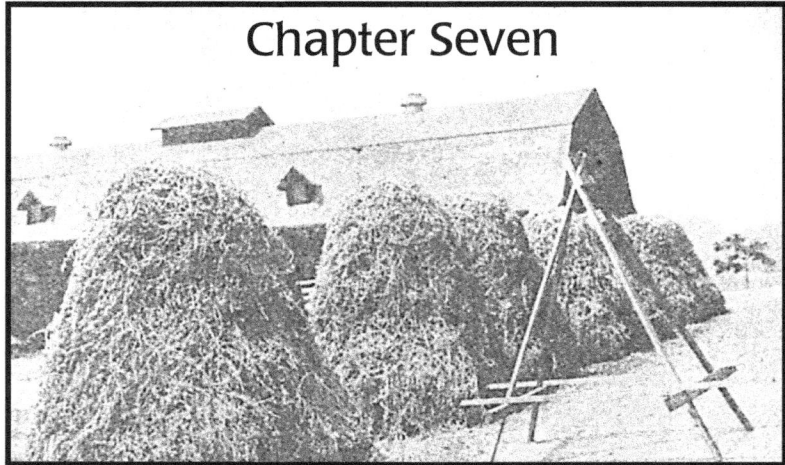

Under this chapter heading we begin by discussing the challenges of trying to put up hay in wet climates where curing weather may be all but nonexistent. Whether it's a lot of rain, absence of strong sunlight and warmth, or heavy fog and dew, there are at least four systems which can be employed with some success if the hay volume is not too great. These ideas fall into four structural plans all of which function in the same way; tripod, pole stacking, curing on the fence, and curing trucks.

Figure 1. Showing a type of tripod frame and hay that has been stacked on them near the barn.

Tripodding

As illustrated, this system (some people believe it originated in Scandinavia) involves draping green hay over a four or three legged standard. The height of

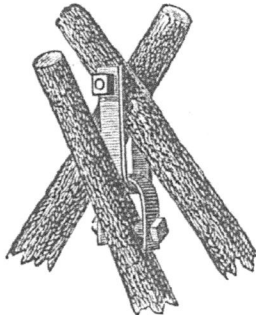

Figure 2. A different frame style and tripod stacks out in a field.

Figure 3. A method for attaching poles.

these standards should be whatever is handy for the farmer to pile hay on or at least 8 feet (some are as high as 10 feet). The hay should be piled high on top and stop just short of the ground. This makes for a hollow area inside helping the hay to air dry. The farmer should be careful not to pile or tightly wind balls of wet hay. Effort needs to be made to fluff out the material as much as possible. Further protection can be had by capping each tripod with a canvas cover to help shed rain. When the hay is cured it is carried by wagon to the waiting barn.

Pole stacking

In this approach, employed in the past with success in the Upper Pennisula of Michigan, trees about three inches in diameter were cut to make poles seven to eight feet long. Stubs of branches were left on to make springs an inch or two in length. Boards four feet long were nailed to the poles about a foot and a half from the larger end, as shown by the diagram below. An iron bar was used to make small holes in the ground in which the larger end of the poles were placed firmly, allowing the attached boards to rest on the ground. This gave the poles sturdy support and allowed circulation of air at the base of the stacks. The other end of the pole was pointed, which aided in proper construction of the stacks. Prongs were left on the poles to prevent the hay from settling too tightly around them. Spikes driven into the poles at alternate places would have served the same purpose. Projections of some sort are essential to keep the center of the stacks from packing too tightly.

Construction of a Pole Stack

Loose hay from the windrow was piled around the pole, each fork load overlapping, requiring four fork loads to make a tier around the pole. (The hay should not be rolled and put into the stack in a tight bunch.) The hay was piled in tiers to a foot from the top of the pole, and from then on it was put directly on top, forcing some hay down around the pole which gave strength to the stack. About two feet of hay was put directly on top of the pole, creating a water-shedding cap. This type of construction allowed the air to circulate through the stack curing the hay.

The stack should not settle below the top of the pole. If it does, rain water will run down the pole, wetting the hay in the center of the stack. It was found that one man could work to better advantage than two. Two men will not place the hay uniformly on the stack and it becomes oversized or settles unevenly. Extra men can be used in bringing the hay to the man doing the stacking.

It was evident from observation and experimental results that a stack should not be built more than 5 to 5-1/2 feet in diameter, and it should be at least 7 to 8 feet in height. The bottom of the stack must be kept narrow as this part of the stack settles into a firm mass. The sides of a well-built stack are straight up and down.

A curing experiment with a second-cutting hay mixture of alfalfa, alsike, and timothy was conducted using pole stacks and average sized cocks. The hay was cut in the morning after the dew had disappeared and was promptly put in windrows with a side-delivery rake. Three hours after it was cut, part of the hay was pole stacked and part was cocked. The moisture content of the pole stacked and cocked hay was 51 and 43 per cent, respectively. Fourteen days, of which seven were rainy, were required to cure the pole stacked hay to a 19 per cent moisture content. None of the hay was spoiled. The cocked hay at the end of the 14-day period contained 27 per cent moisture with a spoilage of 16 per cent.

In the same year, 10 acres of a 14-acre field of first cutting hay was cured in pole-stacks and the remaining four acres in average-sized cocks. Part of the hay which was pole-stacked was free from external moisture, while part was stacked when still wet with rain. The hay that was slightly wilted and free from external moisture, cured in two weeks, resulting in good quality hay with the green color and leaves retained. The hay that was slightly wilted but with external moisture when stacked was not cured at the end of the two weeks. Although the top halves of the stacks were cured, the bottom halves were moldy and wet. It was possible to save the top half of each stack, but the bottoms were discarded as of no feeding value. The four acres of hay which had been cocked was first partially cured in windrows and then put in average-sized cocks and left there two weeks before it was hauled. Much of this hay was black with extensive loss of leaves, moldy and of no feeding value. Seven of the fourteen days during which time the above hays were being cured were rainy.

Conclusion

1. Hay to be cured in pole-stacks may average from 50 to 60 per cent moisture, but should be free of dew and rain.

2. The stacks should be constructed as follows:

(a) Use strong wood poles about three inches in diameter at center, with prongs.

(b) Set the poles firmly into the ground, supported with boards.

(c) Arrange loose hay around the poles.

(d) Keep the stacks 5 to 5-1/2 feet in diameter. Over-sized stacks result in moldy hay.

(e) Keep the sides of the stacks straight up and down.

(f) Build well over the top of the poles so that they will not stick out after the hay has settled.

3. Eight to ten stacks are equivalent to one ton of hay containing 20 per cent moisture.

4. Curing hay in well-constructed pole-stacks eliminates much of the weather hazard at haying time.

5. The cost is small compared with spoilage losses in bad weather.

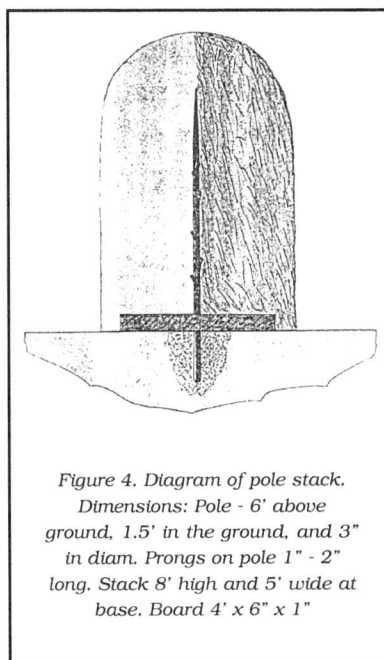

Figure 4. Diagram of pole stack. Dimensions: Pole - 6' above ground, 1.5' in the ground, and 3" in diam. Prongs on pole 1" - 2" long. Stack 8' high and 5' wide at base. Board 4' x 6" x 1"

Curing on the Fence

In terribly wet small farm regions farmers for centuries have draped hay over fences, almost like laundry, for drying. Board, rail or smooth wire works for this purpose. Barbed wire will make hay removal difficult. Obviously such a plan does not work for large fields or large quantities of hay.

The Curing Truck

During favorable weather, or even during transient unfavorable weather when only light showers occur, a fair quality of hay can be made by curing in the cock. If, however, the weather continues unfavorably for a period of a week or more and is marked by frequent heavy rains, the cocked hay will usually become wet through. The bottom of the cock will absorb moisture from the damp ground and soil the hay. After the rains cease it is necessary to spread out the hay and sometimes recock, and when the hay is finally cured sufficiently to be baled or put into the barn or stack it will be of a very poor quality.

Hay caps can be used to advantage to keep the rain from entering the top of the cock, but even then the hay in the bottom will sometimes be damaged from the moisture in the ground unless some special device is used to prevent it. Various types of racks, frames, etc., holding from 100 to 2,000 pounds of cured hay have been designed and used more or less successfully when quality only is considered. Such devices usually require considerable extra man labor in putting the hay on and getting it off the apparatus.

A device was sought that would first permit the rapid handling of the partly-cured hay; one that would keep the hay off the ground and entirely protected from the rain while curing; and, lastly, one that would require but little labor and time in moving the hay after it was cured out. The curing truck fits all the essentials of these ideas.

Description of the Truck

The curing truck is quite similar to the ordinary hayrack. It is 12 feet long and 7 feet wide and will hold from 1,500 to 2,000 pounds of cured hay, depending on how much the hay is cured when put on the truck. The back is supported by two 16 to 20 inch steel or iron wheels running on an axle placed about 4 feet from the rear end of an A-shaped frame. When a loaded truck is not in motion, as when left in the field for the hay to cure or while it is waiting to be baled, the front end is supported by a 6 by 8 inch wooden block or "trigger" of sufficient length to hold the truck level. This trigger is fastened between the two main frame timbers by a heavy bolt, enough play being left

Figure 5. Two curing trucks in tandem the first one being loaded with hay from a bunch.

Figure 6. Right, loaded truck protected by a canvas cover. A well-loaded truck will contain about 1 ton of cured hay. Left, empty truck.

so that it swings easily by its own weight. When the trigger is in use, it is held in a position just past the vertical, by coming into contact with the front coupling casting. When the trigger is not in use, the free end, pointing to the rear, drags lightly on the ground.

The standards at each end run to a point at the top, in which a notch is cut to receive a 2 by 4 inch ridgepole that supports the canvas and keeps it from lying flat on the hay, thus permitting the air to circulate freely at the top. The bottom is made up of seven, 2 by 4 inch pieces, 12 feet long, evenly spaced to allow the air to enter freely. A coupling device is fastened to each end. The front and the back one is shaped in the form of a hook. The couplings make it possible to haul a train of several empty trucks when returning them to the field.

The trucks are moved by means of a two-horse team and a forecart. A seat is provided for the driver, to be used when moving a truck some distance. The forecart is fastened to the truck by means of a long clevis pin. When the truck is being hauled, the weight of the front end rests on the forecart. When the team starts, the forward movements cause the trigger to trip

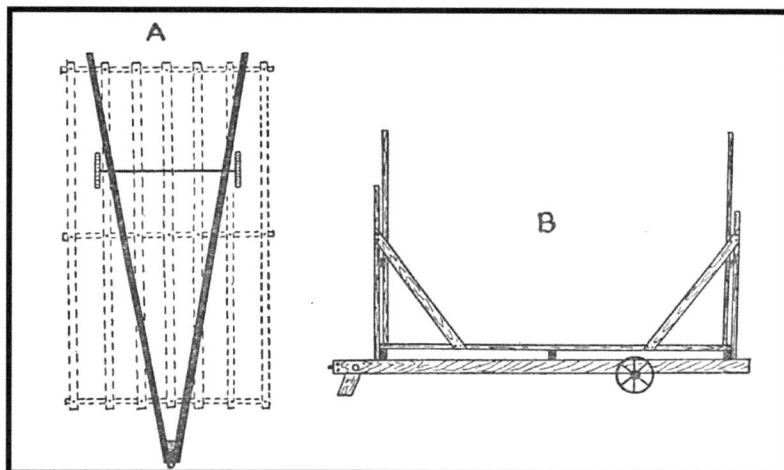

Figure 7. Main frame of truck (A), with rack indicated by dotted lines, and side view (B) showing trigger and position of wheels.

Figure 8. Empty truck, showing details of construction. A 2 by 4 ridge pole is held by the notches at top of standards. Note position of "trigger" which holds the truck level.

as it passes center and to drag on the ground, the weight of the load being thus shifted to the running gears. Upon reaching the destination, the team is backed a step or two, which causes the trigger to assume its upright position and again take its share of the weight of the load.

In order to use the truck successfully, hay must be cured to a certain point before it is put on the truck. If not sufficiently cured, there is danger that it will heat and spoil. Hay put on the truck when almost cured will not have as good a color as hay put on with the proper degree of curing. Hay is in best condition to be put on the truck when it has cured enough to go into the cock or possible just a trifle greener than when it ordinarily is cocked, as it is desirable to do the last third or fourth of the curing on the truck under a canvas cover, where it is protected from the sun and rain.

In order to have the hay in proper condition to be put on the trucks at a set time of day, it is necessary to arrange the time of mowing, tedding, raking, bunching, etc., so the hay will be ready when desired.

When the yield is light, say, about one-half of a ton per acre, the hay is mowed in the morning, raked in the afternoon, and put on the trucks in the evening, or the next morning, after the dew is off. When the yield is about 1 ton per acre, hay is mowed in the morning and tedded the next morning and raked into windrows before dinner. The hay is allowed to lie in the windrow only two hours before being put on the trucks.

Bunching hay to be loaded

Hay may be taken directly from the windrow and loaded by hand onto the truck, but this is not a good practice, because it requires too much time to pitch hay up from the windrow and not leave any scatterings. It also obliges one man to spend much of his time in driving the team hauling the truck, when he might be better engaged in helping load.

The best practice is to bunch the hay after it has been raked into the windrow. This may be done with a dump rake or a buck rake. The dump rake puts a much smaller amount of hay into a bunch than the buck rake, and it is not as efficient an implement as the latter if more than two men are used to load the truck.

Handling and Loading the Trucks

Before the crew starts to work it is customary to bring several empty trucks to the field. Empty trucks are hauled to the field in trains of from 6 to 10. Enough trucks should be brought to the field before loading is begun to last all day or at least half a day. This practice will save time, as the loading crew will not have to wait while trucks are being brought from a distance. If enough trucks are taken to the field in the morning to last until noon, the mower, rake, and buck-rake teams may be used to haul trains of trucks to the field when returning to work after dinner.

The trains are left at different places in the field so as to be quickly accessible. Two trucks are used when loading starts. The front truck is loaded first and the second one is trailed behind to save time, since it does not add enough weight to make any material difference to the team. After the first truck is loaded, the sides of the load are raked down carefully with pitchforks to cause the hay to shed rain more easily, and a canvas is put on and tied at the four corners. This work is done while the team is being hitched to the second truck. The canvas is always put on, even if there is no indication of rain. When the second truck is nearly loaded, the driver unhitches his team and goes for two more empty trucks from the nearest train. While he is doing this, the pitchers, if necessary, carry a few forkfuls of hay from the nearest bunches to finish out the load. The loaded truck is left standing where the last hay is put on. There are two reasons for this practice; the more important one is that it is not advisable to haul loaded trucks very far over the ordinary hay field. The truck wheels are comparatively small and the average alfalfa and Johnson grass hay field is more or less rough, sometimes being very bumpy and uneven; and hauling shakes the hay and causes it to settle and become compact, preventing the circulation of the air, prolonging the curing process, and thereby increasing the danger of loss from heating. Also, if each truck were hauled to a central point, it would be necessary to have a much larger number of teams, drivers, and forecarts in order to keep the loading crews busy all of the time. When the hay has all been loaded, the fact that the trucks are left scattered over the field nearly equidistant from each other indicates that the hay has been hauled no farther than absolutely necessary.

Size of Loading Crews

Loading crews consist of two, three, or four men. A two-man crew can load about 14 trucks in 10 hours. Both men pitch onto the truck until the load is two-thirds on, when one man works on the truck, building and finishing off the top. When three men are used, one stays on the truck all of the time and builds the load and the other two pitch on the hay. They will load about 20 trucks in 10 hours. A four-man crew may work in two ways; three men may be used to pitch on the hay if they are careful in placing the forkfuls on the truck in such a manner as to keep the hay level and fill out the corners. When this is done, one man only is needed on the load. The other method is to use two to pitch and two on the load. The two pitchers are then not concerned about where their forkfuls land on the load. All they need to do is to get as much hay as possible onto the truck, leaving the placing of the hay to the two men building the load. A four-man crew should load, on an average, about 30 trucks in 10 hours. The average time required to load a truck by a two, three, and four man crew is 40, 30 and 20 minutes, respectively.

Loading is not extra labor

Some object to the use of the trucks because of having to load them by hand. This objection is raised only by those who have not compared the amount of man labor required per ton when similarly threatened hay is cured in the cock, with that required when the truck is used, for, as a matter of fact, less hand labor is needed when using the truck than when hay is cured in the cock.

Hay is raked with the dump rake and bunched with the dump rake or buckrake in exactly the same manner whether cocked or put on the truck. If cured in the cock it is necessary to put the hay into carefully made cocks and each forkful must be placed exactly right to form a cock that will be symmetrical. When the

truck is used, it is not necessary to be so careful where each forkful is placed or just how much is taken up with the fork at a time. A man can handle more hay per hour or per day when pitching onto a truck than when building cocks, so that as far as labor requirements are concerned, loading the truck has the advantage over cocking.

Time required to cure on trucks

In good curing weather hay is in condition to be baled or barn stored after curing three days on the truck. If the hay is a little green when put on the truck or the weather is unfavorable, it will take a week or possibly longer to cure out thoroughly. Hay may remain on the truck indefinitely, if well protected by a canvas cover, without injury from sun or rain when the truck is not needed for some time, as is often the case when a cutting is all finished. Some users of trucks put two canvases on each truck when they are to be left standing for a considerable time before being baled. A single canvas cover placed lengthwise covers the top well but does not protect as much of the sides of the load as do two canvases put on crosswise. It is sometimes customary to place two canvases on each truck when there are prospects of a hard wind-driven rain. When hay is on the truck and protected in this manner, the hay grower's mind is at ease; and no matter how hard the storm may be or how long it may last, he is satisfied in the knowledge that his hay crop is safe and will bale up, having a good color and quality.

Economic Considerations

Other uses for the trucks

In addition to its primary use as a haying implement, the truck can be used to advantage in protecting bound grain from the rain until it is ready to be threshed. It is also very handy in hauling corn to the silo, etc.

Life of Truck

The life of a truck depends, as with any other kind of farm machinery, on the care it receives. If a truck suffers hard or careless treatment, such as being overloaded and driven over very rough ground or into small open drainage ditches, it may not last very long. Such handling usually results in a number of trucks being badly injured or destroyed each year.

The truck is, however, a very simply constructed implement. It has no rapidly-moving or delicate parts to wear out, and it is not subject to deterioration by rust. It should last 10 years at least, or longer if a reasonable amount of care is exercised and no serious accident occurs to it. A good canvas will last about 10 years if it is well cared for and never put away when damp.

Limitations of the Truck

A word of caution for those who may be led to adopt the method of curing described. Do not expect all of the hay cured on the truck to be "choice." The reason why curing on trucks will not invariably make "choice" hay is that hay is often damaged more or less by rain before it is in condition to be put onto the truck. It may rain on the hay one or more times between the times the hay is cut and it is ready to be put on the truck.

If the rain comes before the newly mown hay has started to cure, the damage may be very slight. It is the intermittent wetting and drying out that causes the greatest damage to curing hay. When this happens, the curing truck cannot restore the hay to its former color and quality. However, when such hay is finally in condition to go on the truck there will be no further loss.

No matter what other method of field curing is used, the farmer must always run the risk of having his hay damaged or ruined before it can be protected from rain and sun. The point that the writer wishes to emphasize is that the curing truck eliminates all danger of the hay spoiling either by loss of color or quality, after it has been put on the truck in proper condition and is properly protected by a good canvas cover.

Hay Cocks

When curing weather is in question perhaps the most common approach has been to gather in the field, wilted hay from the windrows into small bunches or stacks called hay "cocks". The safest next step, but no guarantee, is to cover each cock with a top canvas or plastic.

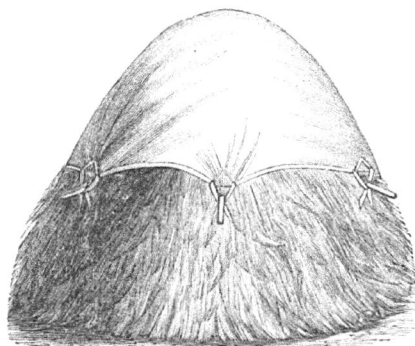

Figure 9. A covered hay "cock".

Roofed Small Stacks

In days gone by some serious small farmers erected poles from which hung sliding roofs as illustrated. This allowed that the top of the stack be snuggly covered. They were commonly six feet square and six to ten feet tall.

Hay in Bundles

This author has bundled hay in shocks with a grain binder and stacked those bundles in the barn as an experiment. The hay was a little green and the object was to get it in the barn in a form that might continue to dry. It worked well and the bundles were very easy to feed out.

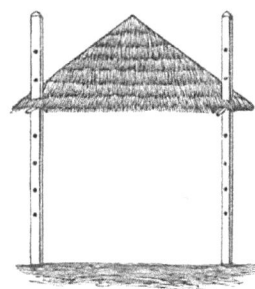

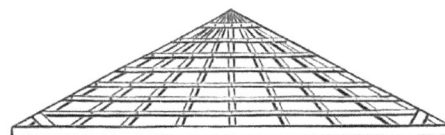

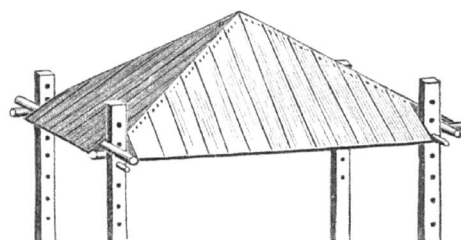

The Right Horses or Mules

Chapter Eight

This chapter is a general discussion on what you will be asking of your animals and a few suggestions to assist if transitions are difficult. As was mentioned in the preface and covered in the introduction, this book does not attempt to cover the mechanics of working horses or mules in harness. That subject has been covered in the *Work Horse Handbook* and *Training Workhorses/ Training Teamsters,* two other books by this same author (see back of this book for more information).

Usually one is in the position to have to make existing horse or mule teams work in the new hay fields. But this author thinks it might be useful to look at the subject from a more direct perspective. Forget for a second that you own work animals (if indeed you do) and try to follow this front door logic.

You're going to make hay with horses or mules. You've decided on the system you're going to use (i.e. baling or loose hay in stacks etc.) and you're shopping for just the right animals to power the venture. You turn to this author and ask *"What's the best animal for haying? Do I need a certain type or size?"*

Figure 1. Singing Horse Ranch colt. Intelligence and attitude show at an early age.
Figure 2. (right) Bulldog Fraser of Montana with a mowing team made up of Belgian and mule.

Figure 3. Chantal Salter with draft Paints in Montana. This young lady has been a student of Bulldog Fraser and learned about mowing the safe way. Crossbreed horses are an excellent choice.

Don't I have to have the bigger animals to pull the mowers and loaded wagons?"

Best animal? That's an easy one. The best animal for the job is the one that excites you. You will hopefully be spending a lot of time with your work mates. And since almost any size of mule or horse can, with ingenuity, get the job done, you owe it to the animals and yourself to be working with ones which get your juices flowing. The draft pony breeds, Haflingers and

Fjords in particular, are ideally suited for hay work. Minor modifications to mowers and care with grades and loads will serve the animals well. Light mules, light horses and light/draft crossbreeds provided the majority of motive power for centuries. This author enjoys draft horses and has raised Belgians and Belgian/Percheron crosses for thirty years. I hope the reader will believe and accept that, whereas I may have a personal preference as to type and size, that view

Figure 4. Justin Miller raking on Singing Horse Ranch with a pair of two year old Belgian fillies. The light work of the raking is an excellent place to get young, quiet, well-started horses accustomed to field work routines.

Figure 5. Tony Miller raking "scatters" with Barney & Molly at Singing Horse Ranch. Sometimes short hay will fall through the buckrake teeth. If there is enough quality and quantity, raking it to pick up with the hay loader may be worthwhile.

Figure 6. A pair of healthy, athletic, intelligent mules can do everything their horse counterparts can do. And some say "more".

should not be taken as a recommendation for others. I believe very strongly in my statement that you ought to have what you like. If you don't, you'll be making excuses and building regrets from the get go.

As to this author's preferences, here goes;
• 16.2 to 17 hands
• intelligent head, well set eyes, alert, confident disposition
• ample heart girth, short back, medium neck length
• strong well set hips, clean wide hocks, medium sized hard hoof

•longer than average cannon bone.

Reasons for the preferences: height translates to good tongue height and length settings for our size of mowers, rakes, and buckrakes. Plus it is a comfortable sized horse for a man, approximately six foot tall, to harness and groom. The head is a strong (not absolute) indicator of how the animal might behave in an extended working relationship. The heart girth is usually an indication of lung capacity and endurance although many is the exception. The long backed horse has, in this author's experience, often had less stamina and fluidity of movement. The neck length does translate to stability and balance which in turn means sure-footedness and a capacity for fluid forward motion. Forward propulsion comes from the hind legs and the primary joints, the hips and the hocks are critical to that transfer of power. The hoof is the contact point and a big foot or a small foot may pose problems but certainly no more than a soft foot.

And finally the relative length of the cannon bone tells us a lot about the stride speeds of that animal. The longer cannon bone means a longer walking stride which means the mower and rake covers more ground with the same amount of effort.

As for Mules versus Horses: It's back to the personal preference thing. There are differences worth noting but there is no way to note them without some people taking issue. Mule people REALLY believe in mules. Horse people LOVE their horses. And that is as it should be. If you want mules don't read any further.

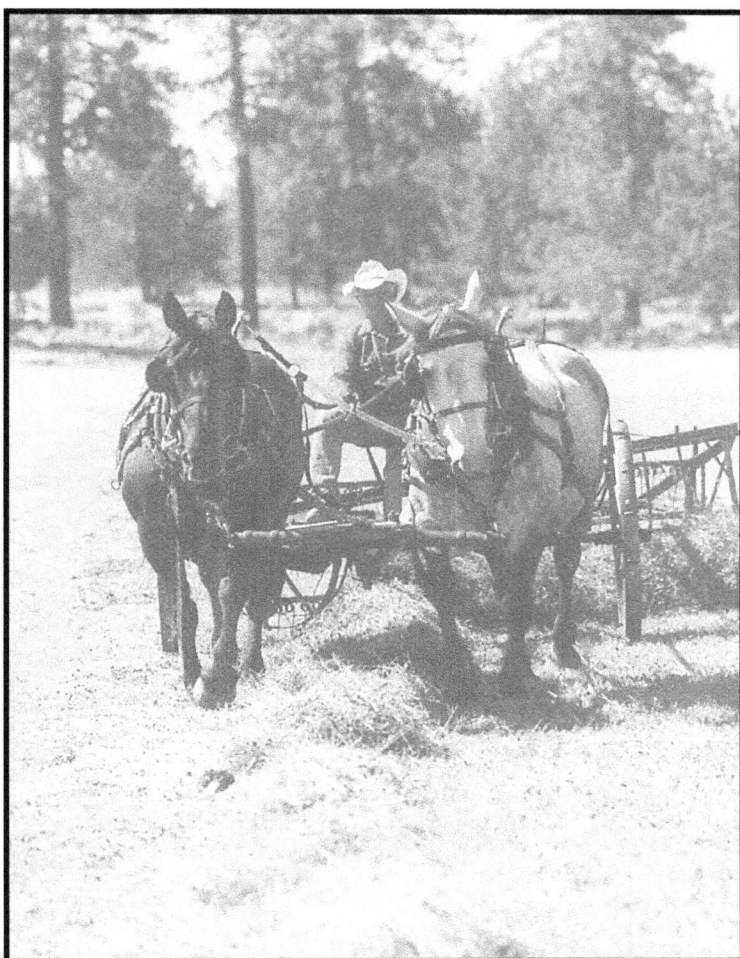

Same thing with horses. If you're a fan of one or the other and just spoiling for a fight, read on.

Mules have a high tolerance for heat, need less water, as a rule are easier keepers, are capable of long periods of hard work, have a phenomenal memory, hold a grudge for a long time, live to get even, are difficult to train, and know that they are beautiful.

Horses have a high tolerance for work, are usually trustworthy, are extremely sensitive, can be as dumb as their owners, seldom hold a grudge, want to please, are easy to train, will eat to excess, seldom forget a nightmare, and don't have a clue how beautiful they are.

(With that this author knows he'll never be invited back to a mule or horse function.)

What You Are Asking of Your Animals

MOWING

If you follow the information in the next chapter on mower setup, mowing need not be an unpleasant job for the animals. Our horses seem to enjoy it. Issues of tongue weight, vibration, side draft, excessive draft all are severely aggravated by inappropriate service, setup, and use of the mower.

With mowing you will be asking your animals to walk straight and at a fluid, comfortable pace. You will be asking of them that they pull a steady moderate load (evenly sharing the

Figure 7. (above) Bud Evers with grade Percherons on side delivery rake. Figure 8. The author delivers a jag of hay to the swinging stacker with a buckrake and pair of Belgian mares. At the stack is Jean Christophe Grosettete of France, on the stack is Carl Leonhardy and Tony Miller.

work) for a long period.[1] Frequency of rest stops are purely at the discretion of the teamster and may be dictated by the conditioning, or lack thereof, of the animals. You may be asking them to do this work for hours at a time and several days in a row. All things with the mower as they should be, there will be little or no tongue weight when the mower is moving ahead and only slight vibration. They'll hear the mower clicking as it cuts the crop. Unexpected obstacles and extreme crop anomalies notwithstanding, side draft should be minimal. They will be receiving a clear sense that something's happening behind and to the right of them.

RAKING

With raking (side delivery) you will be asking your animals to pull a light load straight and in a prescribed pattern. They will be experiencing some mechanical noises and a clear sense that something's happening behind them.

With the dump rake, the horses will feel the spring teeth dragging and the increasing pull until the dump trip clangs and raises to deposit the load. At this point the draft will be released only to start over.

HAYLOADER

With the wagon and hayloader the horses will be asked to pull a long outfit with accumulating draft as the load is built. The vibration from the loader may transfer somewhat through to the wagon tongue, neckyoke and collars. Because of the length of the horses, wagon and hayloader combined, when square corners are attempted, the blindered horses may be able to actually see the machine following. In some

---

1. *The mower is one of the worst places to try to work two animals which are unevenly matched for gait or attitude. It is important that both animals move at the same speed and with the same outlook (i.e. both lazy or both gung ho) otherwise the efficiency required to balance the mower's draft is lost.)*

Figure 9. A pair of Norwegian Fjord Draft Ponies effortlessly pull a new tedder at Indiana Horse Progress Days.

cases this will cause a little surprise panic that the teamster should be prepared for, especially on the first rounds. Of all the different implements this author has used for haying, the hayloader has been the easiest one for the horses to learn (they just need to walk slow) but the hardest one for them to accept.

BUCKRAKE

This is certainly one of the oddest applications of horse power because of the applied dynamics. The Buckrake chapter covers this subject in depth but for here let's just say that the team is split into two individual horses anywhere from 8 to 12 feet apart. They are driven as singles. The animals are expected to push a moderate to heavy load (which is up to the teamster) in front of them. When the load reaches its destination, the animals are then asked to back up and in essence pull the buckrake basket towards them. In this case what is expected of the animals is patience and intelligence. True or false rumors have it that mules won't do this job without counseling. This author has never been able to find for sale any horses or mules which had buckrake experience.[2] We've trained all of ours. In the beginning everything is challenging and odd for teamster and animals alike but most get the hang of it quickly. (See Chapter 12) It is important that any animals put to the buckrake have good attitudes towards backing up as this seems the most difficult concept for them to accept on the buckrake.

BALING

They will have to pull a moderate to heavy load with a curious action on the tongue and necks from the baler's plunger strokes. When actually moving ahead and picking up hay out of the windrow this activity is minimal. It is when stopped and

---

2. *Maybe it's because when animals reach the buckraking level it usually means they've mastered all the other jobs and are therefore too valuable.*

Figure 10. Jess Ross of Oregon mowing with his Percheron mares.

without the resistance of hay that the plunger action will rock the baler back and forth and transfer that action to the horses. Using a baler with a front fly-wheel, such as the older New Holland models, seems to feature an offsetting action which lessens the plunge effect on the horses.

HAUL BACK

Whether on the rope at the barn or the cable at the stacker the animals which are used to pull the hay up have an easy job. It is usually a single animal though sometimes two on a forecart are used. With barn pulley systems there is no counter pull but with many stackers the lowering action actually puts a counter action on the haul back animal requiring, sometimes, a long stretch to back up quietly. This is one job which is excellent for the older horse, the animal which may not be up to all day plowing or mowing.

Training Issues

"So what should I be concerned about as far as the animal's level of training?"

If you're going to go straight to the hayfield, it will be more than a little helpful that your prospective work mates have good solid experience on mowers and rakes and possibly balers.

Barring that, you will find it saves a great deal of time if the animals have experience doing some kind of regular work where it is required that they lean into the collar. This opposed to just light driving pulling a carriage or wagon. Horses with years of experience pulling light vehicles and nothing else are often quick to balk at having to pull a load, especially one which rewards them with vibration and noise.

Be careful about selecting logging or pulling horses for hay equipment without some testing. Some, NOT ALL, pulling and logging horses are made to be high strung and anxious. When they feel a load tighten on their shoulders they think the teamster wants them to put in an all out effort. This can, at best, cause you to be very tired very quickly with having to hold them back. Or, at worst can cause a breakdown of equipment or a serious runaway problem. A runaway is a hazard for you, the animals, the equipment and all other animate and inanimate things in proximity. A runaway is to be avoided at all costs. Runaways are a completely unnecessary evil. This author takes issue, with a certainty born of years of experience and observation, with those who boldly state that run-aways are an inevitable fact of a teamster's life. This does not have to be so. The intelligent horsefarmer has many tools at his or her disposal to safeguard against such occurences. And one of the first and foremost is the employment of intelligent discretion. Good animals will always be worth twice what you pay for them and bad animals will always cost you four times what you pay for them.

This author's first preference, after all these years, is to work with homegrown and trained animals. Second choice is to work with young, green (untrained or just started) animals. And a third choice would be to work with older trained horses. But it is important, in

Figure 11. A light pair of American Cream horses hitched to a forecart and new style rake.

Figure 12. A picture of preference. Dr. Doug Hammill of Montana with his four Clydesdale beauties.

the case of what has just been read, to consider the source. This author has 30 years experience to bring to any training challenge. Even with that qualification, we believe that, in many cases, a well-informed and supported beginner may have an easier time of training new horses than trying to "retrain", "outsmart" or "understand" older broke horses. We recommend a careful reading of *Training Workhorses/Training Teamsters*.

But We've Already Got Animals...

So you've got animals you like and they are trained and you think trustworthy but they have no experience with hay implements or procedures, how do you prepare them? Here's a few suggestions.

MOWING - An ideal situation would be to have one team which is accustomed to mowing, out mowing, and the team in question on a forecart or wagon following that mower as outlined in Chapter Nine, but things are seldom ideal.

So if that's not available: Instead of attempting a full standing hay crop right off the bat find a pasture or light weed patch with lots of room and no nearby obstacles. Hitch to the mower (read Chapter nine first!) and drive around until your team is relaxed. Then set the cutter bar down just ahead of a little pasture or weeds and put the mower in gear (no dogs, livestock, children, or idiot bystanders please). You on the seat, lines in hand, thinking comfortable thoughts about an

upcoming fishing trip but paying close attention to your team, give them the command to go.

If they immediately start dancing on their toes with hips down and noses up, casually stop them and talk to them sweet. Then when they've calmed, give them the command to go again. Repeat if necessary until their hips are up and noses down. When they seem to be taking it well, even though a little nervous, attempt a corner (See Chapter Nine).

OR - If they immediately walk out like its no big deal and show only a passing notice of the mower activity, stop them and talk to them sweet. Then, when you are ready, give them the command to go again. Repeat several times. Attempt a corner. (See Chapter Nine).

The object of all this is to let the animals get accustomed to the mower and its activity BEFORE you attempt to actually mow hay. Because when you do mow you will want to pay attention to the mower and less to the animals. If you are anxious about mowing straight or cutting clean you will be transferring that anxiety to the animals at a time when they need you calm, sure of yourself, and attentive to them.

RAKING, BUCKING, ETC. The individual chapters on these procedures offer suggestions about introducing the animals to the process. Hopefully the reader will understand the objectives and customize the introductions from what is at hand, always with a view towards SAFETY and success.

Mowers & Mowing

Chapter Nine

Figure 1. An 1885 McCormick 'Iron' mower. The sales literature stated "the adaptability of the floating bar to all surfaces, the efficient and rapid stroke of the keen knife through accurately fitting and penetrating fingers, and its established reputation for close and clean cutting over rough or stony ground, or in matted, dead or bottom grass, has made it the peculiar favorite of the progressive farmer.

This will probably be the meat and potatoes chapter for many readers of this book. Mowing and mowers can certainly be the most challenging aspect. The mechanics and idiosyncracies of the mower, to the unintiated, can seem a daunting pile to master. This author has spent thirty years with the machine and the process and is pleased to say that a measure of mastery was accomplished, with excellent help, in the first two years. Since then the machine (in this case a McCormick Deering #9) and the process have been pure pleasure. But we get ahead of ourselves.

Which Mower to Use

If you have elected to mow hay with horses or mules in harness, you have several choices as far as machines and what they might mean to the process. This author has some strong preferences and biases. First a view of the choices:

Machines
- really old rare antique horse mowers
- latest generation (circa pre WWII) horse mowers
- trail-type tractor mowers pulled behind forecarts w/ PTO power source
- combination mower/conditioners pulled by motorized forecarts

Under that first category might fall these manufacturers names and their earliest models;

Acme	**Adriance**
Buckeye	**Champion**
Crown	**Deering**
Dain	**Emerson**
Independent	**Johnston**
Massey Harris	**Milwaukee**
McCormick	**Minneapolis Moline**
Osborne	**Plano**
Thomas	**Wood Bros.**

Figure 2. Two teams mowing under Bulldog Fraser's supervision in Northwestern Montana.

Under the second category would fall these manufacturers and their last models;

Case
Champion
David Bradley
John Deere
McCormick Deering/International
Oliver

The third heading has too many entries and too broad a spectrum to list. And the fourth heading, haybines, includes for our discussions just the smaller models of modern manufacture.

We are concerned, in this book, with the traditional ground drive North American mower. And more specifically with those makes and models manufactured from 1920 through 1945. This is with good

Figure 3. Bulldog Fraser, the infamous Montana teamster with a young pair of Percherons in training on the McCormick Deering No. 7 mower.

Figure 5. Whitely's "New Champion" mower from 1877

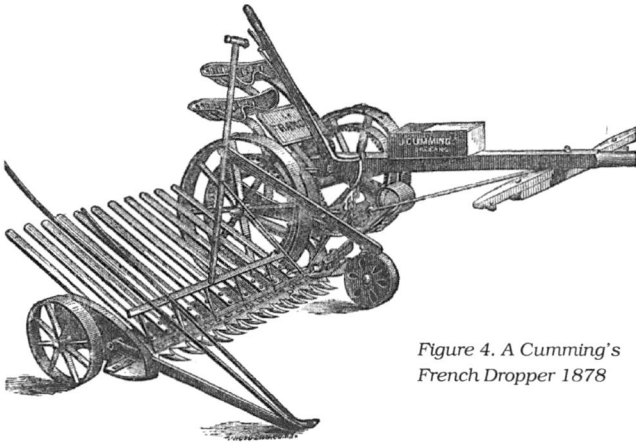

Figure 4. A Cumming's French Dropper 1878

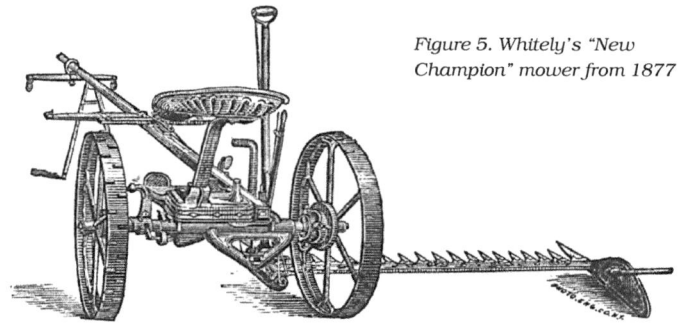

Figure 7. Aultman's "New Buckeye" mower

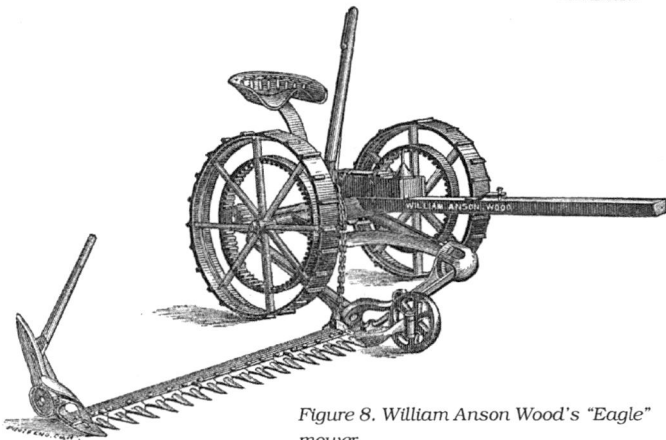

Figure 6. Walter A. Wood's One Horse mower

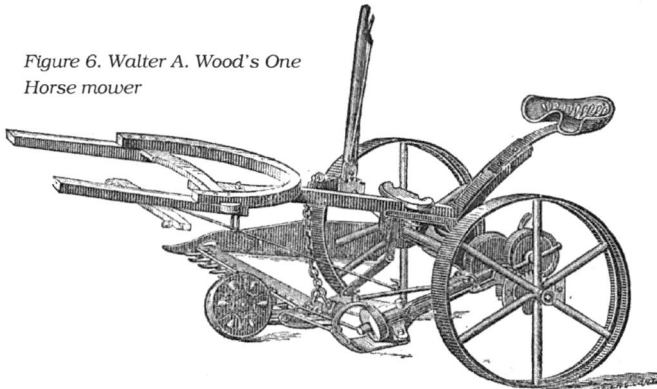

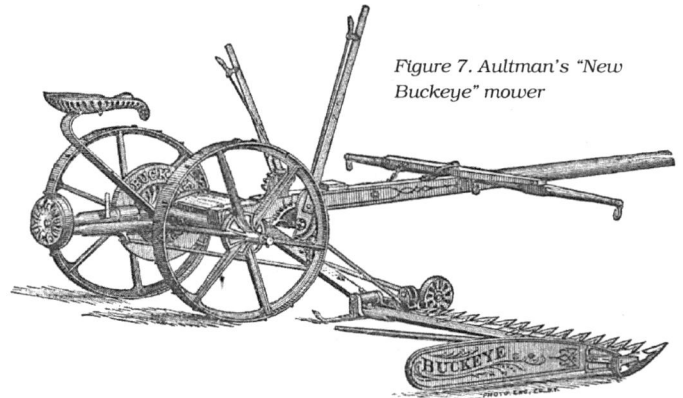

Figure 9. A highly unusual center cut mower. All the makes and models on this page are but a smattering of the hundreds of predecessors of the 'modern' horse mowers.

Figure 8. William Anson Wood's "Eagle" mower

reason which we'll cover very soon.

Although they are pictured and mentioned, primarily through caption, the PTO and motorized mowers are not covered in depth in this book. They are all tractor mowers which have been accomodated for, or modified to work with, forecarts. And as tractor mowers they have ample literature elsewhere on their workings.

Historically speaking, the horse mower is a member of a family of animal drawn technologies which were the pinnacle of 19th century engineering development and had, in this author's opinion, a far greater and more immediate cultural effect on human

development than the other more touted inventions of electricity, radio, telephone and refrigeration. That family of animal drawn technologies included the reaper, the binder, the mower, the combine thresher, the drill and row planters, the straddle row cultivators, and the hay loaders. Though many agricultural historians discount this family of inventions as a mere stepping stone to the internal combustion/hydraulic age, this author disagrees.

These machines, the engineering accomplishments they represent - the immediate effect they had on farm productivity - and the dramatic positive effect they had on rural communities, defined not only an age of agriculture but generations of possibilities and hope. Another discussion another time would have us point to local self-reliance, food, work, churches, schools, institutions like hardware stores, and outside trade expansion and how this all defined the best of

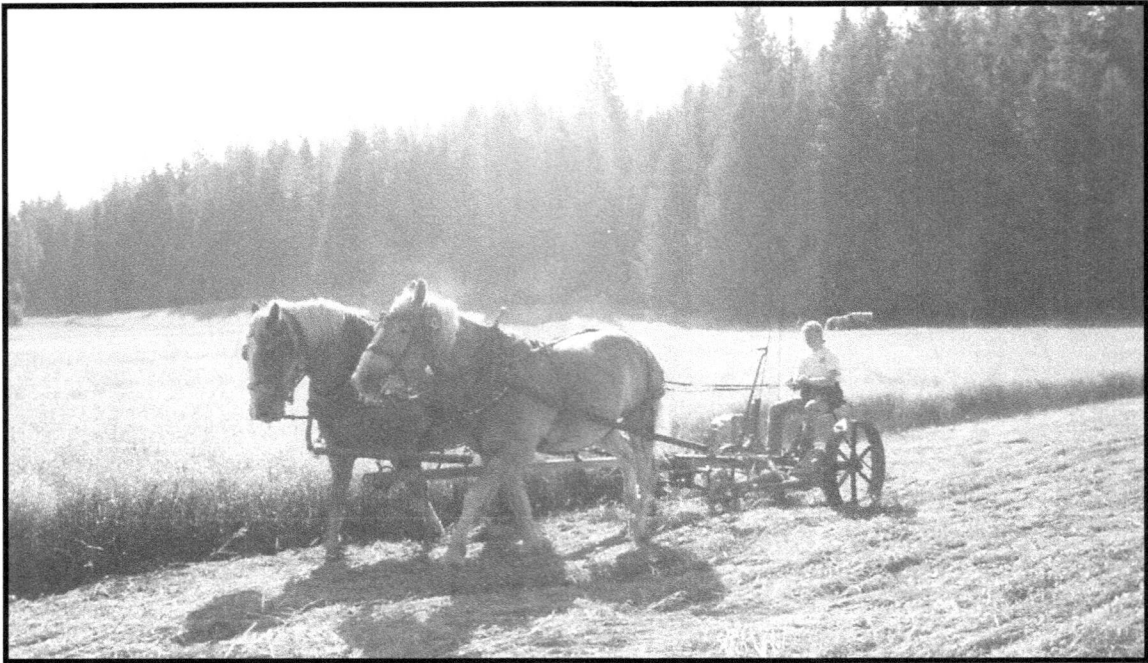

Figure 10. Two McCormick Deering / International model #7s doing excellent work recently in a Montana hayfield. Lynn Goodwin in the forefront.

Figure 11. A John Deere Model #4 with seven foot cutting bar and three abreast setup (see JD section in this chapter). This and the photo above are of recommended makes and models for modern horse haying.

comparison, the next three or four generations of farm technologies were all steps backward into clumsy applications of brute force. In fact, to this day, agricultural engineers have not managed to equal the simplicity, relevance and economy of this marvelous implement. Imagine an implement which, with modest care and a small store of replacement parts, has and will last - with full use year after year - for three or four lifetimes. With a small chunk of information, such as this chapter contains, a person two hundred years from today - upon discovery of a McD #9 or a JD #4 in storage - could put the machine to immediate use. We cannot say the same for most of the modern technologies we depend on (i.e. VCRs, CD players, aircraft, autos, nuclear power, and computers etc.) Our point being that these machines, in their adaptability and suitability, are thoroughly modern.

the twentieth century. And the mower is an appropriate, quiet, undervalued center piece.

The refinement of these animal-powered technologies and especially the mower continued in North America up through the 1930's. In Europe the ravages of war and prolonged domestic uncertainty all but stopped parallel development. It is true that North America was home to the most advanced animal-powered technologies the world has ever seen.

The last models of horsedrawn mowers are marvels of balance, torque, and true efficiency. By

This author recommends without hesitation that anyone considering mowing with horses in the twenty-first century should strongly consider limiting the

*Figures 12 and 13. Two photos of an
upgraded McCormick Deering/International
#9 mower with a six foot bar and special
wheels. This model was displayed at the
Ohio '97 Horse Progress Days. A farm shop
went completely through this unit. The cutter
bar features all new guards, ledgers, knives,
etc. The original wheels on this mower
would have been rubber-tired. The stock
spider assemblies have been used to receive
new wide steel wheels with extreme cleats.
There are several shops around North
America, most are Amish and in the
midwest, which renovate and upgrade
mowers like this. With a little shopping, it is
possible to find them for sale. Prices range
from "DuPont restorations"[1] at $800 all the
way to warranteed work at $3,000 with
many in between.*

selection to two, possibly four, makes.

While it is true that older, more
obscure makes and models are being
used and can be made work-ready
there are challenges which must be
understood especially for the farmer
who is serious about years of haying.

As you will come to understand
in this chapter, the mower has many
moving parts and many removable
parts which are meant to be replaced
at intervals. The serious horsefarmer
will be temporarily crippled by a make
of mower for which there are no
readily available replacement parts
(parts like knives, guards, pitman
bearings, ledger plates, rivets, oil
seals, etc..)

The first two choices of makes
are McCormick Deering/International
(McD/I) and John Deere (JD). The
second choices would be Oliver and
Case. For the McD/I and JD there
exists a vast storehouse of new/old
and aftermarket parts. For the Oliver and Case, many
of the aforementioned cutterbar parts are interchangeable.

So the simple reasons for recommending JD #4
and McD/I #7 & #9 are these:
  • Parts available.
  • More of these models made than any other,
    translating into a vast storehouse of parts
    mowers.

  • They represent the pinnacle of design and
    engineering, the very best.

And outside of an animal drawn mower made in
Poland, no new ground drive mowers are currently
being made for true horsepower.

_____

*1. "DuPont Restorations" is a perjorative which refers to those
implements which have received a coat of spray paint without benefit of
any surface preparation.*

Figure 14. At Horse Progress Days in Ohio this motorized trail mower was demonstrated hooked to a Pioneer forecart.

Utilizing any of a number of forecarts, from basic simple axle and tongue designs to the space age articulated-steering fully motorized units, almost any modern mower can be pulled by horses or mules.

In the case of the one illustrated, the teamster needs to keep in mind that such a set up CANNOT be backed up more than two or three inches and can be unforgiving on the corners.

One farmer, using a similar setup commented that he rigged a kill switch he could pull from the seat of the forecart in case he had to shut it off in a hurry. This one is set up the same way.

Figures 15 & 16 are views of four Percherons pulling a self-powered haybine at the '95 Horse Progress Days in Indiana. Notice how the implement tongue, originally designed for tractor pull, has been outfitted with a set of truck wheels and seat.

The Horse Progress Days were the original idea of Elmer Lapp of Pennsylvania. They have gone through some changes over the years but have settled into somewhat of a pattern. They are organized now, and managed by a committee of Amish as an event to showcase new animal-powered technologies. The event is usually on or near the Fourth of July weekend and rotates between PA, IN, or OH.

Figure 17. A young man drives three Belgians hooked to a simple forecart
which in turn is hooked to a self-propelled mower/conditioner.

Figure 18. This photo gives a good view of the simplicity of the set up.

Figure 19. These photos were taken at the Indiana Horse Progress Days.

Figure 20. The horses are the motive power which, when intelligence comes in to play, can be applied to most any farming technology.

Mowing

After careful thought as to the organization of the next information this author has decided that there is no logical answer as to what should come first. So the reader is encouraged to read this whole segment (indeed the whole chapter!) before attempting anything outlined here.

Laying Out the Mowing

The conventional wisdom would have you mow around any given field until you get it all mowed, period. The way you start out is to mow the first pass around to the left making an opening swath. That round completed you turn your team and mower around and drive on top of that first swath as you mow around to the right. And you proceed this way until the field is mowed.

If your field is small, say five acres or less, this may be a fine way to proceed. What we'd like to suggest however is that some thought go into these aspects of your project;

Why are you mowing this piece?
a. clip a pasture
b. weed control
c. hay
d. green manure
e. root mass management (fertilizer)
f. other

If you are mowing for hay, and that is our primary concern here;
1. Is the hay going into a barn loose, baled?
2. Is the hay going into a bale and elsewhere?
3. Is the hay going loose into a stack nearby?
4. Will the hay require special handling for some reason?

The variables are extensive and interesting. It is perhaps enough to make these notes about what you might consider:

a. If you are short of help and animals consider mowing down no more than five acres of hay at a time with at least a full day between five acre lands, this will prevent you from having too much hay ready to pick up or bale all at once. It may also protect you from losing all your hay to inclement weather.

b. If your field or fields are more than a quarter mile at their widest and you are putting up loose hay, have more than one location to stack hay and try to design your stack yards so that the distance is cut nearly in half. This will save considerable time in bringing hay in and suggest how you might best divide up the mowing.

c. Resist the temptation to drop all of your hay at once unless what you have is less than five acres.

d. If you are stacking hay try to design the location of your stack yards and the layout of the mowed lands so that the travel to and from field is most efficient.

Below is a diagram of how two square forty acre hay fields (quarter mile by quarter mile) were laid out at Singing Horse Ranch with three stack yards. The solid line represents fences, the dotted lines - mowed lands each 4 acres in size. The lower case letters represent lands in the mowing sequence beginning on the left all of **A** went into stackyard #1. Skipping to the fence in the center and working back to the left all of **B** went in to stackyard #2. Then crossing the cross fence and

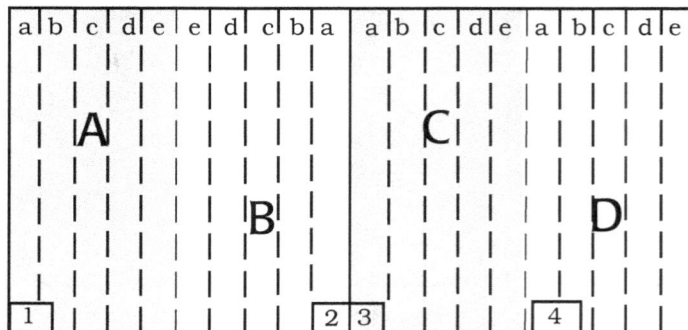

starting with C's 'a' that 20 acres went in to stackyard #3 and so on.

There were plenty of times that all of a given twenty acres was mowed down but it was mowed in sequence which meant, always, that the hay nearest the stackyard had received the most curing time.

The Actual Mowing

Of course, with any complicated process like mowing with horses or mules, there are so many different elements and aspects to draw together that it can be difficult to find a beginning point, especially in a discussion like this.

If the reader can imagine an excellent mower all set up and ready to go, hooked to an experienced healthy team, in a field that has been laid out and opened (a scene one might find with a qualified workshop or instructor), then imagine yourself on the mower seat for the first time with lines in hand. Here's a check list;

• Is cutter bar down and back of the waiting crop at least two inches?

• Visual check of lines and bridles and neckyoke, is everything as it should be?

• Put mower in gear with pedal near gear box.

• Take up slack in driving lines, look at horses ears to make sure both are aimed your way and give the verbal command to go.

• Make sure right horse or mule is walking up against cut edge and that inside heel of cutter bar is in proper position so that there is a full cut of the bar.

• Maintain the team's pace at a steady walk. Don't permit trotting and stalling.

• Occasionally look at cutter bar action to see if the crop is cascading back properly and cutting clean.

Making the Corner

As the team nears the corner, be thinking about the corner. Hold your lines so that the team does not make the corner ahead of time. The animals do anticipate.

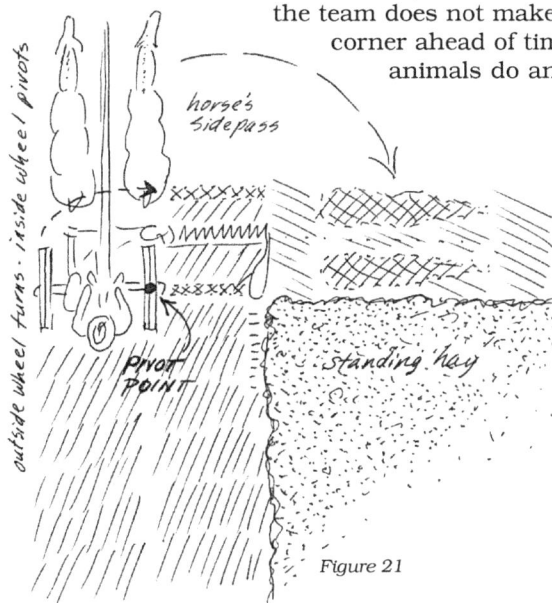

Figure 21

One way to make the corner is to stop the mower just past the edge of cut grass. Press down on the pedal lift and then swing the team to the right without allowing the mower to roll ahead any. (This can be tricky for people new to driving a team.) You may find that you have to back up the mower a step or two to position the cutter bar just back of the mown edge and in proper alignment. All of this can be done with the mower still in gear. All lined up, you can proceed with the mower pass along the end. (Figure 21)

Another way to make the corner is to mow, from the beginning a rounded corner, one which would allow that cutter bar stay down and mowing as the team walks around the corner. (Figure 23)

A third way to make the corner has the teamster mow slightly past the edge, stop the team and back up turning as you go until you are in line to approach the end. (Figure 22)

A fourth and unweildy way is to mow past the edge, lift the cutter bar, and keep going, turning to the left in a clover leaf until back in to the land.

Opening a Land

When a field is divided into chunks we refer to those as *lands*. When mowing an entire field the perimeter is usually established by fences, ditches, pipelines, hedgerows or roadways. When the field is divided into lands the teamster is required to head off across an open expanse of forage without a natural barrier to drive along.

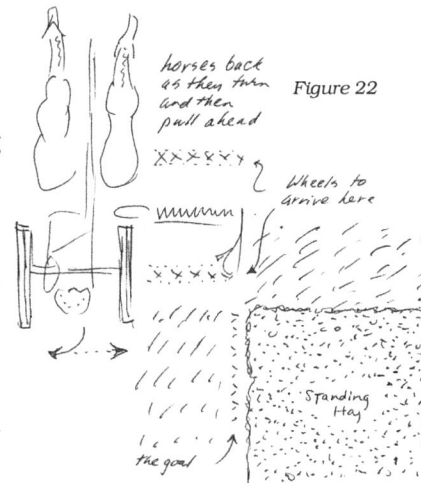

Figure 22

Figure 23

Keep all pets locked up, away from the mowing. Dogs and cats playing in the field WILL end up with leg(s) missing or dead if caught by the cutter bar!

There are certainly exceptions with odd shaped fields, but wherever there is a long straight side tremendous advantage is to be enjoyed when the farmer successfully strikes off the opposing side in perfect straight parallel. It means that when the mowing meets in the center there is opportunity for a last swath the width of the cutter bar. This, as anyone whose mowed with tractor or horses can attest to, is a joy to be planned and wished for.

Pace off, or actually measure, both ends of the land you are going to mow and mark the corners clearly with a post or a flag - something which can be seen from the opposite end. Measure the width to be a multiple of your cutter bar width[1] with hopes that the swaths will come out parallel and even.

When you are ready to open that first 'land', position your mower and team so that the nearest field marker is center of your seated back. Look out across the field to the marker on the other side, have it line up with the mower tongue. Now this can be important if mowing straight is valued, find some tree or object in the landscape that is past where you will be mowing to and in line with the marker. You will be driving your team straight to the further point perhaps occasionally looking to see if the field marker and the further point are still in line. The most important advice we can give you, if your goal is to drive a straight line, is DONT LOOK BACK. When you do, it is a certainty that a swerve will result.

Make that straight first cut across the open field and turn left at the end either heading for a new marker set up or along the field perimeter. Proceed around the land in this direction until a complete round is done. Keep in mind, if you are mowing along a ditch or cluttered fence row that what you cut will become, on the next round, the place where your mower will run. Obviously you can cut over an open ditch, but on the return pass your mower wheel will be in that ditch. Give yourself a couple of feet clearance until you are better at the job.

When you have finished that first opening round you are ready to turn around and proceed with right turns, keeping your mower on the cut grass and the heel of the cutter bar up against the unmown crop.

How to Hitch Horses to Mower

Even with well trained trustworthy horses or mules, extra safety in hitching to, and operation of, the mower is still important. The beginner with the mower and all haying equipment would be well advised to develop a habit of following a safety ladened routine.

First double check the harness. Bridles fit, throat latches fastened, bit straps of good leather? Lines properly set up? (Suggestion; Use billets on the lines and buckle them direct to the bit rather than using snaps. One less thing to come apart and cause misery.) Are the breast straps good leather and ad-

justed to carry the neckyoke high? Do the collars fit properly?

There will be times and reasons why it might not be the first choice, but this author suggests that the team be ground driven over the mower tongue and into position for hitching. This is a good thing to get in the habit with and will eventually lead to being able to do all hitching and unhitching alone. (Remember that with everything you do, you are training your animals. If they get used to being led and held for hitching, they'll expect it.) Side note: if you ground drive your team over the tongue and they don't want to stand there, anxious and spreading out and pawing the ground, make them stand until relatively quiet. Then walk them around in a circle and back over the tongue. Repeat this excercise whatever number of times it takes until they accept it completely and stand quiet. If they are still too anxious and even unruly, ask yourself if you really want to hook them to the mower. If the answer is yes, then make sure you get some experienced help for the hitching and the first mower rounds.

When hitching to the mower, first make sure it's on level ground and out of gear. The cutter bar should be fastened up in the vertical or carrier position. This is for the safety of all people in attendance during hitching. Check the neckyoke on the end of the tongue. If it is a slide type, tie it to the tongue with a piece of wire to keep it from falling off, should the horses step ahead before the traces are hooked. Or should the traces be hooked too long. Always hook the neckyoke FIRST making sure it is holding the tongue up at the proper angle. (The breast straps should have been adjusted to the shortest length before hitching.)

Keep lines in hand all during the hitching process. Neckyoke fastened move back and start with the far right tug and hook trace chains, one tug after the other, until done. It is critical that tugs be tight enough that the neckyoke not slip off when horses step ahead. The proper length will allow a little slack in the trace when the animals step back or a little slack in the quarter straps when they are pulling the load. There is no need to be hooked so tight that the quarter straps rub the bellies when moving ahead. That is just an-

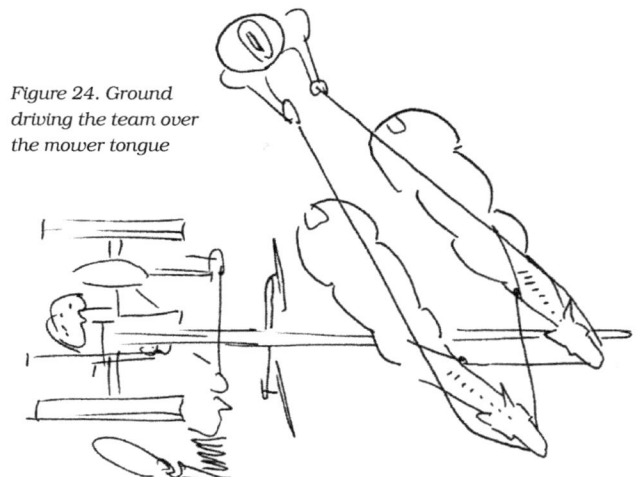

Figure 24. Ground driving the team over the mower tongue

---

1. *If more than one mower is being used and the cutter bar widths are different, it may be impossible to calibrate.*

other discomfort for your working partners. Once the proper length, or hitch point is found, remember it to save time next hitching.

The hitch hardware on the tongue should be adjustable for height with two to four holes in the bracket. It is this author's experience, contrary to prevalent thought, that the animal's are best served if there is a slight lifting action when they pull ahead. If you are unable to move the bracket down, and thereby cause

> *Always hook neckyoke first when hitching and unook neckyoke last when unhitching. If it is a ring style, tie the neckyoke on the tongue cap with a piece of wire.*

a lift, you can accomplish the same thing by fastening a block of wood to the bottom of the tongue between it and the evener shackle. This will cause the same lift. This will translate, in forward motion to little or no tongue weight for the animals and has no appreciable affect on draft.

When unhitching the team from the mower, take care to park in a level area so that the mower, when the traces are unhooked, doesn't try to roll back causing problems at the front of the animals. Lower the cutter bar

> *Keep the lines in hand, and slack, while hitching.*

heel to the ground and, if its not already there, fasten the bar up in the vertical position for your own safety and convenience. ALWAYS unhook the traces before unhooking the neckyoke! Keep driving lines in hand. After traces are unhooked, move to the front of the team and unhook the neckyoke being careful not to let it fall on you or the animals (it's heavy!).

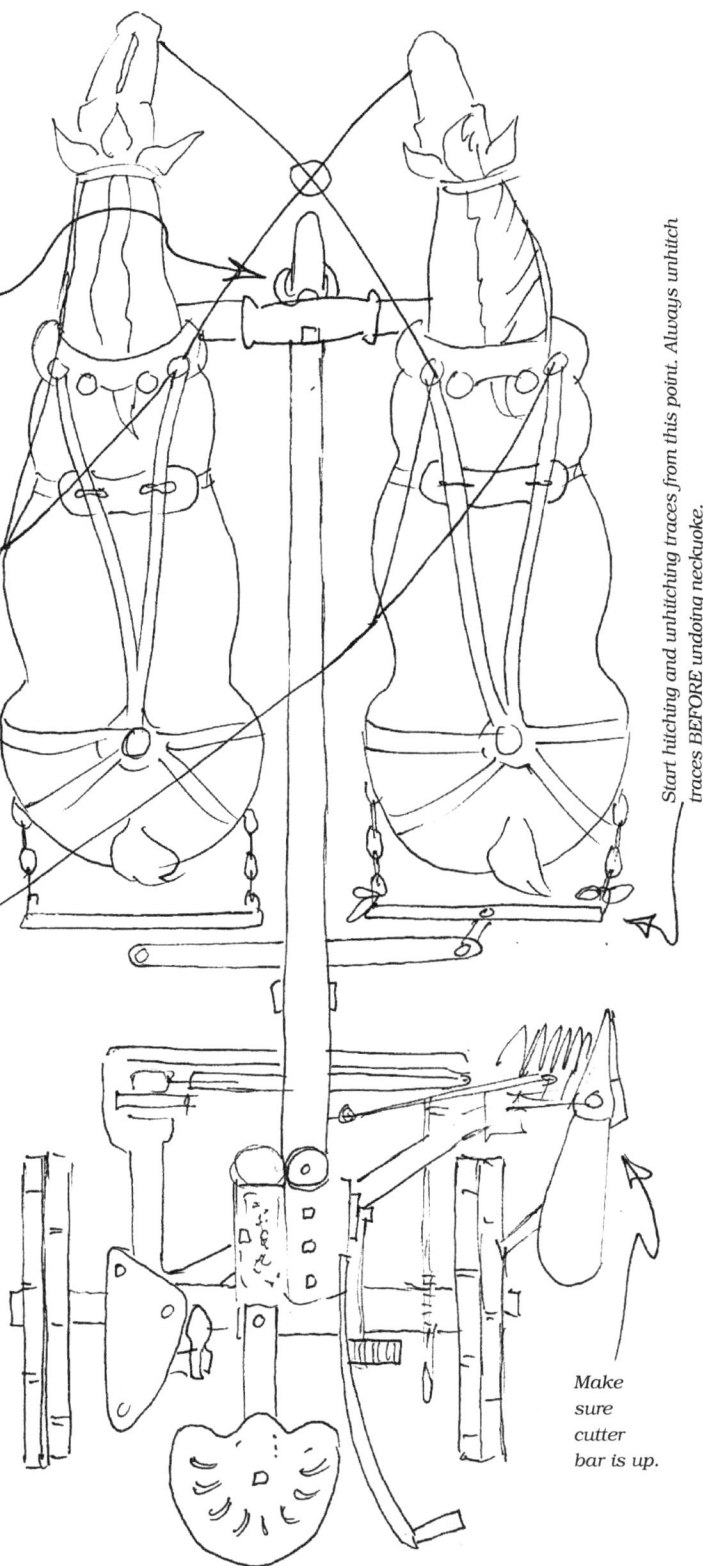

Start hitching and unhitching traces from this point. Always unhitch traces BEFORE undoing neckyoke.

Make sure cutter bar is up.

Make sure mower is out of gear.

Figure 25

Figure 26. A simple, small field pattern, showing direction of mower and also showing the first, opening pass with the side delivery rake.

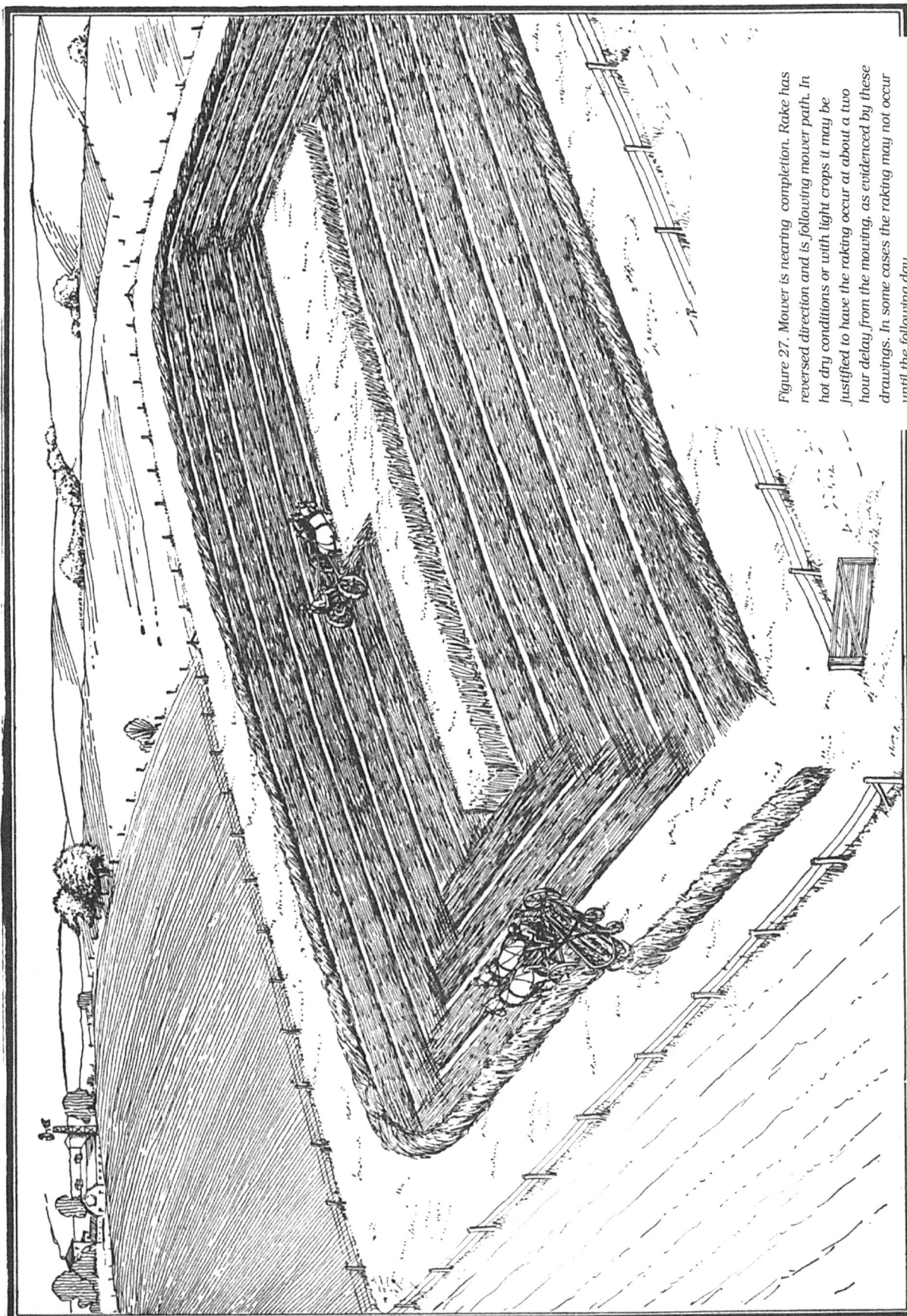

Figure 27. Mower is nearing completion. Rake has reversed direction and is following mower path. In hot dry conditions or with light crops it may be justified to have the raking occur at about a two hour delay from the mowing, as evidenced by these drawings. In some cases the raking may not occur until the following day.

The Understanding and Care of Horsedrawn Mowers

The information which follows is generic to most mowers built after 1915. Following this is information specific to International and John Deere.

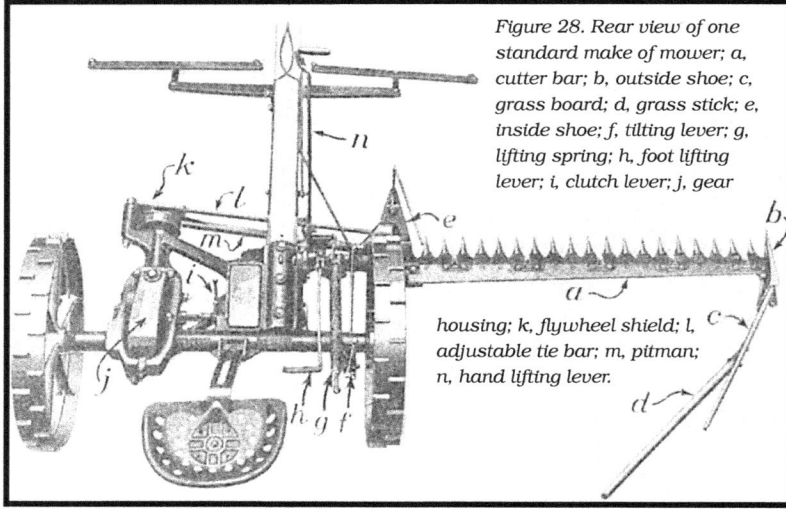

Figure 28. Rear view of one standard make of mower; a, cutter bar; b, outside shoe; c, grass board; d, grass stick; e, inside shoe; f, tilting lever; g, lifting spring; h, foot lifting lever; i, clutch lever; j, gear housing; k, flywheel shield; l, adjustable tie bar; m, pitman; n, hand lifting lever.

A mowing machine kept in good repair and in proper adjustment will last longer and do better work than one which is not. Also the former is less likely to cause costly delay at harvest due to machine failure. Because of the large number of makes and models of mowers it is necessary to limit the scope of the next paragraphs to general information applicable to all machines in common use.

The degree to which a machine may be farm reconditioned depends, among other things, upon available shop equipment. It is not the intention here to encourage farm repair work beyond the possibilities of farm shop facilities or of the ability of the operator. Expert labor or assistance should be employed when there is doubt whether a job can be well done with home equipment.

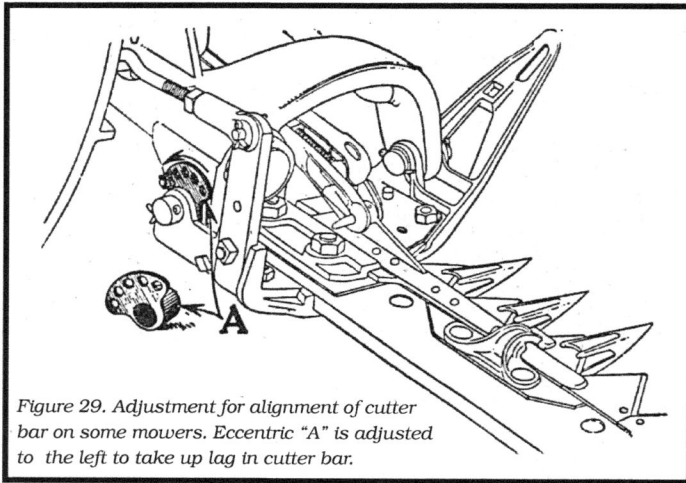

Figure 29. Adjustment for alignment of cutter bar on some mowers. Eccentric "A" is adjusted to the left to take up lag in cutter bar.

HAY AND GRAIN of good quality are most likely to be obtained when the crop is cut at the proper time. Considering the comparatively short period during which the crop may be harvested to best advantage, the time for cutting should be as fully utilized as possible. Putting the machine in the field in the best possible working condition affords some assurance against delays due to machine failure. Further assurance is provided in the ability of the operator to make emergency repairs quickly and intelligently, assuming adequate shop facilities.

Fundamental principles found in the older mowers have been generally retained in the later machines although improvements have been made resulting in more efficient work and less trouble and delay. Lighter and stronger materials of construction,

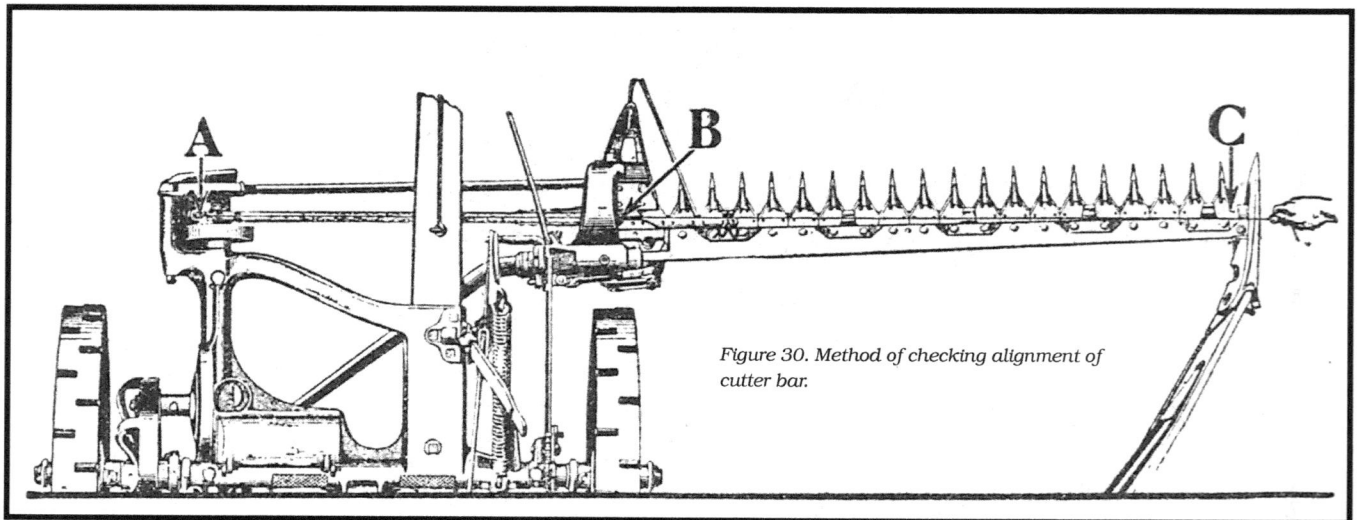

Figure 30. Method of checking alignment of cutter bar.

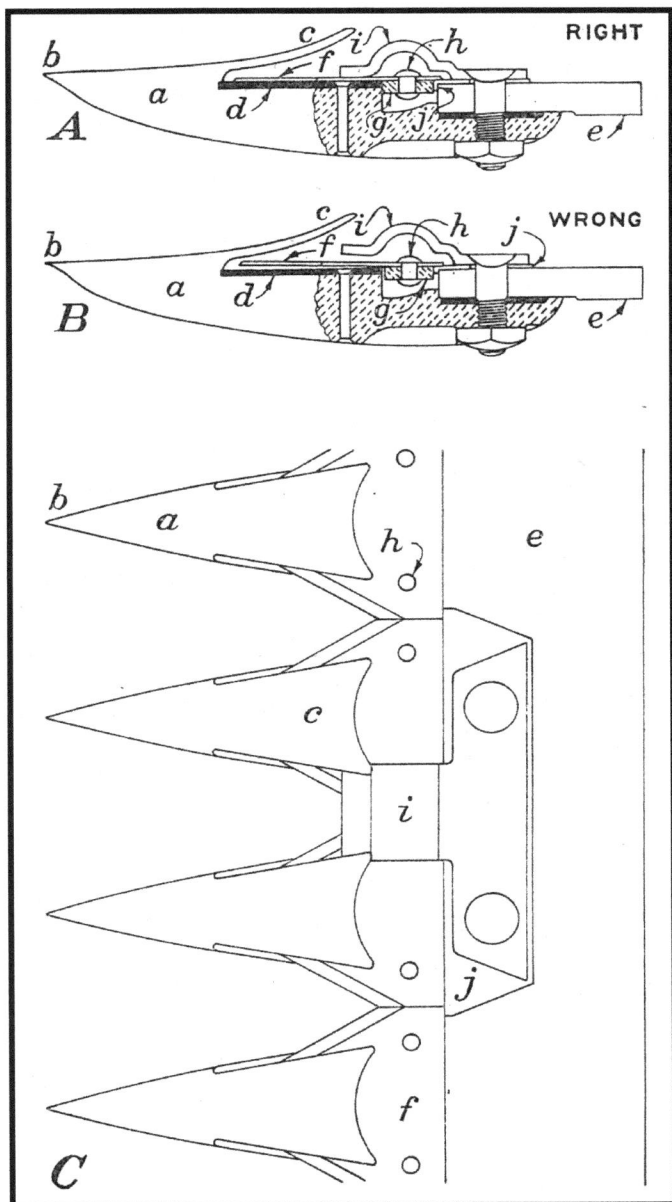

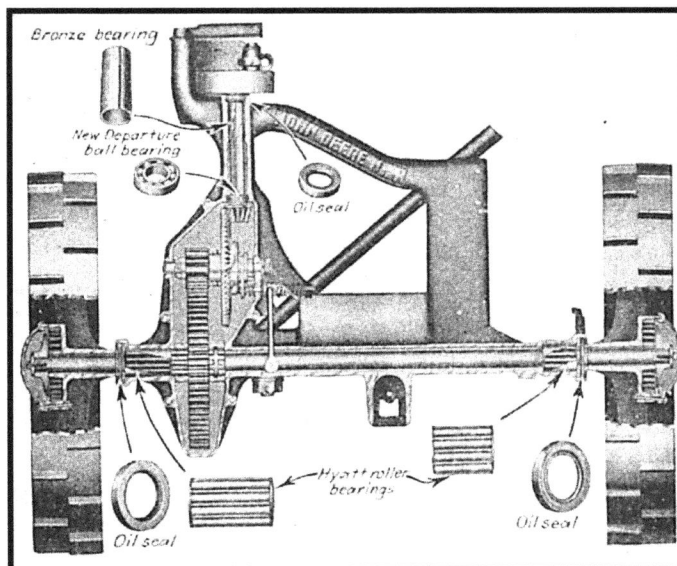

Figure 32. Mower frame showing enclosed gears and automatic lubrication of main operating parts. Note bearing and oil seals on axle line and pitman shaft.

Figure 31. Section and plan of cutter bar of mower: A, Section showing right method of adjusting clips; B, Section showing wrong method of adjusting clips; C, Plan showing relation of guards to knife. a, guard; b, point; c, lip; d, guard plate; e, cutter bar; f, knife section; g, knife bar; h, rivet; i, clip or holder; j, wearing plate.

more general use of roller and ball bearings, improved lubrication, and new or improved units and adjustments have been provided. There is, too, more choice in the selection of the type of machine best suited for the work and in the equipment and attachments available to meet special crop requirements or conditions.

Undoubtedly many machines have been scrapped and many more are likely to be, while still capable of doing several years of useful work if a small sum were expended upon them for repairs. The proper time for overhauling a machine is during its period of inactivity and before the rush of spring work. If overhauling is put off until the machine is needed, failure to obtain repair parts promptly, press of other work, and the hazy recollection of the past season's difficulties may result in costly delays.

The different parts or attachments of the mower are assembled on a main frame, and are readily accessible for repair or replacement (fig. 33). While a comparatively simple machine, many of the adjustments on the mower, particularly on the cutting mechanism, are of an exacting nature. Draft tests on a 5-foot mower showed that when cutting with a dull knife with guard plates in poor adjustment, from 30 to 35 percent more power is required than when the knife is sharp and guard plates are in proper adjustment and that but very little is to be gained by sharpening the knife and leaving the guard plates in poor adjustment.

In overhauling the mower for the next season's work, the cutter-bar assembly is a logical starting place, although the actual repair work should not begin until the entire machine has first been checked over to avoid duplication of work. In doing this, a definite sequence should be followed.

ALIGNMENT OF CUTTER BAR

A cutter bar is in proper alignment when the center of the pitman box, the knife head, and the outer end of the knife bar fall in a straight line when operating. When machine travel is stopped, the pressure exerted on the cutter bar is released, and the bar should come to rest with its outer end slightly in the lead or advance of its inner end. There are several methods of checking the alignment. One is shown in figure 30 and involves the following procedure: With the tongue blocked up to normal position of use and the lifting spring adjusted so that the inside shoe is just

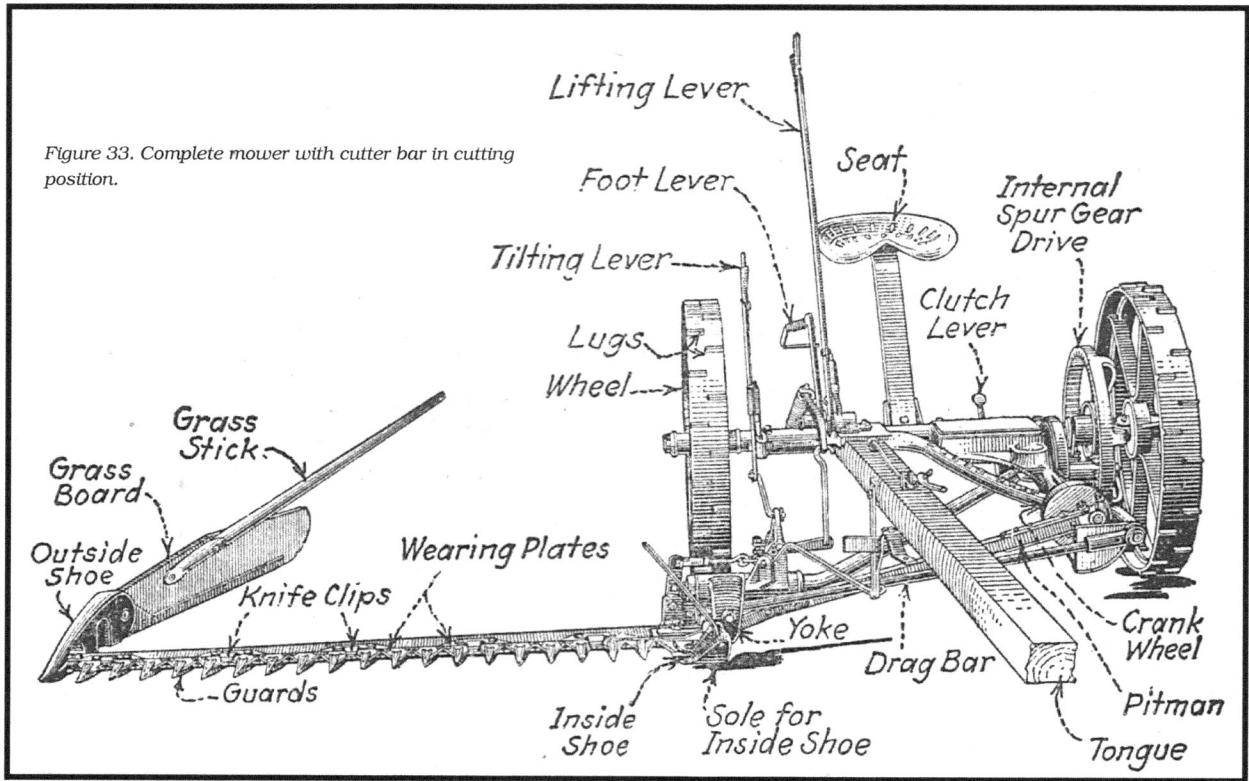

Figure 33. Complete mower with cutter bar in cutting position.

floating, run a straightedge or string parallel to the axle and extending beyond the outer end of the cutter bar. This line should be trued by measuring to it from each end of the axle, and should be held the same distance from the floor or ground along its entire length. After securely establishing the parallel, pull back the outer end of the cutter bar to take up the slack due to wear; then measure from the line to the rear edge of the knife at its outer and inner ends. The lead of the outer end over the inner end can then be determined. Manufacturer's recommendations as to the proper lead are not alike, and the operator of any particular machine would do well to find out and use what the maker recommends. Usually one-quarter inch for each foot of length of cutter bar is recommended.

On mowers of later type an eccentric adjustment is usually provided for aligning the cutter bar (fig. 29). On some machines a step collar is provided in the

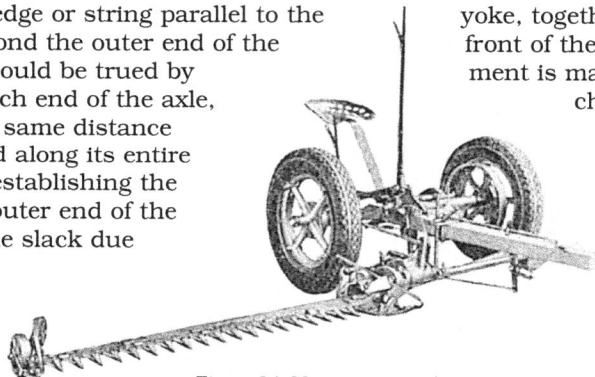

Figure 34. Mower equipped with rubber tires.

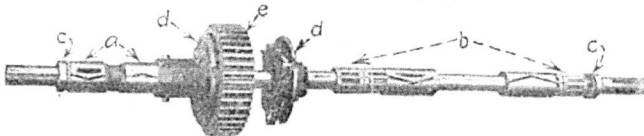

Figure 35. Mower axle and parts: A, B, roller bearings; C, leather oil seal; D, pawl and pawl holder; E, drive gear containing ratchets.

yoke, together with an adjustable tie-bar just in front of the pitman. When this method of adjustment is made the register of the knife should be checked (p.107). Improper lead may frequently be corrected by taking the play out of the yoke pins. Where the holes in the yoke casting have been worn oblong by the pins, which is usually the case, it may be necessary to bore an oversized hole and fit in a larger pin.

When the machine is operating, it is difficult to detect by observation, just when the cutter bar is out of alignment, and therefore it should be checked periodically. A cutter bar out of alignment will do a poor job of cutting and cause heavier draft.

ALIGNMENT OF GUARDS

With the knife removed, guards which are badly out of alignment may be detected by sighting along the guard bar. Take a straight piece of scrap iron or a steel straightedge 15 to 20 inches long and move it along the top of the guard plates d, figure 31, a, noting which plates are high and which are low. In doing this, pay no attention to the points of the guards. Drive the guard which is out of alignment back into place by a sharp blow of a hammer at a point on the guard where the stock is thick. In replacing a broken guard with a new one, the guard plate may be too high. This can be remedied by putting tin shims between the guard and the cutter bar when bolting on.

Figure 36. Gear transmission of a mower, showing spur and bevel gears, clutch, bearings, and bushings. The bevel gears run in an oil bath, which is enclosed and dust proof.

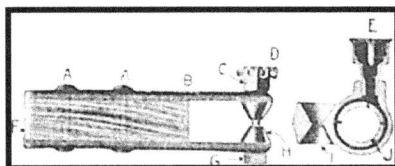

Figure 37. Pitman straps and pitman box: A, rivets; B, pitman straps; C, nut; D, cotter key; E, grease cup; F, pitman; G, pitman bolt; H, conical pitman-strap connections; I, countersunk connection for pitman straps; J, bronze bushing.

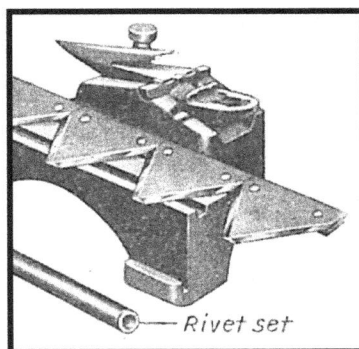

Figure 38. Special anvil for removing and replacing ledger plates and knife

Broken or badly worn guard plates may be replaced and brought into line with shims as for a new guard.

On stony ground the guard wings--the side projections of the guard--may become bent up, down, or sidewise. The function of such a wing is to help hold the guards rigid and to guide the cut grass out of the way of the knife. When bent, they should be straightened to conform as nearly as possible to the position of the other guards.

KNIFE

A sharp, straight, well-centered knife sliding freely without excessive play makes for greatest efficiency. Since the knife is in constant danger of damage, an additional one should be kept on hand. There is, too, the advantage of always having a sharp knife to install when needed.

Examine the knife bar for bends, and if present remove them by straightening on a flat iron surface. If the cutter bar is provided with adjustable wearing plates they should be set to take up any play (fig. 31, A). The same is true of wearing parts in the knife head, and where these parts are badly worn they should be replaced, as excessive wear at this point frequently is the cause of a broken knife.

Knife clips or holders keep the knife flat upon the guard plates, and when the clips become loose or bent upward (I, Fig. 31, B), the front end of the knife is allowed to rise and the result is a pulling rather than a shearing action. These clips should be adjusted only after the knife has been straightened and well centered. Starting with the clip nearest the inside shoe, tap with light blows of a hammer until the clip just begins to tighten on the knife bar. Loosen the clip slightly and proceed with the other clips in the same manner. When all have been set, tighten them down securely.

After this adjustment it should be possible to move the knife freely by hand.

An important adjustment of the knife is what is commonly known as its register. When each section of the knife at the end of its in-and-out strokes does not center on its corresponding guard, the knife is not properly registering. In such a case the cut is incomplete on the one stroke and excessive on the other, resulting in a poor job of cutting, frequent clogging of the knife, and mechanical unbalance. In checking the register tighten the pitman connections at each end, then, after raising the tongue of the mower to normal position of hitch, turn the flywheel first to one dead center and then to the other. If the knife sections do not center in the guards on the dead centers, adjustment is necessary. This is made on some machines at the tiebar, which is threaded at one end, and where this is done it is sometimes necessary to change washers provided in the bar bearing in the yoke from one side of the bearing to the other. Knife-centering adjustments differ with the different machines, and the operator should examine the particular machine to see what provision exists for making adjustments and proceed accordingly.

In grinding knife sections preserve the same angle of cutting edge and bevel as that found on new sections. Many operators grind away too much of the point and too little of the heel of the section, and so change the cutting angle between the section and guard plate. If a section becomes very much shorter than adjacent sections, it should be replaced by a new one.

To remove sections, do not punch out the rivets, thus enlarging the rivet holes and weakening the bar. First remove the knife from the cutter bar, then lightly secure the knife sections in the jaws of a vise, with the knife bar resting on top of one jaw. A section of railroad iron or a similar object will answer if a vise is not available. Strike a sharp blow with a hammer down-

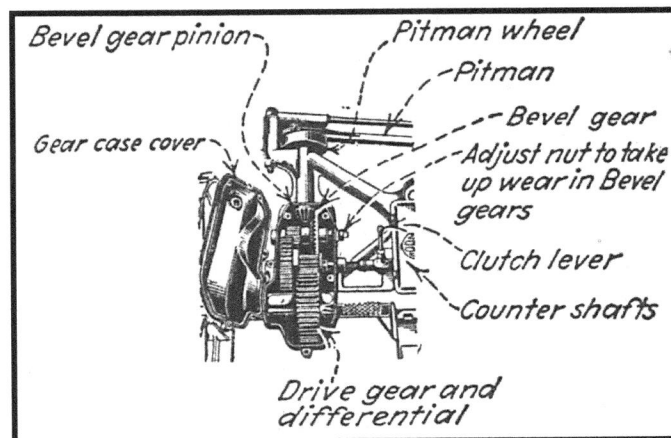

Figure 39. Mower transmission showing triple gear speed step-up.

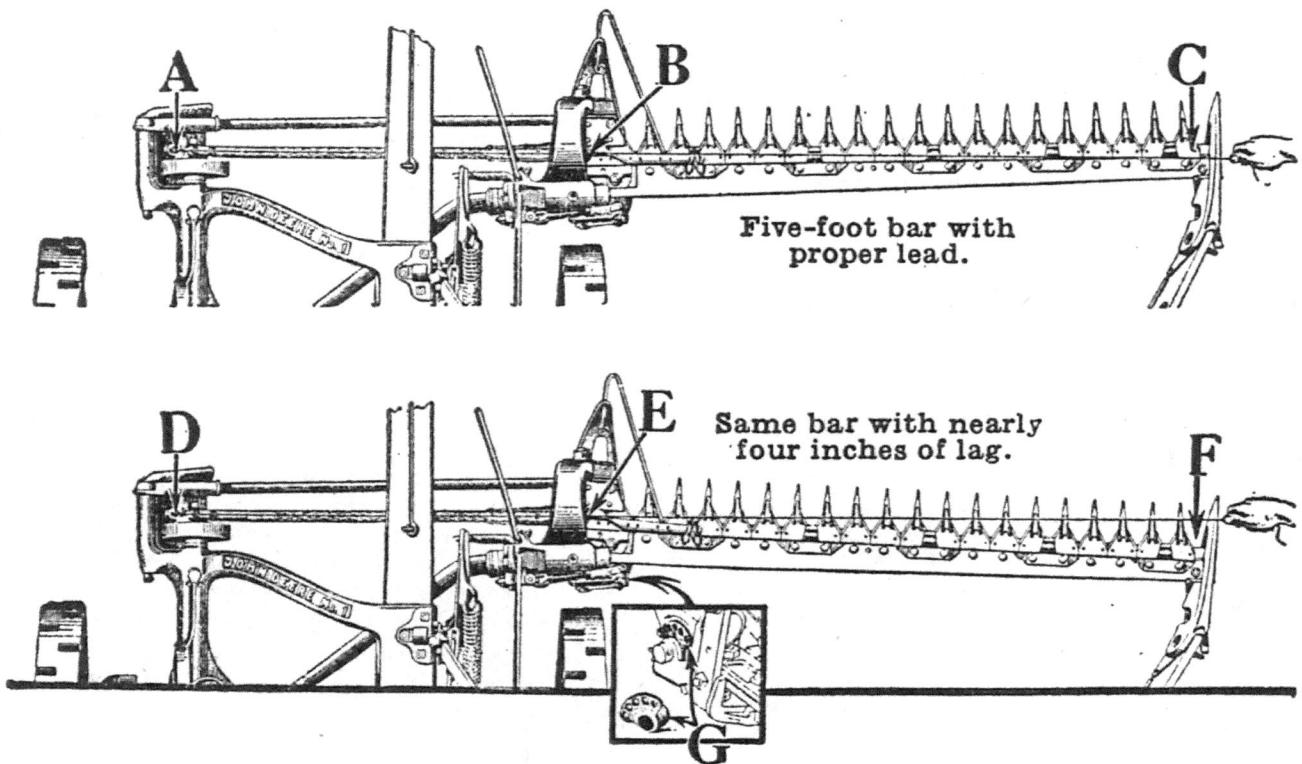

Figure 40. Illustrating how to measure the lead of a cutter bar for proper alignment. Point C should be from 1 to 1-1/4 inches ahead of points A and B. Points D, E, and F show a cutter bar with too much lag.

ward against the back of the section at a point directly above each rivet. This shears the rivet neatly, and the sheared rivet may be removed from the bar with a punch.

In putting on new sections, bear in mind that properly riveted sections seldom become loosened by wear. While a riveting set is desirable, a satisfactory job of riveting can usually be done with the exercise of a little care, using a ball-peen hammer. The center of the rivet should be high and well rounded and the edges worked down flat against the section.

LIFTING SPRING

Proper tension on the lifting spring (Fig. 28, g) transmits most of the weight of the cutter bar to the main wheels of the mower, allowing the bar to float easily over the ground, carrying just enough weight to hold it steady. A bar that drags heavily increases the draft and causes an excessive side draft. When this is in evidence, take up on the lifting spring tension but not enough to keep the end of the bar from following the contour of the ground. Closely associated with the lifting spring is the lifting linkage which should always be kept in proper adjustment to insure even raising of the cutter bar. There is usually provision for taking up wear in the linkage, but any change should be made only after checking for proper tension on the lifting spring.

SHOES, STICK, AND GRASS BOARD

The set of the outside shoe should always be so as to keep the cutter bar level. Replace badly worn soles. When the wheel attachment is used, see that a slight lead away from the grass is maintained, and, if this cannot be done by adjustment, replace the worn bearings. If there is difficulty in keeping a clear path for the inside shoe, examine the grass board and stick. Proper angle of the stick is important, especially in tall grass, in which the stick is raised to better clear the path for the next round. With some mowers, a revolving-stick attachment is available which, when in action, throws the grass away from the end of the cutter bar. Proper adjustment of the grass-board spring should be maintained as otherwise the board may at some time be jammed sidewise and broken. The spring tension should be tight enough to prevent the flapping of the board and loose enough to render the board flexible to side pressure.

PITMAN

End play may develop at either end connection of the pitman and can readily be detected by hand. Be sure the pitman itself is not warped or twisted, a condition sometimes found with wooden pitmans. The knife-head bearing is usually ball and socket, and if by lack of attention the socket has become elongated or the ball worn elliptical so that the lost motion cannot be taken up, the only remedy is to renew either or both the knife head and the pitman bearing. Many models of mowers are equipped with a knife-head connection which automatically takes up the wear as it develops. In tightening pitman bolts be sure there is no binding

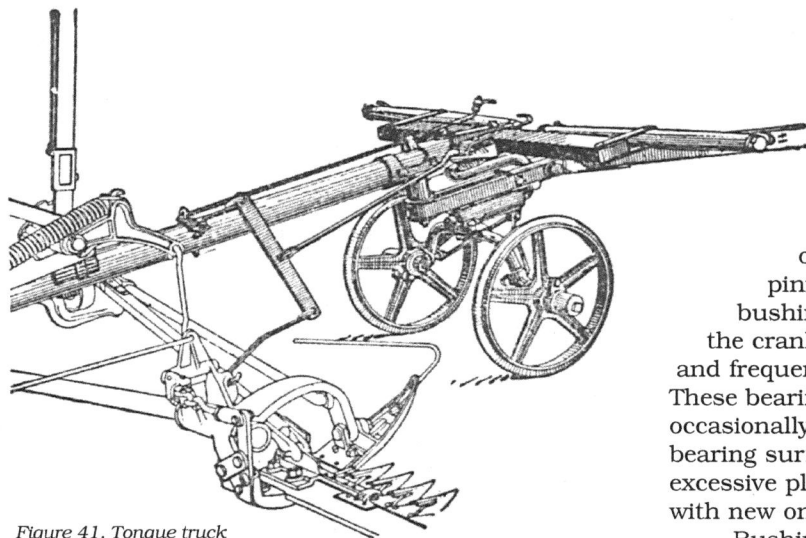

Figure 41. Tongue truck

which would interfere with the up-and-down movement of the cutter bar.

Excessive play or pounding at the pitman-wheel connection is ample warning to the operator that attention is necessary. This usually calls for replacement of the crank pin, the pitman-box bushing, or both. The worn crank pin may be broken off, after starting the break with a hack saw, by striking with a hammer; then the small end of the pin is easily driven from the wheel. This installation should be done with the crank shaft removed from the machine, at which time prelacemtn of the crank-shaft busings may be made if necessary. Be sure the newly installed crank pin is straight in the wheel.

GEARS AND CLUTCH

On many machines provision is made for adjusting the mesh of the gear on the countershaft and the bevel pinion by an adjusting nut on the end of the countershaft. On other machines the main gear or countershaft gear is shifted by an adjustable collar, or by placing washers behind the gear. If washers or gears are worn very badly it may be necessary to replace either or both. Countershaft bushings may be replaced when necessary as indicated under Bearings. See that the gear-control mechanism works smoothly and effectively. Take up wear in the clutch-shifter rod where possible, and check for possible binding of the clutch fork when the clutch is engaged.

MAIN WHEELS AND DRIVING PAWLS

If there is excessive end play in the main wheel bearings this should be taken up by means of the adjustment provided for this purpose. Power to operate the knife is usually transmitted from the main wheels by ratchet and pawls (Fig 35), hence their condition and adjustment are important. Neglecting this adjustment may result in broken pawls and pawl springs. See that the pawls engage at once when the machine starts forward. If the engaging faces of the pawls are worn unevenly, dress them with a file. When pawls are in the ratchets correctly, there should

be a clicking noise when the wheels are turned backward. Any accumulation of dirt in the pawl boxes should be removed.

BEARINGS

Provision for safeguarding the bearing surfaces differs somewhat with different mowers. Generally roller bearings are provided on the main axle, on the countershaft, and sometimes on the pinion end of the crankshaft. Bronze or brass bushings are commonly used on the front end of the crankshaft, in the pitman box for the crank pin, and frequently on the pinion end of the crankshaft. These bearings should be kept well lubricated and occasionally examined for wear. The crankshaft bearing surfaces are especially important, and when excessive play exists, the bushings should be replaced with new ones.

Bushing replacements are neglected frequently because of the time and trouble involved. In removing the crankshaft preparatory to replacing the bushings, bear in mind, when removing the bevel pinion, that the thread of the pinion may be either right or left hand. Determine first how the pinion should be turned, instead of making trial turns in either direction with possibly disastrous results. After taking out the crankshaft, which usually requires the removal of the pitman-wheel guard in addition to the bevel pinion, examine the bushings, and before attempting to drive them out be sure they are not held in place with set screws, and, if they are, remove the screws. The bushings are usually driven out by using a length of shafting slightly less in diameter than the crankshaft and 4 feet long or longer. This is run through the front bushing, and on through the housing until it contacts the near end of the other bushing. By striking on the free end of the shafting, the bushing is driven out. The second bushing may be driven out in a similar manner.

GENERAL

The machine should be gone over to ensure that all bolts and nuts are tight, lock nuts and cotter pins are in place, and parts weakened by wear have been replaced. Attention, especially in hilly sections, should be paid to draft and neckyoke connections. Through wear and exposure to the elements these may be weakened so much that while normally appearing safe, they may give way at an unexpected moment and a serious accident to horses or operator result. Acquire the habit of making frequent inspections of the machine.

Lost motion should be checked periodically, especially in machines that have been used for several seasons. When the knife does not start sliding with the forward movement of the machine, examine the pawls in the hubs, the clutch, the gears, the pitman-wheel bearings, and the knife-head connection. Where lost motion may be taken up at any of these points, it should be done; worn parts that cannot be taken up sufficiently should be renewed.

LUBRICATION

In lubricating the mower, general recommendations of the manufacturer should be followed as nearly as practicable. Lubrication of exposed gears, by increasing lodgment of grit and dirt, may cause excessive wear and grinding out of the parts. This is especially true when operating in sandy soils. Enclosed gears should be kept clean and well lubricated.

Oil cups should be kept clean and care observed that the holes to the bearings are not caked with oil or dirt. Before beginning the season's work it is desirable to remove from the bearings accumulations of all foreign material. This may be facilitated by flooding them with kerosene. Kerosene is an effective cleaning agent, but it should be followed with a bath of lubricating oil as metal washed in kerosene tends to rust. It is best never to use kerosene on metal parts preparatory to storing a machine over the idle season. A wad of wool makes an excellent filter for filling oil cups; unlike cotton waste, it is not easily drawn into the bearings, is not subject to chemical action of lubricating oil, and does not become matted by use. A few drops of oil on the bearing surface of the knife bar when work is started with a new machine or an overhauled one will assist greatly in obtaining smoothness at this point.

Those parts not taken care of automatically should be lubricated little, but often. Fast-moving parts such as the pitman connections, the crankshaft bearings, and the bevel-gear and gear-shift bearings need frequent attention, while the slower moving parts, or those used but occasionally, need less frequent service. Abnormal heating of any part of the mechanism is a certain indication either of the need for lubrication or of improper adjustment of parts. Excessive heating is sometimes accompanied by damage to the part.

STORAGE

At the end of the season's work it is well to make a list of needed repairs and adjustments. If this list is referred to when the mower is being overhauled, necessary repairs and adjustments will not be overlooked. The mower should not be left in a fence corner or other out-of-the-way place for wooden parts to rot and metal parts to rust, which even during a short period may cause more deterioration than a season's use. A weathertight, well-drained shelter with a fairly level floor to prevent machine strain should be provided.

When putting the mower away place boards under the main wheels; clean the knife, wipe it with a greasy rag, and store it in a dry place. Also place a block of wood or other support under the midpoint of the tongue so that it will not acquire a permanent sag. If stored with the cutter bar in a vertical position, place a block of wood under the shoe to take the weight off the frame. All dirt and grease should be removed, and the bright parts oiled or greased.

Wooden parts should be kept adequately painted. The best treatment for the wooden pitman is an occasional application of warm, raw linseed oil. The metal pitman may be preserved by a good grade of paint.

Miscellaneous Updates

If the mower you're going to use is set up with oil cups, we recommend they be replaced with grease zerks so that a modern grease gun and tube grease can be used.

Make sure that the pitman bearing is greased at least every four hours. And do not allow the gear box to run dry. We have been very happy replacing the gear box oil recommendations with 10 weight Hoist Oil, it reduces gear drag but does require a tight oil seal on the pitman shaft. We replace the gear box oil each season.

We have found that the pint sized plastic 90 weight oil jugs found in all hardware stores fit nicely in the oil can holes on the JD and McD mowers. Any oil will work to apply to the cutter bar but we follow this rule: if there is a lot of loose dirt (gopher mounds for instance) the lighter weight oils are better - if there is no loose dirt or the mower is cutting high, the heavier weights seem to last longer. We apply oil to the cutter bar anytime we stop to rest the horses.

Safety note: It will pay you big dividends to carefully check all pins on doubletrees and neckyokes before going to the field. If they look worn or stressed, change them.

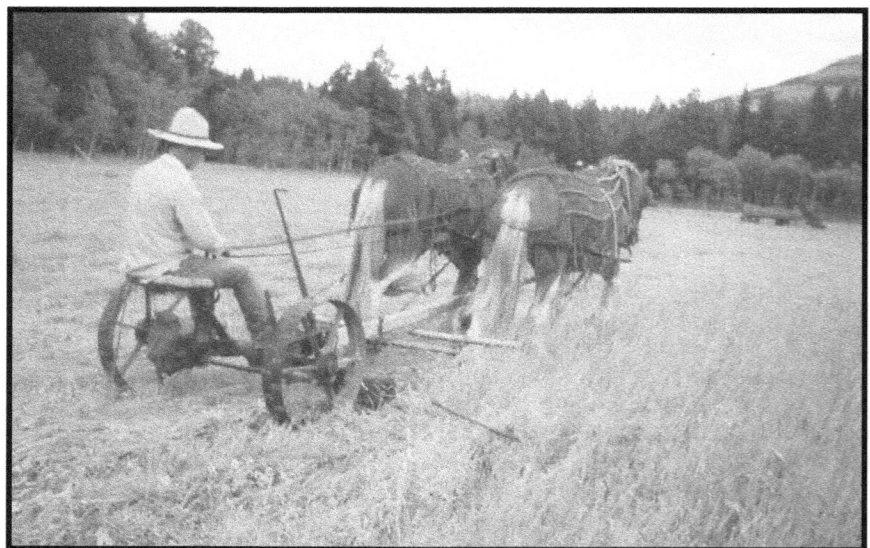

Figure 42. Doug Hammill mowing Montana mountain hay with his Clydesdale team

McCormick Deering/International Mower Service

*In 1945 McCormick Deering/International pro-
duced a booklet entitled **Blue Ribbon Service Training Course Serviceman's Guide** for **Hay Machines.**
The information features the McD/I line of implements
but is presented in somewhat of a generic fashion. The
excellent material which follows is taken from the horse
mower section. Information from this important booklet
on other hay implements is featured in following
appropriate chapters.*

*Immediately following this excerpt is a similar
reprint from the John Deere Company. LRM*

HORSE-DRAWN MOWERS

Mowers are found on practically every farm.
Perhaps it is because they are so common that the
tendency exists to overlook the few chores and mechanical adjustments that are essential for their best
operation.

On the other hand, attending to these details
produces a smooth running, clean cutting mower
which is a real satisfaction to the operator, reduces the
load on the horses or tractor, and materially increases
its operating life.

It is the purpose of this manual to show in
outline form the more vital adjustments which will
enable the serviceman to make a systematic inspection
of the mower in order to apply the remedies found
necessary.

In order to simplify the servicing of the mower, the
information herein presented has been divided into
units. The units consist of the cutter bar, cutter bar
lifting parts and driving mechanism.

CUTTER BAR
IMPORTANCE OF UNIT

The cutter bar and its parts make up the vital
unit in the mower operation, directly effecting the draft
and type of work done. If the parts are properly set
and in good condition, that is, the knife and ledger
plate are sharp and make correct contact, and the knife
registers correctly, the mower should cut a clean swath
with minimum draft.

WEARING PLATES

The wearing plates support the rear part of the
knife or sickle, keeping the knife sections in contact
with the ledger plates. When plates become worn so
that there is no more clearing between ledger plates
(see Fig. 44) the knife section will tip up. This results
in ragged cutting. The plates are provided with slotted

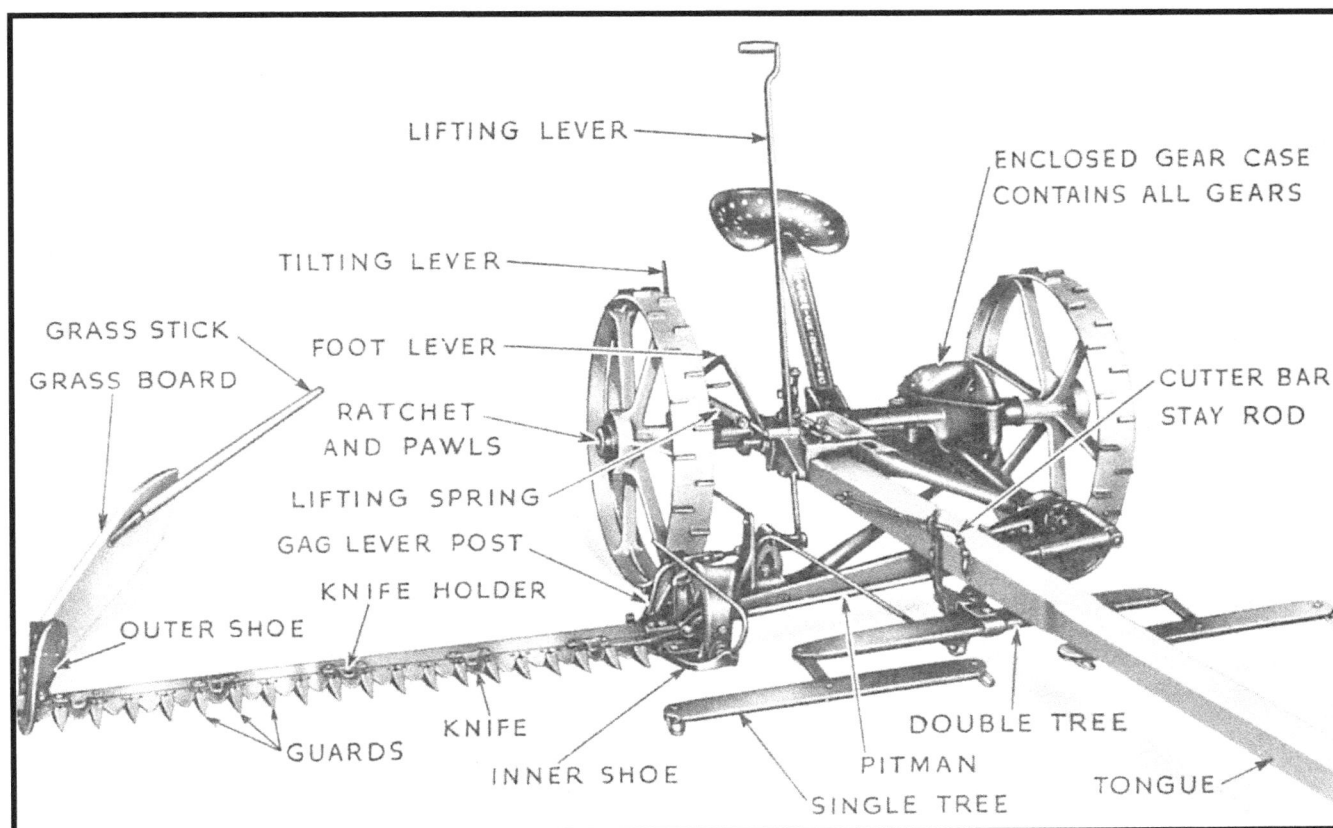

Fig. 43. Principal parts of the No. 9 mower.

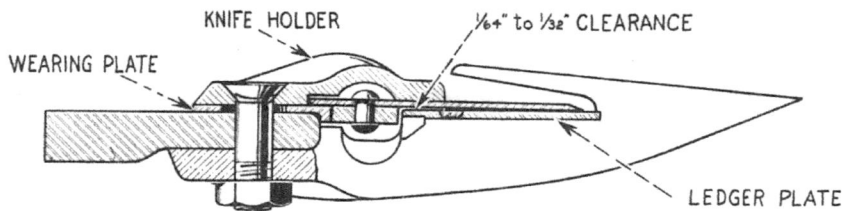

Fig. 44. Cross section of cutter bar showing parts in correct adjustment.

holes to allow adjustment as wear occurs. After all adjustment has been taken up, the plates should be renewed. The wearing plates must line up with each other to give the knife back a straight bearing along its entire length. Be especially careful to line up new plates properly.

KNIFE HOLDERS

Knife holders hold the knife sections down close against the ledger plates (Fig. 46). If they become worn, allowing the knife to play up and down, and making poor contact with ledger plates, the holders should be hammered down.

Knife holder must fit snug on knife without binding. Never bend a knife holder down with knife under it. To fit holder to knife start at outer end of bar, pull knife from under outer holder, then tap holder down at end "A" Fig. 46. Keep trying knife under holder until it sets firmly on knife, yet allows it to work freely. Adjust each holder in the same manner. If holder is too tight on knife strike it on the top of the arch at "B" while knife is under it. A tight working knife causes heavy draft.

GUARDS AND LEDGER PLATES

The guards with the ledger plates "C" Fig. 47 divide the material being cut and protect the cutting units of the cutter bar. The ledger plate "C" acts as one-half of the shear, the knife sections acting as the other half. The sections and plates should fit together when knife is in register with a little clearance as shown in Fig. 44. To insure a clean cut, guard plates "C" Fig. 47 should be replaced when worn dull and the guards carefully set and kept in proper alignment. A special guard repair block (Fig. 48) for removing rivet and plate from any style guard and for riveting new plate is available through the parts department.

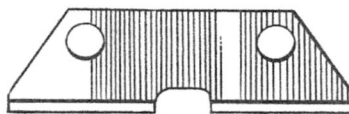

Fig. 45. Wearing plate.

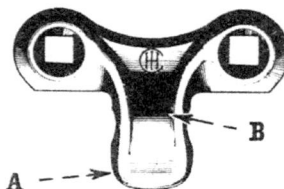

Fig. 46. Knife holder.

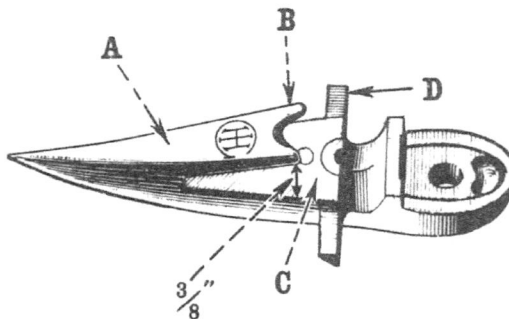

Fig. 47. Guard and ledger plate.

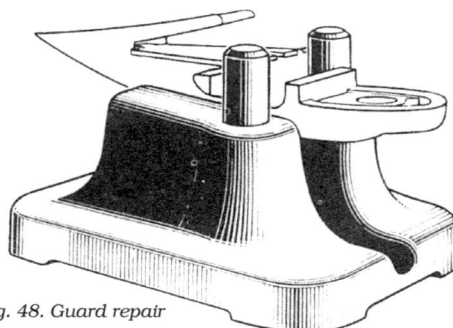

Fig. 48. Guard repair block.

After the guard plates have been replaced, the guards should be aligned to give a sheer cut on every plate. Aligning the guards is an important and exacting operation, and the user of the mower should be instructed to check this alignment periodically throughout the cutting season. A new knife or one that is not badly worn should be used in testing and setting the guards. Begin by pounding down the high guards, first by striking on the heavy section "A" just ahead of the guard plate and then bring up the low ones. Be careful not to drive lip "B" of guard down. The clearance between lip and ledger plate should be no less than 3/8 inch as otherwise clogging may result. Guard wings "D" should also be aligned, to give a smooth surface for knife to work on. Tighten guard bolts before and after aligning guards.

Position of guard points should not be considered; the plates and wings are the important units that must be aligned. Blunt guard points may be repointed with a file or emery wheel, if available.

KNIFE OR SICKLE

Knife and sickle are both names commonly used to designate the moving part of the cutter bar which acts with the ledger plates to form a shear cut. To be technically correct, the knife is equipped with smooth sections and the sickle is equipped with serrated sections. It is essential at all times to have knife sections sharp and replace those that are nicked or broken. The knife back should always be straight; examine it by sighting down bar.

The block is for replacing sections and straightening knife. It may also be used for riveting wrist pin (Fig. 50).

REGISTER OF KNIFE

The register of knife refers to the position of the sections in relation to the guards. The sections should center in the guards when the knife is at extreme outer or inner end of its stroke. (see Fig. 51). Under side

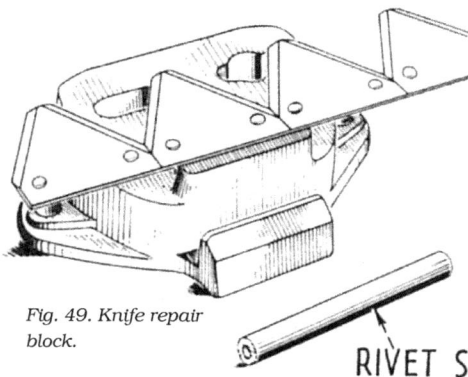

Fig. 49. Knife repair block.

RIVET SET

WRIST PIN

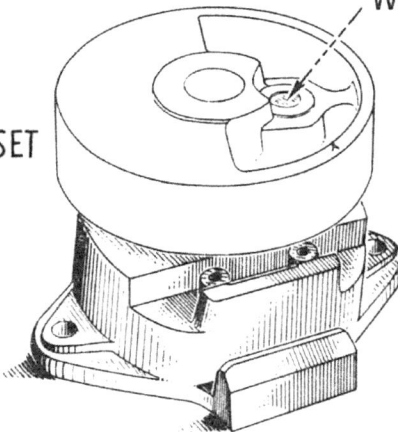

Fig. 50. Riveting wrist pin in flywheel.

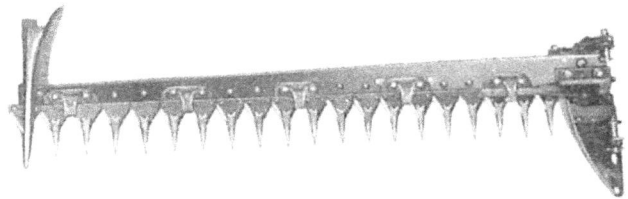

Fig. 51. Regular-type cutter bar showing knife at extreme end of stroke and in register.

of front end of tongue should be 31" from ground when registering knife. (See Fig. 53). To correct for out of register proceed as follows (see Fig. 52): Remove bolt "1". Then turn flywheel shield "2" on coupling bar brace "3" in or out as required to secure proper adjustment. Replace in the reverse procedure.

When knife is out of register, uneven cutting, clogging of knife, and heavy draft result--so it is very important that the knife always be kept in correct register.

CUTTER BAR ALIGNMENT

All mowers should have a certain amount of lead in the cutter bar; that is, the outer end should be ahead of the inner end to offset the backward strain produced by the pressure of cutting. This permits the knife and pitman to run in a straight line. As the mower parts wear, the cutter bar begins to lay back until the knife is running at an angle. This results in excessive wear and breakage of cutting parts. An occasional check should be made for the correct lead. The outer end of the bar should be ahead of the inner 7/8 in. to 1-1/8 in. on a 4 ft. bar; 1-1/8 in. to 1-3/8 in. on 5 ft. bar; 1-3/8 in. to 1-5/8 in. on a 6 ft. bar; and 1-5/8 in. to 1-7/8 in. for a 7 ft. bar.

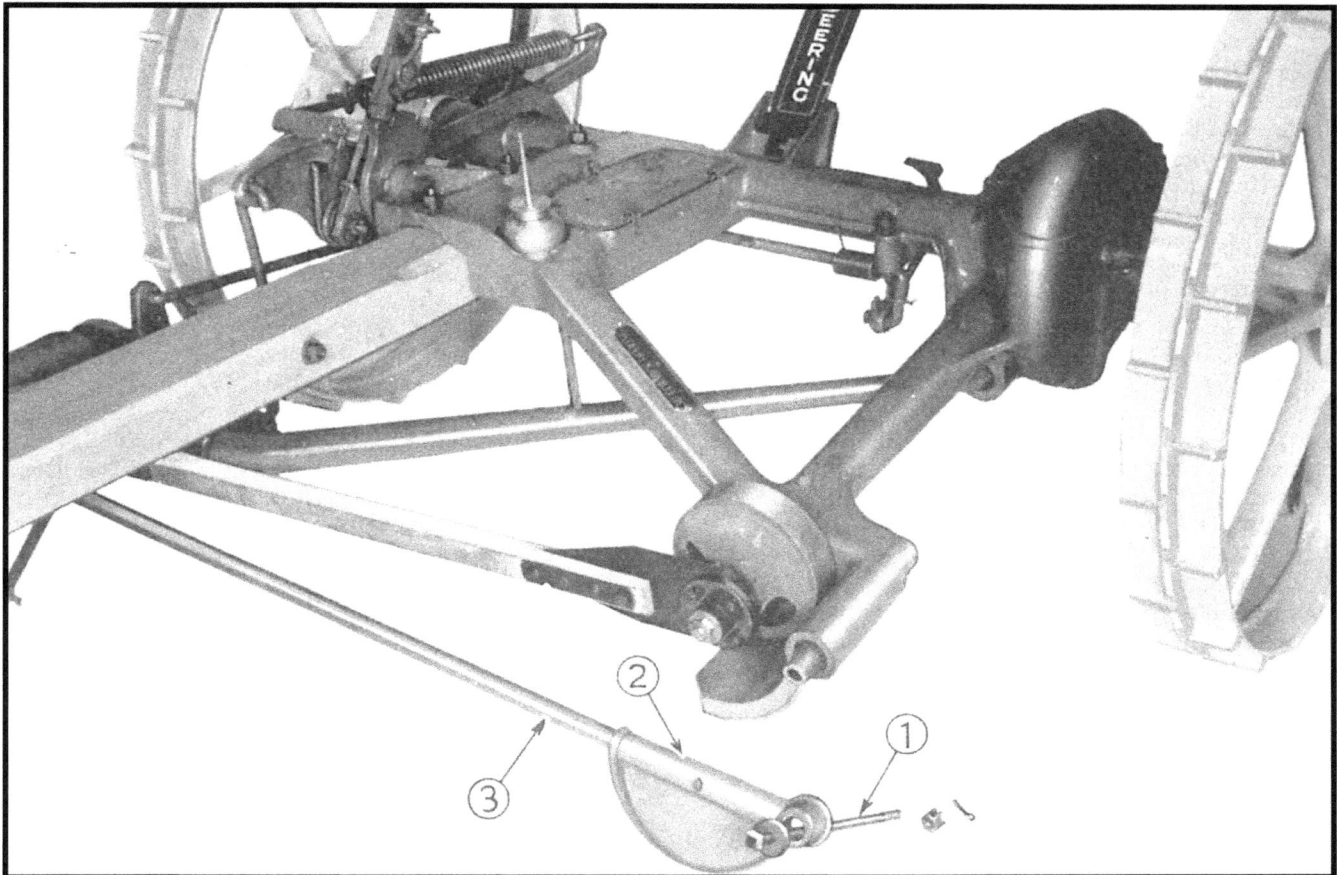

Fig. 52. No. 9 mower showing parts to be removed to adjust register of knife.

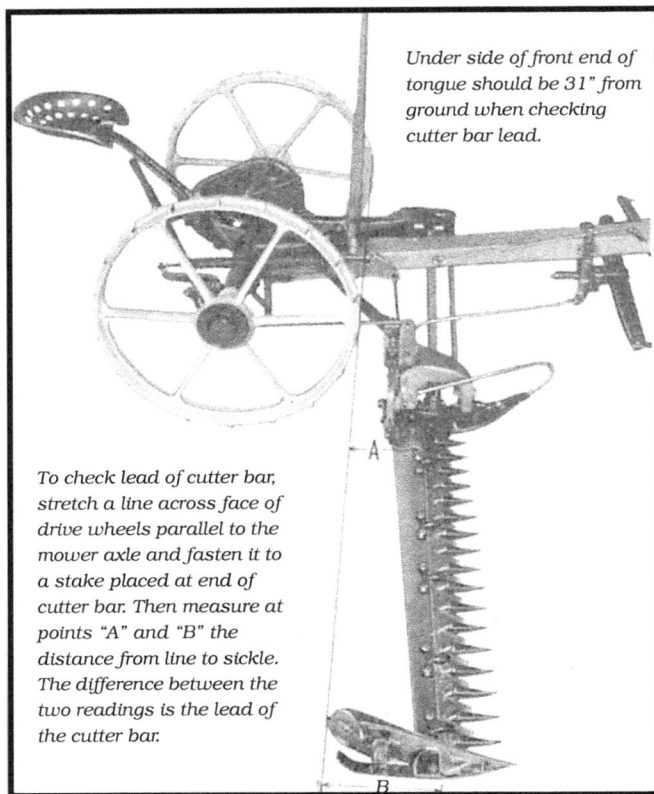

Under side of front end of tongue should be 31" from ground when checking cutter bar lead.

To check lead of cutter bar, stretch a line across face of drive wheels parallel to the mower axle and fasten it to a stake placed at end of cutter bar. Then measure at points "A" and "B" the distance from line to sickle. The difference between the two readings is the lead of the cutter bar.

Fig. 53. One method of checking lead of cutter bar.

To check lead on No. 9 mower, stretch a line across the face of the drive wheels parallel to the axle-- as shown in Fig. 53. Then measure from the line to the back of the knife at the outer and inner end of bar. The difference between the two measurements "A" and "B" is the lead of the bar.

If, after replacing worn hinge pins, the cutter bar should lag rearward, remove the following parts as indicated in Fig. 54 to adjust for the correct lead: lifting spring "1", pin in gag post "2", pin in gag lever hinge "3", tilting rod "4", cotter pin in rear end of coupling bar "5", bolt in flywheel shield "6", pitman "7", and draft rod "8". Then pull the entire assembly out (Fig. 55) and screw the coupling bar "5" out (turn counter-clockwise) of shoe hinge "9" one turn or enough to obtain the correct lead. Reassemble in the reverse order. Check register of knife.

CUTTER BAR LIFTING PARTS
PLAIN LIFT (REGULAR)

The plain lift mower as furnished regularly has a high level lift suitable for clearing all ordinary obstacles with the sickle in operation. When cutter bar is raised by lifting lever to first notch in quadrant, the knife runs parallel to ground. When raised to second notch the inner shoe remains the same and sickle operates at approximately 45 degree angle.

VERTICAL LIFT (SPECIAL)

A vertical lift attachment is available for horse mowers and is recommended for use only on the 4 ½ and 5 ft. cutter bars. The attachment consists of a longer lifting lever connection, a coupling bar stop, and

linkage to the clutch to throw the mower clutch out of mesh. The attachment limits the height of raise of the inner shoe, but allows the cutter bar to be swung to a vertical position by the use of the lifting lever operated from the driver's seat.

When cutter bar is raised by lifting lever to first notch in quadrant, the knife continues to run. The coupling bar should just contact the coupling bar stop at this point. When the cutter bar is raised beyond the first notch, the mower clutch is thrown out of mesh and the knife automatically stops. When further pressure is applied to the lifting lever the cutter bar is swung to a vertical position.

The cutter bar should be held firmly in the vertical position when the latch pawl is engaged in the second notch of the quadrant. If it is not, the coupling bar stop "A" Fig. 58 should be turned counter-clockwise until it contacts the coupling bar firmly. Secure in this position with the lock nut.

LIFTING LEVER

The latch pawl on lifting lever is adjustable to take up wear. To adjust, screw out on nuts "1" Fig. 59. Lower rounded portion on pawl, when engaged with notch in lifting lever lock, should be approximately in line with front edge of lever. Bolts must be kept tight.

LIFTING SPRING

If the cutter bar is too light upon the ground, slacken the lifting spring; if too heavy, tighten spring. This is done by adjusting the bolt in lifting spring "1" Fig. 54. When properly adjusted, the lifting spring reduces the friction of the cutter bar on the ground, thus lessening the tendency toward side draft. The bulk of the weight of the cutter bar should be carried on the wheels, increasing the traction and reducing friction between bar and ground.

GAG POST ADJUSTMENT

If outer end of bar lags behind the inner end in raising, shorten the adjustment as shown in Fig. 60. If outer end is too light, especially on short bar, this done by first removing the lifting spring "1" Fig. 54; then withdraw the pin "1" Fig. 60, which secures gag post to inner shoe and make adjustment as required by lengthening or shortening the gag post adjustment "2"

The gag lever, hinge, and post assembly is adjustable to accommodate different lengths of cutter bars as well as regular or vertical lift. The holes "A" are to be used for 4 ½ and 5-foot bars. Holes "B" are for 6 and 7-foot bars. The pin for holes "A" is ½ inch in diameter and the pin for holes "B" is a 9/16-inch in diameter. Should the length of cutter bar ever be changed, be sure the gag lever is adjusted accordingly.

Holes "C" in the gag post and adjusting link are to be used when the mower is equipped with a plain lift and holes "D" are to be used when it is equipped with a vertical lift.

TILTING LEVER

Tilting the cutter bar to the correct angle is essential to good mowing. If the ground is covered with dead grass or stubble from a previous crop the points of the guards must be tilted upward sufficiently

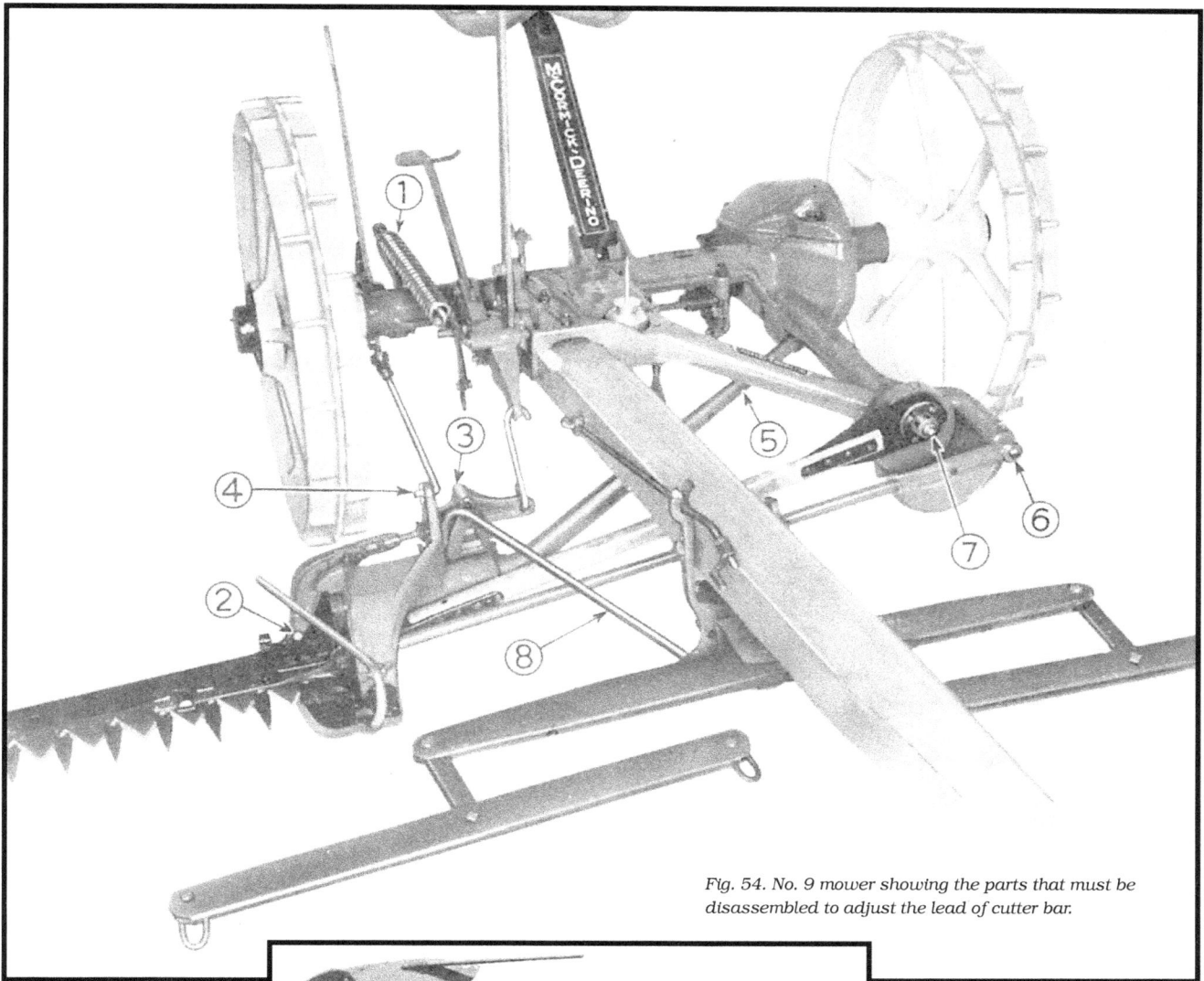

Fig. 54. No. 9 mower showing the parts that must be disassembled to adjust the lead of cutter bar.

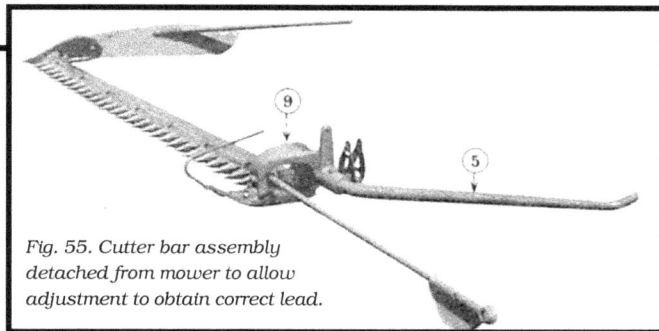

Fig. 55. Cutter bar assembly detached from mower to allow adjustment to obtain correct lead.

Fig. 56. Below: Height of bar on vertical lift mower when lifting lever pawl is in second notch in quadrant.

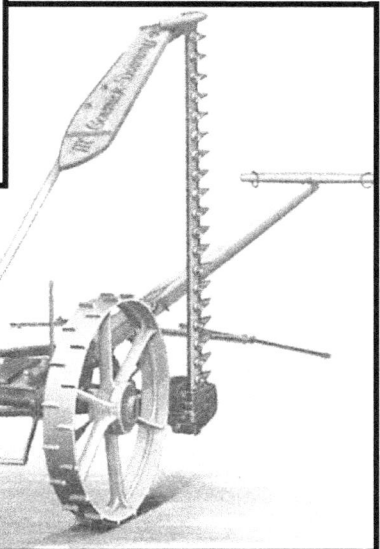

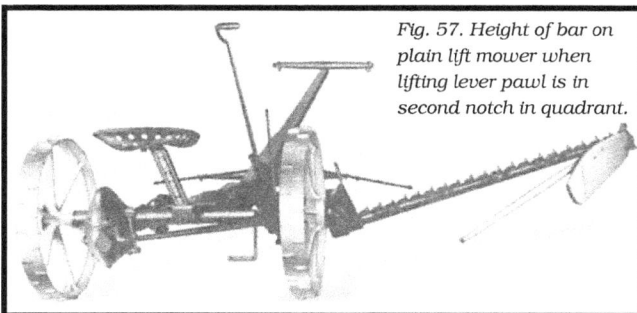

Fig. 57. Height of bar on plain lift mower when lifting lever pawl is in second notch in quadrant.

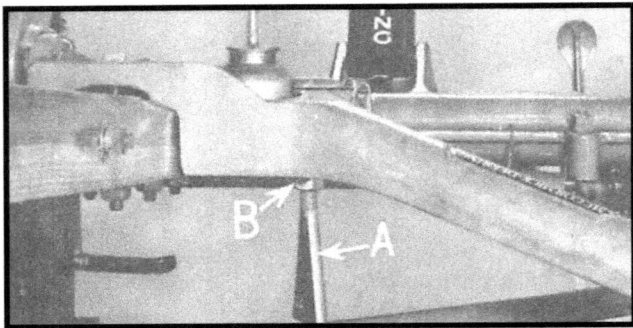

Fig. 58. "A" is coupling bar stop, "B" the lock nut.

Latch pawl releases itself by starting lever forward.

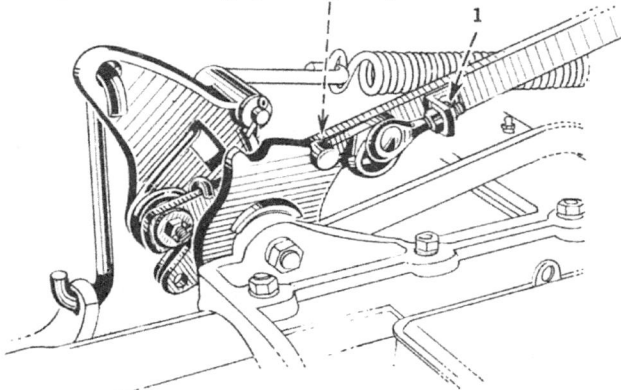

Fig. 59. Unique lifting lever pawl. No hand latch or detent is required. Adjustable for wear so replacement is seldom necessary.

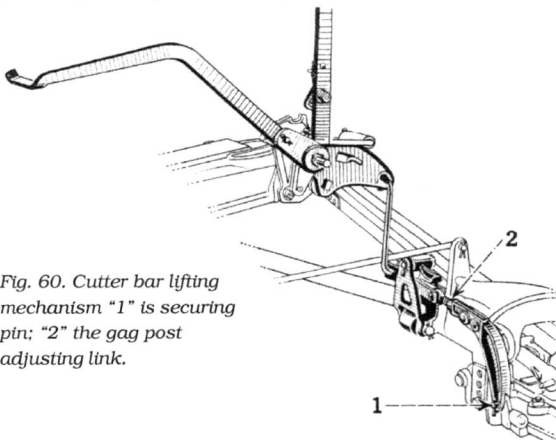

Fig. 60. Cutter bar lifting mechanism "1" is securing pin; "2" the gag post adjusting link.

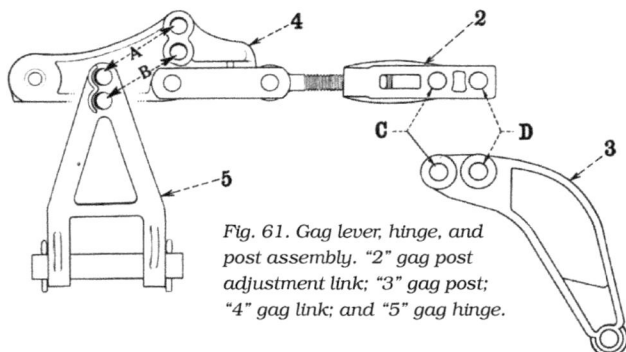

Fig. 61. Gag lever, hinge, and post assembly. "2" gag post adjustment link; "3" gag post; "4" gag link; and "5" gag hinge.

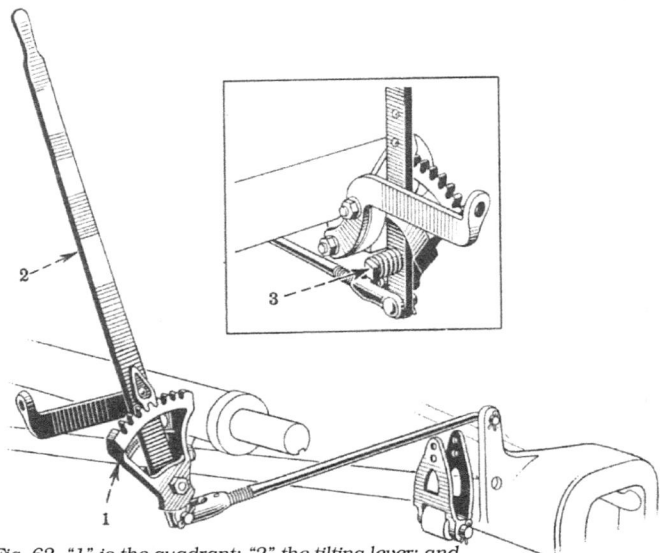

Fig. 62. "1" is the quadrant; "2" the tilting lever; and "3" the spring which holds the lever in position.

to prevent the old stubble from choking the knife. Crops that are down and tangled may require tilting the points down so as to get under the crop and cut clean. The tilting lever on the McCormick-Deering mower has no ordinary latch. It is disengaged from the quadrant teeth by merely pulling the lever out to the side. The bar can then be tilted to the desired position.

HOW THE MOWER IS DRIVEN

Lugs, cast as a part of the face of the wheels, assure positive driving. The wheels turn the axle through ratchet pawls in the wheel hubs. Clamped and keyed to the axle is a cut-steel gear "3" Fig. 63, driving a pinion gear "4", which is cut as an integral part of the pinion shaft "8". On the splined portion of this shaft is the eight-jaw clutch "17" which turns the Zerol bevel gear "1". This bevel gear drives the Zerol bevel pinion "2" on the rear end of the flywheel shaft "10". On the front end of this shaft is the flywheel "13" which drives the pitman and sickle.

The axle turns on roller bearings "9", the pinion shaft on ball bearings "7" & "8", the rear end of the flywheel shaft on a ball bearing "5", and the front end on an extra-long, babbitt-lined bearing "6" which dampens vibration and prevents "whipping." The gears are fully enclosed in a dust-tight, leakproof case. A deep bath of oil envelops the moving parts and assures thorough lubrication.

REMOVING WHEEL AND PAWL HOLDER

To inspect or replace pawls, a wheel, axle oil seal, axle bearings, axle shaft, or main drive gear, it is first necessary to pull one or both pawl holders and wheels. First, remove the groove pin from the pawl holder; then with a puller remove wheel and pawl holder as shown in Fig. 64. The key is tapered, keyway in axle is tapered, and keyway in pawl holder is straight so the key will not wedge as the pawl holder is pulled.

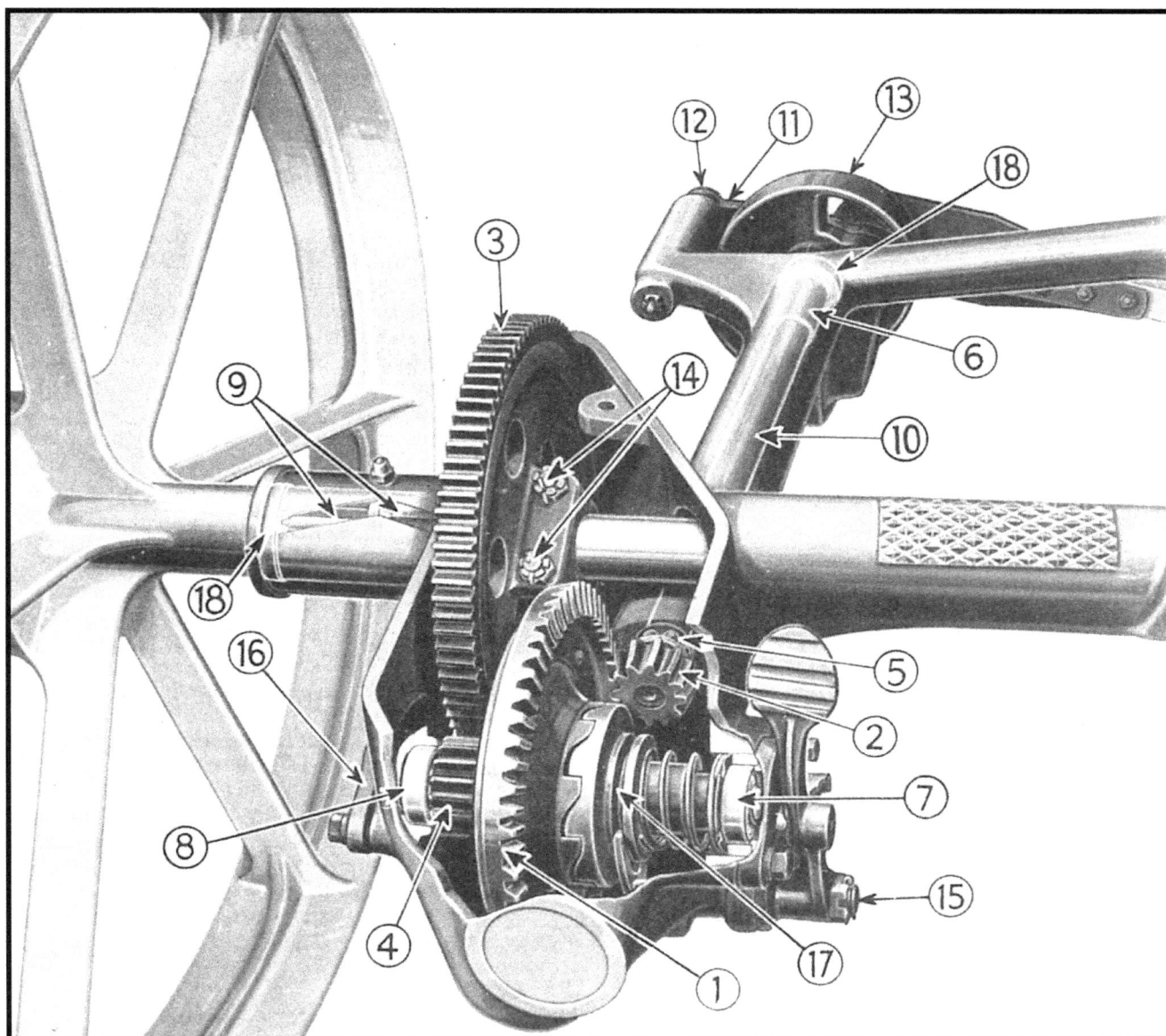

Fig. 63. Driving parts of the No. 9 mower.

1. *Zerol bevel gear.*
2. *Zerol bevel pinion.*
3. *Main drive gear.*
4. *Main drive pinion and shaft.*
5. *Ball bearing.*
6. *Babbitt-lined bearing.*
7. *Ball bearing.*
8. *Ball bearing.*
9. *Roller bearings.*
10. *Flywheel shaft.*
11. *Flywheel shield*
12. *Flywheel shield bolt.*
13. *Flywheel.*
14. *Main drive gear hub clamp bolts.*
15. *Clutch shaft.*
16. *Bearing cap.*
17. *Clutch.*
18. *Oil seal.*

REMOVING AXLE OR MAIN DRIVE GEAR

If axle is being replaced, remove both wheels. If the main drive gear, then it is only necessary to remove the left wheel. Pull the cotters and take the two bolts (see "14" in Fig. 63) out of the main drive gear hub. Insert a wooden block between the hub and the side of the gear case and pull the axle to the right until the key, which secures the hub to the axle, can be removed. Inspect keyway and end of axle, filing off any burrs that might damage the oil seals. The axle may now be withdrawn from the frame.

OIL SEALS

With the axle out of the frame the oil seals "1" Fig. 66 and axle bearings "9" Fig. 63 can be inspected and renewed if sufficiently worn. If oil seals are to be replaced it is recommended that the axle be put in place and the main drive gear keyed and bolted in place on the axle before installing the oil seals.

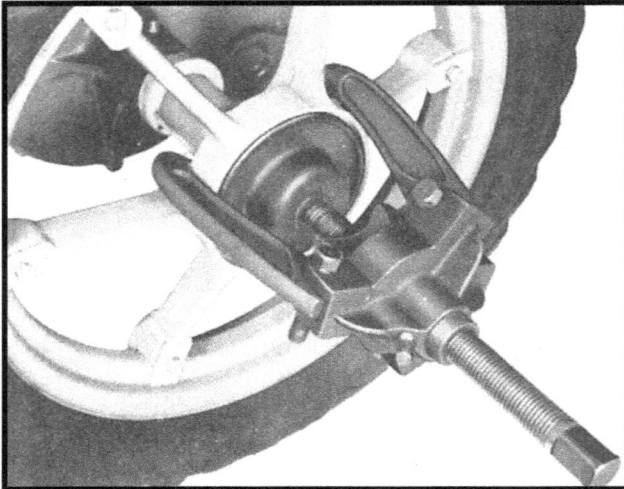

Fig. 64. Pulling wheel and pawl holder.

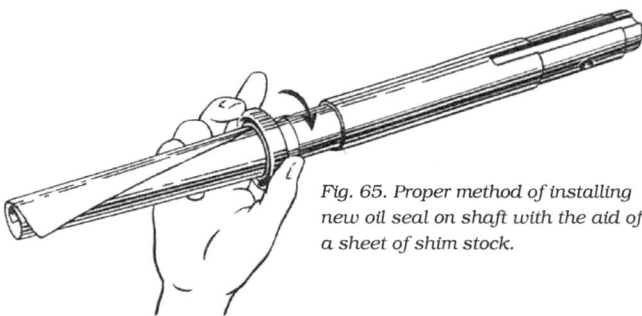

Fig. 65. Proper method of installing new oil seal on shaft with the aid of a sheet of shim stock.

Fig. 66. "1" is the oil seal, "2" the wear washer, and "3" the sand cap in the order in which they should be installed.

Wrap a sheet of shim stock around the axle in a cone shape, slide the oil seal over the cone and onto the shaft, rotating it in the direction the shim stock is wrapped (see Fig. 65). Be sure to install the seal with the lip to the inside. Replace wear washer "2" and sand cap "3" shown in Fig. 66. Wheel and pawl holder can then be replaced.

Pinion "4" Fig. 20 is integral with the shaft. It is mounted on ball bearings "7" and "8" which are carried in bearing caps. Shims are placed between cap "16" and the gear case to provide the proper clearance between the bevel gear "1" and bevel pinion "2". Bevel gear "1" is free on the shaft and turns bevel pinion "2" only when clutch "17" is engaged. The clutch is turned by splines on the shaft so that it is always in motion

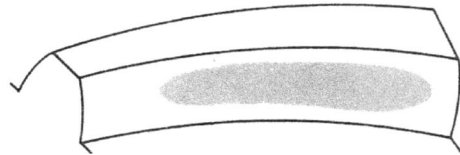

Fig. 67. Tooth contact is represented by shaded area.

when the mower is moving. A foot pedal engages or disengages the clutch and a heavy coil spring holds the clutch teeth in mesh. When the mower is in gear the clutch foot pedal is toward the rear. In this position the clutch fork should work freely in the groove to prevent undue wear on both parts. This clearance can be controlled by adjusting castellated nut on clutch fork shaft "15" Fig. 63.

The clearance between the bevel pinion and bevel gear is properly set at the factory and will require no attention for a long time if the gear case is kept filled with good oil. One of the advantages of the Zerol type bevel pinion and gear is the distinctive tooth shape which reduces tooth friction to a minimum.

Should it be necessary to remove this assembly, proceed as follows: remove the castellated nut on the clutch fork shaft "15" in Fig. 63, and right bearing cap with bearing "7". This bearing is a closer fit in the cap than on the shaft and should remain in the cap. There should be one shim between this bearing cap and the side of the gear case. Next remove the left bearing cap "16" with the shims that determine bevel gear and pinion clearance. Bearing "8" is a close fit to the shaft so does not pull off with the cap. The pinion and shaft can now be pulled out through the left bearing cap opening. Bevel gear and clutch can now be lifted out. Notice that there was a thrust washer on the shaft between the pinion and the bevel gear. Also a flat washer between the right ball bearing and spring retainer.

BEVEL PINION

If the bevel pinion is to be removed, it will be most convenient to do so while the bevel gear is out of the case. Lock the pinion by placing a bar or large chisel between it and the side of the case. Remove the pitman and turn the flywheel counter-clockwise with a bar or wrench until the pinion turns off the threaded shaft. If the pinion is removed while the bevel gear is assembled in the case it will be necessary to slip the flywheel and shaft ahead as the pinion is turned off the shaft to prevent it from crowding the bevel gear.

ADJUSTING BEVEL PINION AND GEAR

Bevel gear and pinion are Zerol type and have curved teeth which combine certain characteristics of both spiral tooth and straight tooth bevel gears. The shape and proportion of Zerol gears are such that tooth contact is always near the center, where the tooth is the strongest and never on the ends or edges of the teeth. Proper tooth contact is as shown in Fig. 67. If replacement is necessary, the tooth clearance should be set as accurately as possible. Shims .005" and .010" thick should be placed between bearing cap

"16" in Fig. 63 and the gear case until the bevel pinion and gear teeth have a clearance of not more than .012" and not less than .008".

FLYWHEEL SHAFT (SEE FIG. 63)

To remove the flywheel shaft proceed as follows: remove flywheel shield "11" by taking out bolt "12", pitman from wrist pin, and gear case cover. Raise front end of mower to prevent oil from running out of flywheel bearing. Next place a chisel between bevel gear "1" and pinion "2", then turn flywheel to the operator's left (counter-clockwise) standing in front of mower.

When removing bevel pinion "2", it is not necessary to withdraw flywheel shaft "10" entirely. However,

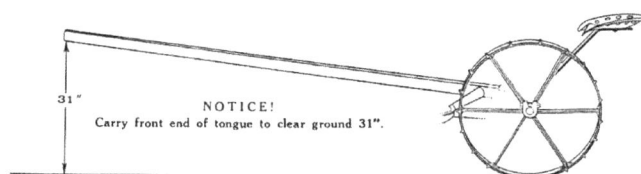

Fig. 68. Carry front end of tongue 31 inches from ground.

if shaft is at any time taken out, be sure to remove burr on shoulder of shaft where it is in contact with bevel pinion. A burr is apt to damage oil seal at front end of flywheel bearing when shaft is removed for replacement.

GUARDS AND BAR ASSEMBLIES

TYPES OF GUARDS

MD-989 MOWER GUARD is a malleable guard 3" spaced, furnished on all regular bars for the horse drawn mowers. The guard is used for average cutting conditions and is accepted as an "all around" guard.

MF-989 is a steel guard similar to MD-989, 3" spaced and built of a material which will withstand greater shock without breaking.

M-2041 MOWER GUARD is a heavy duty malleable guard 3" spaced with a heavier longer body and has serrated ledger plates. The rear or base of the guard when assembled makes contact with the next guard, assuring rigidity. M-2043 is used as the outer guard and M-2042 is used on the inner end of the bar. The guards are adaptable for use in severe ground conditions where obstacles are encountered or for highway use. The guard, because of body thickness, is not as adaptable to heavy grass crops as the MD-989 (guard will not penetrate a heavy crop as readily).

MA-2149 is a 2-1/2" spacing guard and requires the use of a special 2-1/2" spaced knife or sickle and a flywheel with a 2-1/2" throw. The guard is used as a compromise in the two conditions where 1-1/2" and 3" spaced guards are most adaptable. Closer spacing of the guards lessens the ability of the guard to "clean" in the cutting of some crops.

ZA-53 WEED OR BRUSH GUARD is a lipless guard and is adaptable for use in weeds or brush (cardo) or with the canning pea bar. The guard is used in conjunction with a heavy knife and is designed to clean easily under heavy cutting conditions where a 3" spaced guard is adaptable.

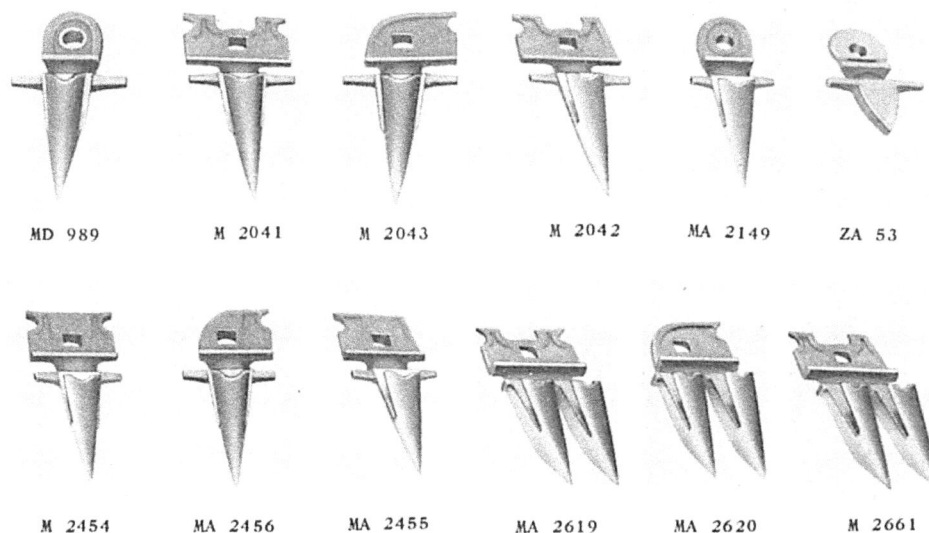

Fig. 69. Types of guards.

M-2454 2-1/2" SPACED GUARD is of the heavy duty type and is adaptable where conditions require a compromise of 1-1/2 and 3" spaced guards and rough and rocky terrain is encountered. M-2456 is used on outer end of bar assembly; MA-2455 on the inner end. M-2454 makes up the balance of the guards on the bar. The guard assembly requires a 2-1/2" spaced knife or sickle and a special flywheel with a 2-1/2" throw.

MA-2619 LESPEDEZA GUARD is specially designed for use in cutting lespedeza. The guard is malleable with under serrated ledger plates and is 1-1/2" spaced using a 3" spaced knife, thus allowing double number of serrations per stroke. The square rear or base of the guards contact each other when assembled assuring rigidity. The lespedeza bar may be used in such places as cemeteries and golf courses where closer cutting than the regular guard will afford is desired. MA-2620 guard is used on the outer end of the bar and M-2661 is used on the inner end. The balance of the guards on the lespedeza bar are MA-2619 guards. Guards may be bolted or assembled on the regular bar.

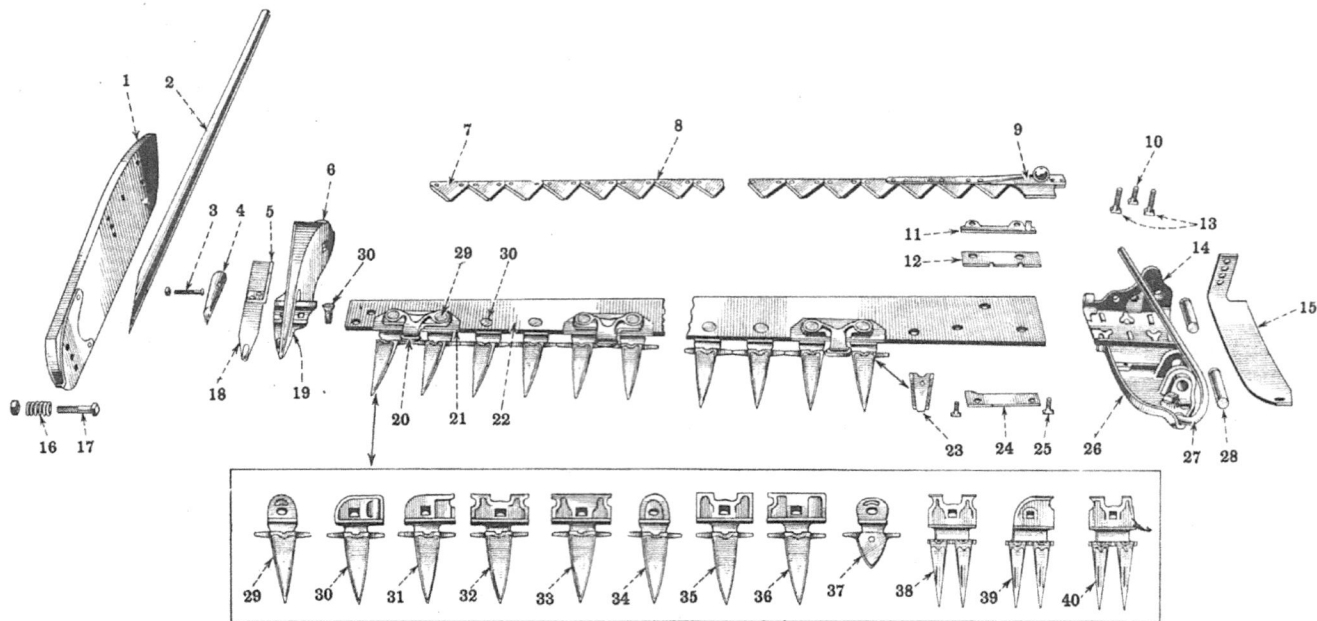

Fig. 70. 1. Grass board, 6. Outer shoe, 12. wearplate, 15. Inner shoe sole, 18. Outer shoe sole, 20. knife clip, 22. Cutter bar, 23. Ledger plate, 24. Knife head cap, 27. Fender rod, 28. Inner shoe pin, 29.(inset) Guard, 30.(inset) Outer guard.

MOWER ATTACHMENTS

REAPING ATTACHMENT

A reaping attachment for cutting grain and special seed crops can be supplied for use with mowers having 4-1/2 and 5-foot cutter bars. Attachment includes an extra seat over the right wheel for the operator, who reels the crop onto the platform with a rake and retains it until a gavel of the right size is secured, then dumps it off upon the ground. This attachment converts mowers into serviceable grain reapers for small acreages, and is excellent for harvesting clover and small seed crops.

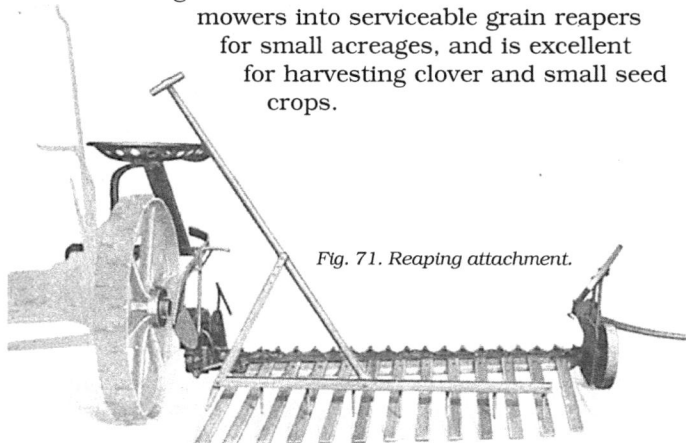

Fig. 71. Reaping attachment.

BUNCHING ATTACHMENT

This attachment is designed for gathering very short hay which cannot be raked easily. It is used also for gathering seed crops. The hay is gathered upon the slatted platform back of the cutter bar, and when a sufficient amount has accumulated, the driver dumps it by raising the shield as shown in Figure 72. This is accomplished with a foot lever. Buncher attachments can be furnished for 4-1/2, 5 and 6-foot mowers.

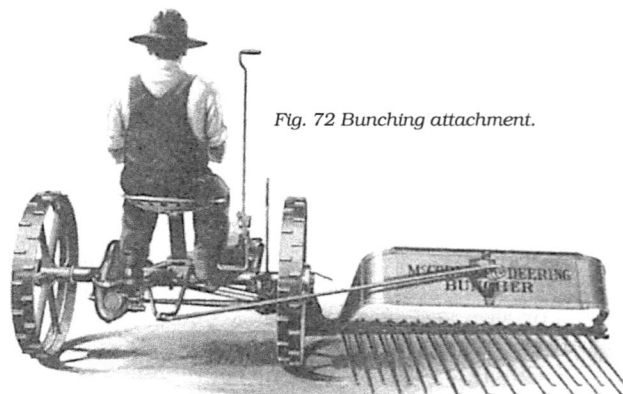

Fig. 72 Bunching attachment.

Fig. 73.

MOWER TONGUE TRUCK

Under certain conditions owners sometimes prefer to equip their mowers with a tongue truck. The tongue truck reduces side draft and takes the neck weight off the horses. The truck is provided with a casting permitting the attachment of a draft rod running from the inner shoe hinge on the cutter bar to the tongue truck. The tongue truck can be supplied with a long pole or stub pole or without either pole when so ordered. Pneumatic-tired wheels are also available.

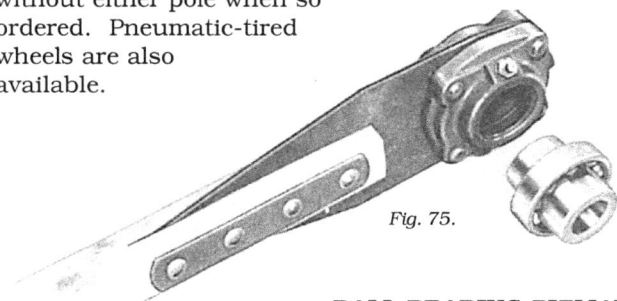

Fig. 75.

BALL BEARING PITMAN

A ball bearing pitman is furnished as regular equipment on the #18 mower and as special equipment on the #9, #16 and #25 mowers. The bearing consists of a wide cone with single row ball bearings, lubricated and protected from dust and dirt by a felt washer.

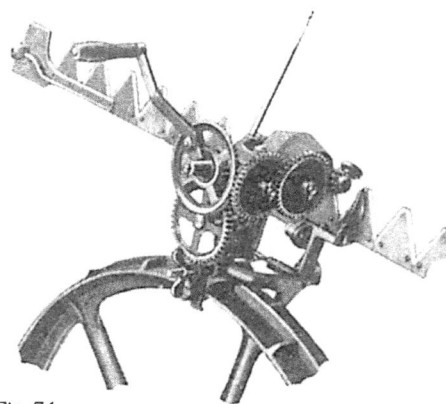

Fig. 74.

KNIFE GRINDERS

The regular hand knife grinder is designed to meet the demands for a handy, portable machine with which the farmer can sharpen his mower knives easily and quickly, and still retain the correct bevel edge and cutting angle of each section. It will grind one edge of each of the two sections at the same time. The knife is then moved along and the next two sections are ground. A spring maintains the pressure of the stone against the knife and the stone is of the proper shape to give the sections the correct bevel. The knife is held in the machine by clamps and simply turning the crank moves the stone over the edges of the sections. A handle on the frame permits holding the stone at any desired point so that nicks can be ground out.

Foot Power Attachment:

This quality-built knife grinder is regularly supplied for hand power. At small extra cost, a stand, seat, pedals, chain, and sprocket will be supplied. The grinder is then clamped to the frame and operated from the seat by means of the foot pedals.

Regular Equipment:

Crank for hand power and bevel stone for grinding mower knives (3-inch spacing). Weight, including stone, 20 pounds.

LEAD WHEELS

Lead wheels are available for #9, #25, and #16 mowers. (Illustrated wheels are for the #9 mower). The attachments are used where high cutting is required for the bar to clear small obstructions, or it is desired to cut over the top of a young crop. Wheels are furnished in either steel or pneumatic tired.

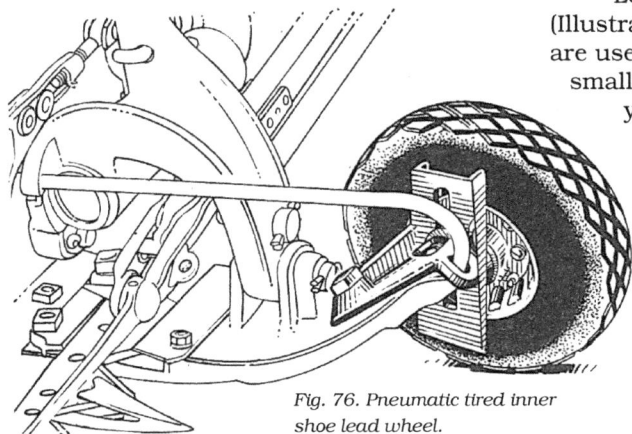

Fig. 76. Pneumatic tired inner shoe lead wheel.

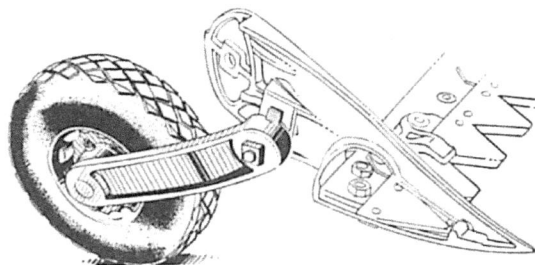

Fig. 77. Pneumatic tired outer shoe lead wheel.

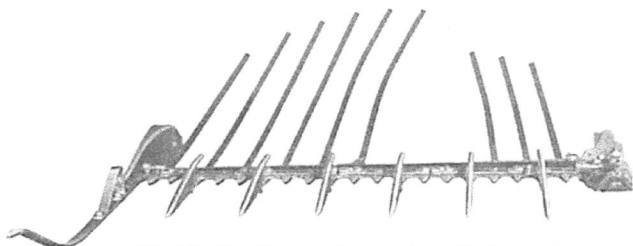

Fig. 78. Five-foot canning pea bar attachment is available for the no. 9 horse-drawn mower.

Fig. 79. Lifter guard for pea bar. The lifter is provided with a tension spring.

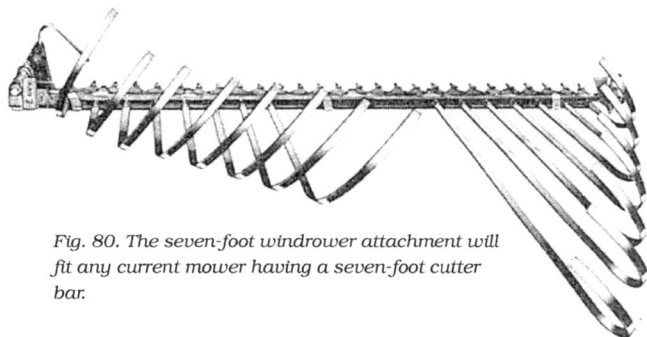

Fig. 80. The seven-foot windrower attachment will fit any current mower having a seven-foot cutter bar.

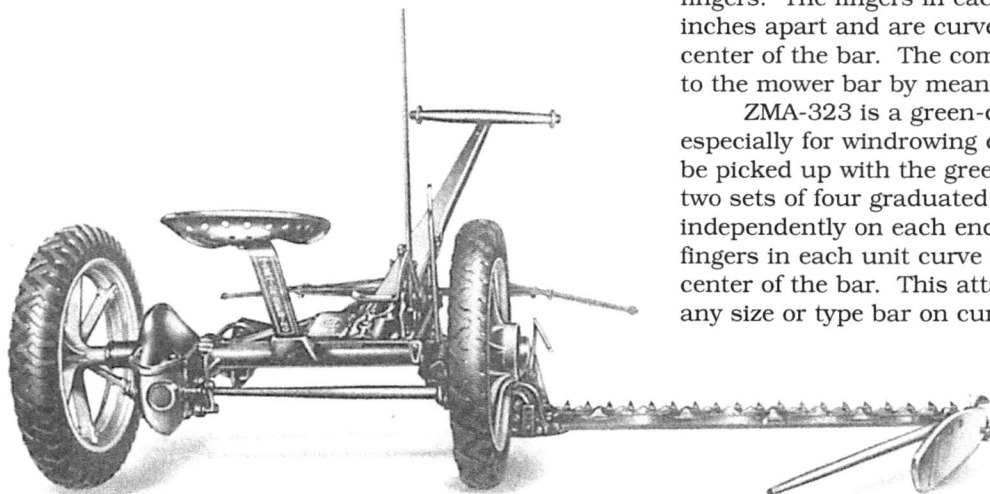

Fig. 81. No. 9 mower equipped with 5.00 x 21-inch pneumatic-tired wheels.

PEA BAR ATTACHMENTS

These bars are especially designed for harvesting green canning peas in the most efficient manner. They are equipped with stub guards, vine lifters, special outside divider, and windrower fingers. The vine lifters are of special design, smoothly finished and streamlined to give the most efficient action. The lifters are hinged and are provided with a tension spring so that the points will follow the ground closely. They pick up all the vines and raise the pods above the sickle so that none will be cut or wasted.

The stub guards between the lifters are lipless and will not clog easily. The knife sections extend beyond the ledger plates and cut readily through heavy growths. The windrower fingers turn the cut vines gently into a windrow with the pods mostly inside so that the peas remain fresh. The windrow on the five-foot bar has a center delivery while the six-foot bar has end delivery. This assures proper windrow spacing so that the windrow is deposited out of the way of the horses or tractor wheels when cutting the following round.

WINDROWER ATTACHMENTS

M-22817 is a 5-foot (center-delivery) windrower available for all current mowers having 5-foot cutter bars. It consists of a steel bar to which graduated flat steel fingers are secured. There are six fingers at the outer end and three fingers at the inner end. This is the same type of windrower as used on the 5-foot canning pea bar attachment. (Figure 78)

M-32611 is a 6-foot (end-delivery) windrower available for all current mowers having 6-foot cutter bars. It consists of eleven flat steel fingers graduated in such a way as to deliver the windrow from the inner end. This is the same type of windrower as used on the 6-foot canning pea bar attachment

ZMA-333 is a 7-foot windrower attachment (center-delivery) applicable to any current mower having a 7-foot cutter bar. It consists of a flat steel bar to which are bolted groups of graduated flat steel fingers. The fingers in each group are spaced three inches apart and are curved upwards and toward the center of the bar. The complete assembly is attached to the mower bar by means of clevices.

ZMA-323 is a green-crop windrower designed especially for windrowing crops which are intended to be picked up with the green-crop loader. It consists of two sets of four graduated flat steel fingers mounted independently on each end of the cutter bar. The fingers in each unit curve upwards and toward the center of the bar. This attachment can be used with any size or type bar on current mowers.

JOHN DEERE NO. 4 BIG ENCLOSED GEAR MOWER

The information in this section comes, verbatim, from the John Deere Operator's manual. It covers, in great detail, one of the finest horsedrawn mowers ever designed and built. The other two, in this author's opinion, were the McCormick Deering No. 9 just covered, and the McCormick Deering No. 7.

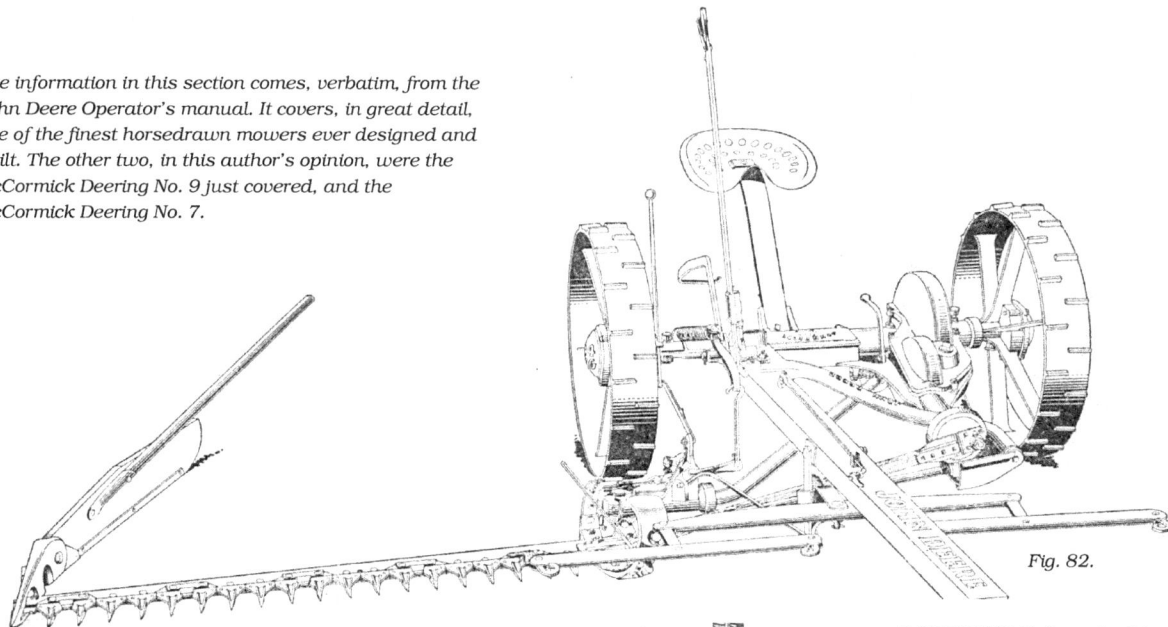

Fig. 82.

Fig. 83.

IMPORTANT: Before attaching cutter bar to Mower, GET BAR UP where all wearing surfaces can be oiled freely before putting in Knife. Run Knife back and forth and oil until Knife works freely (see instruction No. 4 below).

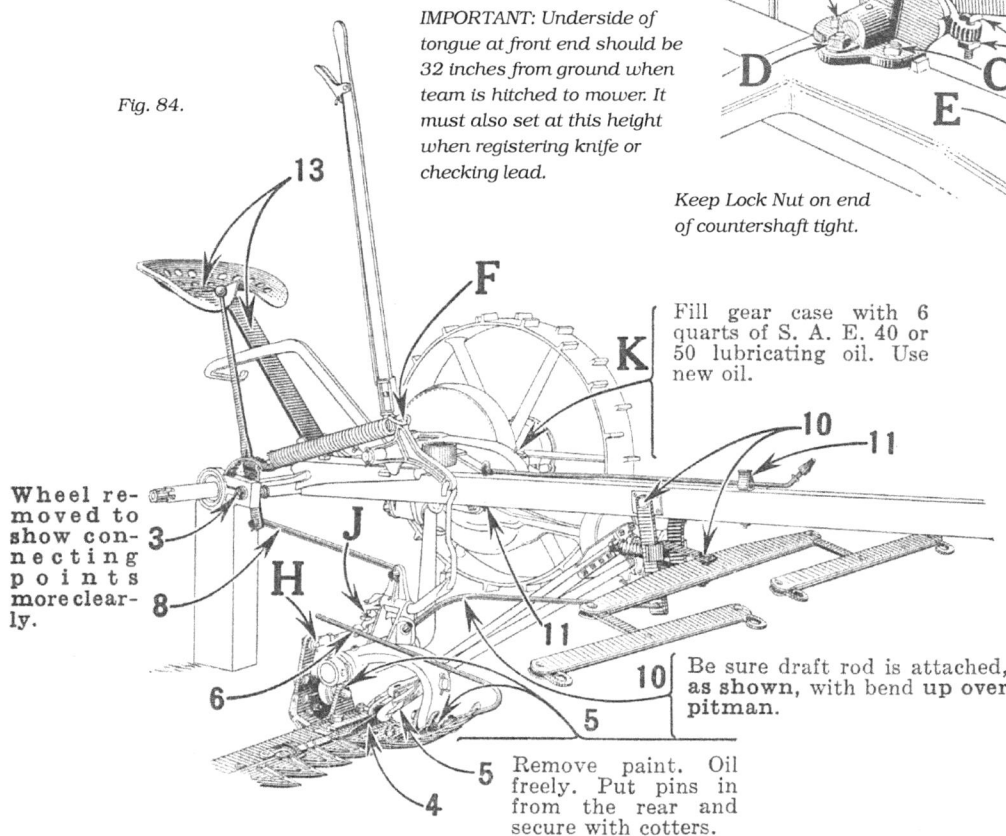

IMPORTANT: Underside of tongue at front end should be 32 inches from ground when team is hitched to mower. It must also set at this height when registering knife or checking lead.

Keep Lock Nut on end of countershaft tight.

Fig. 84.

Wheel re-moved to show con-necting points more clear-ly.

Fill gear case with 6 quarts of S. A. E. 40 or 50 lubricating oil. Use new oil.

Be sure draft rod is attached, **as shown,** with bend up over pitman.

Remove paint. Oil freely. Put pins in from the rear and secure with cotters.

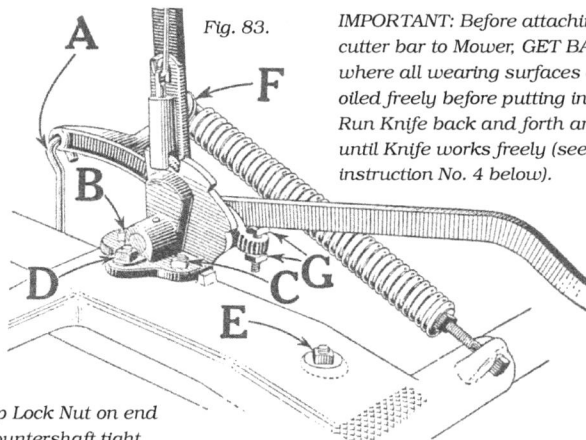

1. (Fig. 83.) Put block under flywheel bowl to raise frame in position to receive pole. Bolt pole loosely at "E". Attach lift rod at "A", and bolt sector at "B" and "C", at same time attaching oil can holder at "C". Bolt securely at "D", then at "E". Be sure to use heavy washers under head of pole bolts. These bolts must be tight.

2. Screw Zerk fittings into hubs of main wheels.

3. (Fig. 84.) Bolt tilting lever ratchet and lever to main frame.

4. KNIFE. Before putting knife in bar, oil guard plates,

wearing plates, and knife head guides liberally. Put in knife, oil again. Move knife back and forth until it works freely.

5. CUTTER BAR. (Fig. 84.) Scrape paint from hinge pins and pin bearings. OIL. Pine and cotter cutter bar to yoke. Connect pitman to knife head. (See Fig. 85 and Instruction No. 10.)

6. (Fig. 84.) Bolt Fender Rod to inner shoe, as shown.

7. (Fig. 84.) Pin and cotter swivel at "H" with head of pin to the rear (see Instruction No. 13, page 8).

8. (Fig. 84.) Cotter tilting rod to tilting lever and to yoke, as shown.

9. (Figs. 83 and 84.) Remove adjusting bolt from lifting spring. Hook front end of lifting spring onto support arm with loop pointed down, as shown at "F", Fig. 84. Slide adjusting bolt through lug on frame and screw into lifting spring. See Instruction No. 14, for adjusting tension of spring.

Note: For all Canning pea cutter bars and all 7-foot cutter bars, bolt lifting spring bracket to main frame.

10. (Fig. 84.) Hook draft rod in yoke, as shown; then draft bracket to rod. Bolt bracket to pole, then doubletree to clevis.

11. (Fig. 84.) Bolt cutter bar latch rod and holder to pole.

12. (Fig. 86.) Attach grass board and stick. Stick goes to inside of board.

13. (Fig. 84.) Attach seat and seat spring (see Instruction No. 6).

14. Loosen bolt through seat spring and main frame. Insert Gun Holder plate (Z7256 H) between main frame and seat spring. Tighten bolt.

15. Attach gun holder spring (Z7257 H) to seat spring. Locate Z7257 H on seat spring so that nozzle of grease gun will rest in Z7256 H plate.

16. (Fig. 84.) Put six (6) quarts of S.A.E. 40 or 50 lubricating oil in gear case at "K". Put one (1) pint of oil in each pawl plate housing. Use new oil. Do not use burned-out crankcase oil (see Instruction No. 3.)

INSTRUCTIONS FOR ADJUSTING AND OPERATING

Read every word of these instructions. You will like this John Deere No. 4 Mower. With proper care, it will give years of satisfactory service. You will agree that careless setting up, careless operation and neglect are the causes of most troubles. This machine will continue to cut like a new Mower if properly oiled and kept in good repair. Cutting parts must be kept sharp, knife head guides, knife holders, and wearing plates must be replaced when badly worn, and carefully set; guards must be kept in alignment; adjustments to restore alignment of knife and pitman and for registering knife sections in guards should be used if ever necessary; lifting spring should be properly adjusted and pole hitched at correct height. Proper attention to these essentials insures clean cutting, light draft,

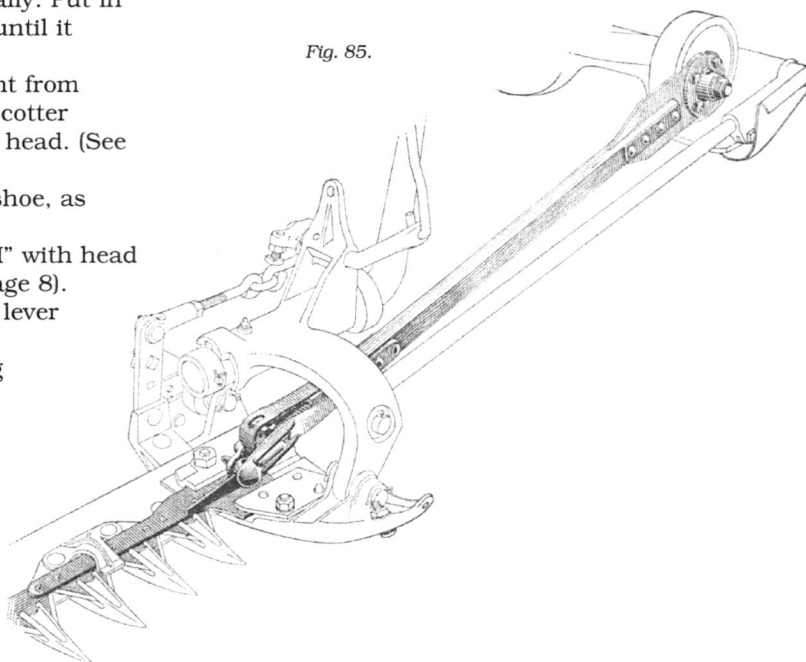

Fig. 85.

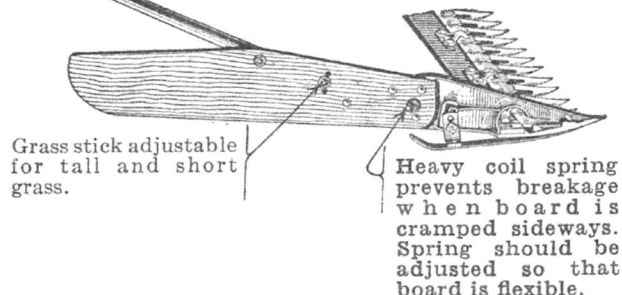

Fig. 86.

Grass stick adjustable for tall and short grass.

Heavy coil spring prevents breakage when board is cramped sideways. Spring should be adjusted so that board is flexible.

continuous operations and low upkeep cost. The responsibility for this kind of service rests with the owner and operator.

Upkeep cost will be reduced by good storage and by replacing worn parts promptly.

1. Be sure the mower is properly set up, according to these directions, and that it is properly adjusted and oiled.

2. See that every moving part works freely before putting machine in field. Keep all nuts tight and cotters spread.

3. Oil from Gear Case lubricates Main Axle, Crankshaft, Countershaft, Gears, and Bearings. Keep 6 quarts of S.A.E. 40 to 50 lubricating oil in gear case. With tongue at operating height (see Instruction No. 5) oil must cover countershaft at all times to insure proper lubrication of Pitman Shaft Bearings. To test: Remove filler plug and wipe gauge. Insert gauge, but do not screw plug into case--oil should show up to

When aligning guards, PAY NO ATTENTION TO POINTS OF GUARDS—LINE UP SURFACES OF GUARD PLATES.

When setting knife holder DOWN, PULL KNIFE FROM UNDER HOLDER.

Keep shoe bolts tight.

TO ALIGN, STRIKE GUARD ON THICK PORTION JUST AHEAD OF LEDGER PLATE.

If knife holder binds knife, with knife under holder, strike holder on flat surface between bolts to relieve.

Fig. 87.

SLOTTED HOLES IN WEARING PLATES provide adjustment to set plates ahead to TAKE UP LOOSENESS OR WEAR.

DO NOT POUND DOWN LIPS OF GUARDS—choking will result.

Knife back guided in neck of Guards and rubs on hardened wearing plate at rear.

To insure a shear cut, POINT OF SECTION MUST SET ON GUARD PLATE.

KNIFE HOLDER MUST FIT DOWN SNUG on knife but without binding.

SET WEARING PLATES AHEAD IF THERE IS TOO MUCH "PLAY" IN NECKS OF GUARDS FOR KNIFE BACK.

Z7961H spacers must be inserted between wearing plates and Z7798H knife clips when knives equipped with extra heavy sections are used.

Fig. 88.

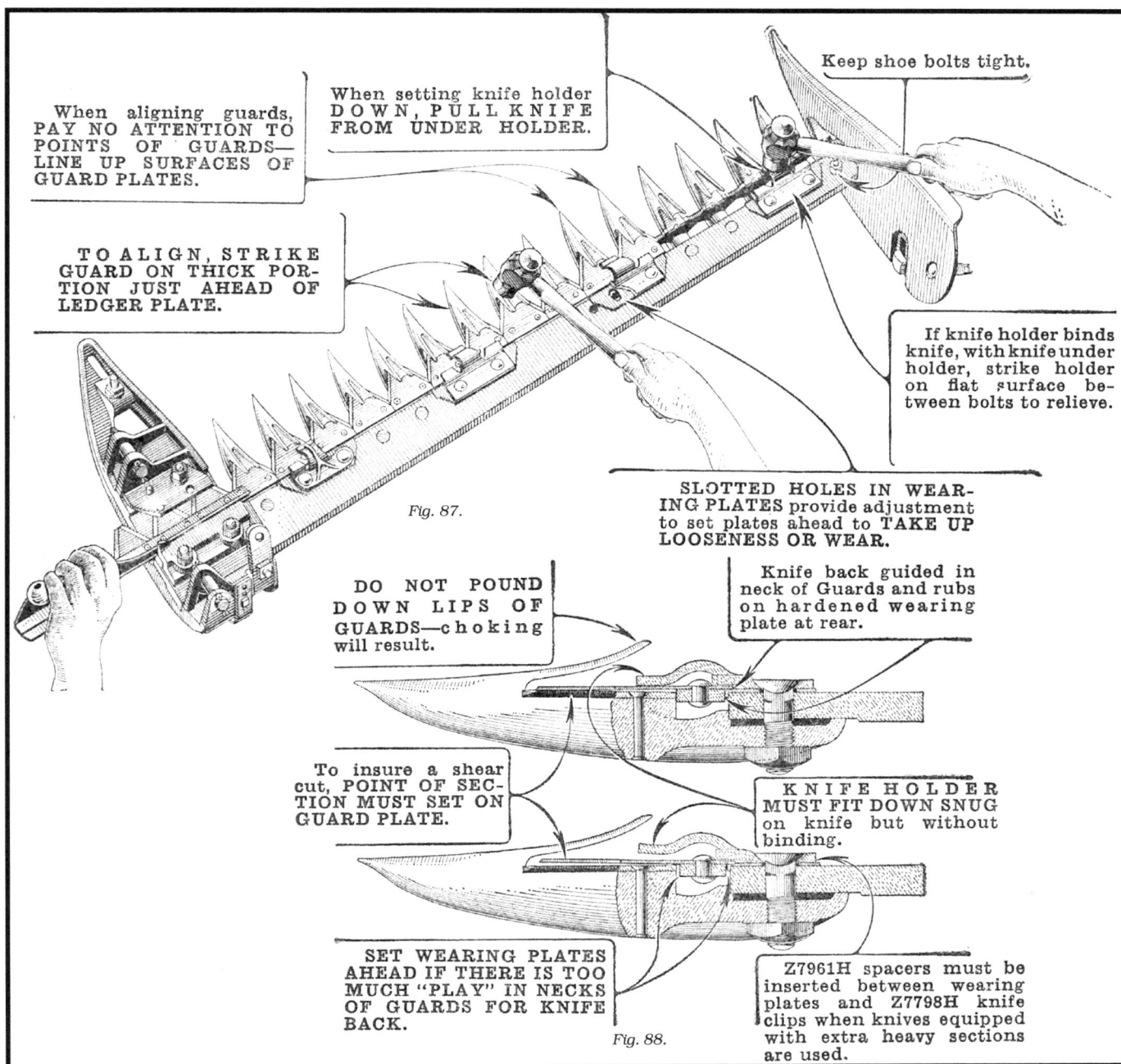

mark on gauge at all times, with front end of tongue at the correct height. Never permit oil to get so low that it does not register on gauge.

Drain and strain the oil after the first week's run. Put this oil back in case after it is carefully strained. It is not necessary to change the oil until the end of the season. New oil should be put in case and wheel caps at the beginning of each year's cutting. Do not use burned-out crankcase oil from automobile or tractor.

Use one (1) pint of oil in each pawl plate housing.

4. Use a good grade of heavy machine oil in oil can. Lubricate pitman box and pitman connections often. Oil lifting mechanism bearings and cutter bar hinge pins occasionally. Keep cutting parts oiled except

in dry, dusty, and sandy conditions, where cutting parts usually work the best without oil. Use water to remove gummy trash that packs on wearing plates and guards. Do not let it harden.

5. TONGUE. Front end of tongue, measuring from underneath, should be 32 inches from the ground when team is hitched to Mower. This is important. At this height, the Mower is in the correct working position.

6. SEAT SPRING. Two holes at lower end of spring provide adjustment in main frame for convenience of operator and for reducing neck weight.

7. CUTTER BAR. (See Figs. 87 and 88.) The cutter bar of the Mower, in principle, is nothing more than a

multiple set of shears--the blades of shears, to cut properly, must be sharp and have a shear cut--likewise, the sections and guard plates of a mower must be sharp and have a shear contact. Cutting edges of the ledger plates must line up the full length of the bar if ledger plates and sections are to have a shear cut, and knife holders must be set to hold front part of sections down against guard plates and heel of sections against wearing plate but must permit knife to run without binding. Wearing plates under knife holders should be replaced when worn enough to cause sections to raise from contact with ledger plates at points. Always look for cutting troubles in the cutter bar--guards out of alignment, worn wearing plates, bent knife back, dull knife and ledger plates, and worn, improperly-set knife holders.

Never bend a knife holder down with the knife under holder. Start at outer end of bar by pulling knife from under outer holder and tapping holder down. Keep trying the knife until holder sets so that knife works freely and at the same time is down on knife. Then work each holder in the same manner. If holder is too tight on knife, strike holder between the bolts while knife is under holder. A Knife working tight in the bar will cause heavy draft. Knife holders should not be set until after guards are aligned. Be sure all guard bolts are tight. After setting knife holders, try the knife, put oil on guard plates and be sure knife is working perfectly free.

Z7961 H spacers must be inserted between wearing plates and Z7798 H knife clips when knives equipped with extra heavy sections are used.

Do not pound down lips of guards--choking will result--the lip of the guard is the portion that covers the guard plate. Repoint blunt guards.

Wearing plates should be set ahead or replaced to take up wear on knife back, and reduce play of knife back in neck of guards--in setting wearing plates ahead, there should be enough clearance left at front of sections so that sections do not strike the guards. Turned-down edges of wearing plates must line up with one another to give knife back a straight bearing along its entire length. Be especially careful to properly line up new plates.

Knife Head Guide, Front, Z6010 H, should be reversed if worn badly.

A dish is built into the 6-foot and 7-foot cutter bars (Z6698 H, Z7049 H and Z7091 H, 6-foot Bars--Z6439 H and Z7603 H, 7-foot Bars) to compensate for the added weight of the outer shoe and guards. The 6-foot bar has a 9/16-inch dish in the center--the 7-foot bar has 15/16-inch dish in the center. This automatically straightens out when the cutter bar is attached to the mower and the proper lifting spring tension is applied.

The repairing of the Mower Cutter Bar is generally put off too long--the Mower does not give satisfactory service, is abused more than necessary, and the horses are overworked. This neglect is generally due to the lack of proper tools and the necessary repairs to

make a quick, easy job. Many times, new guards are put on when new plates only are needed. Note the helpful suggestions for repairing in Figs. 87, 88, and 115.

Run Bar Level--Keep guard and shoe bolts tight (see Fig. 116).

If knife becomes bent in rough cutting or storage, straighten it on a level pole or block.

A Guard Block and Knife Repair Anvil (Z763 H) provides the easiest and quickest way for replacing guards, replacing worn or broken sections, and straightening knives (see Figs. 114 and 115).

9. ADJUSTABLE SUBSOLES. These soles under inner and outer shoes should be adjusted to regulate the height of cut for different field conditions. They will set bar to cut as high as 3 inches. Be sure to have cutter bar same height at both ends. On rough or stony land, the cutting parts should be protected by adjusting the subsoles to raise cutter bar.

10. PITMAN. The knife head connection of the automatic pitman requires no attention from the operator to keep in proper adjustment. Wear on knife head is automatically taken up by spring tension, see Fig. 89.

Pitman can be quickly attached or removed from knife head without aid of a wrench.

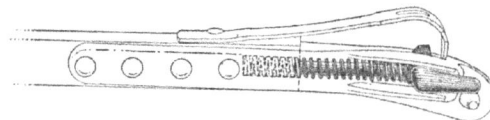

Fig. 89.

To Attach Pitman

Hold pitman down with foot and use punch to pry up the spring, allowing pitman straps to close over knife head. (Fig. 90.)

To Remove Pitman

Insert punch through hole in flat spring and into yoke plunger, and force yoke plunger back, and the flat spring down between the straps, to spread them and release the knife head. (Fig. 91.)

11. REGISTERING KNIFE. The knife registers if sections center in guards when pitman is at the inner or outer end of its stroke. A knife off register will not cut clean. Adjustable (forked) washers are provided at inner and outer ends of drag bar bearing in yoke. Knife is registered by transferring one or more washers from one end of yoke to the other and by lengthening or shortening the Brace Bar "A" at Flywheel Bowl (see Fig. 92). One turn of Brace Bar gives bar 1/8-inch movement. When checking knife for register, be sure that pitman connections are adjusted as directed (see Instruction No. 10.) Be sure that tongue is 32 inches from the ground at front end (see Instruction, No. 5). Be sure that lifting spring has proper tension (see Instruction No. 14).

To Make Registering Adjustment.

Unhook lifting spring from lifting lever support arm. Disconnect brace bar from front side of yoke. Remove pin and collar from outer end of drag bar. Move yoke out just far enough on drag bar to permit transferring washers from one end of yoke to the other as may be necessary to reregister the knife. Adjust brace bar in socket at "A", Fig. 92, at flywheel end to correspond to the amount or thickness of washers transferred. When Mower is old or has had severe use and is badly worn, the brace bar socket connection can be shortened one or two turns after all other adjustments have been made. Reassemble parts and recheck knife making sure it registers. Looseness of head on brace bar may be taken up with adjusting bolt. Do not tighten too tight. Brace bar must swivel in socket when tilting cutter bar.

12. ALIGNMENT OR LEAD OF CUTTER BAR. (See Fig. 93.) "Lead of Cutter Bar" is built into John Deere Mowers and realigning cutter bar to regain the original lead helps make an old mower cut like new, reducing the chances for knife head breakage so common in old machines with lagging bars.

The cutter bar should have a lead from inner to outer end of 1 to 1-1/4 inches on 4-1/2-foot; 1-1/4 to 1-1/2 inches on 5-foot; 1-1/2 to 1-3/4 inches on 6-foot; and 1-3/4 to 2 inches on 7-foot. This amount of lead keeps knife and pitman in line when Mower is cutting. If, after long use, the outer end of bar lags back, bring it forward by turning eccentric "A" to the left in yoke. This restores alignment of knife and pitman, and prevents wear and breakage. Do not use eccentric to "register" knife.

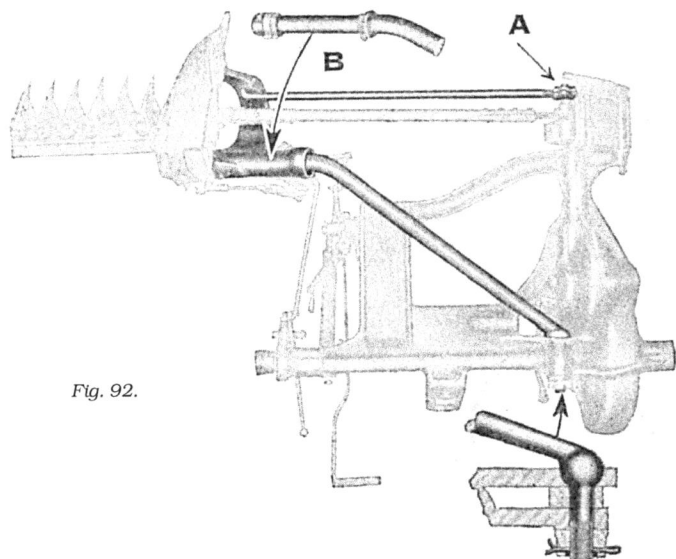

Fig. 90.

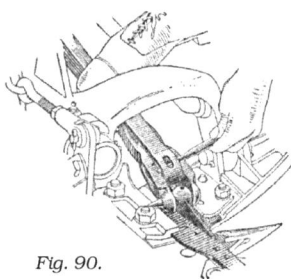

Fig. 91.

When making "lead" adjustment, have lifting spring properly adjusted (see Instructions Nos. 13 and 14). Be sure front end of tongue is 32 inches from ground; cutter bar level; knife at outer end of its stroke, and wheel blocked securely.

Move outer end of cutter bar forward as far as slack permits and mark this location with a stick at point of outside shoe. Next, move cutter bar back as far as the slack will allow. The distance from point of shoe to stick will show the amount of slack or lag in the bar at the outer end and this much play can be taken up by adjusting the eccentric "A" to the left.

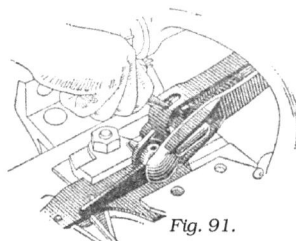

Fig. 93.

Adjust eccentric "A" to LEFT to take up wear and increase lead.

Keep nut on rear end of drag bar just tight enough to take up end play. Drag bar must swivel freely.

Cutter Bar properly aligned, knives centered properly in the guards, and careful lubrication, lighten draft and insure satisfactory work.

13. LIFTING CHAIN, ADJUSTING BOLT IN BELL CRANK, AND CHAIN SWIVEL ADJUSTMENT. (See Fig. 84.) Inner and outer end of bar should leave the ground at the same time. If outer end of bar rises too fast, screw adjusting bolt down in bell crank at "J". If outer end rises too slow, screw adjusting bolt up enough to make both ends of bar rise evenly at the start. Chain swivel adjustment, "H" can also be shortened to take up wear and to make outer end of bar rise faster, or lengthened to make outer end of bar rise slower. Be sure that lifting spring has proper tension (see Instruction No. 14).

14. LIFTING SPRING. There should be enough tension on lifting spring so that bar will rise easily and make the bar float rather than drag, and yet move steadily over the ground. If there is not enough tension, the bar will ride on the ground and increase side draft. Too much tension will not allow the bar to follow the uneven ground, and may hold it up after it has risen over a mound or other obstruction. When properly adjusted, the lifting spring carries the bulk of the weight of cutter bar on the wheels, increases the traction, and reduces the friction of bar on ground.

15. FOOT LEVER. Foot lever is adjustable at "G" (fig. 83). It can be raised or lowered for the convenience of the operator and to raise the cutter bar higher without using hand lever.

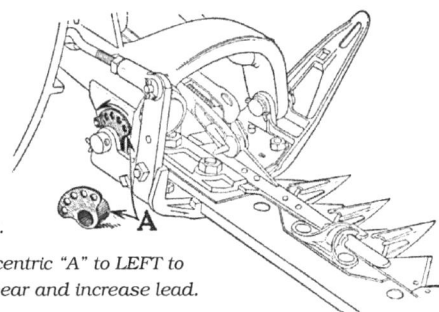

Fig. 92.

16. CLUTCH. (See Fig. 94.) The clutch shifter rod is adjustable at "B" to keep clutch properly engaged with bevel gear. When shifter lever is up and clutch meshes full depth with clutch jaws on gear, the clutch shifter yoke should be free in clutch and not bind against either side of groove in clutch. When shifter lever is down, clutch jaws should clear each other 1/8-inch. If the mower does not go out or stay out of gear when shifter lever is moved down, shorten the rod adjustment.

To Adjust: loosen bolt at "A" and turn clutch throw-out adjusting sleeve at "B". Turn sleeve forward (R. H.) to lengthen the throw of shifter lever, and rearward (L. H.) to shorten. After adjustment is made, be sure to tighten bolt at "A" and secure with cotter.

17. REMOVAL OF GEARS, AXLE, AND SHAFTS: It is not necessary to take all gears from Mower to replace individual parts.

(A) Main Axle and Main Spur Gear. Block up Frame, Drain Gear Case and remove Cover. Remove pin and bolt locating Main Spur Gear on Axle. Remove R. H. Wheel Cap, Pawl Plate (Sq. Hd. Cap Screw in end of axle has R. H. Thread) and Wheel (see Instruction No. 18)--then remove L. H. Wheel with axle attached. Main Spur Gear may then be lifted from case.

(B) Countershaft Assembly. Drain Gear Case and remove Cover. Remove Pin at "F" (Fig. 94) and lock washer under nut at "D", replace nut loosely on shaft to protect threads. Strike countershaft to drive out expansion plug--then remove nut and slip countershaft out through expansion plug hole. Lift bevel gear and clutch parts from case.

(C) Pitman Crankshaft. Block up Frame, Drain Gear Case and remove Cover. Remove flywheel shield, brace bar and pitman. Unscrew bevel pinion (R. H. thread) and remove flywheel with shaft attached.

If ever necessary to remove bearings--tap rear

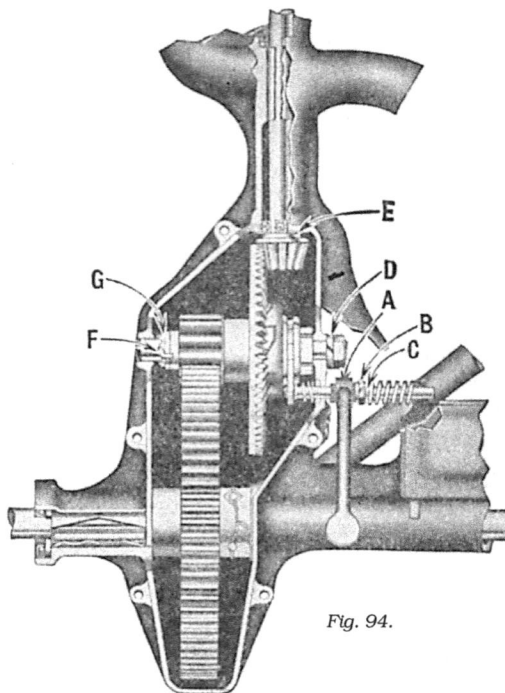

Fig. 94.

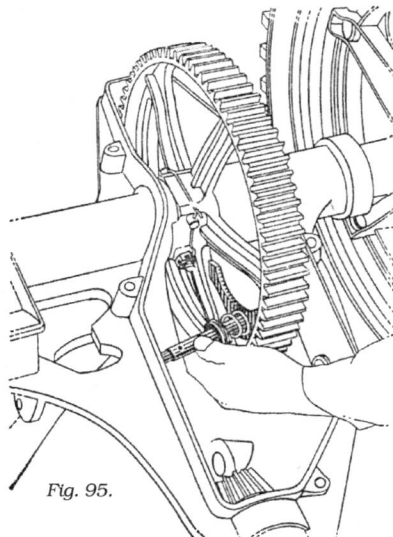

Fig. 95.

ball bearing out toward gear case. To remove front (bronze) bearing, place a 9/16-inch carriage bolt in bearing from rear. With head against end of bearing and with a heavy rod or shaft and hammer, drive out bearing. It may be necessary to file edge of bolthead to fit barrel of frame.

To replace bearings remove flywheel--from pitman drive shaft. Slide pitman drive shaft into barrel at front of main frame. Slide rear ball bearing over rear of pitman drive shaft and screw bevel pinion (R. H. thread) onto rear end of shaft. Fit ball bearing and bevel pinion snugly against shoulder in main frame. Slide bronze bearing onto front end of pitman drive shaft. Drive bearing into barrel at front of main frame using pitman shaft as guide.

Be sure bearing is driven straight into main frame to avoid damage to bronze bearing. Do not batter end of bearing when driving into main frame. If bearing is battered or damaged, the bearing will heat when operating Mower. Replace oil seal and flywheel.

(D) Clutch Shifter. Remove countershaft assembly as described under (B) countershaft assembly. Loosen bolt "A" (Fig. 94) in throw-out lever. Remove cotter "C" through throw-out shifter rod. Slide shifter yoke toward large drive gear, as shown in Fig. 95, being careful not to damage leather washer and cupped washer in frame. Turned-down end of shifter rod allows play enough in hole in gear case to permit removal of shifter between spokes of large drive gear. Bushing "A", Fig. 96, is inserted into lug in main frame to make up for difference in size of turned down end of shaft and hole in lug.

Replace parts, as shown in Fig. 94. Should leather washer become damaged, it should be replaced or an oil leak might result. After shifter has been replaced, adjust as described under "Clutch".

Care must be taken when making repairs to keep dirt and foreign particles out of gear case;

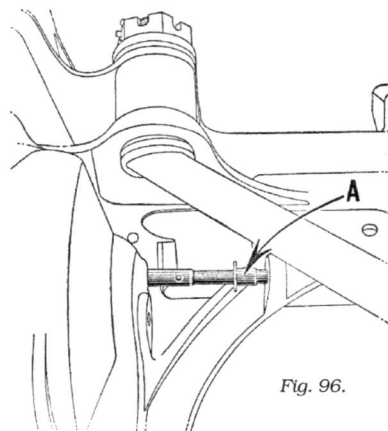

Fig. 96.

also to keep oil free from dirt and grit. Strain oil if it is to be used again.

18. WHEELS AND PAWL PLATES. To Remove: Block up frame. Remove pawl plate cover and pawl plate washer (cap screw in end of axle has R. H. thread). Screw cap screw with lock washer tightly into axle. Strike head of cap screw with a heavy hammer at same time pulling out on wheel, until pawl plate is flush with head of cap screw. Then bump pawl plate off with wheel.

To Replace: Set wheel at outer end of axle. Place pawls and plate in wheel hub and drive pawl plate onto axle at same time holding wheel against pawl plate. Keep pawl plate cap screws tight. Be sure lug on washer fits into slot in pawl plate.

Pawls are in wheel ratchets correctly if they click when wheel is turned backwards. The end of the dog that engages the ratchet in wheel must point opposite to the direction of wheel travel.

19. BEVEL GEAR AND PINION at rear of pitman crankshaft can always be kept in proper mesh by adjusting notched collar at "G", Fig. 94, and by using Z5406 H washers between collar on crankshaft and pinion at "E". Be sure gears are properly meshed. Gear teeth should not bottom, but must have about 1/16-inch clearance so that gears work freely.

20. STORING FOR THE WINTER. If space is at a premium, the tongue can be removed without disturbing other parts. If tongue is not removed, place block under center to keep it from warping. If Mower is stored with cutter bar in vertical position, place block under shoe to relieve lifting spring and parts--if stored with cutter bar down, unhook lifting spring. Be sure bar is not tilted so that pitman is under strain--it is likely to take a permanent twist and later cause trouble. It is best to remove pitman when storing.

Clean off the dirt, tighten loose bolts, spread cotters, grease the knives and wearing parts on cutter bar, paint the wood parts, and order the repairs that will be needed before mower is used again.

If you do not order the needed repairs when mower is put away, be sure to make a list of them and

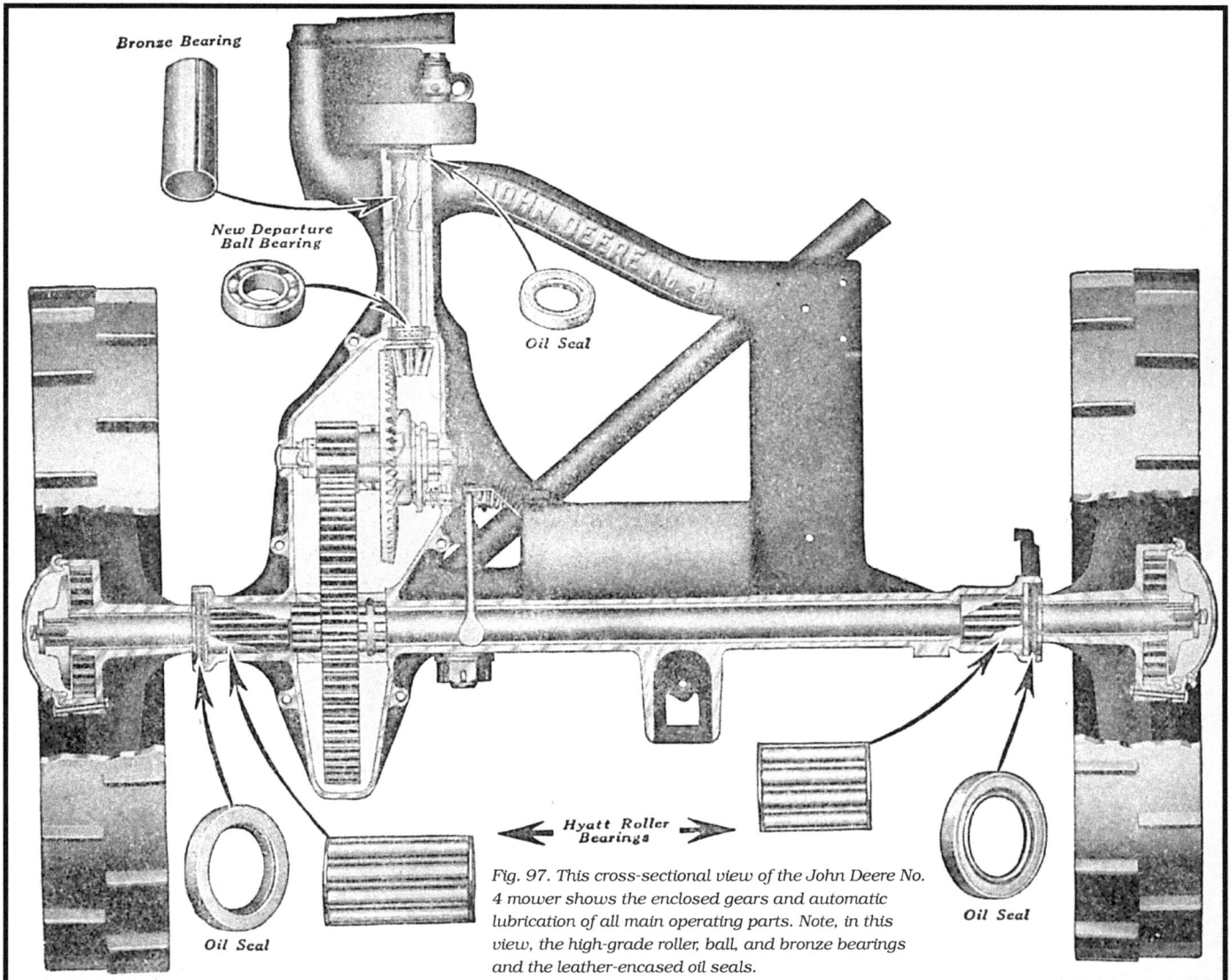

Fig. 97. This cross-sectional view of the John Deere No. 4 mower shows the enclosed gears and automatic lubrication of all main operating parts. Note, in this view, the high-grade roller, ball, and bronze bearings and the leather-encased oil seals.

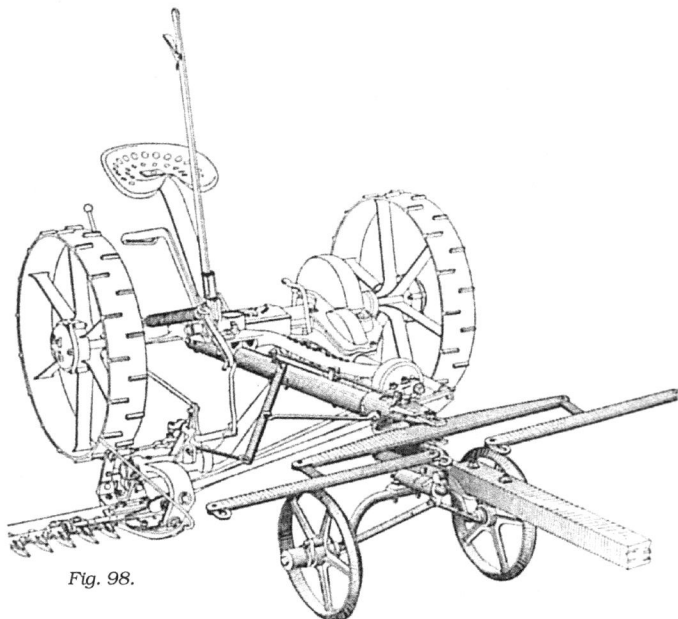

Fig. 98.

give the order to your dealer early--he can give you better service if you do.

Note: Keep an 8-inch block under flywheel bowl of mower when not in use.

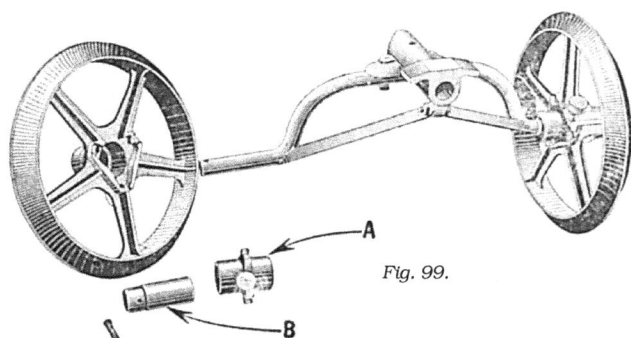

Fig. 99.

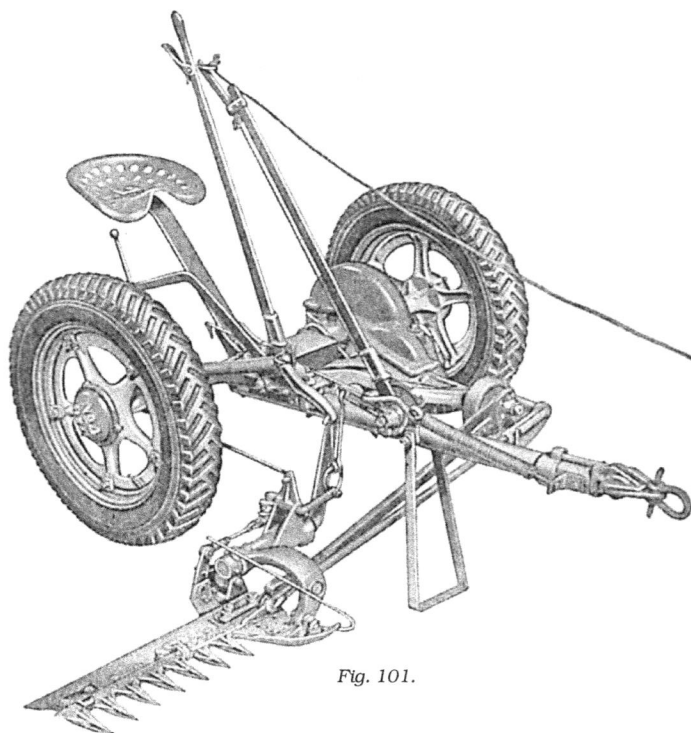

Fig. 101.

turning faster than the pole, make square turns without "backing up". It is easier for driver to control the team and to operate the mower. More comfortable riding. The Quick-Turn Tongue Truck saves the horses' necks, regardless of operator's weight--no whipping of pole. Flexible axle permits truck wheels to penetrate soil in all ground conditions, resisting side draft.

Tongue Truck Wheel is equipped with a new type removable white iron box "A". Box turns on a chilled sleeve "B" that is secured to axle. All wear is on box and sleeve. Zerk fittings provide lubrication.

QUICK-TURN TONGUE TRUCK FOR JOHN DEERE HIGH-LIFT MOWERS

A Forward Step in Mower Building

This Quick-Turn Tongue Truck keeps the pole (stub tongue) at the correct height, insuring proper setting of the drag bar and cutter bar, and a smoother, easier-running mower. Truck wheels,

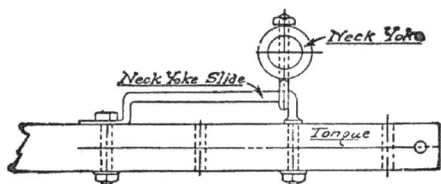

Fig. 100. Notice--Hitch team so that neckyoke is on forward end of neckyoke slide when team is going forward. This will give the horses more clearance for turning.

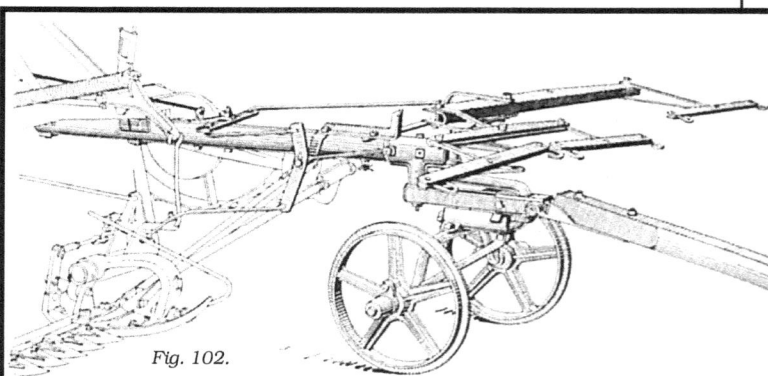

Fig. 102.

3-HORSE HITCH FOR MOWER TONGUE TRUCK

The 3-Horse Hitch can be furnished for mowers equipped with tongue truck. It is required only where cutting conditions are severe or when the horses used are small.

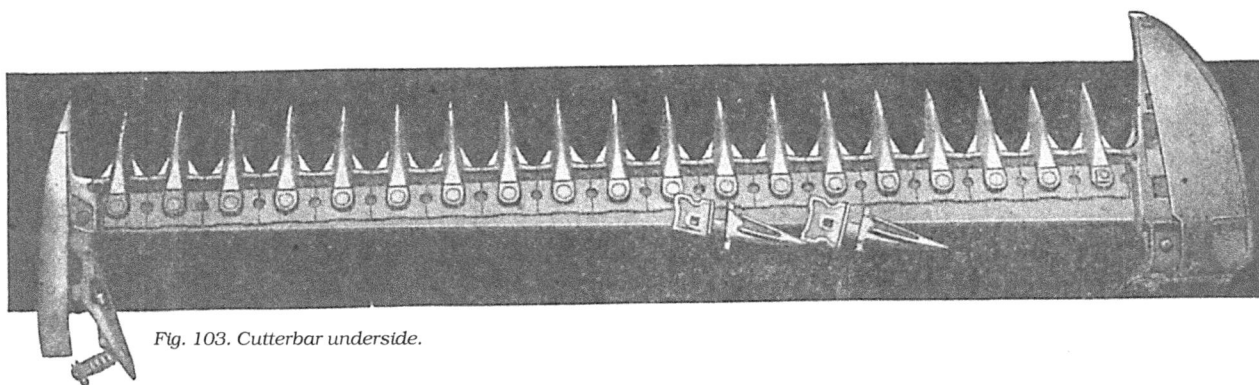

Fig. 103. Cutterbar underside.

SPECIAL STEEL ROCK GUARDS AND HEAVY KNIFE FOR EXTRA HEAVY DUTY

This Steel Rock Guard Cutter Bar with heavy duty knife is recommended for stony fields, road work where there is crushed rock or stone, and for heavy brush cutting. This bar is made up of heavy forged steel guards.

Rock Guards can be furnished in sets and put on regular bars. Extra Rock Guard Cutter Bars, or mowers equipped with these bars can be furnished.

When these steel Rock Guards are bolted to the cutter bar, be sure to attach the inside guard (next to inner shoe) with the head of the bolt in the steel cutter bar; that is, put the bolt through from the top side. If this bolt is put in from the bottom, the nut will interfere with sickle head. The nut should be turned with its oval surface against the countersunk hole in the guard so that it can be drawn down tight into the countersunk hole.

Order: AZ2131 H Steel Guard (16 used 4-1/2 foot; 18 used 5-foot; 22 used 6-foot; 26 used 7-foot).

AZ2132 H Steel Outside Guard (1 used on all sizes).

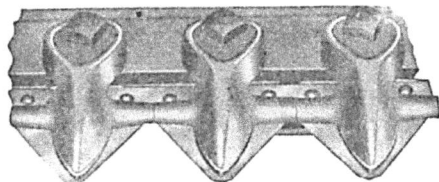

Fig. 104. Section of underside of pea bar. Note the knife sections extend beyond heavy guards.

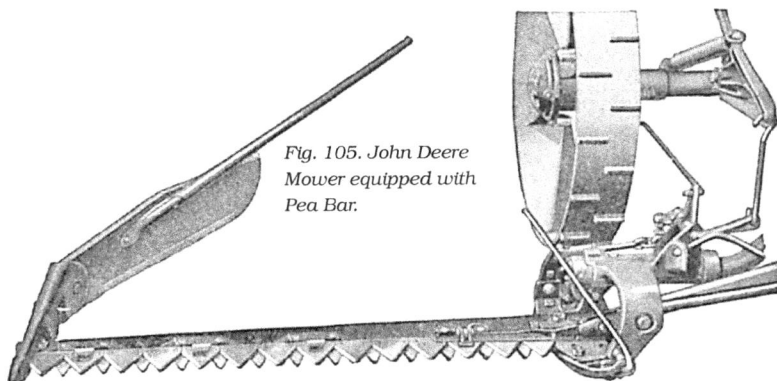

Fig. 105. John Deere Mower equipped with Pea Bar.

Special Pea (Cowpeas) Bar, Special cutter bar for cowpeas and beans can be furnished in 2-foot, 3-1/2-foot, 4-1/2-foot, 5-foot, 6-foot, and 7-foot sizes. The guards are extra short and without lips. They will not clog readily, as they allow the heavy stalks to bend sideways. The sections are made of heavier steel than the regular and extend beyond the special guard plate so as to readily cut their way through the heavy growth.

WEED CUTTING ATTACHMENT

Sweet Clover Bar-Wheel High-Cutting Attachment

A wide range of adjustment is possible with this bar-wheel attachment made especially for cutting sweet clover at just the right height (from 3 to 9 inches) to insure a good stand and prevent killing. Has no projection above shoe to catch and drag the crop. Also suitable for roadside or weed cutting. This attachment consists of a wheel and adjust-

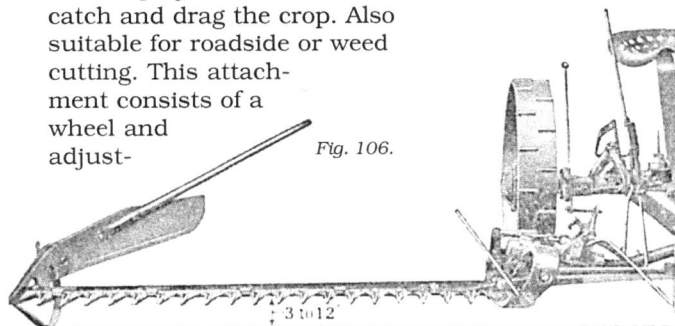

Fig. 106.

able bracket at the outer end of the cutter bar and a hook bolt which attaches to the pole to hold up the drag bar and the inner shoe. Hook bolt is illustrated in Fig. 132, page 140.

Weed Attachment. AZ2007 H weed or high-cutting attachment is furnished at small additional cost. It consists of a hook bolt attached to pole to hold up drag bar and inner end of cutter bar, and a special subsole for outer shoe. Both parts are adjustable so as to cut from 3 to 12 inches high. It can be used to excellent advantage in road work and cutting weeds and alfalfa tops. The regular subsoles under inner and outer shoes can be adjusted to cut as high as 3 inches above the ground.

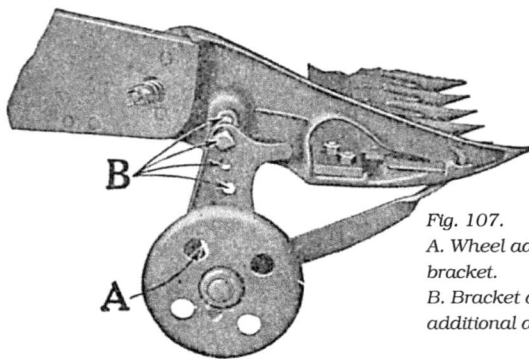

Fig. 107.
A. Wheel adjustable in
bracket.
B. Bracket and shoe provides
additional adjustment.

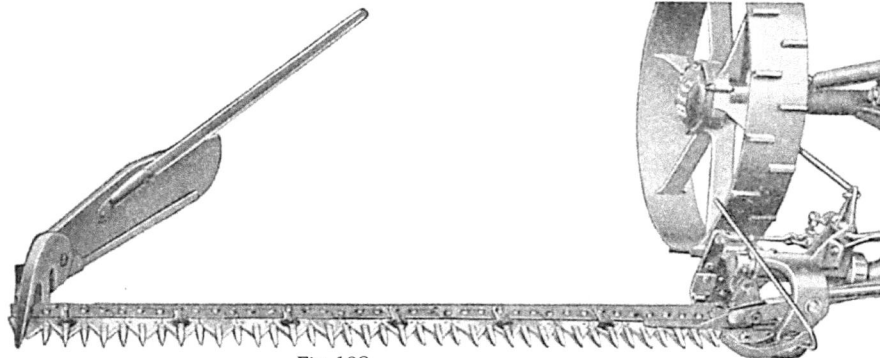

Fig. 108.

LESPEDEZA CUTTER BAR

A low-cut cutter bar is best adapted for cutting Lespedeza, either for hay or seed, because of the tendency of Lespedeza to mat. The John Deere Lespedeza Cutter Bar cuts approximately 1-inch closer to the ground than our regular Cutter Bars. The guards are spaced 1-1/2 inches instead of 3 inches. They are malleable, cast in pairs, with removable plates.

Smooth, serrated, or underserrated sections, as desired.

Furnished in 3-1/2 foot, 4-1/2 foot, 5-foot and 6-foot sizes.

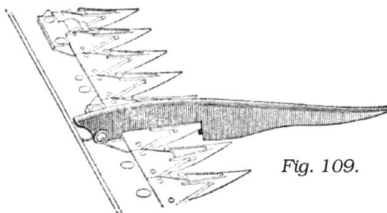

Fig. 109.

LIFTING GUARDS

Lifting Guards Can Be Furnished for Cutting Peas and Beans. These lifting guards fit over the regular guards and extend somewhat below the points. They are flexible and follow the uneven ground. They raise the vines over the knife so that very few pods are cut.

No. Used: 5 used 4-1/2-foot; 6 used 5-foot; 7 used 6-foot; 8 used 7-foot.

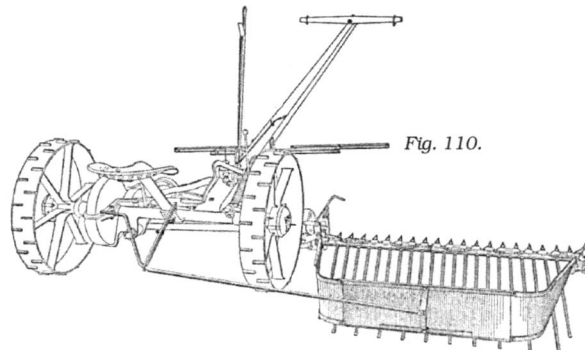

Fig. 110.

BUNCHER ATTACHMENT

Bunchers are used with profit in soybeans, short grain and hay, seed clover, and other seed crops. It saves a lot of seed by bunching with the bulk of the seeds on top where they dry quickly. Most any short crop that cannot be raked easily may be bunched to advantage. The crop is gathered on the fingers, and operator dumps by raising gate or shield with a foot lever.

Furnished in 4-1/2-foot, 5-foot, 6-foot, and 7-foot sizes for this mower. A special 5-foot buncher attachment for Lespedeza can also be furnished.

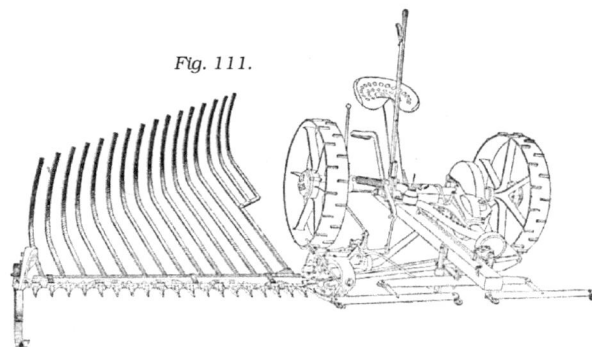

Fig. 111.

WINDROWER ATTACHMENT

In many parts of the country, windrowers are used quite extensively in harvesting clovers, soybeans, canning peas, cowpeas, grains used for hay, and other hay crops. The John Deere moves the crop off to the rear of the mower in light, fluffy windrows that dry and cure readily. A clear path is left for the horses and mower for the next round, and much seed is saved when cutting ripe crops for seed.

Furnished in 4-1/2-foot, 5-foot, 6-foot, and 7-foot sizes for this mower.

CANNING-PEA BAR AND WINDROWER

This cutter bar and center delivery windrower is built to handle the pea harvest quickly and economically. The lifting guards are so designed that they get under the down and tangled peas and raise them

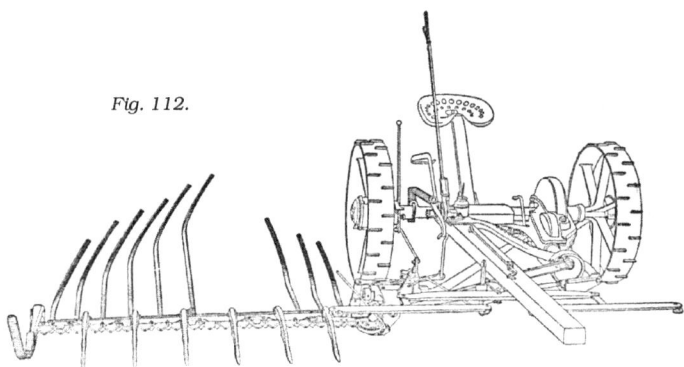

Fig. 112.

gently over the cutting parts, saving the pods from injury. Clean-cutting without clogging results from the stub guards and exposed sickle construction. Windrower delivers peas in clean-cut, narrow wind-rows with pods inside, protected from the hot sun.

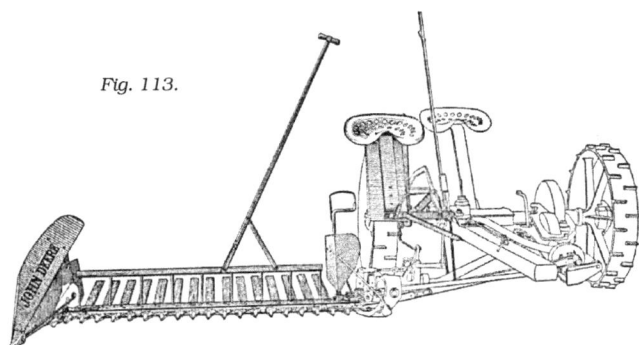

Fig. 113.

REAPING ATTACHMENT

Reaping attachments are used for cutting grains where there is only a small acreage, and for seed crops.

This attachment includes an extra seat mounted over the right-hand mower wheel for the operator that handles the hand rake for the reaping attachment. This operator reels the crop and keeps it on the reaping platform by holding platform up with foot lever until a load is accumulated.

Furnished in 3-1/2-foot, 4-1/2-foot, and 5-foot sizes for this mower.

ROTARY SWATH-CLEARER

It turns the tall hay and makes a good, clear track, greatly reducing the labor cost and saving some of the crop. This is the kind of swath-clearer you have been wanting for cutting tall or down and tangled hay--

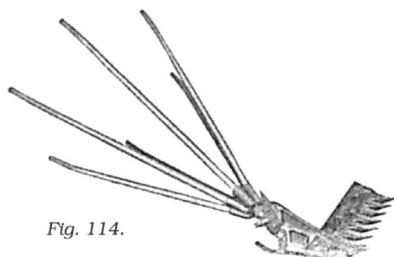

Fig. 114.

vetch, Johnson grass, peavines, and small grains not harvested with the binder.

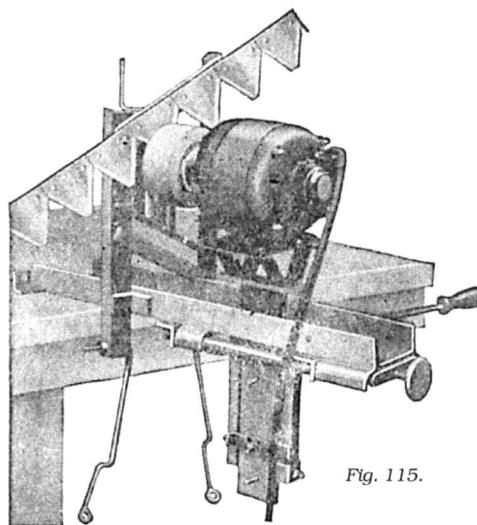

Fig. 115.

JOHN DEERE KNIFE GRINDER

The only Mower Knife Grinder that maintains the original bevel and angle of cutting edge of the sections. It sharpens sections under knife head. Insures better work and longer life of knives. Will grind all makes of Mower Knives. It can be attached to bench, mower wheel, or wall. It can be operated by hand power, foot power, electric motor, or gas engine.

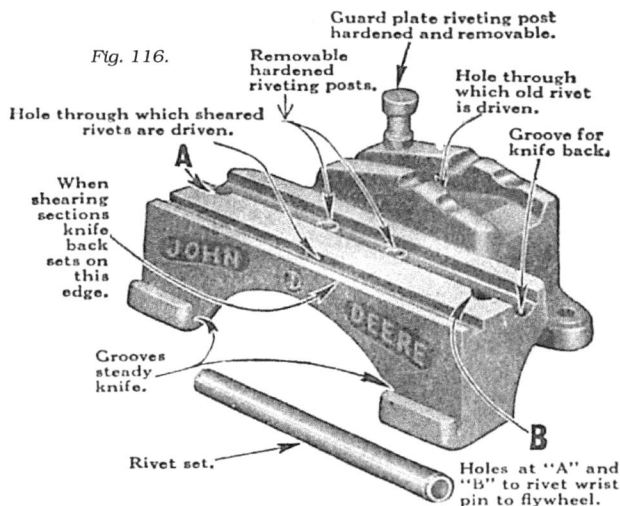

Fig. 116.

Guard plate riveting post hardened and removable.

Removable hardened riveting posts.

Hole through which old rivet is driven.

Hole through which sheared rivets are driven.

Groove for knife back.

When shearing sections knife back sets on this edge.

Grooves steady knife.

Rivet set.

Holes at "A" and "B" to rivet wrist pin to flywheel.

MOWER GUARD BLOCK AND KNIFE REPAIR ANVIL

The Improved Mower Guard Block and Knife Repair Anvil is a convenience that every farmer should have--fits any guard or knife--can be attached to workbench or carried to field. This convenient block and anvil makes a quick, easy job of replacing mower guards, replacing knife section, flywheel wrist pins, and for straightening knives. It saves time, hard work, and black smith bills. Mower is kept in better shape because it is easy to make repairs.

Fig. 117 shows operator replacing guards without removing them from bar.

A SOCKET WRENCH THAT FITS MOWER GUARD BOLTS

Z969 H socket wrench is made for tightening bolts for Z463 H guards on John Deere Mowers. It is inexpensive and you are sure to have guards tight, without skinning your hands.

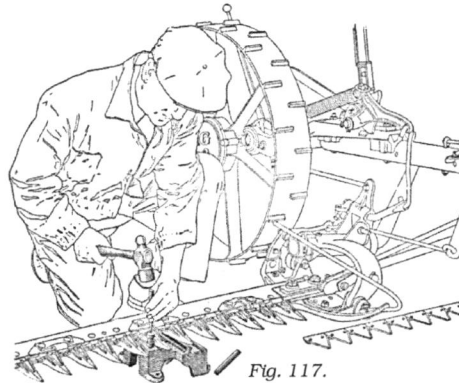

Fig. 117.

Fig. 118.

MAIN FRAME, WHEELS AND SEAT *Fig. 119*

1. Oil Plug (2 used)
2. Hub Cap (2 used)
3. Gasket (2 used)
4. Pawl Plate
5. Pawl Spring

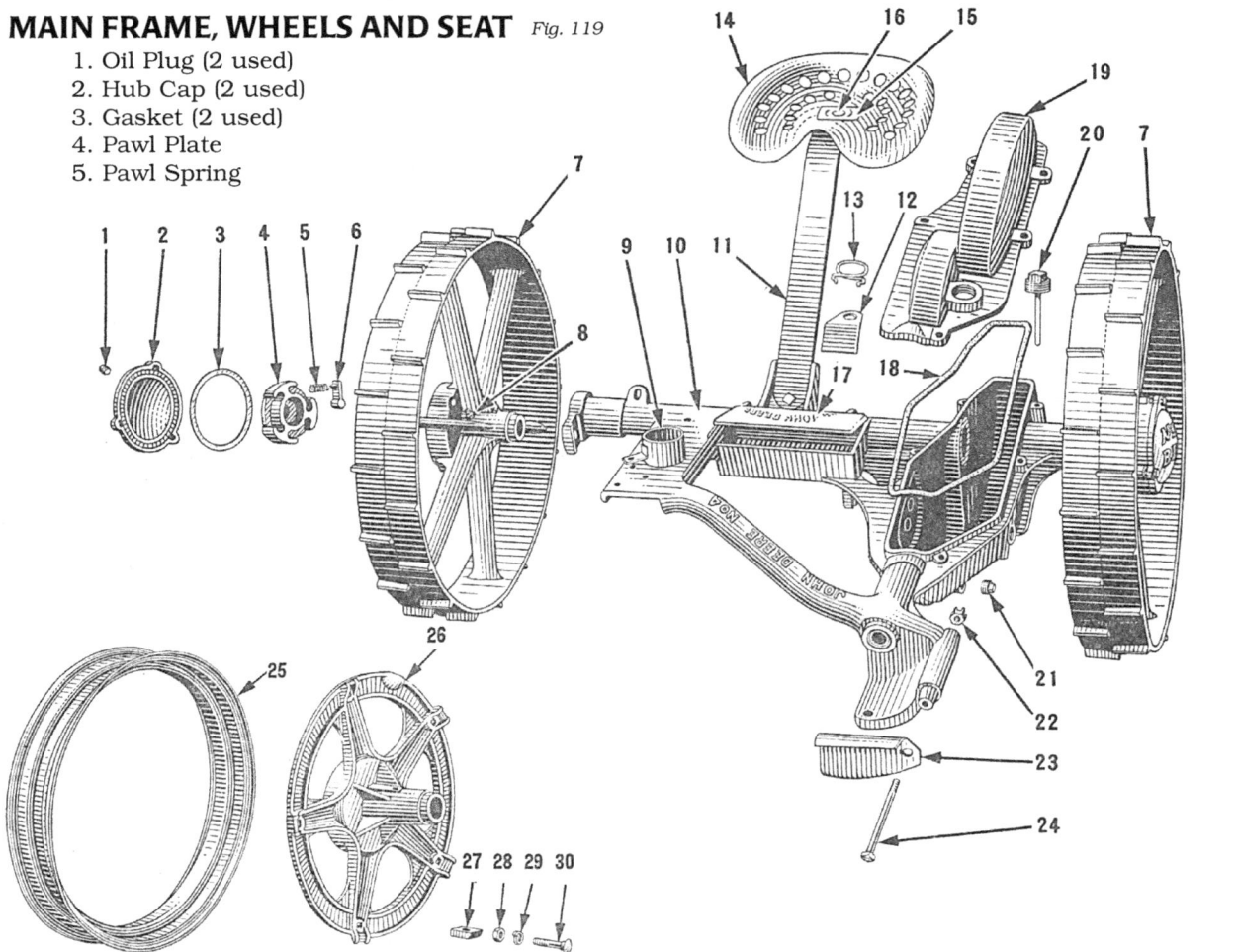

6. Pawl (6 used)
7. Main Wheel Regular
   Main Wheel Center Lug
8. Alemite Fitting
9. Oil Can Holder
10. Main Frame (Z854 H, Complete)
    Main Frame (Order AZ1399 H)
11. Seat Spring
12. Alemite Gun Holder, Lower
13. Alemite Gun Holder, Upper

14. Steel Seat
15. Seat Plate
16. Seat Bolt
17. Cover (tool box)
18. Gasket
19. Cover (gear box)
20. Filler Plug
21. Drain Plug
22. Horned Nut
23. Shield

24. Bolt
25. Wheel Rim (2 used)
26. Wheel (2 used)
27. Lug
28. Nut
29. Lock Washer
30. Bolt

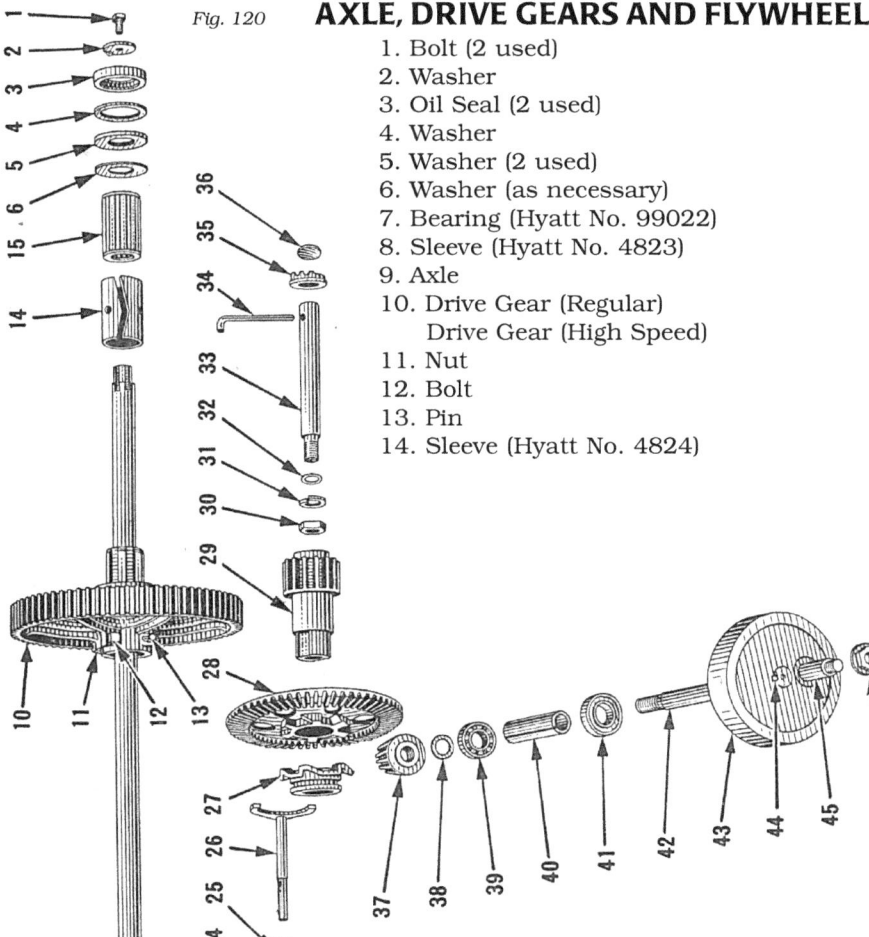

Fig. 120

AXLE, DRIVE GEARS AND FLYWHEEL

1. Bolt (2 used)
2. Washer
3. Oil Seal (2 used)
4. Washer
5. Washer (2 used)
6. Washer (as necessary)
7. Bearing (Hyatt No. 99022)
8. Sleeve (Hyatt No. 4823)
9. Axle
10. Drive Gear (Regular)
    Drive Gear (High Speed)
11. Nut
12. Bolt
13. Pin
14. Sleeve (Hyatt No. 4824)
15. Bearing (Hyatt No. 99023)
16. Bushing
17. Spring
18. Washer
19. Sleeve
20. Bolt
21. Nut
22. Clutch Lever
23. Leather Washer
24. Cup Washer
25. Spring
26. Throw-Out Yoke
27. Clutch
28. Bevel Gear
29. Spur Pinion (Regular)
    Spur Pinion (High-Speed)
30. Nut
31. Lock Washer
32. Copper Washer
33. Countershaft
34. Pin
35. Collar
36. Expansion Plug
37. Bevel Pinion
38. Washer
39. Bearing
40. Bushing
41. Oil Seal
42. Crankshaft
43. Crank Wheel with Pin & Nut
44. Key
45. Pin and Nut
46. Nut

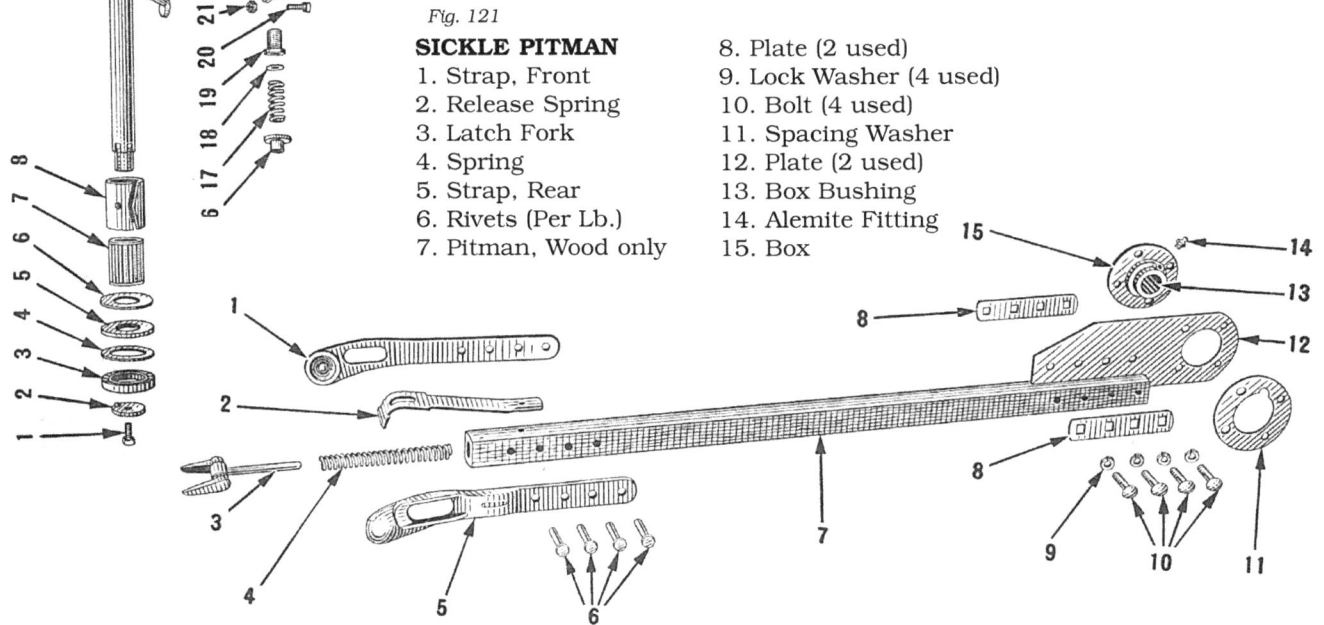

Fig. 121

SICKLE PITMAN

1. Strap, Front
2. Release Spring
3. Latch Fork
4. Spring
5. Strap, Rear
6. Rivets (Per Lb.)
7. Pitman, Wood only
8. Plate (2 used)
9. Lock Washer (4 used)
10. Bolt (4 used)
11. Spacing Washer
12. Plate (2 used)
13. Box Bushing
14. Alemite Fitting
15. Box

Fig. 122

DRAG BAR, BRACE BAR AND YOKE

1. Adjusting Nut
2. Washer
3. Drag Bar
4. Yoke Washer
5. Yoke Washer
6. Yoke
7. Alemite Fitting
8. Yoke Washer
9. Yoke Washer
10. Collar
11. Pin
12. Collar
13. Eccentric
14. Brace Bar
15. Connection

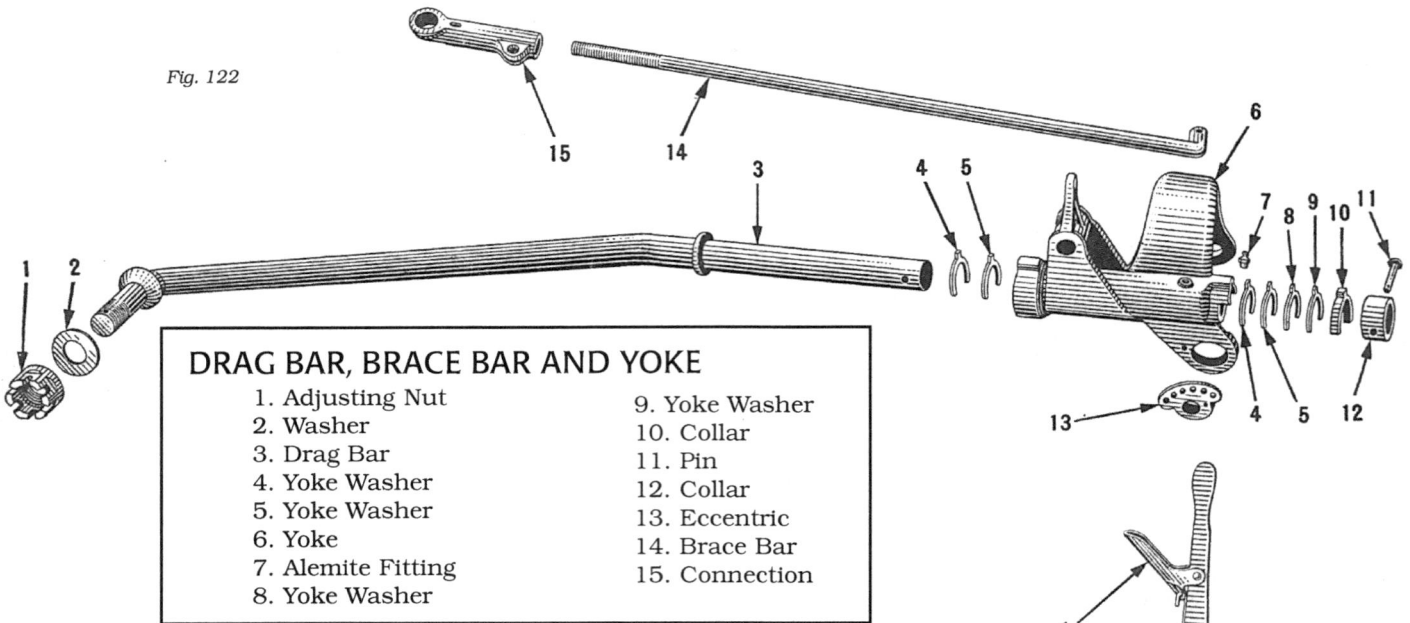

LIFTING LEVER   Fig. 124

1. Hand Latch
2. Detent Rod
3. Lever only
   Lever Complete
4. Lever lunger
5. Spring
6. Spring Stop
7. Sector with Stud
8. Foot Lever
9. Pin (7/12 x 1-3/8 inches)
10. Bolt
11. Nut
12. Spring with Nut
13. Bolt (7/16 x 2 inches)
14. Spring Support Arm
15. Lifting Rod
16. Lifting Crank
17. Lifting Crank Arm
18. Bolt (½ x 1-1/2 inches)
19. Chain Hook
20. Chain Link
21. Chain Eyebolt with Link
    Chain Link with Hook and Eyebolt
22. Chain Connection

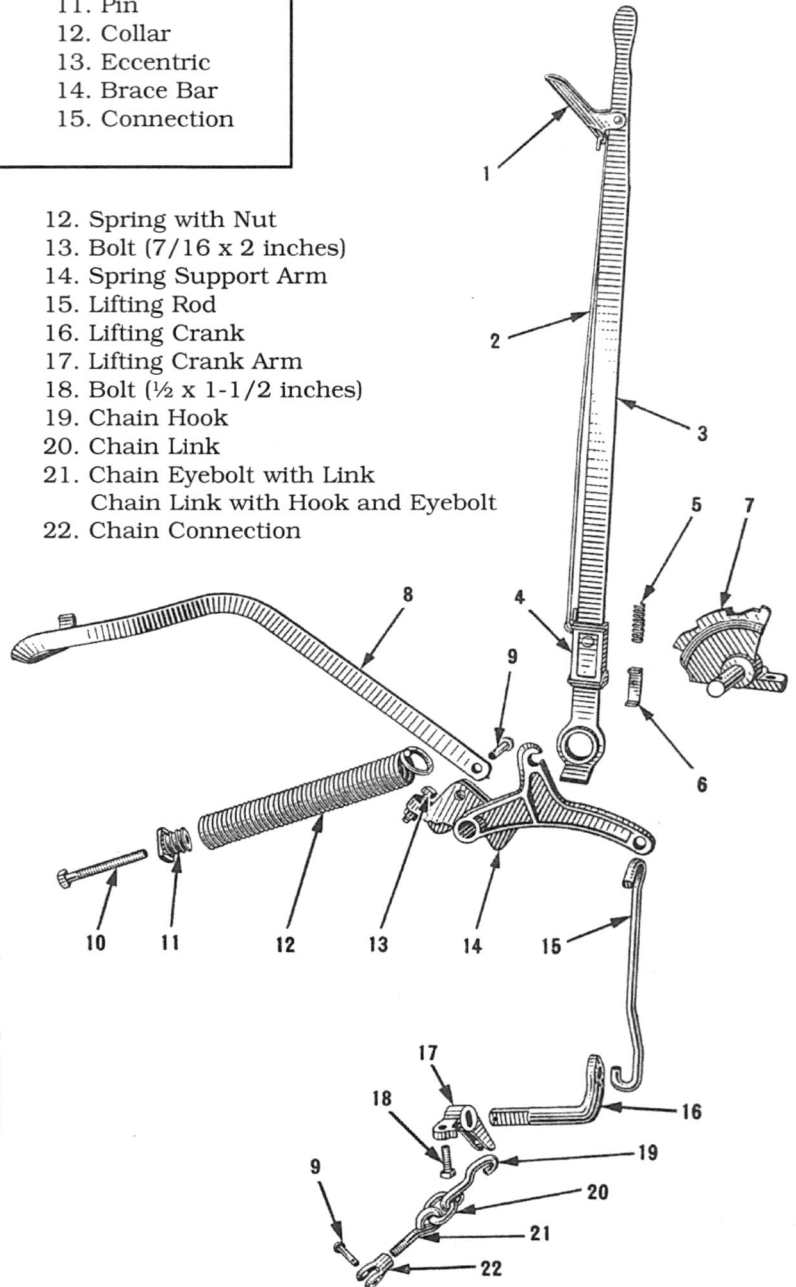

TILTING LEVER   Fig. 123

1. Tilting Rod
2. Bolt (½ x 3 with 1-7/8 inches Thread)
3. Spring (19/32 x 1-7/8 inches long, 9 Coils)
4. Tilting Lever
5. Tilting Sector, Complete

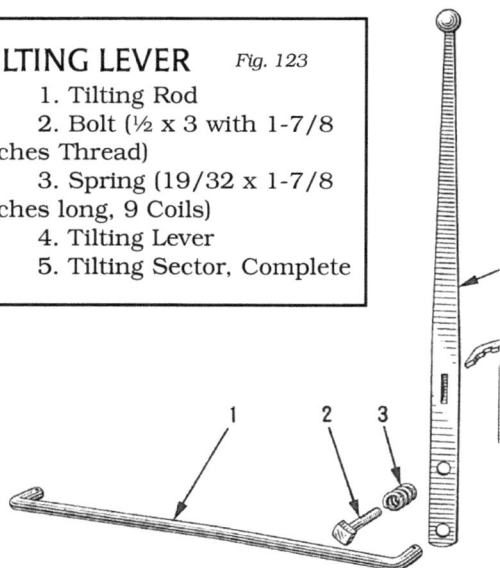

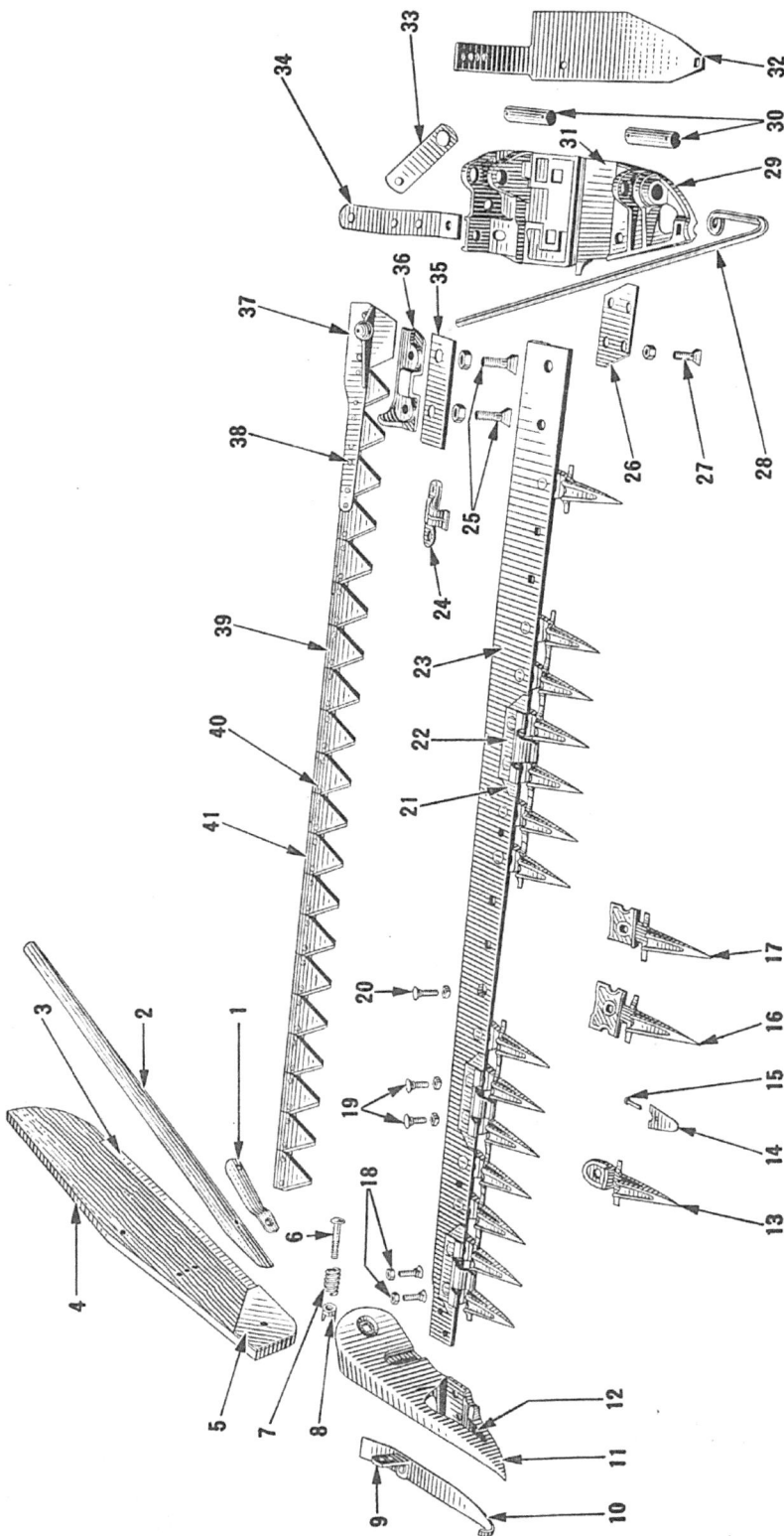

CUTTER BAR *Fig. 125*

1. Stick Socket
2. Grass Stick
3. Wearing Strip
4. Grass Board
5. Plate
6. Bolt
7. Spring
8. Horned Nut
9. Subsole Bracket
10. Subsole
11. Outside Shoe
12. Ledger Plate on outer Shoe
13. Guard
14. Ledger Plate
15. Ledger Plate rivets, per Lb.
    Ledger Plate Rivets, per ½ Lb.
16. Guard Steel
17. Outside Guard Steel
18. Bolt (7/16 x 1-1/4-inch Rd. C. S. Hd.)
19. Guard and Clip Bolt, Reg. Guard
    Guard and Clip Bolt, Steel Guard
20. Guard Bolt, Reg. Guard
    Guard Bolt, Steel Guard
21. Wearing Plate
22. Knife Holder
23. Cutter Bar, came in reg. & hvy. dty.,
lengths 4 ¹/2' through 7'
24. Inner Knife Holder
25. Bolt (5/8 x 2-1/4-inch, Sq. C. S. Hd.)
26. Head Guide, Front
27. Bolt (½ x 1-1/2-inch, Sq. C. S. Hd.)
28. Fender Rod for Shoe
29. Inner Shoe
30. Shoe Pin (2 used)
31. Ledger Plate
32. Sole for Shoe
33. Brace, 4-1/2, 5-foot Brace, 6' and 7'
34. Chain Standard, 4-1/2', 5', 6' and 7'
35. Wearing Plate
36. Knife Head Guide, Rear
37. Knife Head
38. Knife Head Rivets, per Lb.
39. Knife
40. Section Rivets, per Lb.
41. Section, Smooth
    Section, Top Serrated
    Heavy-Duty, Smooth
    Section, Reg., Underserrated
    Section, Heavy-Duty Underserrated

POLE, NECKYOKE, DOUBLETREE, DRAFT ROD AND BRACKET

Fig. 126

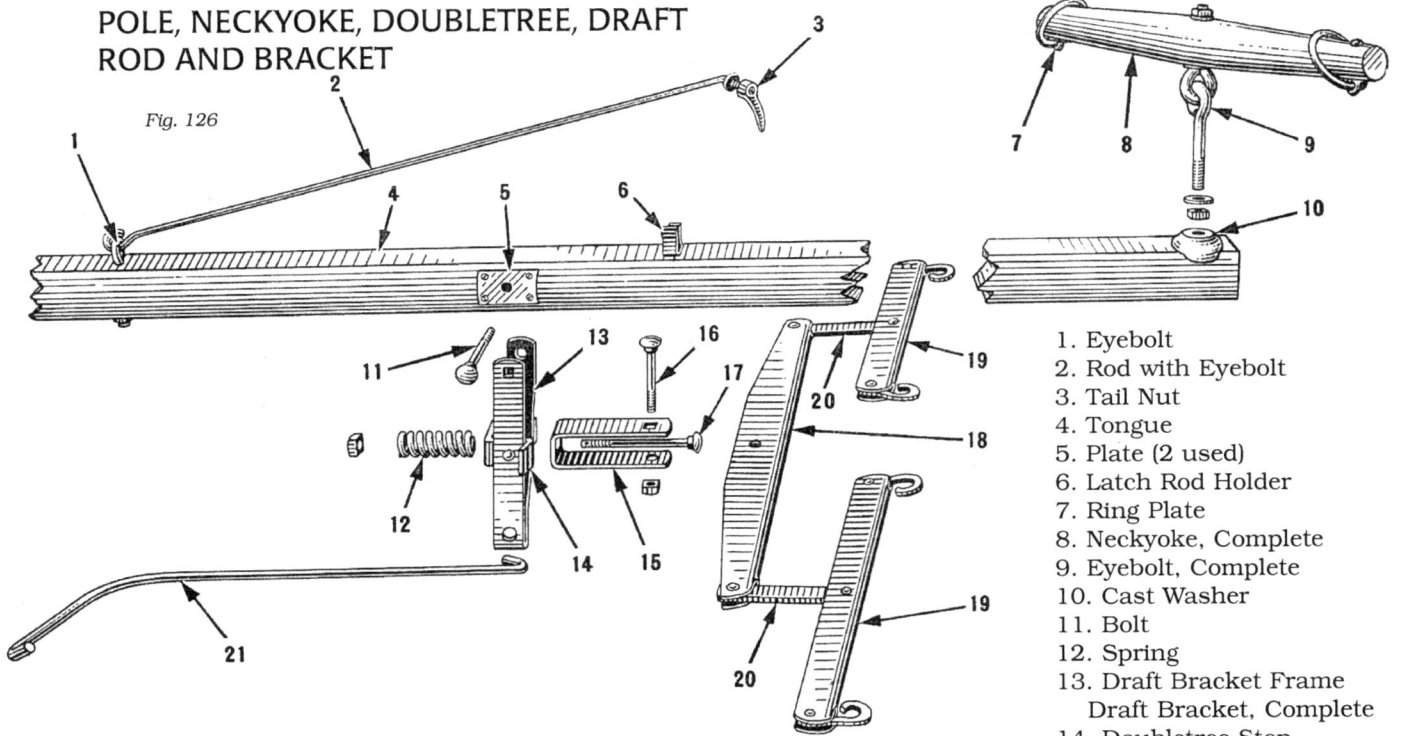

1. Eyebolt
2. Rod with Eyebolt
3. Tail Nut
4. Tongue
5. Plate (2 used)
6. Latch Rod Holder
7. Ring Plate
8. Neckyoke, Complete
9. Eyebolt, Complete
10. Cast Washer
11. Bolt
12. Spring
13. Draft Bracket Frame
    Draft Bracket, Complete
14. Doubletree Stop
15. Clevis
16. Bolt
17. Bolt
18. Doubletree Bar
    Doubletree, Complete
19. Singletree (2 used)
20. Strap
21. Draft Rod

TONGUE TRUCK   *Fig. 127*

1. Stub Tongue
2. Pin (2 used)
3. Stub Bracket, R. H. (2 used)
4. Stub Bracket, L. H. (2 used)
5. Holder
6. Stub Bracket
7. Bolt
8. Horned Nut
9. Axle Crank
10. Crank Box (2 used)
11. Crank Yoke
12. Axle Bracket
13. Wheel (2 used)
14. Wheel Bearing (2 used)
15. Grease Cup (2 used)
16. Washer
17. Pin
18. Axle Bearing (2 used)
19. Axle, Complete
20, Hinge

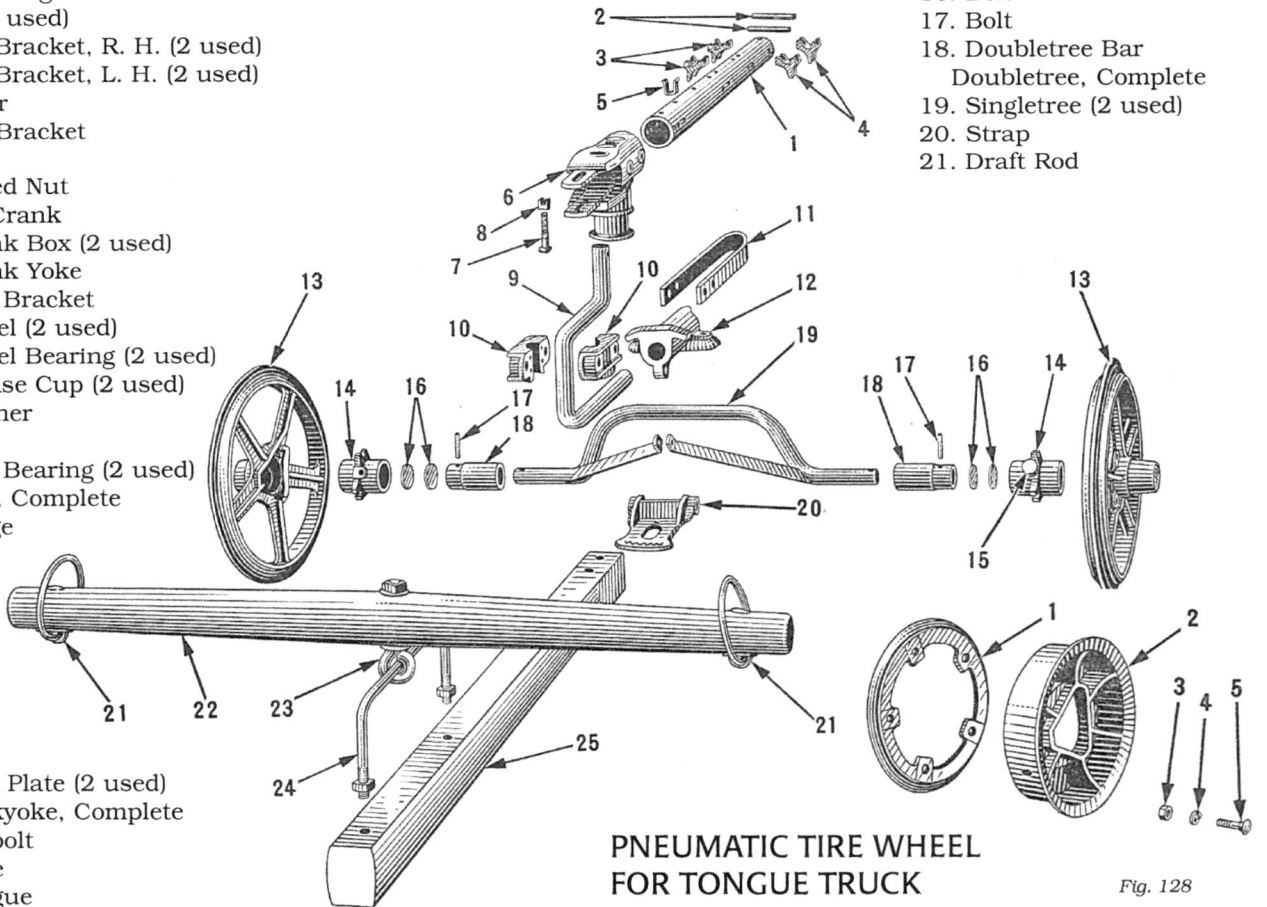

21. Ring Plate (2 used)
22. Neckyoke, Complete
23. Eyebolt
24. Slide
25. Tongue

PNEUMATIC TIRE WHEEL FOR TONGUE TRUCK

Fig. 128

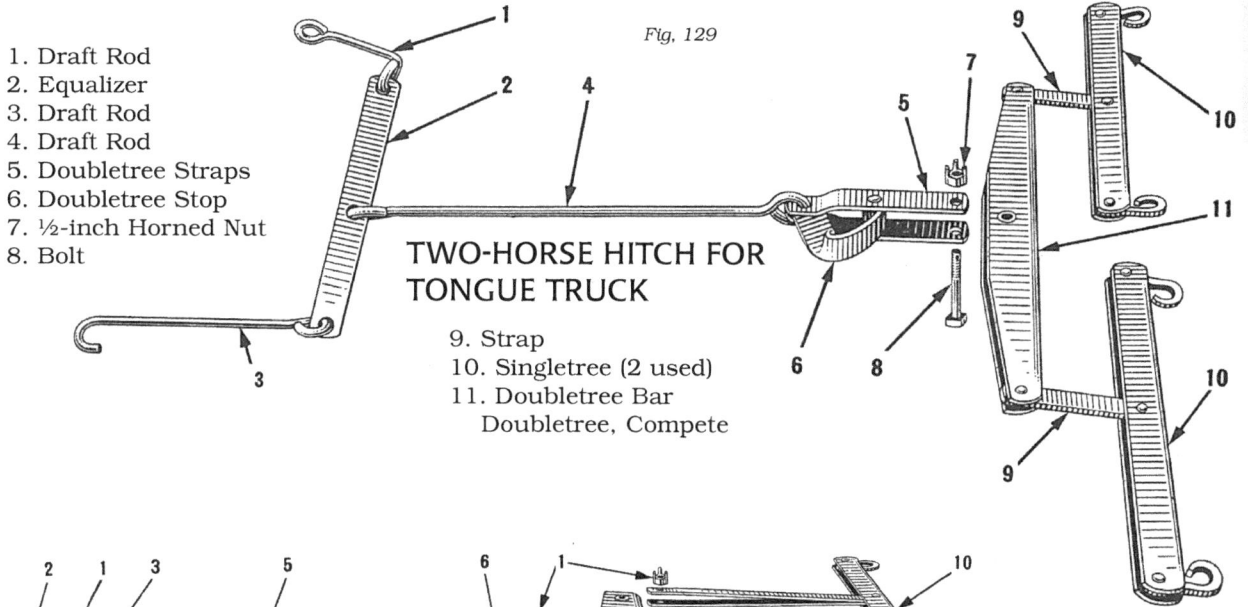

Fig, 129

1. Draft Rod
2. Equalizer
3. Draft Rod
4. Draft Rod
5. Doubletree Straps
6. Doubletree Stop
7. ½-inch Horned Nut
8. Bolt

TWO-HORSE HITCH FOR TONGUE TRUCK

9. Strap
10. Singletree (2 used)
11. Doubletree Bar
   Doubletree, Compete

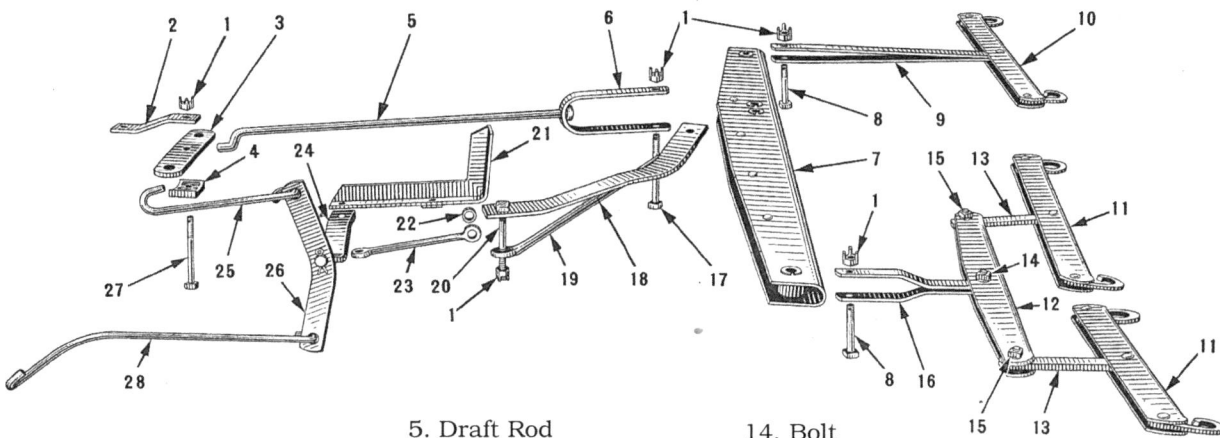

Fig. 130.

THREE-HORSE HITCH FOR TONGUE TRUCK

1. Horned Nut
2. Hammer Strap
3. Equalizer
4. Spacer

5. Draft Rod
6. Clevis
7. Evener Bar
8. Bolt
9. Straps
10. Singletree
11. Singletree (2 used)
12. Doubletree Bar
   Doubletree, Complete
13 Strap

14. Bolt
15. Bolt (2 used)
16. Strap
17. Bolt
18. Evener Support
19. Brace
20. Bolt
21. Stop angle
22. Trunnion

23. Brace
24. Support Pivot
25. Draft Rod
26. Equalizer
27. Bolt
28. Draft Rod

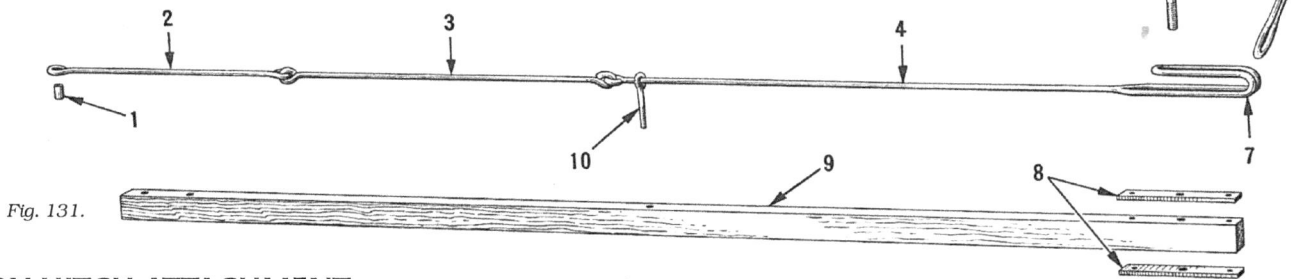

Fig. 131.

OX HITCH ATTACHMENT

1. Pipe Spacer
2. Rod Rear (Stiff Pole)
   Rod Rear (Tongue Truck)
3. Rod, Middle

4. Rod, Front
5. Front Eyebolt
6. Link
7. Complete Rod (Stiff Pole)
   Complete Rod (Tongue Track)
8. Plate (2 used)

9. Tongue (Stiff Pole)
   Tongue (Tongue Truck)
10. Rear Eyebolt

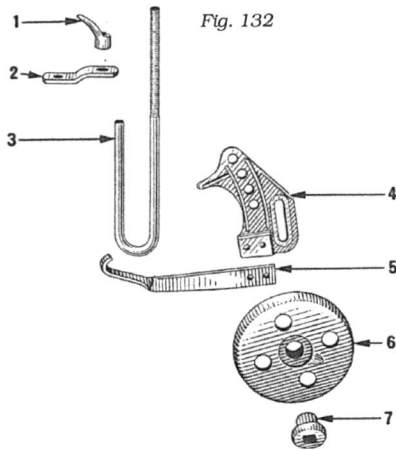

Fig. 132

BAR WHEEL HIGH CUTTING ATTACHMENT

1. Nut
2. Bracket
3. Hook Bolt
4. Bracket
5. Subsole
6. Wheel
7. Bearing
   Attachment

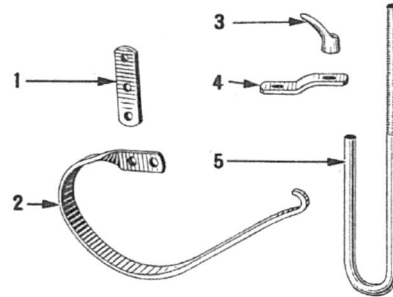

RUNNER TYPE HIGH CUTTING ATTACHMENT

1. Bracket
2. Subsole
3. Nut
4. Bracket
5. Hook bolt
   Attachment

Fig. 133.

PARTS SPECIAL FOR COWPEA CUTTER BAR ATTACHMENT

1. Knife, Smooth, 4-1/2-foot
Knife, Smooth, 5-foot
2. Section

3. Guard
4. Guard Plate
Bar, Complete, 4-1/2-foot
Bar, Complete, 5-foot

Fig. 134.

Fig. 135.

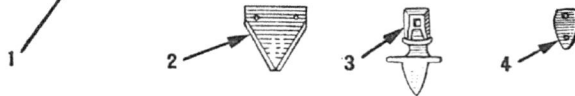

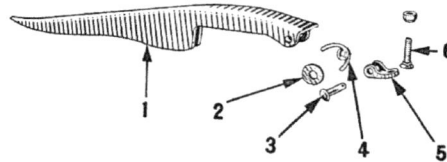

Fig. 136.

LIFTING GUARD

1. Guard, Complete
Guard, only
2. Spacer
3. Pin (3/8 x 1-3/4-inch
4. Spring
5. Guard Hinge
6. Bolt (7/16 x 1-3/4-inch Rd. C. S. Hd.)

SPECIAL LIFTING SPRING BRACKET

1. Bracket
2. U-Bolt

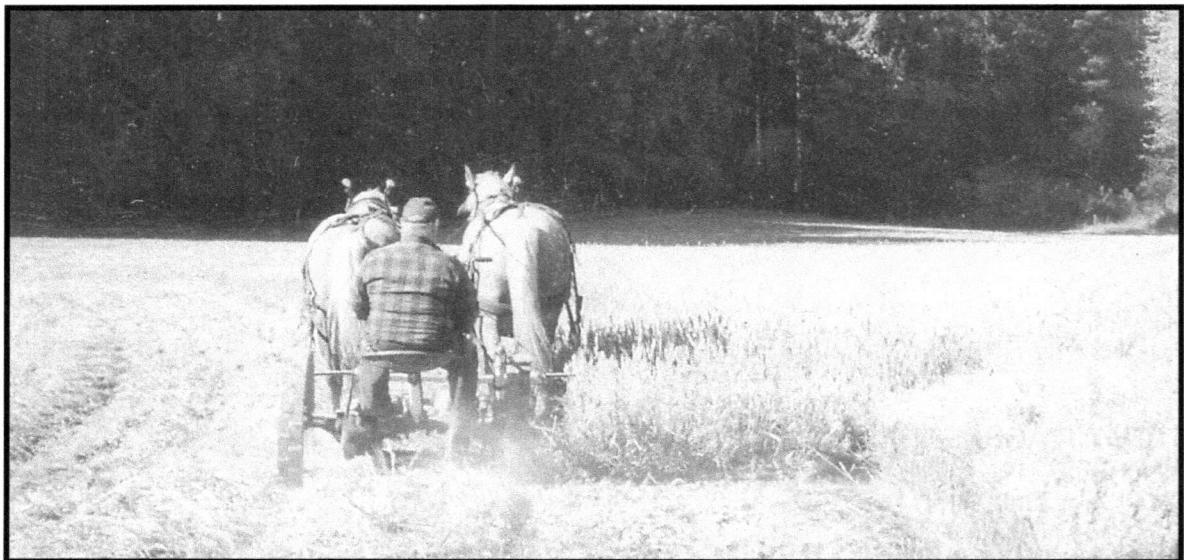

Fig. 137. Montana's Bulldog Fraser with two Percheron colts in training on a McCormick #7 mower.

REPAIRING AND ADJUSTING THE CUTTING PARTS OF A MOWER

The information below comes from Shopwork on the Farm, by M. M. Jones and gives a more general repair view applicable to most every make of mower cutting bar. Some of this information is a slight repeat of what has appeared in the manufacturer's information just presented, but in the interests of being thorough and having this book be usefully accessible to the farmer it appears under this heading also.

Heavy draft, ragged cutting, and excessive wear and breakage can often be avoided by a few simple adjustments and replacement of parts of the cutting mechanism of a mower.

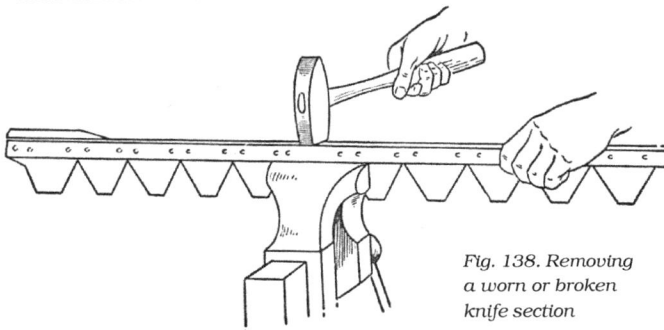

Fig. 138. Removing a worn or broken knife section

Replacing Knife Sections

To remove a broken or worn knife or sickle section, support the knife rib or bar firmly, and strike the back of the section one or two sharp blows with a

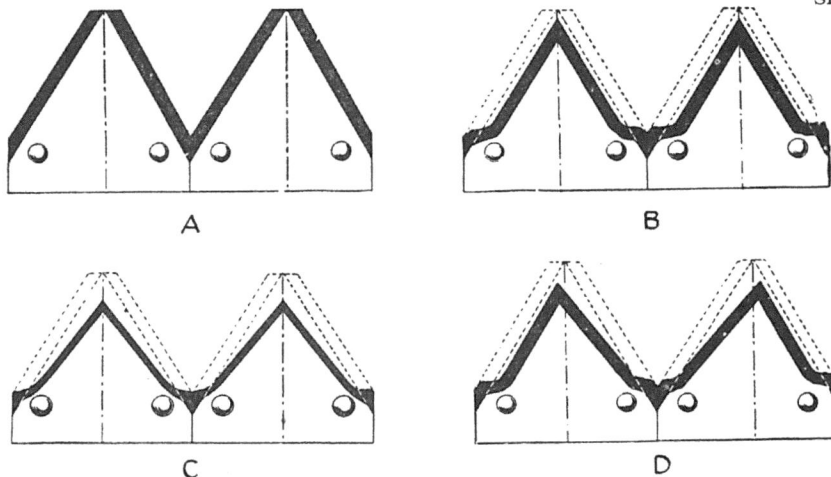

hammer. A good way to support the knife is to clamp it loosely in a vise, pointed end of the sections down (see Fig. 138).

After the broken section is removed, then punch the sheared rivets from the holes in the knife bar, being careful not to enlarge the holes. Then put the new section in place, insert the rivets, and rivet them down. Strike one or two heavy blows first to swell the rivets, and then form the heads by light peening with the ball peen of the hammer, or with a rivet set.

Sharpening a Mower Knife

A dull or improperly ground knife causes ragged cutting, rapid wear, and extremely heavy draft. Three points are very important in grinding a knife:

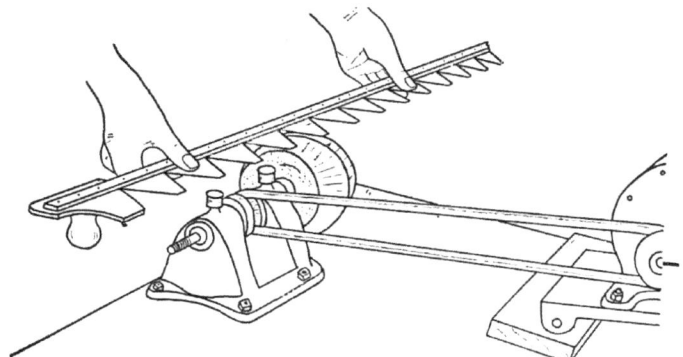

Fig. 140. Mower sickles are ground easiest on special grinders made for that purpose, although thye can be satisfactorily sharpened on an ordinary grinding wheel or with an electric hand grinder.

1. Maintain the original width of bevel (see Fig. 139). A narrow, blunt bevel does not cut easily; and a wide, keen bevel nicks easily.

2. Maintain the original angle of shear, or the angle between the cutting edge of the section and the guard plate. Otherwise, the grass will tend to slip away and not be cut.

3. Do not over heat and draw the temper.

Mower knives can best be sharpened on special grinders or grinding wheels made for that purpose (see Fig. 140). With such a grinder, it is easy to maintain the original bevel and angle of shear. With a little practice and patience, however, mower knives can be ground satisfactorily on regular grinding wheels. Grinding may be done on either the flat vertical side or on the regular curved grinding surface. Motor-driven grinders are much faster and require less work than hand--or foot--operated grinders.

Straightening a Mower Knife.
If a knife bar is bent, either edgewise or flatwise, it will bind as it works back and forth in the

Fig. 139. Right and wrong ways to grind mower knives. Dotted lines show outlines of new sections. A. New sections with proper width of bevel and angle of shear. B. Sections properly ground. Proper width of bevel and angle of shear are retained. C. Sections improperly ground. The bevels are too narrow, and the angle of shear is wrong. D. Proper width of bevel, but improper angle of shear.

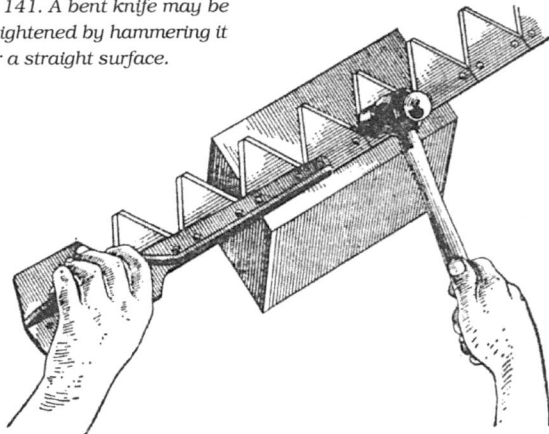

Fig. 141. A bent knife may be straightened by hammering it over a straight surface.

cutter bar and cause both rapid wear and increased draft. To straighten a knife, sight along the knife bar to locate the bend. Then place it on some straight surface, as a bench top or mower tongue, with the bend or bulge up, and strike with a hammer (see Fig. 141). Sight again and hammer more as may be required. Be sure to check the knife bar for bends both edgewise and flatwise.

Replacing Guard Plates.

A guard plate, also called *ledger plate* serves as one blade of the shears and should be kept in good cutting condition just the same as the knife section. When guard plates become worn and dull or nicked or

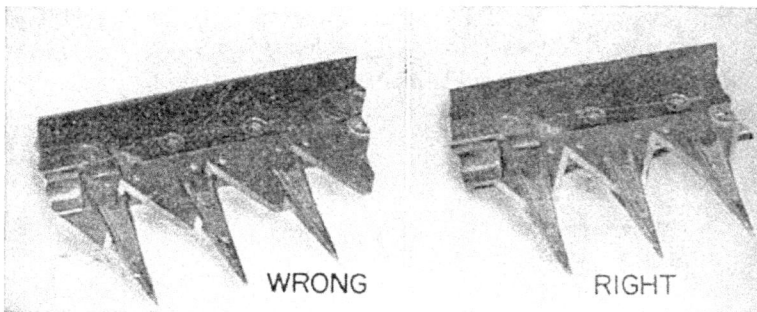

Fig. 142. Knife sections should register or center in the guards at the ends of the pitman stroke.

broken, remove them and install new ones.

Guard plates may be removed with the guards either on or off the cutter bar. To remove a guard plate, firmly support the guard from beneath. A special guard-repair anvil (see Fig. 117, page 134) is excellent for this purpose. Drive the guard rivet down from the top, using a stout punch to start the rivet and a slim punch to finish removing it. Then insert a new guard plate and rivet, and securely brad the end of the rivet. If there is any part of the rivet projecting above the guard plate, trim it off smooth with a sharp cold chisel. New rivets may be inserted from the bottom or from the top of the guard.

Guards should be aligned frequently, for they often strike stones, sticks, or other obstructions and become bent. Even a nearly new mower is likely to need some guards bent back in line. Probably the best

way to align guards is to insert a straight knife and hammer the guards that are out of line, bending them up or down as may be required. Strike on the thick part of the guard just ahead of the guard plate. Pound the high ones down first, and then bring the low ones up (see Fig. 87, page 125). Be sure the guard bolts are tight, and remember in hammering to make the guard plates line up. It is now important if the points are somewhat out of line

Adjusting Other Cutter Bar Parts.

The cutter bar is the heavy steel bar to which guards and other parts are attached. It should not be confused with the knife bar, which is the small bar or rib to which sickle sections are riveted.

The parts of the cutter bar form a sort of groove or trough in which the knife works back and forth. Not only should the knife be straight, but these parts on the cutter bar should be aligned and form a straight place in which the knife can work. Also, they should be adjusted to fit the knife. They should not fit too tightly and therefore bind. Neither should they fit too loosely and allow the knife to bounce or flop about. The knife sections and the guard plates should fit together snugly and form sharp shearing edges, just the same as the two blades of a pair of scissors should fit together reasonably tight.

The knife hold-down clips should almost but not quite touch the knife when it is resting on the guard plates. To adjust the clips, simply hammer them up or down, but be sure the knife is not in place under a clip when it is being hammered down (see Fig. 87). When a thin piece of tin can be just slipped under the clip, it may be considered in good adjustment.

The wearing plates, which support the back edge of the knife, are replaceable. When they become worn, replace them with new ones. Always adjust new wearing plates so their front edges just touch the back of the knife bar. The wearing plates are held in place with guard bolts, and the holes through which the bolts go are slotted. It is therefore a simple matter to loosen the bolts and adjust the plates forward or backward until they all line up.

Worn knife-head guides are a common cause of knife breakage, as well as poor cutting near the inner end of the knife. These parts should therefore be adjusted or replaced whenever looseness develops.

It is important that all guards and other cutter-bar parts be kept tight. Tight-fitting strong wrenches, such as socket wrenches, are best for tightening cutter-bar bolts.

Aligning a Cutter Bar

A cutter bar is in proper alignment if (1) the pitman is square with, or at a right angle to, the pitman drive shaft, and (2) if the pitman pushes and pulls straight on the sickle.

To offset the backward strain when cutting and to make the pitman and knife run straight, the cutter bar is given a certain amount of lead. That is, when the mower is standing still, the outer end of the cutter bar is slightly ahead of the inner end. The proper amount of lead is about 1/4 in. per ft. of cut.

To check the pitman angle, place a square or other straight edge against the front face of the pitman wheel. If the pitman is parallel to the edge of the square, it is square with the pitman drive shaft. To change or adjust the pitman angle, adjust the tie rod in front of the pitman or the diagonal push bar behind it so as to move the inner shoe of the cutter bar forward or backward as may be needed. This adjustment will also affect the register of the knife. Therefore, check the register (Fig. 142) before making this adjustment. If some parts have been sprung and thus allow misalignment of the pitman with the pitman drive shaft, these parts may have to be straightened or replaced.

To check the lead of a cutter bar, place the mower on level ground, block the wheels, raise the end of the tongue 31 in. from the ground, and pull the outer end of the cutter bar back as far as it will go. Hold a string against the front of both wheels, stretch it tight and straight on out to the end of the cutter bar (see Fig. 53). Note you measure to the back edge of the knife (not the cutter bar) from the string at the inner end and the outer end of the knife.

An eccentric busing is provided on some mowers for adjusting cutter-bar alignment (see Fig. 93). On many mowers no adjustment is provided. On these mowers, and also frequently on mowers having an eccentric busing, it will be necessary to determine just what causes the lag in the cutter bar, and then remove the cause. In many cases, it is wear on the hinge pins of the inner shoe. New hinge pins may need to be installed, or possibly the holes drilled oversize and oversize pins installed. Sometimes parts of the mower have been sprung, and these must be straightened or replaced. Another cause of misalignment is worn bolt holes and bolts that fasten the cutter bar to the inner shoe. In such a case, the bar and inner shoe may be welded or brazed in proper position.

Lengthening the diagonal push bar behind the pitman or shortening the tie bar in front of the pitman is not satisfactory for restoring lead. Neither of these adjustments will improve the angle between the pitman and the sickle. Furthermore, they will change the angle between the pitman and the pitman drive shaft.

Adjusting Knife Register.

By register of the knife is meant the centering of the knife sections in the guards at the ends of the knife strokes (see Fig. 142). If the knife does not register, the mower will do an uneven job of cutting, it will choke easily, and the draft will be heavy. If a knife is out of register, first check to be sure the pitman straps are properly tightened and that there is not excessive play in the bearings at the ends of the pitman. Then if the knife is still out of register, move the whole cutter bar in or out as may be necessary. To do this, shorten

or lengthen the tie bar in front of the pitman, and also move the back of the inner shoe yoke in or out on the diagonal push bar the same amount. Various methods are provided for adjusting the position of the yoke on the diagonal push bar. On some mowers, washers may be shifted; on others, screw threads are provided; and on other mowers, still other methods are used.

Restoring Mowers

This author has worked over a dozen years restoring HD mowers for use and resale. We choose to share a few pictures of the McD #9 tricks and goals. To determine if an old mower is worthy of restoration we first put the frame up on jacks or blocks so that the wheels may turn free.

If you grab the top of the wheel and rock it and there is no free play, that is good. If there is free play you must try to determine if it is the wheel loose on the axle or the axle loose in the frame. Wheel loose on the axle may be remedied, whereas axle loose in the frame is bad news. In our high-grading process, if we found a mower with axle loose in the frame it would be parted out, considered not worthy of rebuilding.

The next thing to inspect is the gear box. Remove the lid and look inside. If it is full of gunk and garbage or dry you will need to clean it and fill it with oil before you can do the next checks. Use small brushes and wash with gasoline or diesel fuel. Then replace drain plug and fill to axle level with oil (we prefer to use 10w hoist oil). Remove pitman stick from pitman flywheel, put the mower in gear and turn the wheel. The pitman flywheel should spin free and easy. Now grab a hold of that flywheel and see if you can feel any play. It should be snug. Some people think that the replacement of the oil seal is all that is necessary to correct any play. Not so. If the mower has been drug behind a tractor at high speeds and/or without sufficient oil the pitman shaft housing in the frame can get wallowed out or the pitman shaft may be scored, in either case it is too expensive to machine and is a candidate for parting out.

If the mower has passed these tests, the next thing to check are the yoke and pin clearances for the cutter bar. Take hold of the end of the cutter bar, when it's flat on the shop floor, and gently push and pull it, You may have to have another person hold the left wheel to keep the mower from falling off the jacks. If the cutter bar attachment feels loose this will need attention because it can throw off any timing. It may be as simple as replacing a pin with a slightly larger one.

At this point you've checked the big things and should be able to determine whether or not you have a keeper or a junker.

If you're just going to take it to the field the information already presented in this chapter can help you to get things right.

If your goal, however, is to restore the mower to pristine condition here's a suggested outline of additional steps following what has already been suggested

Fig. 143. Mower on jacks with gear box lid removed.

Fig. 144. Pedal position indicates mower is in gear, A view of the uncleaned gear box.

Fig. 145. The cleaned gear box.

Fig. 146. Making a template to use to drill holes in a new tongue.

in the diagnostics.

Take a written inventory of what parts are broken or missing from the mower so you may start to gather those up.

Next disassemble the mower completely with the exception of the frame, axle, gears, and pitman shaft assembly. Remove the cutter bar from the frame. Then, using a heavy duty gear puller, remove the wheels. On the #9 the wheels slide over the axles and are kept in place by the wheel hub and pin. To remove the wheel you must remove the hub. (Sometimes it is necessary to strike the puller head with a sledge hammer or heat the wheel hub with a torch. Remember it is cast iron and can break or change shape from excessive heat.

When you have the wheel hubs off you'll see the spring and pawl assemblies. These hubs are right and left. Mark them to make sure they go back right. Check the springs to see if they are okay and replace if necessary. Clean the inside of the wheel and the wheel hub assemblies. After they are painted you will be coating their insides with wheel grease.

Next remove the lifting and tilting handle assemblies. And finally the seat.

Now, carefully restrict the turning of the pitman shaft by pulling up a bent screwdriver end between the ring gear and pinion gear teeth in the mower gear box. While someone holds that in place put a big pipe wrench on the flywheel end of the pitman and unscrew the pinion gear. Watch for the thin washer(s). You may now carefully, gently, pull the pitman shaft forward and free of the frame. You must try to avoid disturbing the oil seal(s) because they may be alright to use again or in bending them you may score the oil seal seat and create a place for a leak. Now's the time to check the oil seal. If it looks good and you had no oil leaks, put the pitman back in and screw the pinion gear back on.

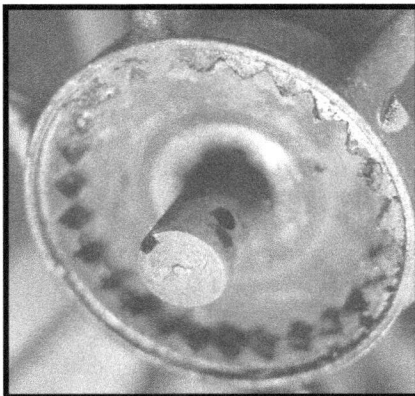
Fig. 145. Inside of wheel showing a view of uncleaned axle end.

Fig. 146. Cleaned wheel and axle.

Fig. 147. McD #9 wheel hub as it came off, uncleaned.

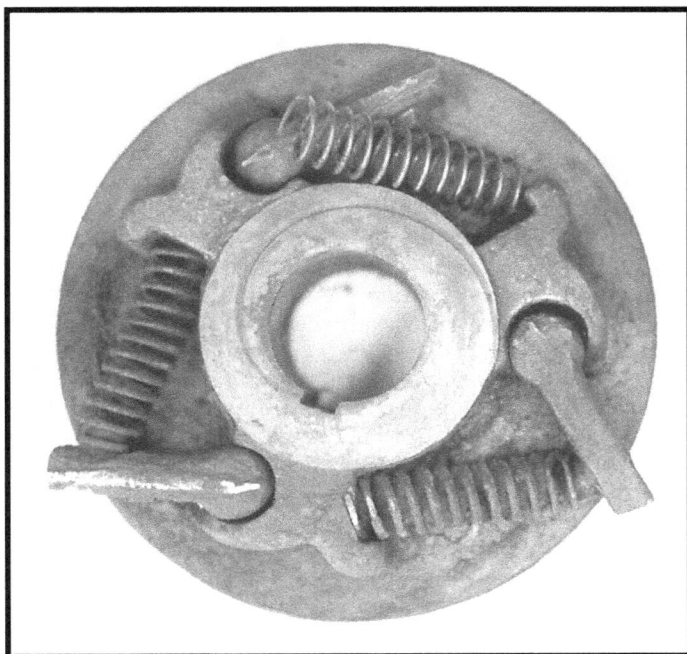
Fig. 148. Same hub cleaned, one spring lays on top to allow view of mechanism.

If it looks bad or was leaking, you will need to replace the oil seal. With either the old one or the dimensions of shaft and housing you can get aftermarket oil seals for your mower. In some cases these new oil seals are thin enough that it is possible to put two in where one once was. This, although not strictly necessary, is advisable.

With help you can hold up a piece of wood into the tongue chamber and mark it from above to give you a template for drilling a tongue. When you are under a McD mower look for raised numbers which will tell you the date your mower came off the assembly line.

Wash all the disassembled parts with gasoline or diesel. Remove most of the paint if there is any, paying close attention to where paint may have been applied over grease or dirt. If your mower is very rusty you may want to consider applying a coat of rust neutralizer. There are lots of products on the market. We use

something called *Ospho* which is a liquid equivalent of naval jelly. It is a toxic compound and care must be taken with its use. When it is painted on the rusty metal and allowed to dry, a milky white primer surface results ready for paint!

Apply two or three coats of paint sutiable for metal. And reassemble the mower. This is an excellent time to upgrade the cutter bar parts, put in a new tongue and pitman and shop around for neat decals. It might be fun to punch in your initials and the date of your restoration.

Mower Review Checklist

What is the general principle of horsedrawn mower operation?

Motion is transferred from the wheels to the main axle to the knife by gears, shaft and pitman. Rotary motion of the ground wheels is converted into reciprocal motion of the knife.

What is the arrangement of gears?

There are three gears. A large internal gear mounted on the main axle drives the one-piece spur and bevel gear, which in turn drives the pinion on the flywheel shaft. The design is such that the thrust of the internal drive gear and spur pinion is balanced by the thrust of bevel gear and bevel pinion.

Where is the flywheel shaft bevel pinion located?

Between the bearings in the main frame with the large end of the pinion to the rear.

Why?

It gives a steady motion and reduces friction. Such tendency as exists for the pinion and flywheel to work forward and cause end thrust is overcome.

Why is the yoke built so strongly?

Because it takes the heavy strain of the cutter bar.

How is the yoke designed?

It is made extra wide, and is held between double lugs at the front and rear. There is a close fit between the large steel pins and bored holes of the yoke and shoes.

How is the mower put in or out of gear?

By means of a tooth clutch. The driving member of the clutch is mounted on and turns with the axle, the latter receiving its motion from both drive wheels. While in gear the clutch teeth mesh with teeth in the hub of the larger internal gear. Control of the clutch is by means of a short lever.

What are some points in design of the cutter bar and knife?

The cutter bar is reinforced its entire length, for strength, with a heavy rib. Holders and wearing plates keep the knife in the proper cutting position so as to make a clean shear cut. The knife sections travel back and forth through the guards, the latter being provided with raised ledger plates.

What is a ledger plate?

It is that part of the guard against which the knife cuts the grass. The edes of the plate are "serrated," which means notched or toothed somewhat as a saw.

What is meant by the proper alignment of the cutter bar?

When the knife, cutter bar and pitman are all on the same straight line.

Why is the cutter bar alignment so important?

To maintain a free and easy motion of the knife, by keeping it in line with the pitman, to keep the draft low, reduce wear, and to do a good job of cutting.

How is the cutter bar put in proper alignment?

The cutter bar is re-aligned by turning an eccentric collar on the rear inner shoe pin. This draws the outer end of the cutter bar forward, so that all lag due to unusual strains or wear in the shoe pin bearings is taken up. Knife and pitman are thus placed in alignment.

How are the guards aligned?

The knife is removed and by the use of a straight edge those guards that are either high or low can be seen. Those guards not too badly out of line can be driven back into place with a hammer.

Can knife sections be easily replaced?

Yes. Each section is held to the knife bar by two soft steel rivets. These rivets are readily cut off so that new sections can be attached.

What is meant by "lead"?

When standing idle the outer end of the cutter bar should be set ahead of the inner end a short distance, called the "lead".

How much lead should the cutter bar have?

Approximately 1-1/4 inches for each five feet of cutter bar length.

How is the proper lead determined?

Directions for this operation are as follows:

With the wheels blocked, place the pole, or as it is often called, "tongue" at the normal height, 32 inches from the ground, measuring underneath the front end of the pole. Run a cord across the face of the two ground wheels about 5 inches from the floor, straight out to a point beyond the end of the cutter bar. Place the knife on the extreme outward stroke. Pull the cutter bar back, to take up all slack. Measure from the cord to the rear straight edge of the knife at the sickle head end and at the extreme outer end.

When is a knife said to register?

When each knife section centers in its guard. That is to say, the sections are properly centered when each section is directly under a guard with the pitman at either end of its stroke.

If the cutter bar is "off register" how can the proper adjustment be made?

There are adjusting washers at both ends of the drag bar bearing in the yoke. To adjust for register, place more or less washers at the inner or outer end of the yoke. At the same time lengthen or shorten the brace-bar connection at the flywheel bowl.

Adjustment should not be necessary except for wear. This adjustment is usually taken up outwardly or by moving the washers from the outer to the inner end of the drag bar bearing and, of course, lengthening the connection at the flywheel bowl. "Off register" may also be caused when the pitman has been repaired and not kept at the right length. Such a pitman should be repaired for length or repalced, rather than adjusting the parts to make the knife register.

How is the high lift arranged?

The bar is raised at both ends with either foot or hand lever. In this position the machine can be operated; that is to say, it is not necessary to throw it out of gear.

How high does the foot lift raise the cutter bar?

From 8 to 11 inches at the inner end and from 25 to 35 inches at the outer end. This height is for turning, and to provide clearance for passing over ordinary obstructions.

How high does the hand lift raise the cutter bar?

The inner shoe is raised 13 inches and the outer shoe 44 inches.

What is the purpose of the lifting spring?

The large coil lifting spring not only assists in raising the cutter bar, but when properly adjusted, makes the cutter bar float. Because the weight of the bar is carried by the main frame through the lifting spring, the traction of the wheels is increased as well as the draft being reduced. Being spring carried, the cutter bar rests lightly on the ground.

What is the purpose of the tilting lever?

It raises or lowers the points of the cutter bar and so makes it possible to cut high or low as desired. The lever is made to use where stubble has to be run over, or where cutting over rough ground. By tilting the bar upward at the front, low obstacles are easily passed over. On the other hand, by lowering the guards, matted or tangled grass can be picked up and cut.

What is the purpose of the adjustment for the sub-shoes?

The sub-shoes under the inner and outer shoes may be raised or lowered to regulate the height of cut.

Why is the pole attached underneath the frame?

So that water cannot collect around the butt of the pole and cause it to rot.

Raking & Rake Setup

Chapter Ten

After the forage crop is mowed, to assure a timely and proper curing, it needs to be fluffed into a standing position to shield the leaves from too much solar bleaching and to allow drying air to pass freely through. The most common practice is to rake the hay in what are called windrows. The first way this was done was by hand with pitchforks. Various different designs of appartus were pulled by animals and bunched hay either in small stacks (cocks) or in long rows. These implements evolved into the Sulky or Dump rake as is shown below. Even to this day the dump hay has its place in certain operations but has largely been supplanted by the side delivery rake (seen above).

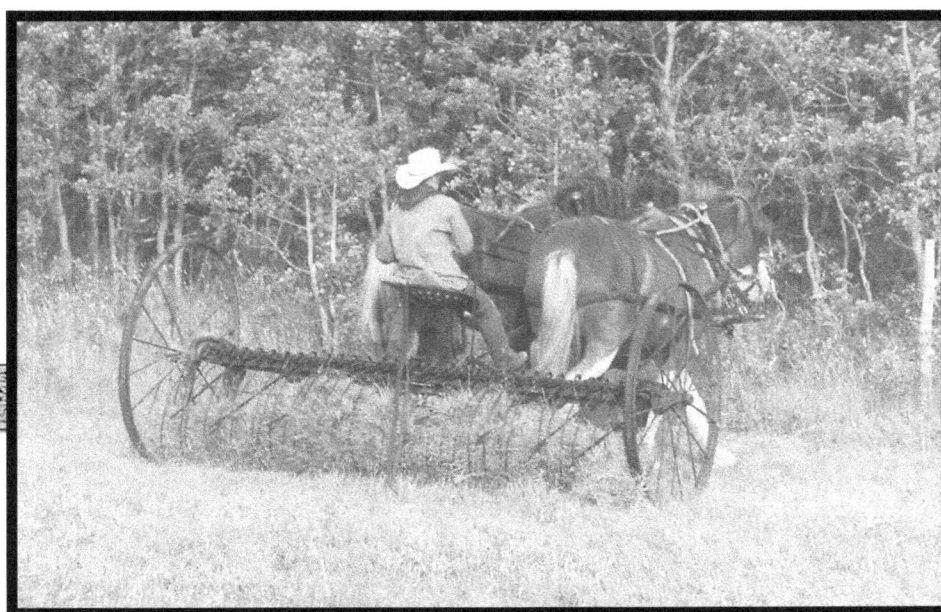

Laurie Hammill with two Clydesdales on sulky rake.

There is a third category of rake which might be called an UNrake but which is commonly called a tedder. This machine is like several pitchforks fluffing and spreading hay out rather than making windrows. Nowadays its primary value or utility is to spread out rained-on hay to help dry it. Although ingenious farmers have found a variety of uses for the tool.

This chapter will be far less complex than the mower chapter because the rakes and tedders featured are, though elaborate, much simpler in basic design and certainly in function.

Rakes are far easier to operate than mowers but they do present some issues which must be respected.

Before we get to those we should discuss what's available and what you might reasonably use. To our knowledge no dump or sulky rake is currently being manufactured. However, during their "hayday" (pun intended) or during the turn of the last century, there were at least 40 million dump rakes scattered around North America. Though

To this day farm equipment companies continue to make ground drive rakes and tedders in all sorts of new designs. These implements can be used with horses or mules in harness either by hooking them to a forecart or by modifying the front end with tongue

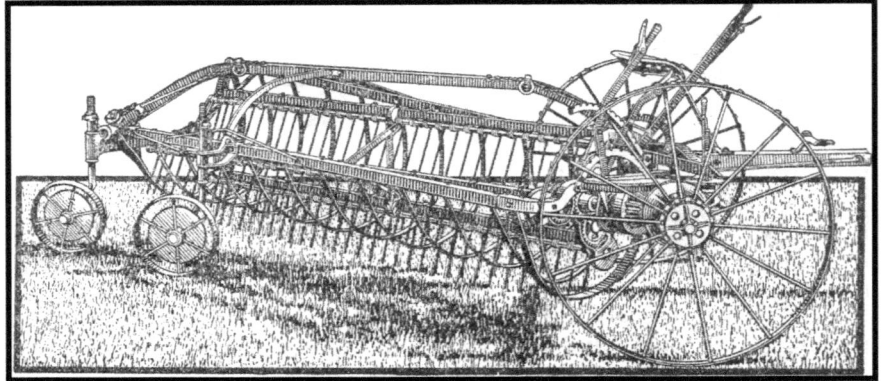
Figure 3. A John Deere Side Delivery rake

trucks or crazy wheels.

As for the front line of old original rakes there are several makes which are extremely service-able. In fact from region to region through-out the U.S. and Canada some makes might be better to have than others. This relates simply to the fact that when the rakes were originally sold they may have found greater popularity in certain regions. It makes good sense, if the older rake is your choice, to select a make and model which was numerous in your area in its day. In this way you can be certain of a better chance at finding either parts or second and third rakes which can be used as parts. We neglected to mention in the mower chapter that it is extremely useful to pick a make/model and stick with it. If you have a Case, a New Idea and a John Deere rake in your lineup it is more difficult to keep them running than if they were all the same make.

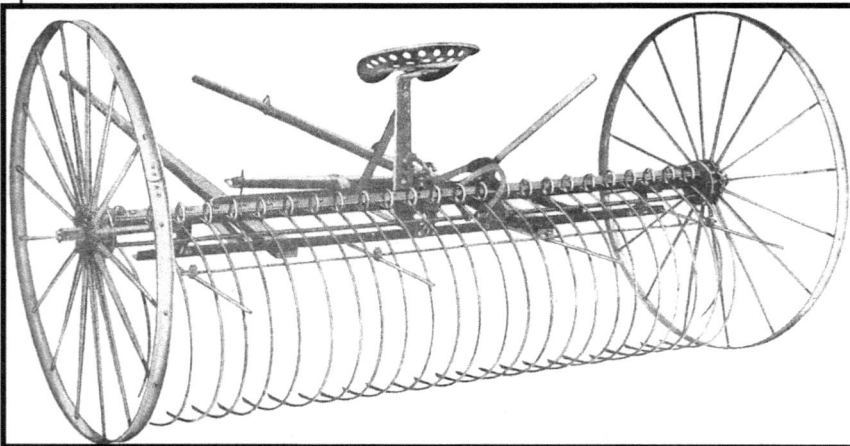
Figure 4. Wood Brothers Dump Rake

many of those have gone to scrap iron, hundreds of thousands still exist in various states of repair. And, unless something changes soon they are one of the older implements which have remained relatively inexpensive at yard sales, implement yards, and auction sales. So, with the dump rake the reader is stuck with the older pieces. Keep in mind that the bigger companies (i.e. Massey, New Idea, Case, McCormick Deering, John Deere etc.) kept making these rakes as well as the side delivery and tedder models, equipped for horses up until WWII. And the ground drive rakes they offered for a generation after that were the same identical machines with the exception that tongues were set up for tractors.

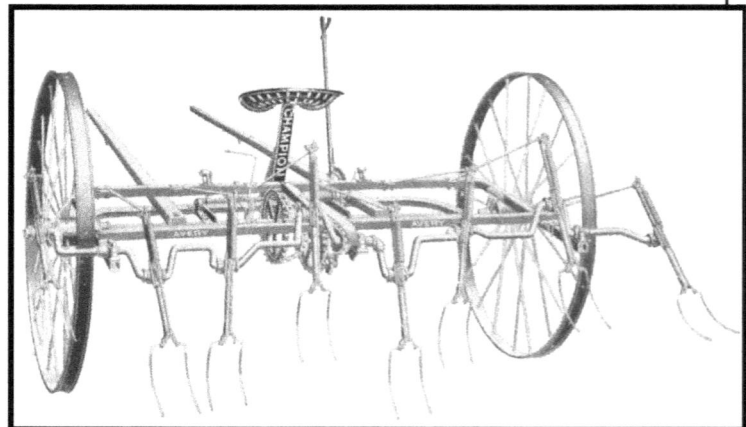
Figure 5. Champion Tedder

In our region of the Pacific Northwest the list of popular makes looks something like this and in this order;

John Deere
New Idea
Case
McCormick/International

Your area may be different. Best place to find out would be to ask at the used farm implement dealerships.

Hitching to Rakes

The hitching procedure outlined for the mowers will work well for rakes. Obviously there is no cutter bar to be concerned with but there is the width of the implement and it may catch someone unawares if the animals should move ahead unexpectedly and quickly while hitching.

Also it is important to note, as evidenced by Figure 8, that the teamster is sitting atop an implement that is difficult to get off of in a big hurry. There is no escape hatch from the conventional side delivery rake. (This author, 28 years ago, was involved in a runaway on a side delivery rake which resulted in a broken wrist.)

For these reasons it is important that only well broke and dependable animals are hitched to rakes. This is doubly so if the teamster is a novice.

If your animals are new to rakes,

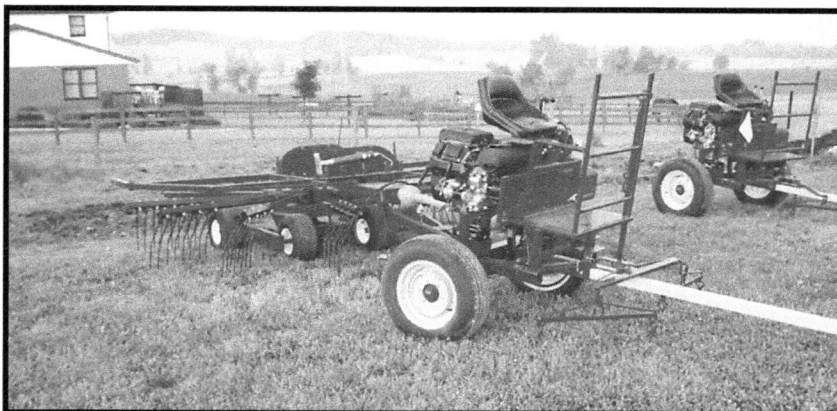

Fig. 6. The next several photos were taken at Horse Progress Days and show off modern horsedrawn raking apparatus. Above is a new motorized forecart providing PTO and hydraulics to a new tractor finger rake.

Fig. 7. Another new style of rake hooked to a motorized forecart

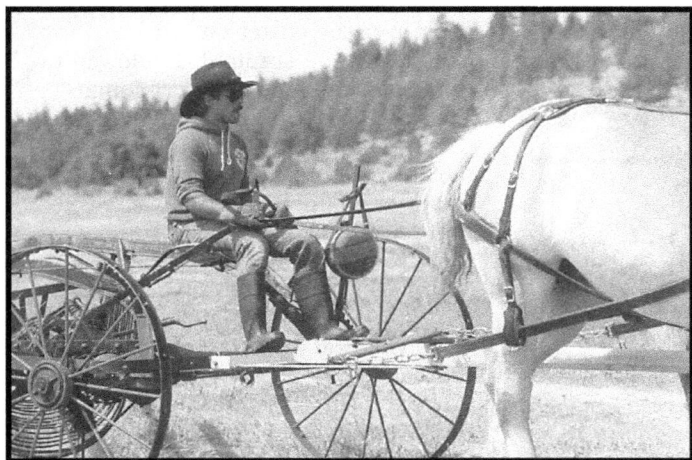

Fig. 8. Tony Miller sits on a New Idea side delivery rake. Note his position and how getting off will take some finesse.

Fig. 9. A new style of ground drive tedder.

you should find creative ways to introduce them, safely, to the sounds, and actions. A system was suggested with mowers. It is certainly important to note that in this author's thirty years all problems occurred early on, before experience and discrimination were available to assist. And that, in retrospect,

with qualified help ALL problems could have been avoided. Better put, they SHOULD have been avoided.

Expect no problems because your expectations shape your days. Watch for problems in the making because your watchfulness will give comfort to your best expectations.

Fig. 10. A new New Holland ground drive tractor rake which works well behind any forecart.

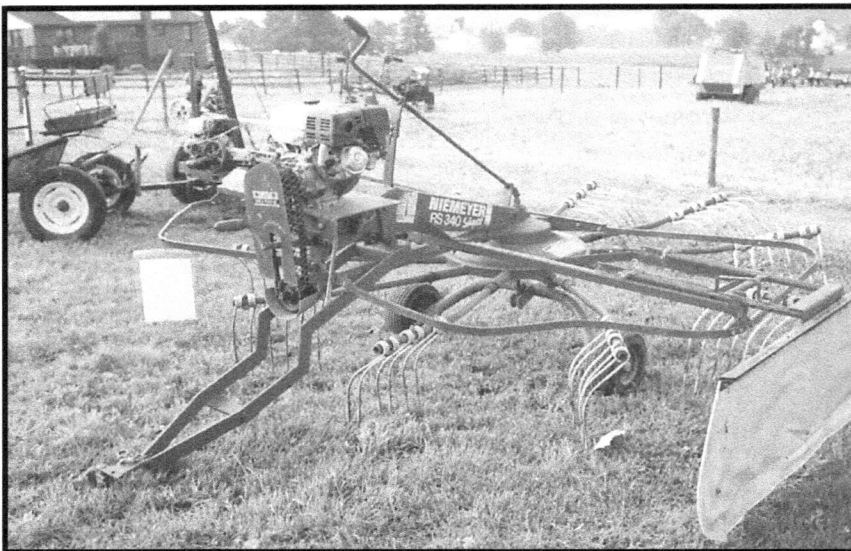

Fig. 11. All manner of new rake designs can be made to work behind horses or mules. This model features its own motor for power.

Fig. 12. Demonstrating the White Horse Machine Shop power forecart which, with four Belgians, was used to successfully pull two side delivery rakes as seen in Figure 13.

Raking

The popular practice formerly was (it prevails still in some sections) to let the hay lie in swaths after being cut, or to kick it around with a tedder until dry. Swath-curing causes the hay to bleach, and the leaves, being dry, are easily shattered by the tedder. When the leaves are lost, 65 per cent of the feeding value -- the bran of the hay -- is sacrificed.

When the plant is growing, it is constantly drawing nourishment and moisture from the ground through the roots. The plant food thus absorbed is retained and the excess moisture passes out through the leaves. One ton of hay will siphon between 300 and 400 tons of moisture from the ground during its growth. To gain an idea of the amount of rain required to mature one ton of hay, consider that one inch of rainfall over an acre of land weighs approximately 113 tons.

When the hay is cut, the flow of ground moisture is shut off; but the plant is full of water and the problem then is to reduce the moisture to a safe percentage for storing the hay, and to do this in the shortest possible time.

The leaves, or tops, as they fall back over the cutter bar, are left exposed to the sunlight. If allowed to remain in this position very long, the leaves dry up and shatter. When this happens, the natural flow of moisture from the stems to the leaves is stopped and the water is "bottled up" in the stems. This results in unevenly-cured hay that is unfit for feeding or storing and that grades low on the market.

The first step to be taken after the hay is cut, therefore, in order to prevent loss, is to get the hay out of the swath and into the windrow where the leaves will be in the shade while they are yet green.

To accomplish this, the hay is raked as soon as reasonable after it is cut, the rake being driven in the same direction as the mower travels. The curved teeth, working against the tops of the hay, lift the hay and

Fig. 13. Two New Holland rakes behind a four abreast, rake large amounts quickly at Horse Progress Days.

place it in loose, fluffy windrows with most of the green leaves inside, protected from the sun's rays. The leaves, shaded by the stems, cure rapidly by the free circulation of air through the loose windrows.

Turning the Windrow

Windrow-turning is an important operation, one that is a big factor in preserving the quality of the hay and saving time by hastening the curing. As a general rule, the best time to turn a windrow is when the hay feels dry about one-half of the way down.

In the turning operation, the top of the windrow which has been exposed should now be placed on the bottom, next to the stubble, with the damp, or bottom, side up.

The windrow-turning operation should be repeated as often as necessary in rainy weather, until the hay is properly cured.

Curved Teeth are Important. The teeth of the rake must be curved to make airy windrows. Curved teeth lift the hay. Straight teeth make what is commonly called "roped windrows" which exclude the air, lengthen the curing process, and produce inferior hay.

The advantage of the curved teeth on a rake over straight teeth could be best appreciated if you were to attempt to pitch hay with a straight-tined fork instead of one with curved

Fig. 14. Doug Hammill serenely raking with his two Clydesdales and dump rake in Montana Rockies.

Fig. 15. Bob Oaster on a New Idea Rake, with Singing Horse Ranch team.

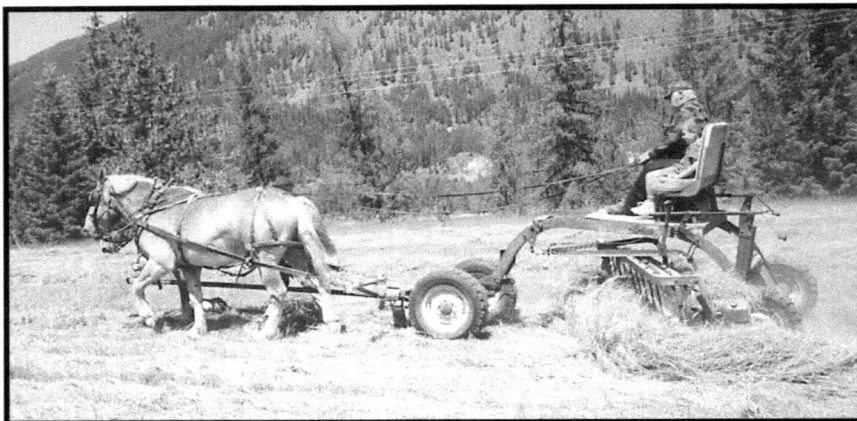

Fig. 16. Bulldog Fraser of northwestern Montana sits atop a gooseneck ground drive tractor rake which has been adapted for horse use. Note the rubber-tired truck wheels to carry the tongue.

Fig. 17. This side view closeup shows how the bench seat has been mounted atop the rake.

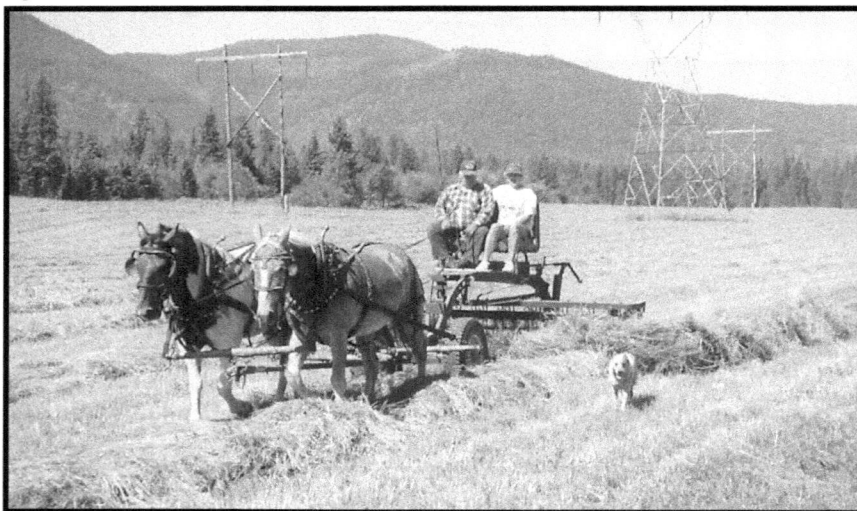

Fig. 18. As this young team in training evidences, raking is not a difficult job. Bulldog's passenger is there to keep him awake. So much for the myth of horsefarming being hard work!

tines.

Operation is Simple

See that your rake is properly set up and adjusted before you go into the field with it.

The rake usually has three levers. The outside lever on the right-hand side controls the height of the inner end of the reel. This lever should be set in the center notch for ordinary conditions.

The inside lever usually controls the angle of the teeth. When raking, throw this lever back to angle the teeth as much as possible and still get all the hay. The farther back this lever is set, the looser and fluffier the windrow will be.

The left-hand lever controls the height of the outer end of the reel. This end of the rake usually can be carried from one to two inches higher than the inside end. It should be just low enough to get all of the hay.

To get a parallel and uniform windrow, and at the same time prevent the horses from trampling the hay, adjust the pole on the rake so the horses will walk in the swath board space between the swaths.

Changes for Turning Windrows

On some rake models the left-hand wheel can be set in on the axle to give better results when turning windrows. When the left wheel is run alongside the windrow, the teeth protrude just far enough beyond the wheel to cause the windrow to be turned upside down.

The tongue can be adjusted to prevent the horses from trampling the windrow or bean rows, and also to equalize side draft.

Caster Wheels

The side delivery rake can be furnished with either one or two caster wheels. Two caster wheels are always recommended because they distribute the weight properly and insure smooth operation in going over rough ground or ditches. When using a rake with two caster wheels, one wheel carries the weight in case the other wheel should enter a furrow or low spot, so that the reel will not dig into the ground. Steel caster wheels with removable bearings, practically dust-proof, insure long life and satisfactory operation.

Side Delivery Rake Review Questions

(from John Deere Co.)

What is the horse powered way of making hay?

The hay is cut with a mower. A left-hand side delivery rake is driven in the same direction, raking the hay into fluffy windrows while it is still green. When the hay is cured in the windrow, it is loaded on the wagon with a hay loader and hauled either to the barn or stack.

What part in this haying system does this rake play?

The most important part because it saves the leaves. This rake is not cylindrical, but the frame tapers, the small end being at the right. By using the left-hand rake, the teeth work against the heads, the crop is picked up as laid down by the mower and

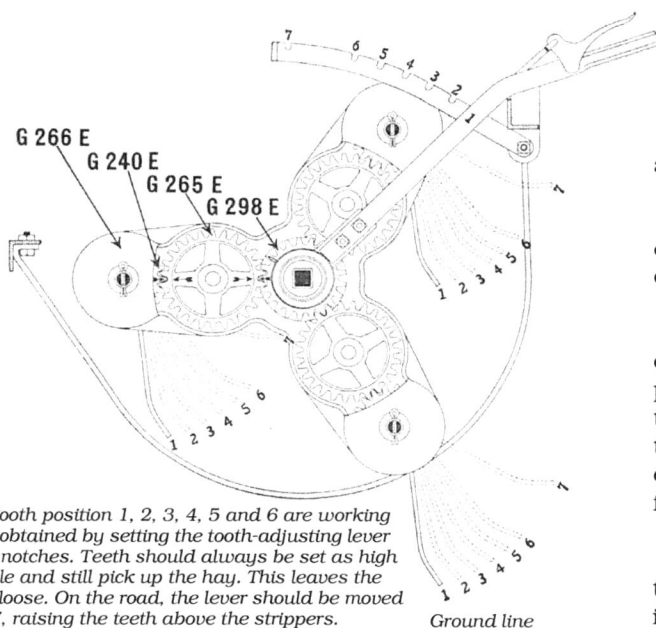

Fig. 19. Tooth position 1, 2, 3, 4, 5 and 6 are working positions obtained by setting the tooth-adjusting lever in the six notches. Teeth should always be set as high as possible and still pick up the hay. This leaves the windrow loose. On the road, the lever should be moved to notch 7, raising the teeth above the strippers.

Ground line

Fig. 20.

rolled in a loose windrow with approximately four-fifths of the stems out. The crop is raked as soon as cut, and cures in the windrow in one or two days.

Why is the function of this rake so important?

Because it deals with the actual curing of the hay.

What is the greatest cause for loss in hay quality?

Improper curing and handling, resulting in excessive loss of leaves. Hay not cured properly is low in feeding value.

What happens when hay is cut and not promptly raked?

When the hay falls in the swath the stems which contain the bulk of the moisture are covered. The leaves on top are exposed to the direct rays of the sun. The hay is in such position that it cannot cure evenly. The leaves, under these conditions, dry quickly and shatter, while the stems may still contain moisture.

How should the hay be handled after cutting?

The quick way to get moisture out of the stems, when cut, is through the leaves by evaporation into the air. The leaves, therefore, have to be preserved on the stems in an active condition. This means the leaves have to be kept from the sun's rays, yet where there is free circulation of air, while they are yet green and actively evaporating moisture.

What is the function of the side delivery rake?

It picks up the hay swath and places the bulk of the leaves in the shade in a loose windrow upon clean stubble.

Why are the rake teeth curved?

To make the windrows loose and fluffy, which condition can only be accomplished by actually pitching the hay into the windrows. The reason for the curved teeth can best be explained by recalling that the tines on a hand pitchfork are curved. One could not successfully use a straight-tined pitch-fork.

Why is the frame inclined?

The frame is low at the front end and increases toward the rear. This arrangement takes care of the increased volume of hay toward the rear. Because the frame is low at the front end and set slightly ahead of the tooth bars, the hay is caused to come against the frame and pitch forward into a loose coil, with the leaves inside and the stems outside.

Name and explain the lever controls?

There are three levers. The front lifting lever on the right hand side controls the height of the inner end of the reel. This lever is ordinarily set in the center notch.

The inside lever controls the angle of the teeth. When raking, this lever is thrown as far back as possible and still rake all the hay, in order to make the windrow loose and fluffy.

The left-hand lever is the rear lifting lever and controls the height of the outer end of the reel. In operation this end of the reel is generally carried one or two inches higher than the inside end.

Is it a good idea to turn the windrows?

It is if the windrow is turned upside down, so the wet hay is placed on top and the dry top hay on dry stubble. Windrow turning is an important operation, particularly in wet weather, because by so doing the quality of the hay is preserved, and the curing process hastened with a resulting saving of time.

Is the side delivery rake adapted to windrow turning?

Yes. And I this operation this rake is so arranged that the left-hand wheel can be set in on the axle so that by driving with the left-hand wheel next to the right-hand edge of the windrow enough of the rake comes in contact with the windrow to turn it upside down.

What is the gearing arrangement on this rake?

The axle is at right angles to the line of travel, while the rake reel is at an angle to the axle, the left end of the reel being back, the right end nearly touches the axle. Bevel gears transmit the power from axle to the reel through an internal gear and pinion. The internal gear is one-piece with the driven bevel gear. The cast reel heads, with chilled bearing surfaces, are mounted on the square reel shaft. At the right-hand reel head are three trains of gears, each mounted on a reel arm. The middle gear is held from turning by the tooth-adjusting lever. As the reel revolves, each set of rake teeth turn on their respective shafts. The timing of each train of gears is such that, as the reel turns, the rake teeth point in a general direction downward as they come in contact with the hay.

Explain the tooth-bar attachment.

The teeth on each tooth bar are coiled around a shaft and retained by a wood bar. The pressure of the hay is not taken by grasping the metal bar, but by the tooth bar as a whole with a solid casting that forms part of the gear. The metal bar is not used as a bearing in the inner reel head. The wear is taken up by the large bearing in the gear and the chilled bearing surface in the reel head.

Fig. 21.

How is the turning motion of the axle derived from the wheels?

Through dogs and ratchets in the hub of each wheel. A special advantage of this design is the fact that the wheels have releasing connections for turning corners.

How is the rake thrown in and out of gear?

By means of a jaw clutch located near the right-hand wheel on the axle.

Why is there an additional wide band and brace at the front end of the reel?

The purpose of these parts is to press down the heavy hay in the swath so that it can be engaged by the teeth and not pushed to the right or clog the rake.

What is the caster-wheel equipment on this rake?

The rake can be equipped with either one or two caster wheels. The best arrangement is for two caster wheels because two wheels divide the weight and provide smooth operation in going over rough ground. The caster wheels are of steel with removable bearings, so arranged as to be practically dust-proof.

Fig. 22.

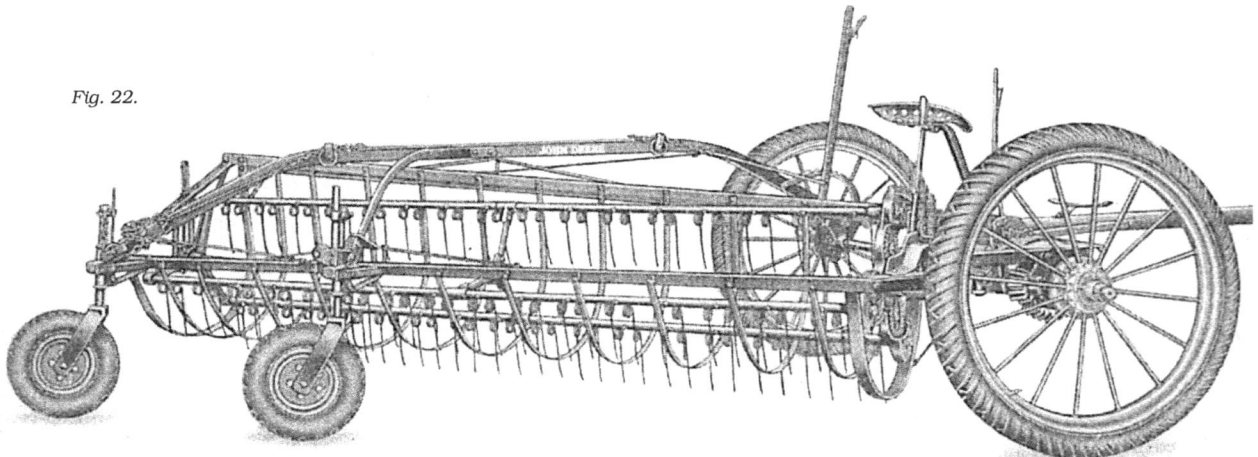

Fig. 23. As was shown in the previous chapter, the mown hay is raked first in the same direction as the mowing. When, and if, it is necessary to turn the hay again it should be done in reverse direction.

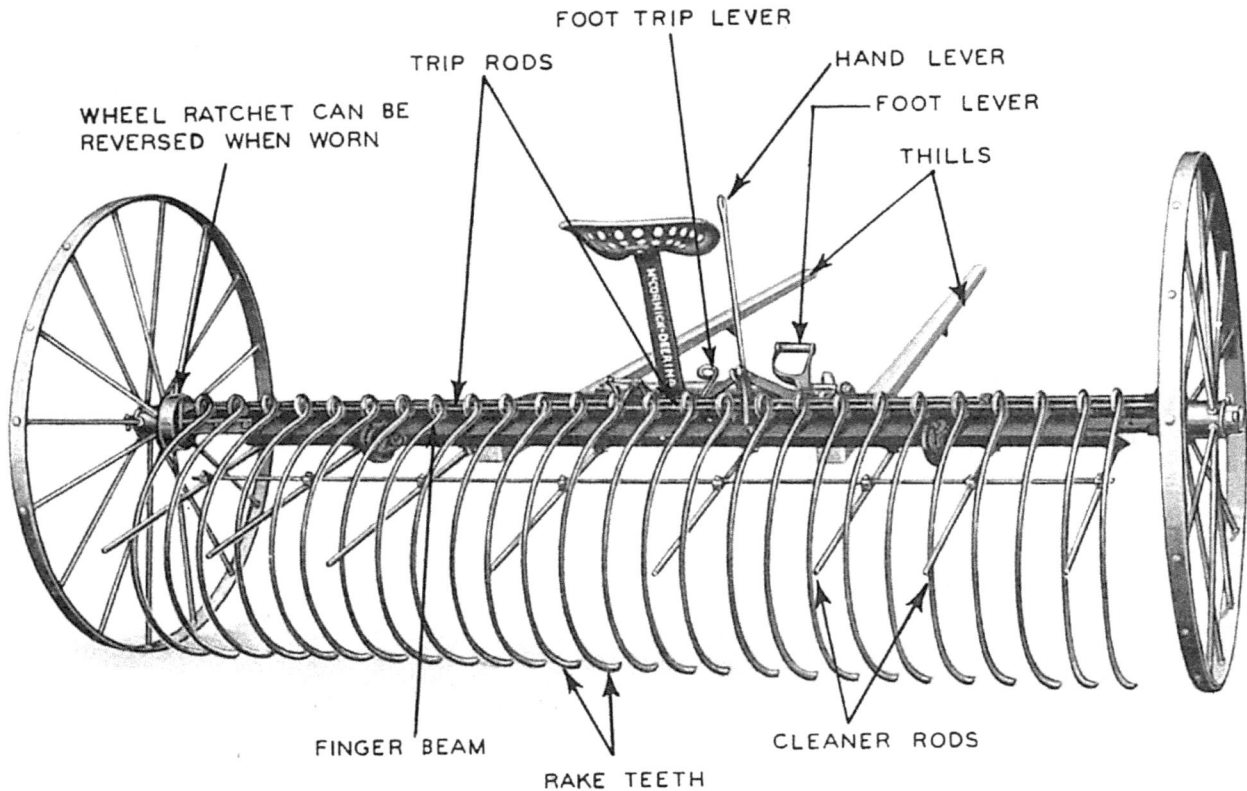

Fig. 24. McCormick-Deering type M self-dump rake, 10-foot width, with various parts identified.

SELF-DUMP RAKES

Type M self-dump rakes are furnished in 8, 9, 10, and 12-foot sizes as horse or tractor drawn types. They are extensively used for raking crops into windrows or convenient piles for hand loading.

DETAILS OF OPERATION

The self-dump rake includes a finger beam to which is attached the various parts that make up the rake, such as thills, axles, rake teeth, and trip mechanism.

The rake is self-dumping. All the operator has to do to dump the rake is step on the foot trip lever. This engages pawls at the ends of trip rods into ratchets in each wheel. The forward movement of the wheels raises the rake teeth to a height regulated by an adjustable trip stop which releases the pawls from the ratchets and permits the teeth to drop back into raking position.

A hand lever is provided to manually raise the teeth. The foot lever is used to hold the teeth down when raking heavy crops. The foot lever also permits holding the teeth in a raised position when turning corners or when passing over obstacles.

For traveling on the road it is necessary to lock the teeth in the up position. First, swing trip stop around so that either arm "1" or "2" Fig. 26 is in the control position. This will permit rake teeth to be raised to a sufficient height to lock hand lever with hold-up hook "1" Fig. 25.

ADJUSTMENTS

A--Trip Stop: (See Fig. 25 and Fig. 26).

The height of tooth rise, when dumping, is regulated by an adjustable trip stop. This stop can be raised, turned, and set in any one of four positions. This gives four adjustments as to the height the teeth will rise when dumping. Fig. 26 illustrates the posi-

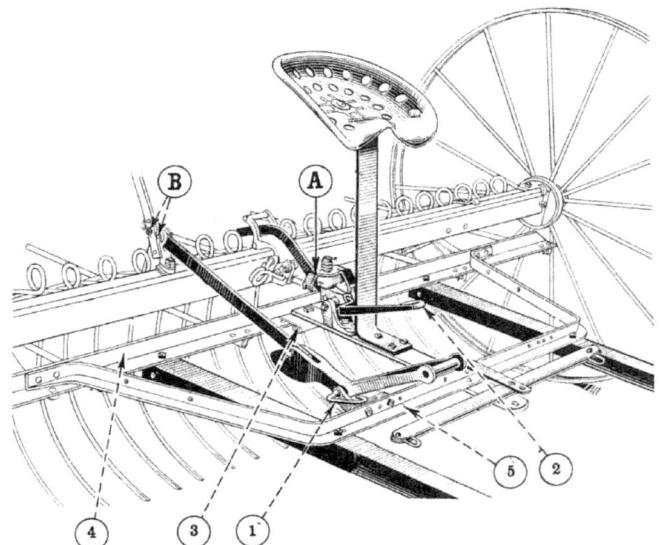

Fig. 25. Adjustments on type M self-dump rake are indicated by letters "A" and "B". "1" is the hold-up hook; "2" the foot trip lever; "3" curved metal piece; "4" cleaner bar angle; and "5" thill frame.

tions of the trip stop. With arm "1" in control position, teeth can rise high for heavy crop or for bunching. Arm "2" in control position allows teeth to rise not as high as arm "1". Arm"3" in control position allows teeth to rise not as high as arm "2". Arm "4" allows teeth to rise only a short ways; this position is used for short, light crops.

B--Tooth Position:

The front end of thills should be carried 42 inches from the ground and the front end of pole should be carried 32 inches from the ground. When operating at these heights, pin at "B" Fig. 25 should be located in the center hole in hand lever bracket for best raking performance. Whenever these heights are changed, on account of the height of the horse or horses, the pin should be moved. Placing pin in forward hole brings the teeth forward which will compensate for carrying the thills or pole lower than the above dimension.

SERVICING INFORMATION

Wheels and wheel ratchets are reversible; thus, when the ratchet surface becomes worn in one direction, the wheels with ratchets can be reversed left for right, giving double life to the ratchets A split cotter pin is all that holds the wheel on the axle.

Trip rods are reversible, end for end, but the right-hand rod is not interchangeable with the left hand rod. To reverse trip rod end for end proceed as follows: use a drift punch as leverage to release spring tension in trip "2" Fig. 27 (right) and take out pin "3". Remove trip rod holder "4" and rake wheel. Slide trip rod "1" out of axle holder. Turn trip rod "1" end for end and assemble parts by reversing above procedure.

The self-dump rake is regularly equipped with *thills* which can be spread apart for one horse or brought together to form a tongue for two horses. For one horse, bolt thills to under side of cleaner bar angle

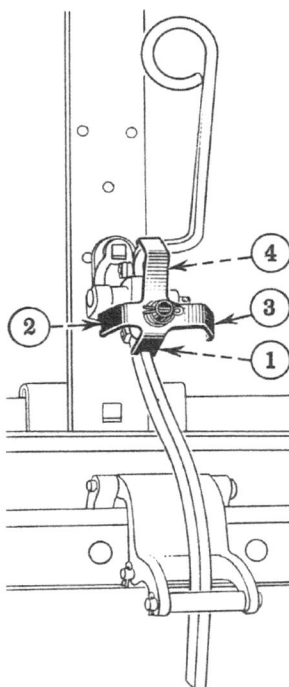

Fig. 26. Adjustable trip stop in position for highest tooth rise when dumping.

"4" Fig. 25 and thill (shafts) frame "5". For two horses, bolt thills to upper side of cleaner bar angle "4" and under side of thill frame "5".

A *balancing spring* is regular equipment on the 12-foot rake and is available as special equipment for all other sizes. This spring assists in raising the teeth when dumping and eases the shock when the teeth return to normal raking position.

SPECIAL EQUIPMENT

Special teeth available are 1/2-inch no coil teeth, 25/64-inch double coil teeth, and 7/16-inch double coil teeth. Teeth are furnished with round or flat points (see Fig. 29).

Mountain wheels, with either plain or roller bearings, are available for rough terrain. They have a wider and heavier tire as shown in Fig. 30.

A remote control rope trip attachment enables the tractor operator to trip the rake from the tractor seat by pulling a rope. A pressure spring mechanism is also embodied in this attachment which will allow the rake teeth to pass over an abrupt rise in the ground without bending or distorting supporting numbers.

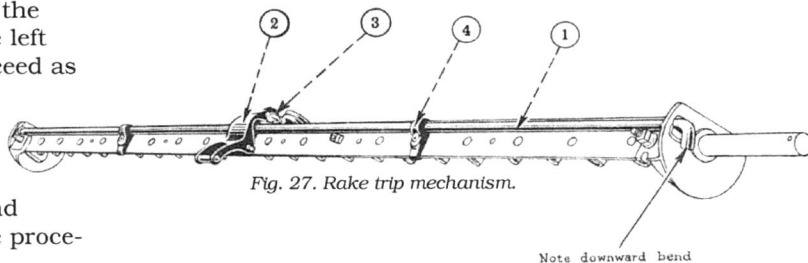

Fig. 27. Rake trip mechanism.

Note downward bend

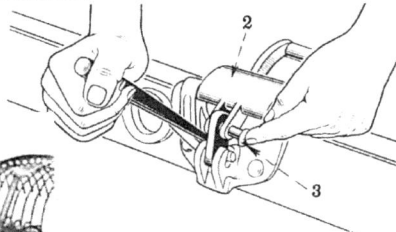

Fig. 28. Using drift punch for releasing spring tension. "1" trip rods; "2" trip; "3" pin; "4" trip rod holder.

Fig. 29. (left) teeth have single or double coils, round or flat points. (Right) two-wheel tongue truck.

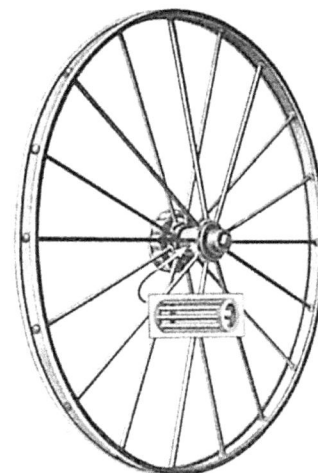

Fig. 30. Mountain wheel with roller bearing.

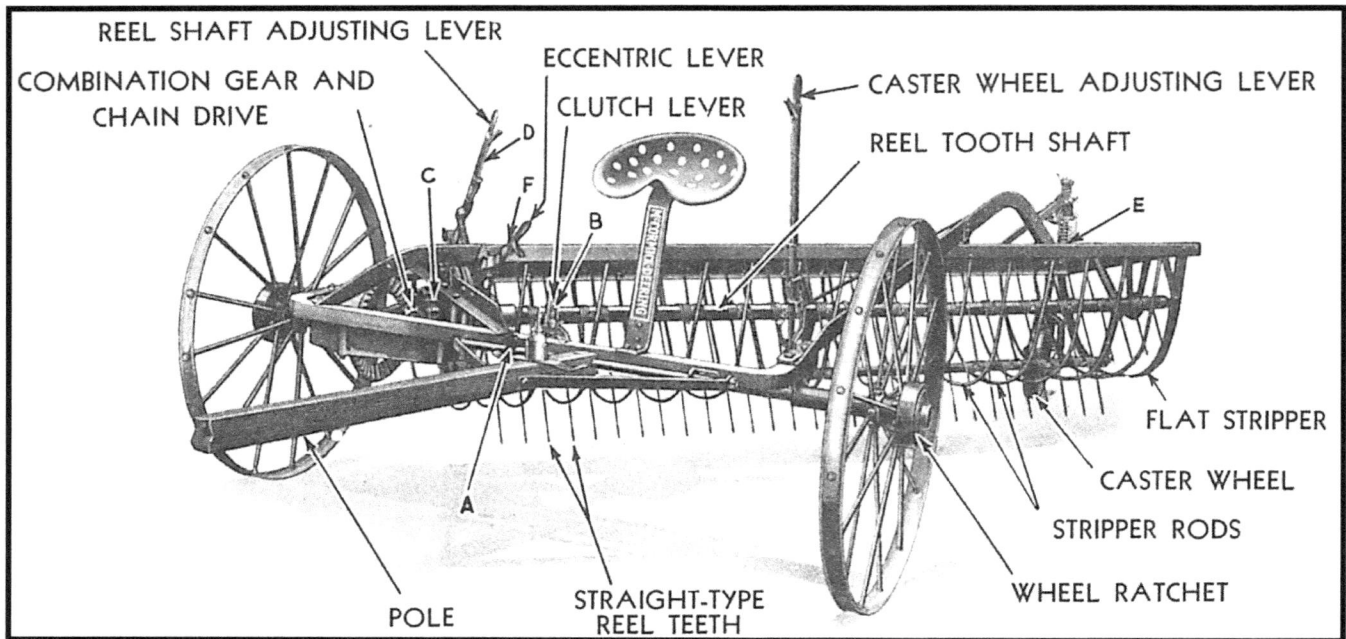

Fig. 31. Combination side-delivery rake and tedder with various parts identified. Letters "A", "B", "C", "D", "E", and "F" indicate adjustments.

SIDE-DELIVERY RAKES

Side-delivery rakes are used to place the mowed crop in loose, fluffy windrows, with the bulk of the leaves turned inward and the stems outward. The stems are thus subject to faster drying by the sun while the leafy portion of the plant is in the shade, protected from bleaching and overcuring. Hay cured in this manner will have better color and less leaf loss. There is also a combination side rake and tedder which is a machine designed to serve a dual purpose, operating as a side rake or as a tedder. Tedding fluffs the crop into loose, airy bunches which allows free air circulation and hastens curing.

The side-delivery rake or tedder is operated in the same direction as the hay was cut, against the heads of the mowed crop.

The following side-delivery rakes are available: combination side rake and tedder (regular and special bean); enclosed-gear side-delivery rake (three-bar reel); and enclosed-gear tractor side-delivery rake (four-bar reel).

COMBINATION SIDE RAKE AND TEDDER--REGULAR

DETAILS OF OPERATION

The combination side rake and tedder includes a three-bar straight-tooth reel which rotates in a diagonal position to the direction in which the rake is pulled. The reel is carried by a heavy steel frame which is supported at the front end by two main wheels and at the rear by a caster wheel. Each main wheel is ratcheted to a main axle at its hub; this causes the axle to turn with the wheel that has the greatest forward rotation. The main axle turns in three roller bearings. The reel is also carried on roller bearings.

The reel is driven by a combination of bevel gears and chain drive. Two bevel gears and a sliding drive clutch are mounted directly on the main axle. The sliding clutch can be moved to engage the proper gear for raking or tedding, or set in a neutral position by means of a clutch lever "B" Fig. 32. This lever is short and purposely located where it is necessary to reach down and make the change, thus eliminating any likelihood of the operator accidently reversing the reel while the machine is in operation, which might result in breakage.

The ends of the reel are individually adjustable

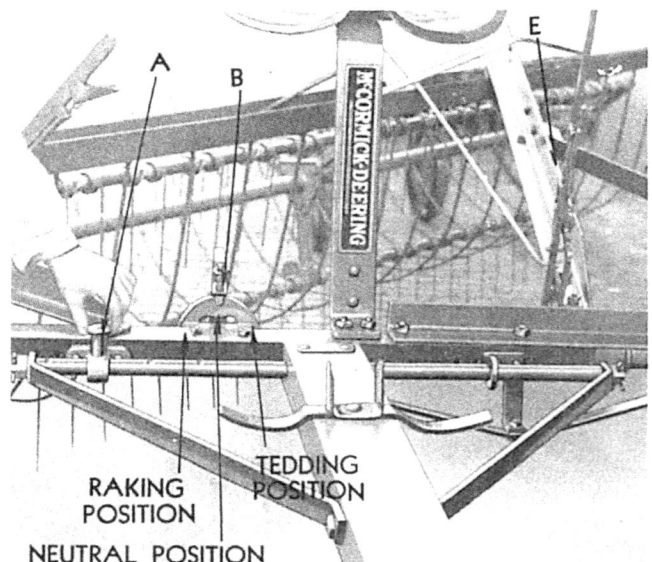

Fig. 32. "A" is pole shifting adjustment; "B" clutch adjustment; and "E" caster wheel adjusting lever.

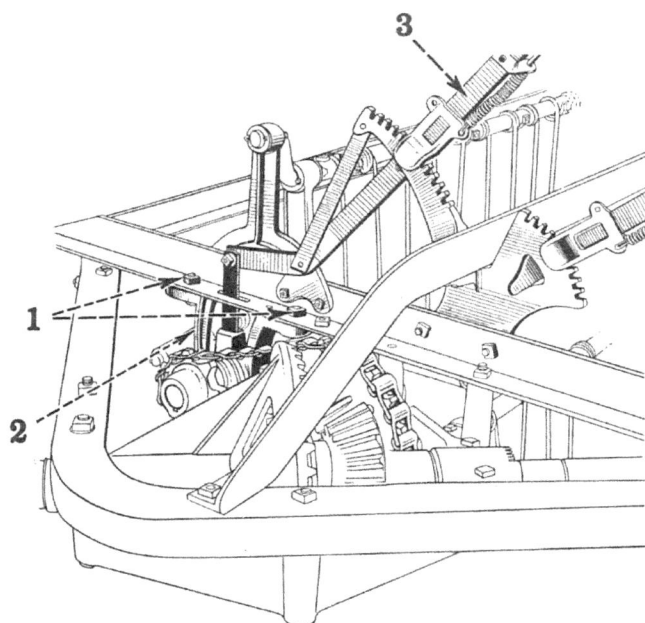

Fig. 33. Chain adjustment and reel shaft adjustment. "1" nuts; "2" reel shaft bearing support; and "3" reel shaft adjusting lever.

for height from the ground and can be set in a desired position for raking or tedding. The height at the forward end of the reel is controlled by the reel shaft adjusting lever, and at the rear end by the caster wheel adjusting lever.

The working angle of the teeth can be changed and set by means of an eccentric lever.

ADJUSTMENTS
A--Pole Shifting:

The pole is adjustable laterally for different raking conditions by lifting the spring plunger "A" Fig. 32 and shifting pole to desired position. Releasing

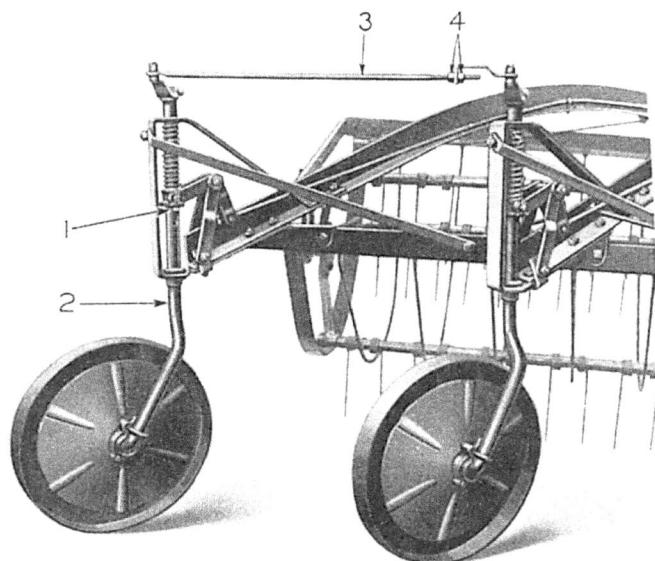

Fig. 34. Caster wheel assembly with extra caster wheel attachment installed. "1" is cotter pin; "2" caster wheel axle; "3" caster wheel adjusting rod; and "4" lock nuts.

plunger which is spring-actuated will lock the pole position.

This adjustment permits the operator to compensate for raking side draft when operating under various crop conditions.

B--Clutch:

When clutch lever "B" Fig. 32 is in neutral position as shown, the clutch should turn freely between bevel drive gears, and should engage bevel gears when in tedding or raking position. The position of clutch can be changed by loosening lock nut "2" Fig. 33 and turning clutch rod "1" in or out as needed in clutch rod adjuster "3". Tighten lock nut.

C--Chain:

Chain is tightened by loosening nuts "1" Fig. 33

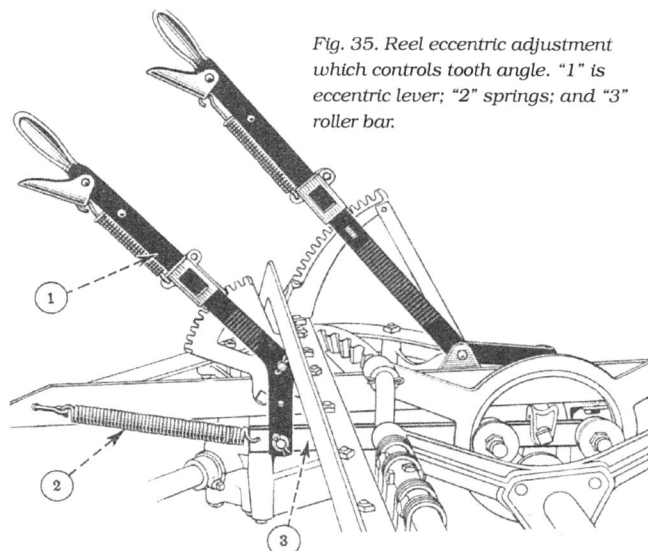

Fig. 35. Reel eccentric adjustment which controls tooth angle. "1" is eccentric lever; "2" springs; and "3" roller bar.

and moving reel shaft bearing support "2" backward to take up chain slack; then tighten in place. Slots are provided in reel shaft bearing support "2" to permit adjustment. Washers on the lock bolts are serrated on one side which engage serrations on the bearing support. To tighten chain, it is necessary to loosen lock bolts sufficiently to release serrations and reengage in new position.

The curved surface in the reel shaft bearing support "2" must curve around the drive sprocket. This is to keep a uniform chain tension throughout the raising or lowering movement of the front reel shaft bearing.

D--Reel Shaft:

The forward end of the reel can be raised or lowered with relation to the ground by means of reel shaft adjusting lever "3" Fig. 33. For raking, the front end of the cylinder should be slightly lower than the rear end. For tedding, the cylinder should be level. Do not set the cylinder teeth closer to the ground than is necessary to gather all the crop.

E--Caster Wheel:

The rear end of the main rake frame which supports the reel is carried on a caster wheel and is spring-mounted for flexibility when traveling over rough ground. The height of teeth from the ground at the outer end of the rake is adjustable by means of the caster wheel adjusting lever "E" Fig. 32.

Three holes are provided in caster wheel axle "2" Fig. 34 for cotter pin "1". Use the hole which will give the desired range of elevation in conjunction with caster wheel adjusting lever "E" Fig. 32. The center hole is used for normal conditions. The caster wheels are provided with stationary sheet steel shields which prevent crops winding around the hub.

F--Eccentric:

The angle of the teeth in working position can be regulated by means of eccentric lever "1" Fig. 35. Lowering the lever causes the ends of the teeth to move ahead. Raising the lever moves them back. For ordinary use, the teeth should stand slightly ahead of vertical, but the operator should at all time use his best judgement in keeping the teeth at the proper angle for best results in raking or tedding. For transporting on the road, the teeth should be moved back as far as possible. Slack in the eccentric mechanism is taken up by the pull of spring "2" on roller bar "3", resulting in a smoother running machine.

G--Main Wheel:

The left main wheel can be moved in approxi-

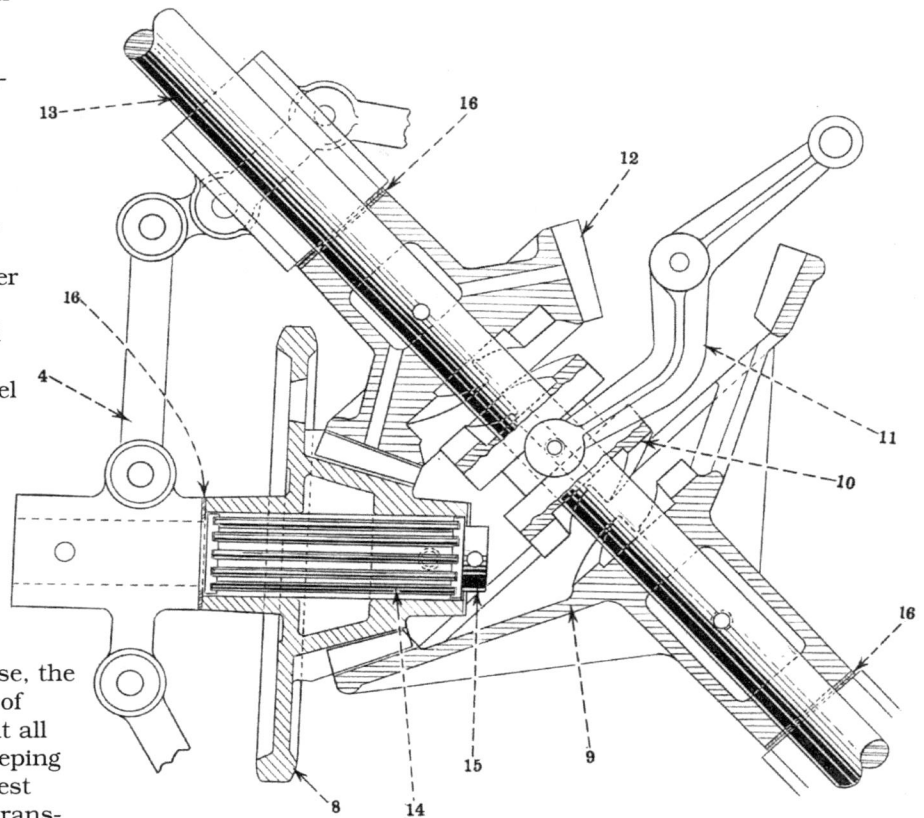

Fig. 36. Cross-sectional drawing of transmission on side rake and tedder. "4" is gear frame; "8" drive sprocket and gear; "9" internal clutch gear; "10" clutch; "11" clutch shifter fork; "12" clutch bevel gear; "13" main axle; "14" roller cage for short shaft; "15" short shaft; "16" washers.

Fig. 37. Transmission of the combination side rake and tedder. "1" bolts; "2" reel shaft sprocket; "3" chain; "4" main frame; "5" bolts; "6" gear frame; and "7" bearing cap bolts.

mately 4 inches by removing a spacer and pinning the pawl plate to the axle in the second hole provided.

SERVICING INFORMATION

To remove the *drive chain*, "3" Fig. 37, loosen the two bolts "1" holding reel shaft support. Move reel shaft sprocket "2" ahead until chain can be slipped off teeth. Disconnect link in chain and remove chain. Replace chain with slotted sides of links facing out and loop ends facing forward in direction the chain travels when raking shown by arrow, Fig. 40.

To dismantle *transmission*, support main rake fram "4" on wooden horses. Remove two main wheels and drive chain and disconnect clutch control rod. Take out six bolts "5" holding gear frame"6" to rake frame and two bolts holding main axle bearing box to rake frame. The axle with gear drive assembly can then be removed from under the rake frame on opposite side (not shown) of transmission. Take off bearing cap which is held by two bolts "7"; then axle with two drive bevel gears and bearings can be lifted from gear frame. Drive sprocket and bevel pinion can be readily removed. Bevel gears and pinion when installed properly should line up even at the tooth ends and should revolve freely without excessive backlash.

To remove *teeth*, it is first necessary to remove

the tooth bar. This is done by taking out three bolts holding crank to the tooth bar. Remove strap at center bearing support and single bolt holding short shaft in tooth bar at end opposite crank end. Using a punch in the single bolt hole, loosen the short shaft and pull it out of tooth bar. Tooth bar can then be pulled off of crank and removed.

To remove an *eccentric*, it is first necessary to remove the reel with eccentric mechanism as an assembly. Take off all strippers, remove two bolts holding reel shaft support to frame, and slip chain off small sprocket. Disconnect reel shaft adjusting lever from reel shaft bearing strap. Disconnect tension spring and roller bar from eccentric lever. Front end of reel will drop free of rake frame and rear end will slide free of rear reel shaft bearing box. Eccentric can then be readily overhauled. Replace reel by reversing the removal procedure.

SPECIAL EQUIPMENT

Quick attachable teeth can be furnished on the rake in lieu of regular teeth from the factory if specified.

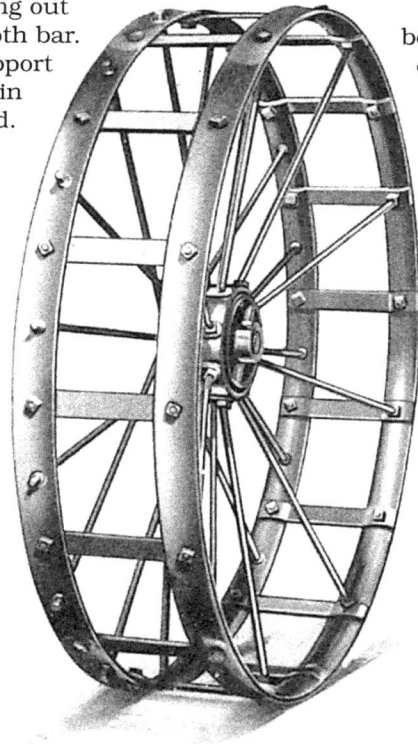

Fig. 38. Special extension rim wheel.

When using the side rake and tedder, better results are sometimes obtained by changing the cylinder speed for various crops. For beans or light dry hay, a slow cylinder speed is desirable for gently building the windrow; for heavy green crops, the cylinder may be speeded up. For this purpose special sprockets are available to replace the regular RA 809 (8 tooth) reel shaft sprocket.

R 5020 sprocket (9 teeth) or R 5021 sprocket (10 teeth) will replace RA 809 for reduced cylinder speed, and R 5033 sprocket (7 teeth) will replace RA 809 for increased cylinder speed.

These special sprockets will replace RA 809 on machines in the field without the use of additional parts.

A supplemental caster wheel is available for use where the ground is rough. Fig. 34 shows this caster wheel installed. These caster wheels must run parallel to each other; two lock nuts "4" are provided on the threaded end of caster wheel adjusting rod "3" for setting caster wheels parallel. Use a ruler to measure the distance between the caster wheel rims at the front and at the rear in checking wheels for parallel alignment. These dis-

ENCLOSED—GEAR SIDE—DELIVERY RAKE (THREE-BAR REEL)

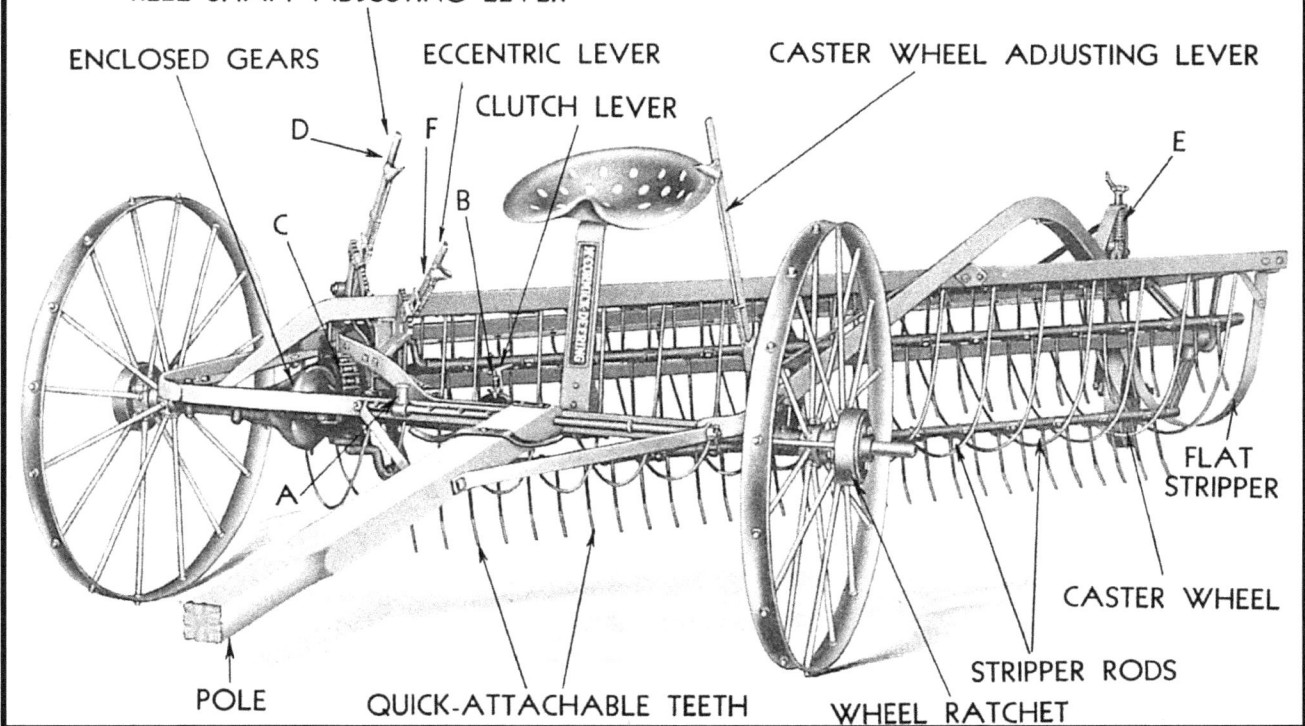

Fig. 39. Enclosed-gear side-delivery rake (three-bar reel) with various parts identified. Letters "A", "B", "C", "D", "E" and "F" indicate adjustments.

tances should be exactly the same if wheels are parallel. When wheels are parallel there will be no noticeable chatter during field operation.

It is advisable to have the supplemental caster wheel removed when the rake is to be used for tedding. This caster wheel interferes with the tedding operation and also packs a track in the crop already tedded.

An extension rim wheel can be obtained for use in soft ground or irrigated sections. This wheel consists of an extra rim which is fitted with cross bars to a regular wheel rim and has a greater number of spokes (see Fig. 38)

Other special equipment available for the combination side rake and tedder includes two horse evener and neckyoke, tractor hitch, main axle extension with truss, 4.00x36-in. 4-ply implement tires on main wheels, and 4.00x9-in. 4-ply implement tire on caster wheel.

COMBINATION SIDE RAKE AND TEDDER-- SPECIAL BEAN

DETAILS OF OPERATION

The special bean side rake and tedder is similar to the regular side rake and tedder with the exception of the raking cylinder or reel. It is longer on this machine and will rake a swath 9 inches wider than the regular rake. All preceding information on the regular side rake and tedder will apply to the special bean rake.

The enclosed-gear side-delivery rake is a horse-drawn rake, and is equipped with a three-bar reel that revolves on roller bearings. It is practically identical in

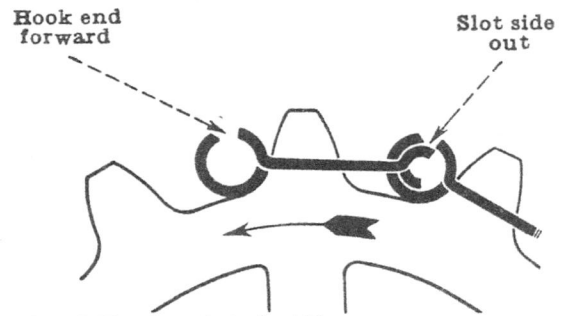

Fig. 40. How drive chain should be installed on sprocket.

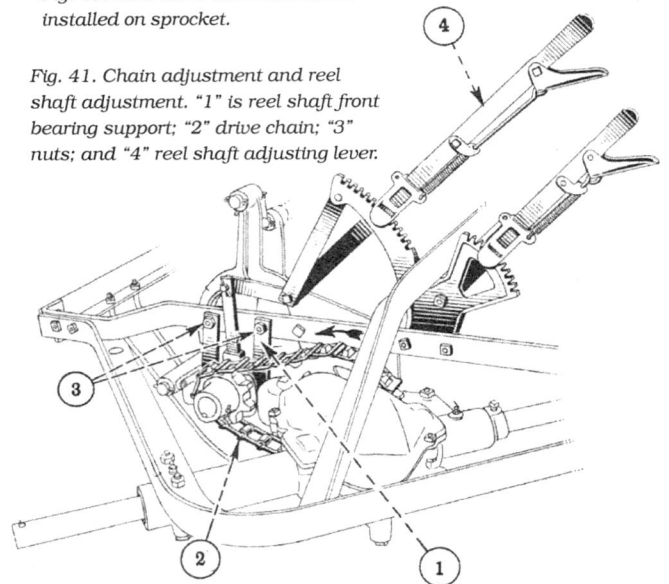

Fig. 41. Chain adjustment and reel shaft adjustment. "1" is reel shaft front bearing support; "2" drive chain; "3" nuts; and "4" reel shaft adjusting lever.

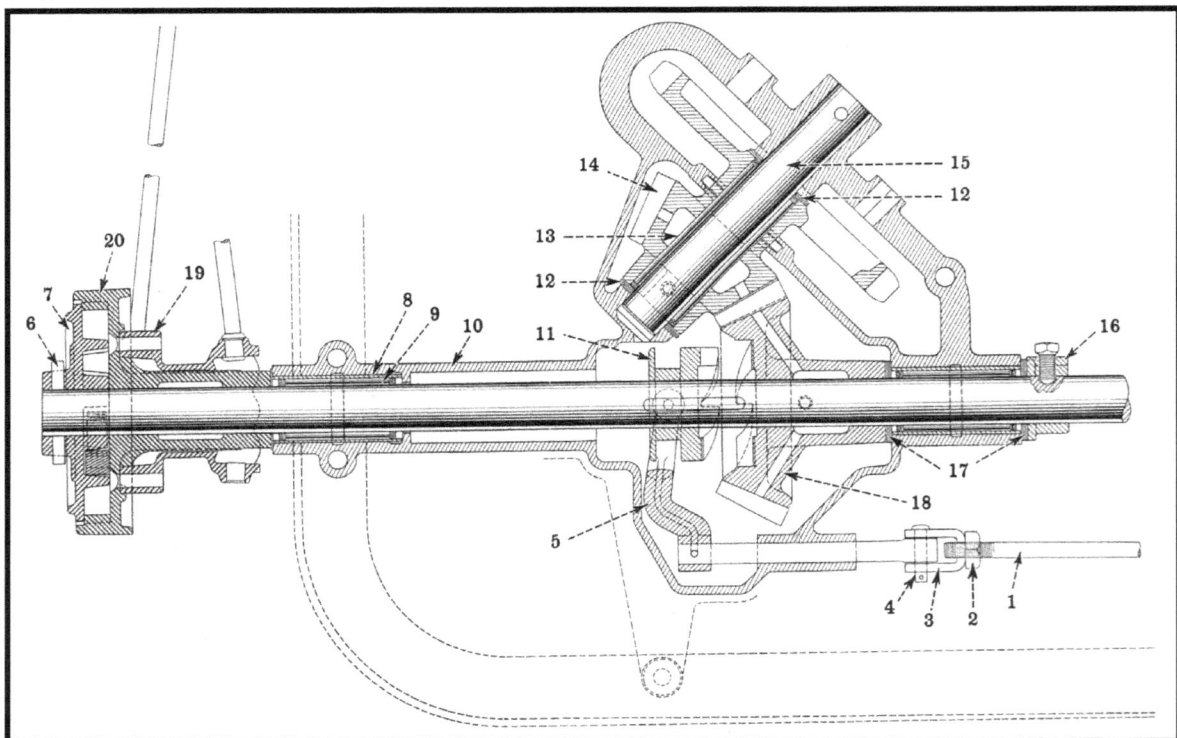

Fig. 42. Cross-sectional view of gear drive on enclosed-gear side-delivery rake. "1" is clutch rod; "2" lock nut; "3" clutch rod adjuster; "4" pin; "5" clutch shifter fork; "6" pin; "7" pawl plate; "8" roller bearing outer race; "9" roller bearing; "10" gear housing; "11" clutch; "12" thrust washers; "13" roller bearing for short shaft; "14" reel drive sprocket and gear; "15" short shaft; "16" collar on main axle; "17" thrust washers; "18" clutch bevel gear; "19" main wheel; "20" main wheel ratchet.

EXTRA CASTER WHEEL

REGULAR CASTER WHEEL

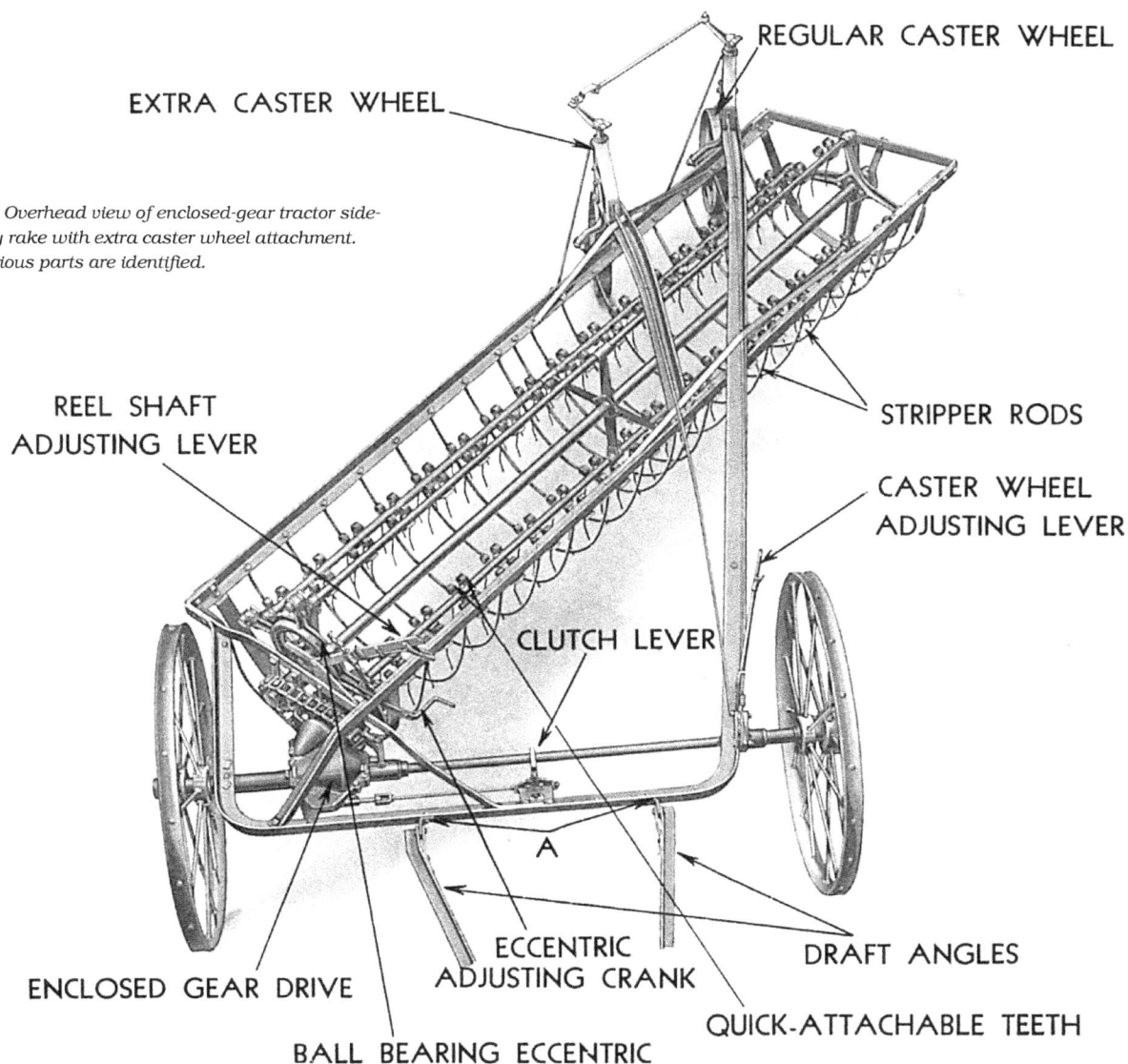

Fig. 43. Overhead view of enclosed-gear tractor side-delivery rake with extra caster wheel attachment. The various parts are identified.

REEL SHAFT
ADJUSTING LEVER

STRIPPER RODS

CASTER WHEEL
ADJUSTING LEVER

CLUTCH LEVER

ENCLOSED GEAR DRIVE

A

ECCENTRIC
ADJUSTING CRANK

DRAFT ANGLES

QUICK-ATTACHABLE TEETH

BALL BEARING ECCENTRIC

construction to the combination side rake and tedder, previously covered, except that it can only be used for raking crops in windrows and cannot be used for tedding. The enclosed-gear side-delivery rake, as the name indicates, has its drive gear and pinion completely enclosed in a housing to protect them from dirt and wear.

The reel is equipped with quick-attachable double teeth which can be removed independently when replacement is necessary. A keeper fastens on each double tooth to retain the tooth and prevent it from falling into the crop, should accidental breakage occur.

ADJUSTMENTS

A--Pole Shifting:
Refer to page 159, pole shifting adjustment "A", for information covering this adjustment.

B--Clutch:
This adjustment is the same as clutch adjustment "B", page 159, except that the clutch lever cannot be set in tedding position. The clutch should turn freely when the clutch lever is in the neutral position and engage the bevel drive gear when in the raking position.

C--Chain:
Drive chain should be installed with link slot side facing out and hook end facing ahead in direction of rotation of sprockets. See Fig. 37, also arrow indicating direction of chain travel in Fig. 40.

Drive chain "2" Fig. 41 may be tightened by loosening the two nuts "3" and sliding reel shaft front bearing support "1" back; then tighten nuts "3" securely. The curve in reel shaft front bearing support "1" should curve around the drive sprocket. This will maintain a uniform chain tension throughout the raising and lowering movement of the reel shaft.

D-Reel Shaft:
The forward end of the reel can be raised or lowered with relation to the ground by means of reel shaft adjusting lever "4" Fig. 41. The front end of the reel should be nearer to the ground than the rear end

for best raking. Do not set the reel teeth closer to the ground than is necessary to gather all the crop.

E-Caster Wheel:

Refer to page 159, caster wheel adjustment "E" on the combination side rake and tedder. This adjustment is the same for the enclosed-gear side-delivery rake.

F-Eccentric:

Refer to page 160, eccentric adjustment "F" for information.

G-Main Wheel:

The left main wheel can be moved in approximately 4 inches by removing a spacer on the axle in back of the wheel and pinning the pawl plate to the second hole provided in the axle.

SERVICING INFORMATION

To dismantle the *transmission*, support the rake main frame on wooden horses. Remove drive wheels and chain. Disconnect the clutch control rod. Take out six bolts holding the gear housing to the rake frame and three bolts holding the left main axle box to the rake frame. The axle and gear housing assembly will then drop free of rake frame. Remove gear housing cover which is held by two bolts with wing nuts and lock washers. Loosen collar setscrew, slide axle in until key clears drive bevel gear, then remove key with aid of pliers. Axle will then slide out of housing releasing drive bevel gear.

To remove bevel pinion, take out bolt holding short shaft to housing and slide short shaft back until drive sprocket and pinion are released. Reverse above procedure for reassembling. Bevel gear teeth should line up even at their ends with mating bevel pinion teeth and should revolve freely with no excessive backlash.

The eccentric used on the enclosed gear side-delivery rake is the same as that used on the combination side rake and tedder. Therefore the servicing information covered on page 160 will hold for this rake.

SPECIAL EQUIPMENT

An extra caster wheel is available for use where ground is rough. When the extra caster wheel attachment is assembled on the rake, it is necessary to set the caster wheels so they run parallel to each other. See instructions on page 161 for making this adjustment.

Main wheels with extension rims are available in lieu of regular wheels for use on soft ground or irrigated sections. See Fig. 38, page 161.

Main wheels with 4-7/8-inch wide tires in place of regular 2-1/2-inch tires can also be furnished as original equipment from the factory.

Other special equipment available for the enclosed-gear side-delivery rake includes doubletree and neck yoke, tractor hitch in lieu of horse hitch, main axle extension frame and truss braces, and 4.00x36-in. 4-ply implement tires on main wheels and 4.00x9-in. 4-ply implement tire on caster wheel.

ENCLOSED-GEAR TRACTOR SIDE-DELIVERY RAKE (FOUR-BAR REEL)

DETAILS OF OPERATION

The enclosed-gear tractor side-delivery rake is especially designed for fast, efficient raking behind a modern tractor. It is equipped with a four tooth-bar reel which provides greater raking capacity to compensate for higher travel speed of the tractor. The main drive bevel gear and pinion are completely enclosed to protect them from dirt and wear. The working angle of the teeth is controlled by a ball-bearing eccentric. Quick-attachable teeth are furnished as regular equipment.

ADJUSTMENTS

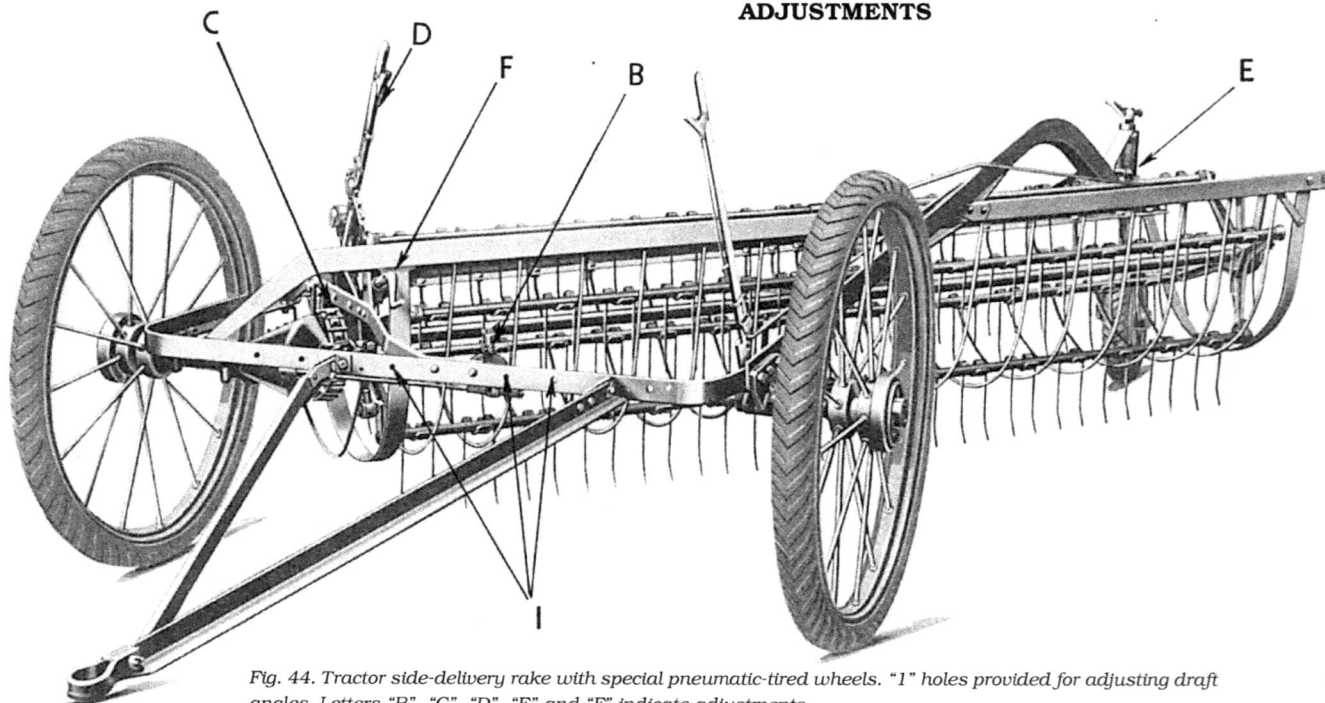

Fig. 44. *Tractor side-delivery rake with special pneumatic-tired wheels. "1" holes provided for adjusting draft angles. Letters "B", "C", "D", "E" and "F" indicate adjustments.*

A--Draft Angles:

The draft angles may be shifted to the right or left to obtain the desired line of draft, by using the series of holes "1" Fig. 44 provided in draft frame.

B--Clutch:

This adjustment is the same as clutch adjustment "B", page 159, except that the clutch lever cannot be set in tedding position. The clutch should turn freely when the clutch lever is in the neutral position and engage the bevel gear when in the raking position.

C--Chain:

Refer to chain adjustment "C" explained on page 49. The procedure for adjusting drive chain on the tractor side-delivery rake is the same as the horse side-delivery.

D--Reel Shaft:

Reel shaft adjusting lever "D" Fig. 44 controls the raising or lowering of the forward end of the reel with relation to the ground. For best raking, the teeth should work as far from the ground as possible and still gather all the crop. For best raking the forward end of the reel should also be nearer to the ground than the rear end.

E--Caster Wheel:

This adjustment is the same as caster wheel adjustment "E" covered on page 159 for the combination side rake and tedder.

F--Eccentric:

A screw-type crank Fig. 45 provides a means of adjusting the working angle of the teeth. For ordinary use, the teeth should stand slightly ahead of vertical.

SERVICING INFORMATION

The *eccentric* is of ball-bearing type and requires frequent greasing. Greasing helps prevent dirt entering the bearing and tends to force out dirt which does enter. Although this eccentric is a ball bearing type the same procedure for removing, as outlined on page 160 for the combination rake and tedder is followed.

Tooth bars can be removed from the reel by taking off three bearing straps (two around reel head bearings and one around eccentric bearing) and bolt holding center bearing. The tooth bar is then freed from the reel. The crank is held to the tooth bar by three bolts.

SPECIAL EQUIPMENT

An extra caster wheel is available for the tractor side-delivery rake where rake is to operate on rough ground. Refer to Fig. 34, page 159 for picture of the attachment and to page 161 for instructions for adjusting caster wheels to run parallel.

A horse hitch is available as extra equipment for tractor side-delivery rake, Fig. 46. It consists of a seat, spring, pole, and attaching parts. The pole can be shifted left or right to obtain the desired line of draft, the same as with the tractor draft angles.

A special main axle extension is available for use in irrigated sections and wherever special wheel spacings are required. It moves the left hand wheel out 14 inches, giving a thread of 7-feet, 9-inches.

Pneumatic tires and wheels are available in the following sizes: 4.00x36-in. 4-ply implement tires for

Fig. 45. Eccentric adjustment controls angle of teeth.

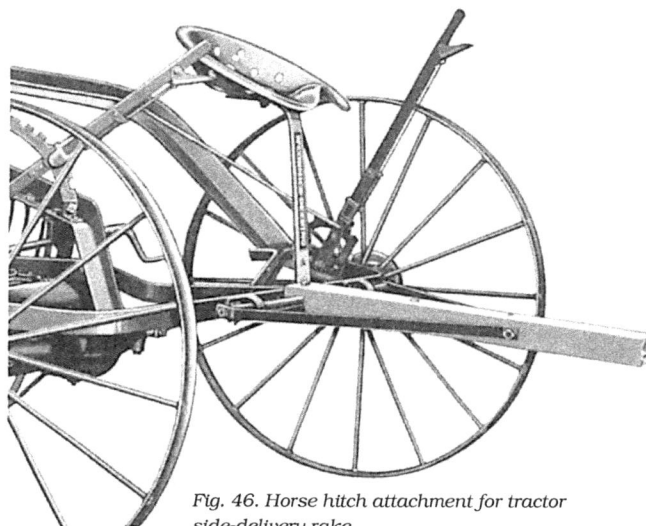

Fig. 46. Horse hitch attachment for tractor side-delivery rake.

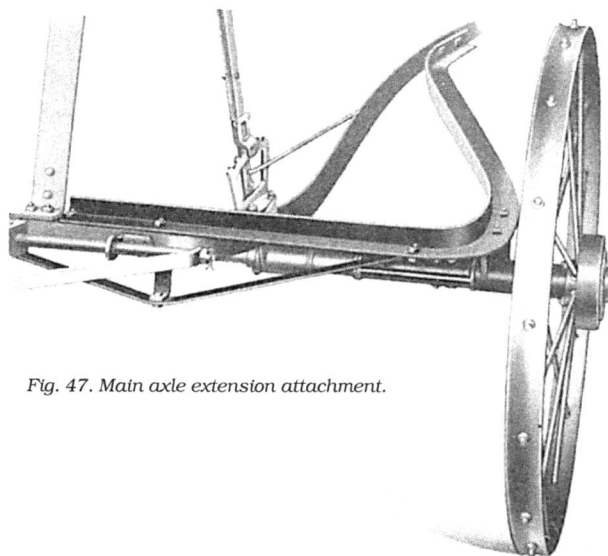

Fig. 47. Main axle extension attachment.

main wheels and a 4.00x9 in. 4-ply implement tire for caster wheel.

Main wheels can also be furnished with 4-7/8-inch wide rims in place of 2-1/2-inch wide rims furnished regular.

JOHN DEERE SIDE DELIVERY RAKES
OPERATION AND ADJUSTMENTS

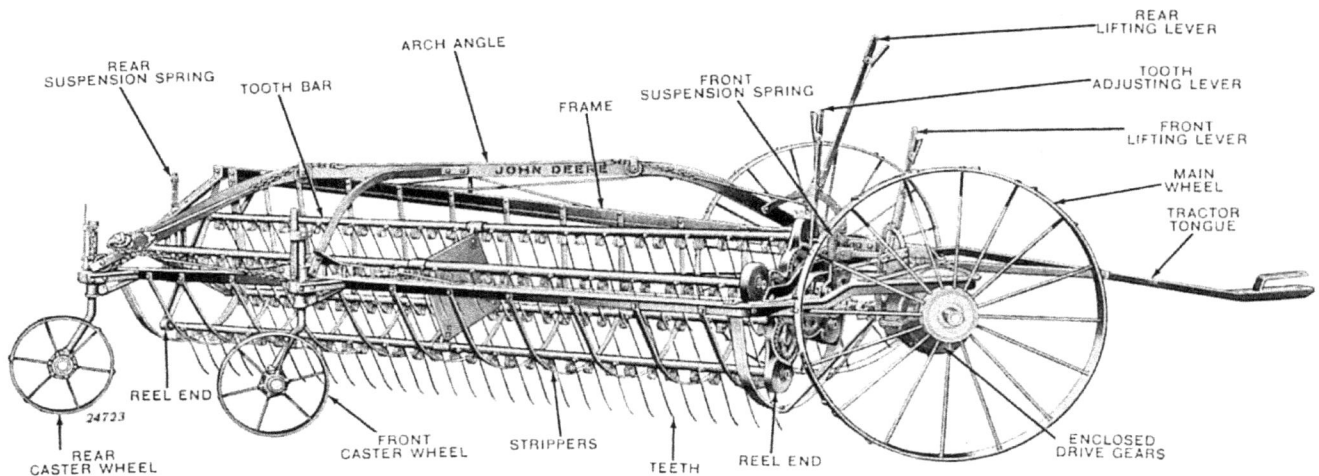

Fig. 48. John Deere No. 594 Side Delivery Rake.

Before starting the John Deere Side-Delivery Rake, make sure that all bolts are tight, cotter pins are spread, and machine has been properly set up.

Be sure to fill gear case with the proper grade of oil and lubricate.

When starting a new (or any) side rake, turn the reel by hand to be sure it revolves freely and the teeth do not strike the stripper bars. Then throw the rake in gear and turn the wheel by hand to see that the tooth bars and gears run free. Breakage of parts, which causes serious delay and additional expense, can be avoided by taking these precautions before entering the field.

An occasional thorough inspection for loose nuts, worn bolts, and other parts will add to the efficiency of your rake.

TOOTH-ADJUSTING LEVER

The most important adjustment is the angle of the teeth in relation to the surface of the ground. This adjustment regulates the raking of the teeth for loose or tight windrows.

Under average conditions the normal position for the tooth-adjusting lever will be in the center of the rack, at Notch 3. (Fig. 50) Moving the lever to

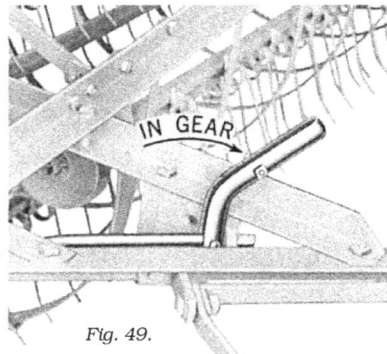

Fig. 49.

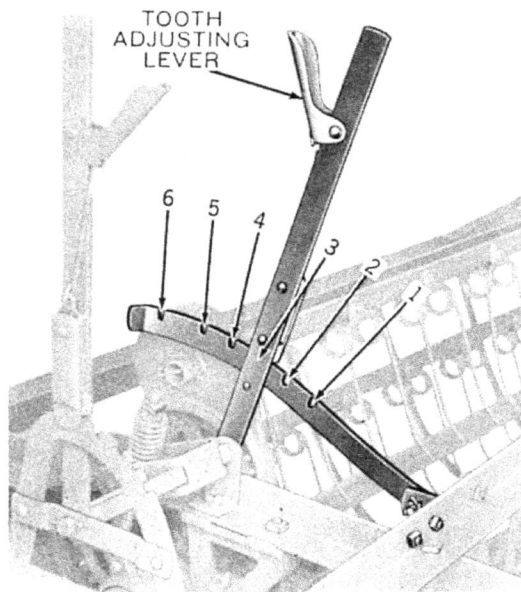

Fig. 50.

the rear toward Notches 4 and 5 increases the forward angle of the teeth to produce a loose, fluffy windrow. Moving the lever forward toward Notches 1 and 2, will decrease the tooth angle to produce a tighter windrow. The Sixth Notch is used when transporting rake.

FRONT LIFTING LEVER

The teeth should always be set as high as possible and still pick up all the hay. This setting causes the teeth to pitch the hay into loose windrows permitting free circulation of air. A trail in the center notch of the Front Lifting Lever will give an indication as to the position in which it should be set.

REAR LIFTING LEVER

The Rear Lifting Lever is properly set when the rear end of the reel is slightly higher than the front end. This aids in making the windrow loose and fluffy.

TRANSPORTING

In traveling on the road, the Tooth Adjusting Lever should be moved to Notch 6. In this position, the teeth are raised above the strippers, out of danger of being bent by hitting obstructions. Raise both ends of the reel as high as possible by moving the front and rear lifting levers into the extreme forward position.

When transporting the machine on a public road at night or during other periods of poor visibility, use a warning lamp in socket provided on the extreme left-hand side of the rake.

A warning lamp, that also may be used with other implements, can be purchased from your John Deere dealer.

MAKING HAY THE JOHN DEERE WAY

For proper method of cutting and side raking hay the John Deere Way, see illustrations below.

In mowing, enter the field as shown in Fig. 23, page 155, making one round to cut hay along the fence. Reverse direction of travel and continue around the field making right-hand turns until the entire field is cut.

Drive the John Deere Side Delivery Rake in the same direction the mower traveled. Working against the heads of the plants, the John Deere places the majority of the leaves inside the windrow. The leaves, shaded from the direct rays of the sun by the stems, are cured rapidly by the free circulation of air.

To hasten curing of especially heavy crops, or to preserve the quality of hay dampened by a shower, turn the windrow upside down by simply driving alongside the windrow with the left rake wheel just at the edge of the hay. This causes the windrow to be placed with the dry side down on dry stubble.

TIMING TOOTH PIPES AND MOUNTING STRIPPERS

1. Set tooth-adjusting lever in first notch, and move one arm of reel head into horizontal position toward rear frame angle. NOTE: It may be necessary to turn pipe gear so arrow will be visible for timing.

Place one intermediate gear in reel head, so that the arrows will line up with the arrows on the tooth pipe and center gear. Attach the intermediate gear to reel head with a bolt and washer. Revolve reel in normal operating direction 1/4 turn and repeat this timing operation until all four tooth pipes are timed.

NOTE: The timing arrows should align themselves on every thirteenth revolution of the reel.

2. Lay out strippers in numerical order. They are stamped on the back, 104 to 114. No. 104 bolts in the first hole of the frame next to the front reel guard. The others are mounted in every third hole thereafter; that is, leave two holes between each stripper. Attach the

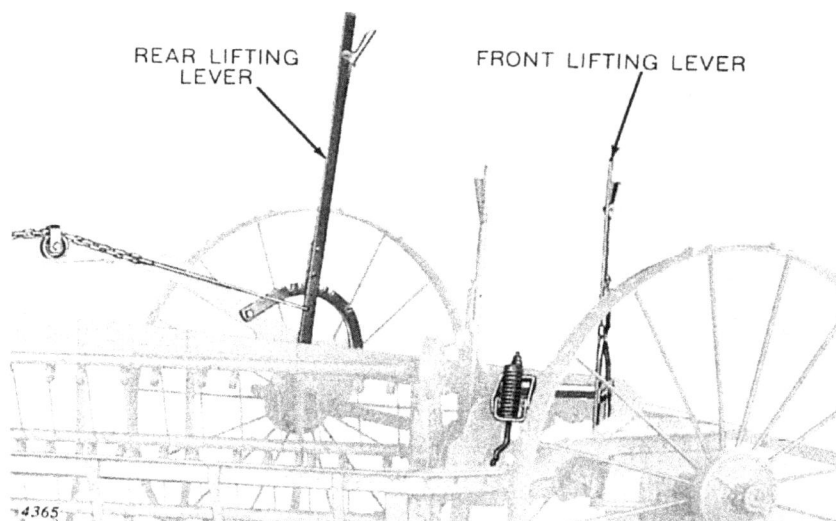

REAR LIFTING LEVER   FRONT LIFTING LEVER

Fig. 51.

back end of stripper first to the back side of rear frame angle and then the front end to the back side of the front frame angle with a 3/8" bolt. The heads must be on the inside next to the reel so the nuts will not snag material.

3. It is necessary to set the back ends of all strippers with a hammer and chisel, then finally tighten.

When setting up Mint Rake, use one 104 Stripper and three of all the others.

4. Attach rear reel guard in the last set of stripper holes in the front and rear frame angles. Attach rear reel guard brace as shown.

5. Attach the stripper support to the reel guards and each stripper with the 5/16" bolts. The heads of these bolts must be on the inside next to the reel.

Fig. 52. Bud Evers with Percheron horses on a Case rake.

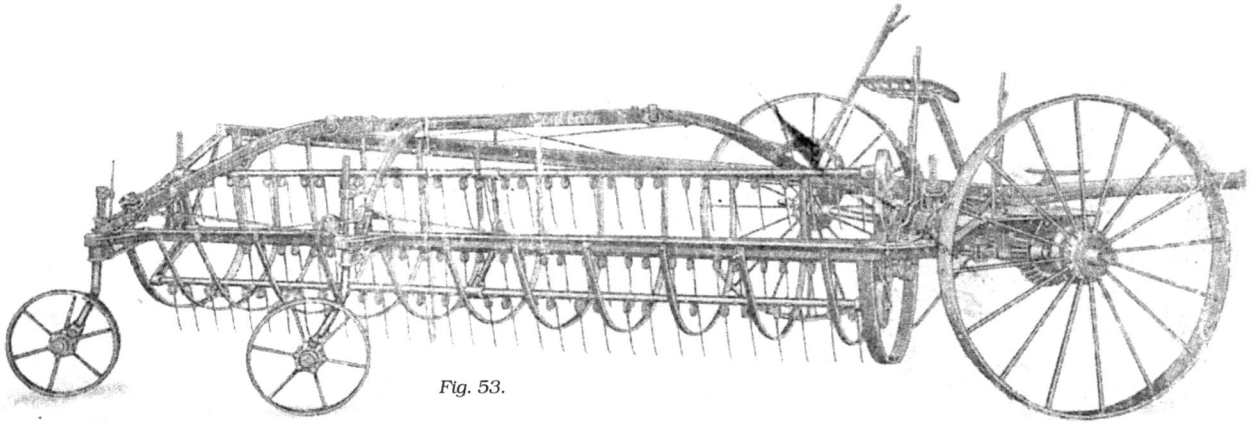

Fig. 53.

JOHN DEERE SIDE-DELIVERY RAKES

The material which immediately follows came from an Operation and Setup manual for JD models 553, 554P, 563, 564, 674, 574P, 583, 584, 593, 594, 594B, 594P and 594M. The diagrams were drawn to help a person work through the assembly process for new rakes. They are equally helpful to anyone trying to repair or reconstruct an older rake.

The John Deere Side Delivery Rake is set up as illustrated on this and the following pages. The darkened portions in the progressive illustrations show clearly the parts to be assembled and attached in proper order. Where the instructions or the connecting points are numbered, follow closely the order in which they are numbered. Arrows are also used to point out important adjustments or parts that need special attention in setting up.

Practically all the trouble with new machines is due to improper or careless setting up and lack of oil.

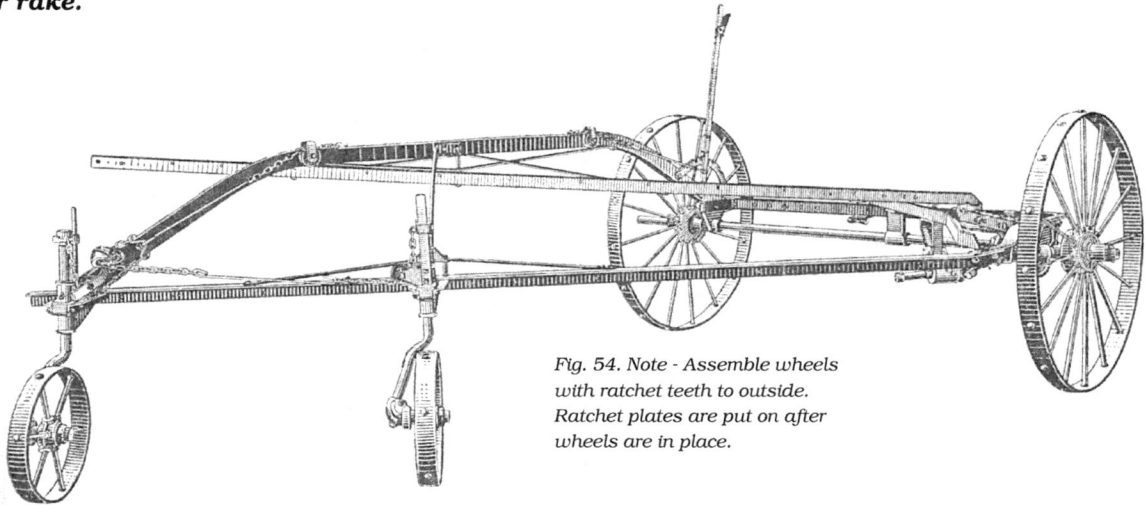

Fig. 54. Note - Assemble wheels with ratchet teeth to outside. Ratchet plates are put on after wheels are in place.

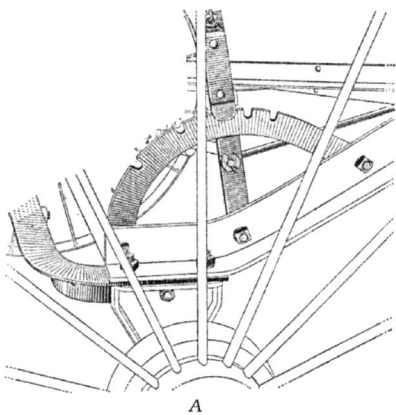

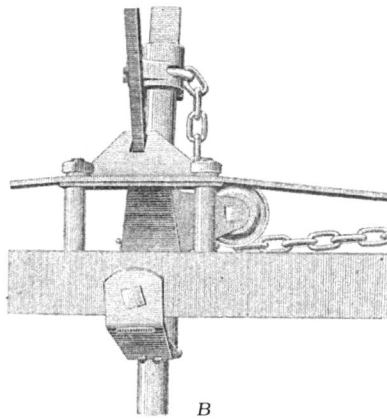

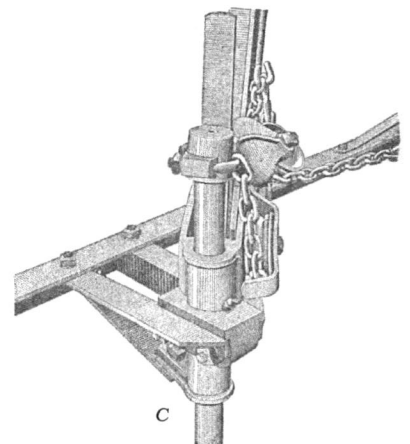

A

B

C

Fig. 55. A, B, C - Assemble frame as shown.

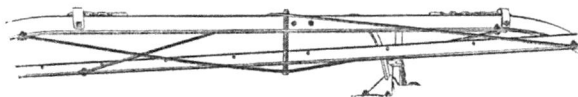

Fig. 56. Portland arch angle brace.

Fig. 57. (Left) Remove tooth pipe and intermediate gears from reel head and assemble front stripper, reel head and levers as shown. (Right) Shows front stripper and lever on three-bar rake.

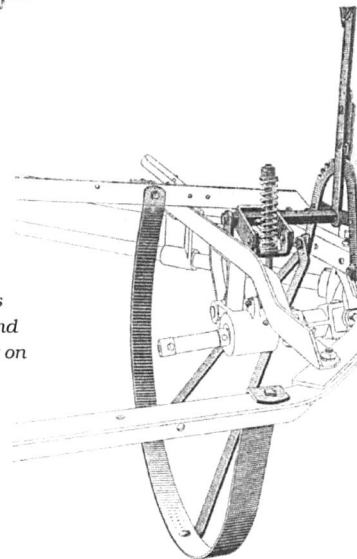

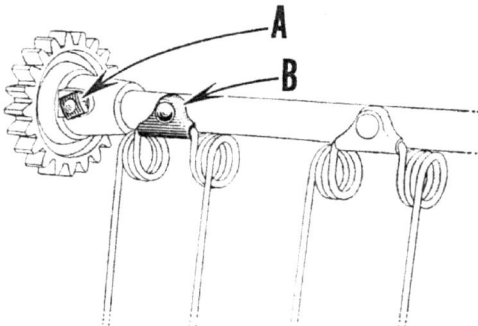

A

B

Fig. 58. Bolt tooth pipe gears on tooth bars as shown. Be sure gear is put on pipe so that nut "A" is on same side of pipe as tooth clips "B" as shown in above cut. The bolthead fits into the square hole in the gear on the opposite side from that shown.

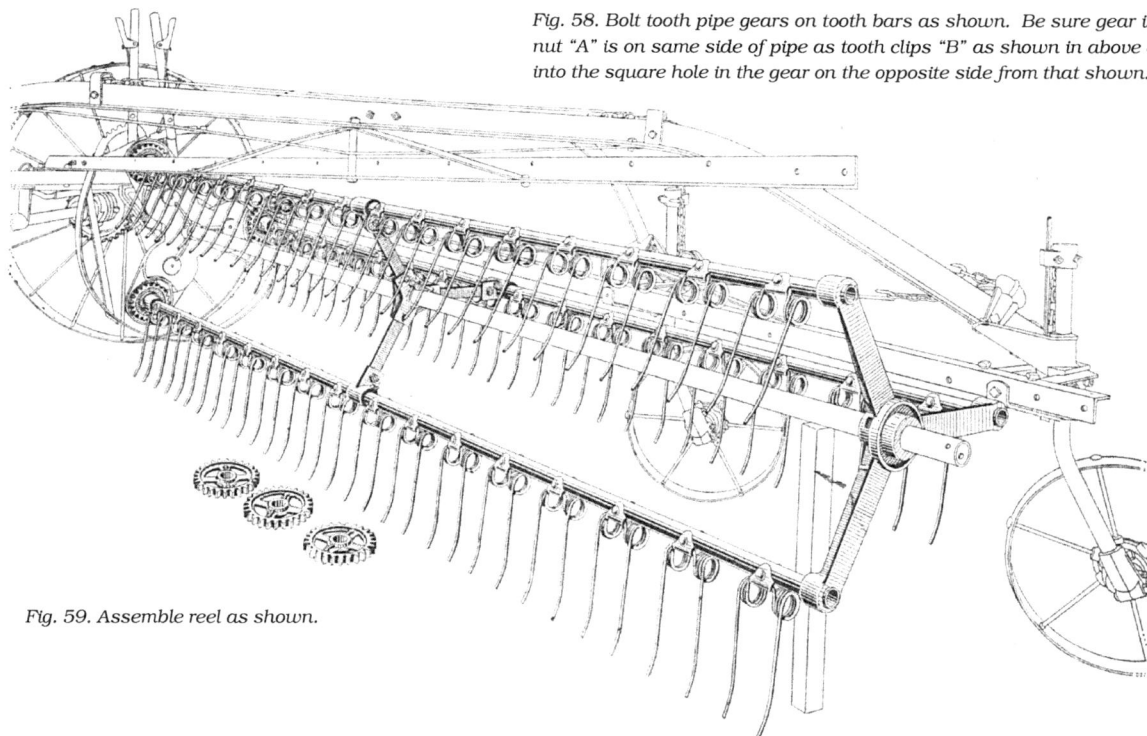

Fig. 59. Assemble reel as shown.

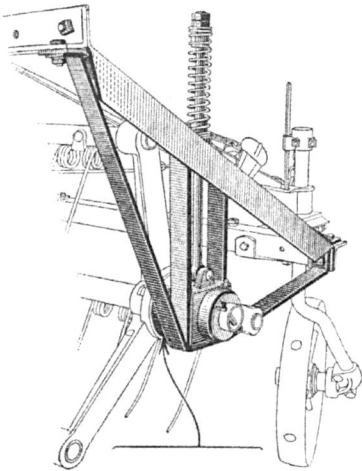

Fig. 60. Attach rear cross angle, bearing support, and brace as shown. **Make sure that Lugs in Bearing fit into offset in hardened retainer washer.**

Fig. 61. Attach strippers as shown. Bolt strippers on rear frame angle first. Note: Numbers are stamped on back of strippers; No. 104 stripper goes next to gears, then No. 105, etc. **Note that strippers bolt on rear of rear frame angle.**

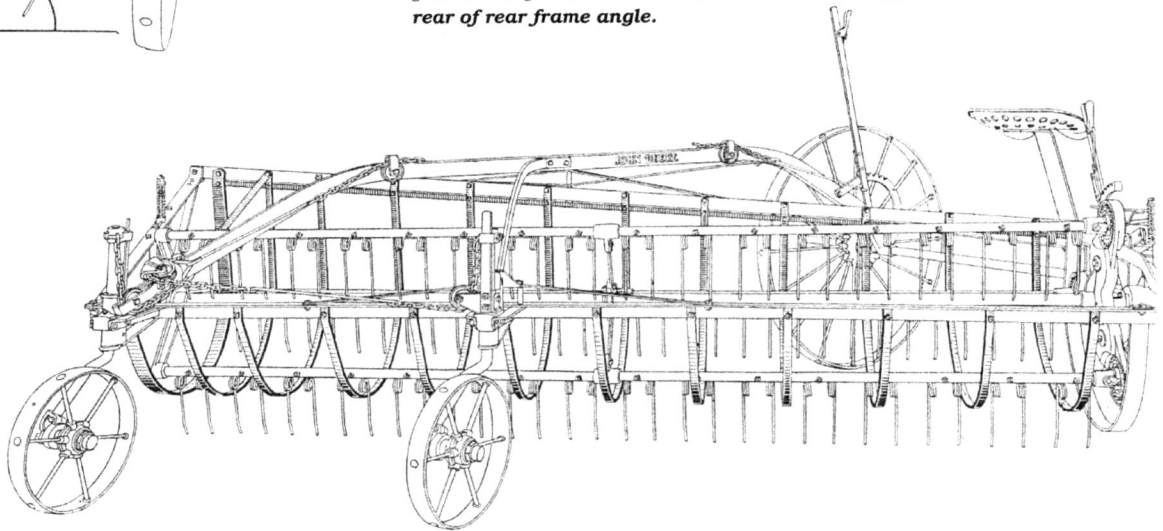

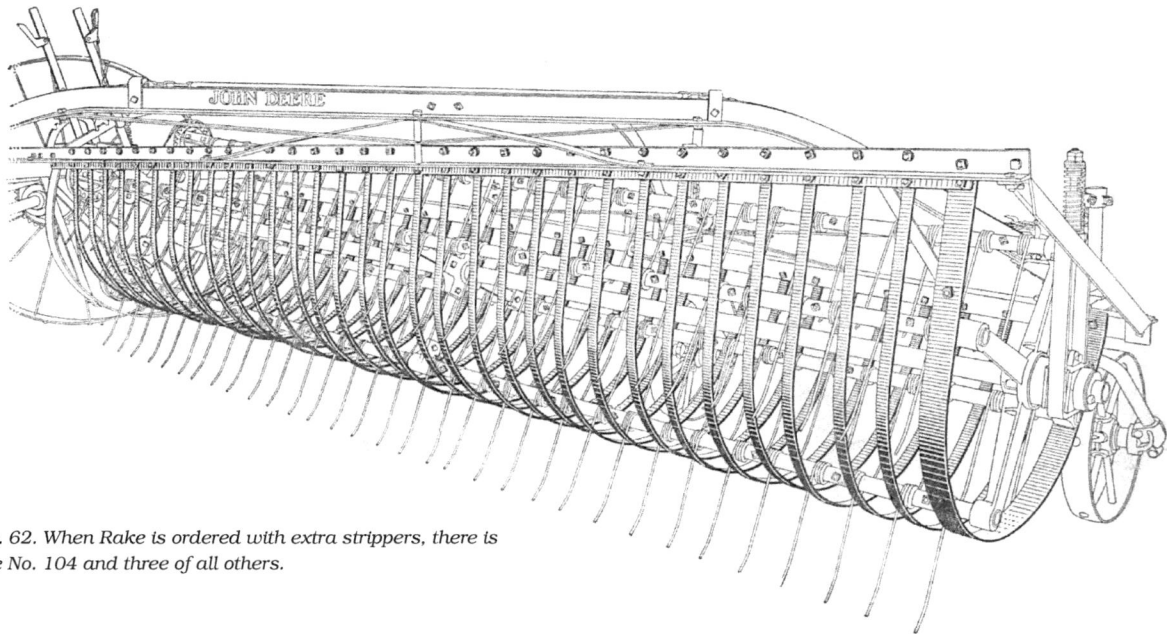

Fig. 62. When Rake is ordered with extra strippers, there is one No. 104 and three of all others.

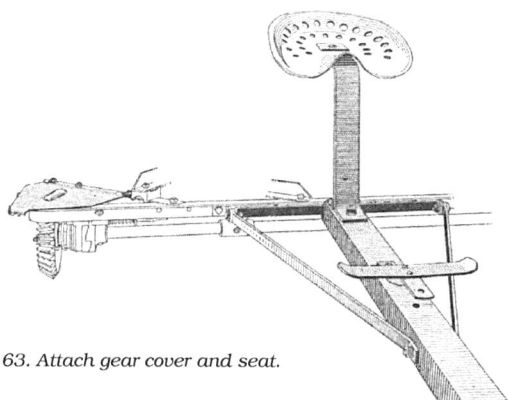

Fig. 63. Attach gear cover and seat.

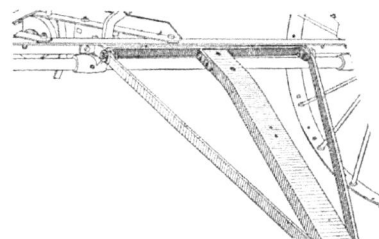

Fig. 64. Shows tractor tongue attached.

Fig. 65. Set tooth-adjusting lever in lowest notch and move one leg of reel head into horizontal position toward rear frame angle. Put intermediate gear in place so that the arrows will point to the teeth marked with an arrow on center gear and gear on end of tooth pipe. Bolt intermediate gear to reel head with bolt and washers. Repeat the above operation with the other two intermediate gears. Use same procedure on 4-bar rakes.

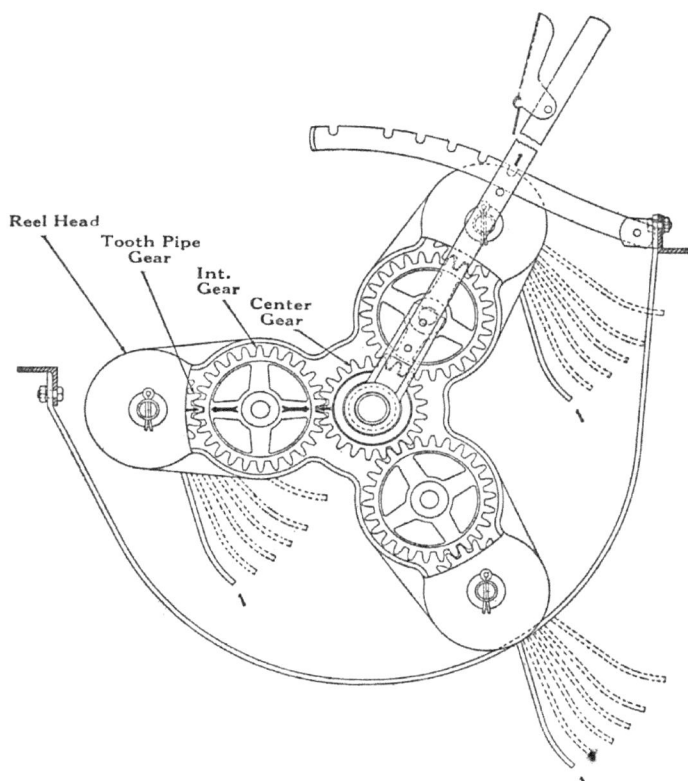

Reel Head
Tooth Pipe Gear
Int. Gear
Center Gear

Fig. 66. Right-hand extension axle used when raking bean rows varying in width and when raking hay in irrigated fields. When it is necessary to rake close to the borders, the wheel can be set out 16 or 20 inches.

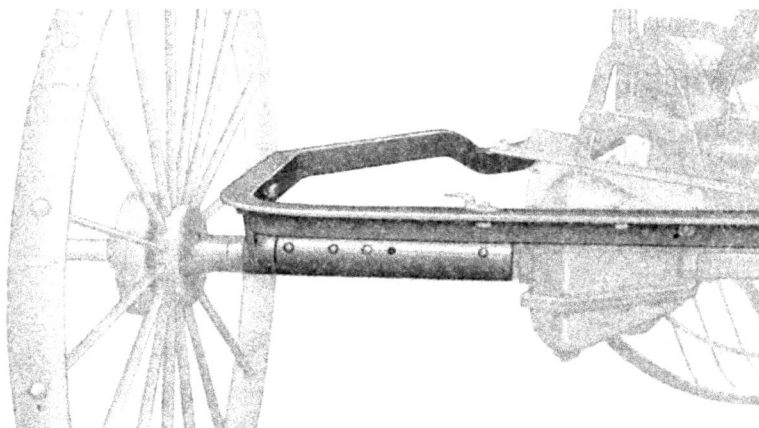

To Operate

A. Oil all wearing parts regularly and keep bolts tight. See Fig. 71 for oil instructions for gear case.

B. Turn reel by hand to be sure it revolves freely and teeth do not strike strippers.

C. Throw rake in gear and turn reel by hand to see that axle and gears also revolve freely.

D. Tooth position 1, 2, 3, 4, 5 (Fig. 50) are working positions obtained by setting the tooth adjusting lever in the 5 notches. Try it in the center notch first. The lever should be placed in notch No. 6 when traveling on the road. Teeth should always be set as high as possible and still pick up the hay. This leaves the windrow loose, permitting free circulation of air.

E. Adjust rear lifting lever so that rear end of reel is higher than front end of reel.

F. To turn windrow halfway over or upside down, rake should be driven alongside of windrow to be turned instead of straddling it.

G. Always drive in the same direction the mower has run if possible.

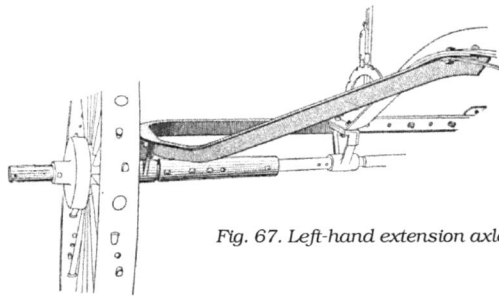

Fig. 69. AG302 E wheel scrapers for side rake caster wheels.

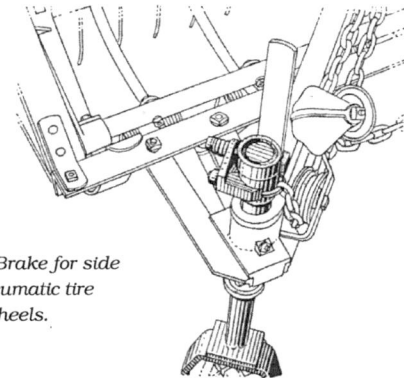

Fig. 70. Brake for side rake pneumatic tire caster wheels.

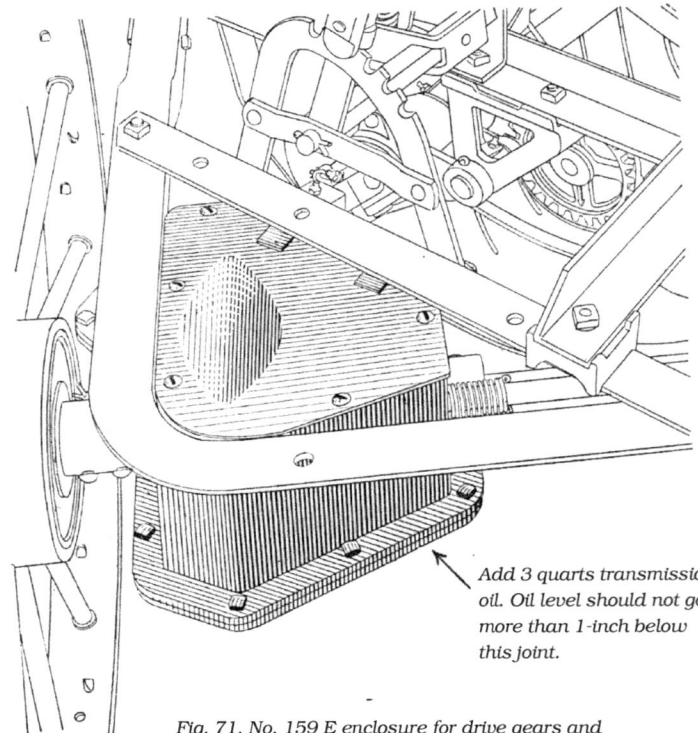

Fig. 67. Left-hand extension axle.

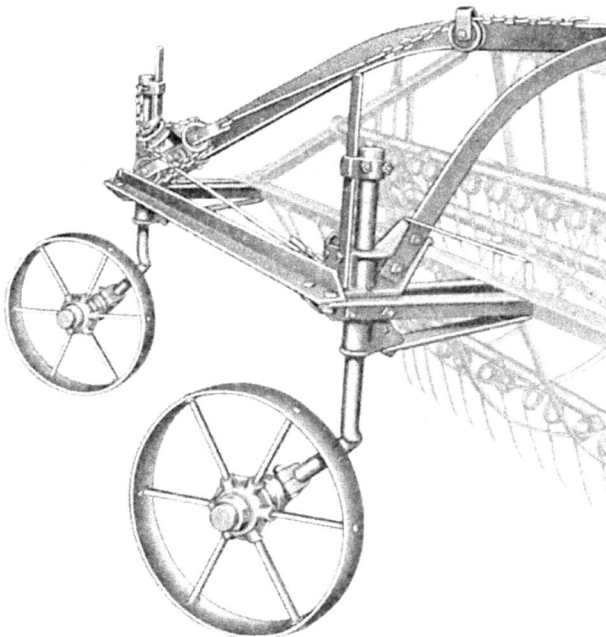

Fig. 68. The Portland style caster wheels give the rear of the rake better support in irrigated land. This equipment is furnished on special rakes for these territories.

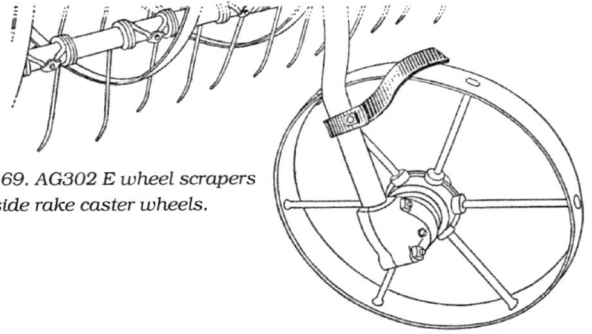

Add 3 quarts transmission oil. Oil level should not go more than 1-inch below this joint.

Fig. 71. No. 159 E enclosure for drive gears and clutch which protects these parts from dirt and allows them to operate in oil.

THE JOHN DEERE SULKY or DUMP RAKE

All the equipment manufacturers came out with promotional literature on their implements. Most of it is laced with sales talk as to be expected. Some of the brochures actually provided good information as well. What follows is the text from the John Deere brochure on its last model of dump rake. LRM

In light or heavy hay, while raking into windrows or bunching--in all conditions--the John Deere Self-Dump Sulky Rake does a clean job of raking and dumps easily. Even in exceedingly rough ground, it is giving remarkable satisfaction to users the country over. Its better performance makes a place for it on every farm.

One trip across the field on the John Deere will convince you that there's a big difference in sulky rakes. It will also make you realize, beyond all doubt, that the John Deere has the features for which you have long been looking--clean raking and ease of operation in varying conditions, and a sturdiness that means long life.

Improved Foot Lever Control

The outstanding ability of this new rake to fulfill the sulky rake needs of farmers in all sections of the country is due in no small part to the new foot control lever (See Fig. 73). This important John Deere feature makes it an easy matter to hold teeth down to their work or up when bunching from the windrow or turning at the end of the field.

The notch in the bottom of this foot lever makes possible long-wanted advantages. When the teeth are raised and the foot lever is pressed down, the notch catches over the stud in center hinge, thus holding the rake teeth up (See Fig. 74)

The unusual length of the foot lever and hinge

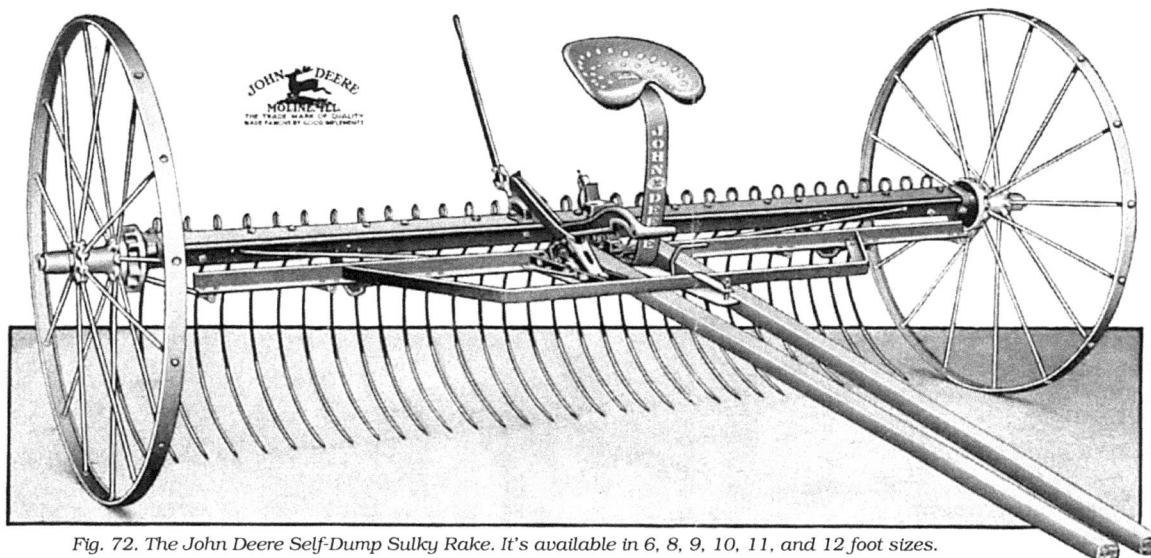

Fig. 72. The John Deere Self-Dump Sulky Rake. It's available in 6, 8, 9, 10, 11, and 12 foot sizes.

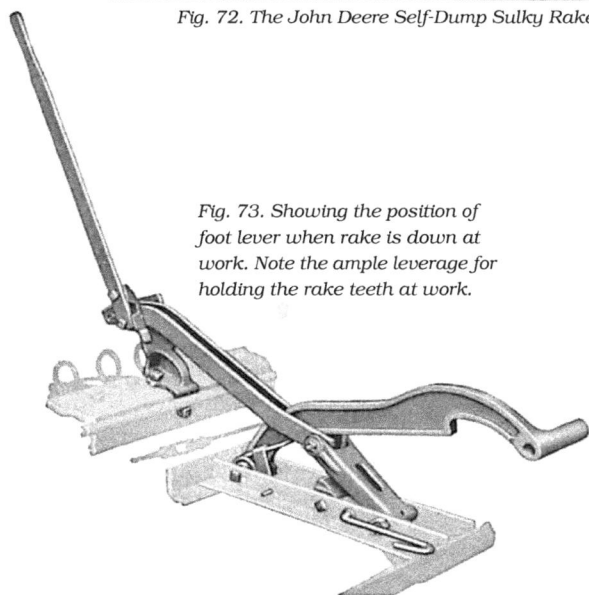

Fig. 73. Showing the position of foot lever when rake is down at work. Note the ample leverage for holding the rake teeth at work.

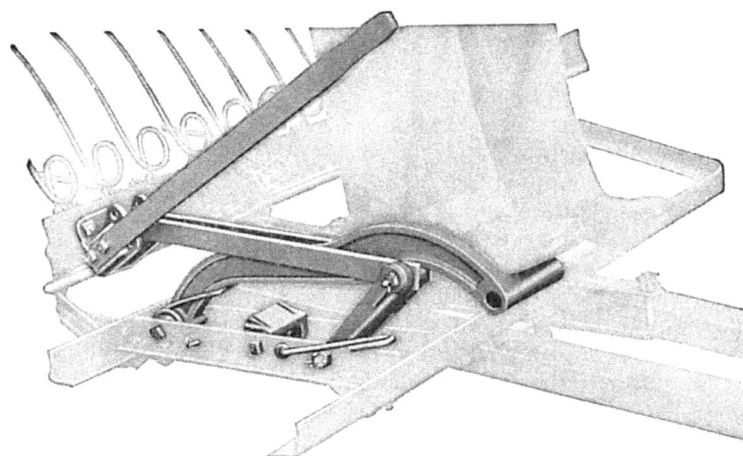

Fig. 74. Here is the rake with teeth in raised position with the operator holding the foot lever down.

gives extra leverage, which aids in making this operation extremely easy. This feature also enables the operator to hold the teeth down to their work with little effort. Note the large foot rest and ample leverage as illustrated in Figs. 73 and 74.

Foot Lever Self-Releasing

The self-releasing feature is a real convenience. The instant the foot is removed from the foot lever, a spring forces the center hinge upward (See Fig. 73). As a result, there is no undue strain on any part of the rake as the teeth start upward. The foot control is always in proper adjustment.

Long life is assured in these important working parts because the wearing surface on the bottom of the foot lever, the hook, and the stud over which it works, are hardened...

The transport link fastens over hand lever, thus holding teeth up while in transit.

Positive Tripping Device Affords Many Advantages

With the new positive tripping mechanism (See Fig. 75) the operator has constant and complete control over dumping and tripping...

When raking, dump rods are positively locked out of the wheel ratchets by the trip plunger (See Fig. 76). This makes it impossible for rake to dump until pressure on trip lever engages and locks dump rods in wheel ratchets. When tripped, dump rods are securely locked in wheel ratchets by trip plunger (See Fig. 77). They cannot release until forced out of engagement by trip plate coming in contact with snubbing block (See Fig. 75).

This lock-in-and-lock-out feature of the John Deere Sulky Rake makes premature dumping or tripping impossible under any condition.

Tripping Adjustment Changed Quickly

Without the aid of a wrench, the snubbing block may be turned to provide four different positions for tripping (See Fig. 75). In this way the rake can be adjusted to dump quicker or slower without stopping the team.

It is a real advantage to be able to quickly make adjustments so the teeth return to working position at the proper time.

Rake Has Correct Balance

To insure proper balance, the frame hinges and axle supports are located so the weight of operator not only assists in dumping but also forms a cushion-like action as the teeth return quickly to working position.

The seat spring, by reason of its backward tilt, helps to distribute the operator's weight to balance the rake. It also reduces neck weight on the horses and

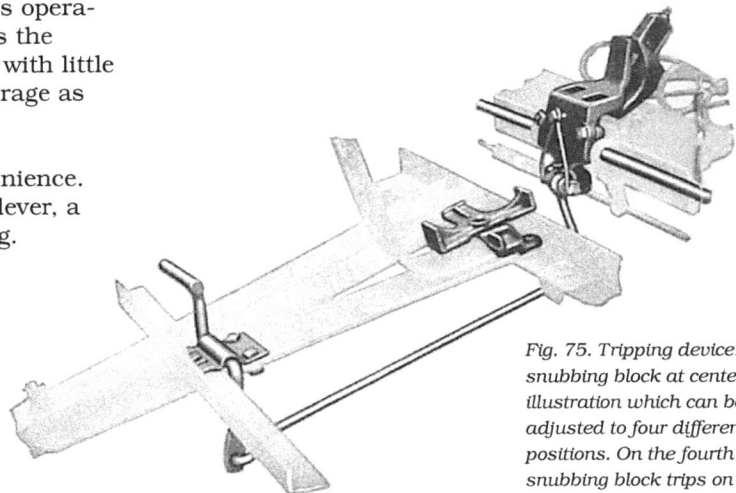

Fig. 75. Tripping device. Note the snubbing block at center of illustration which can be adjusted to four different positions. On the fourth position, snubbing block trips on rake head angle.

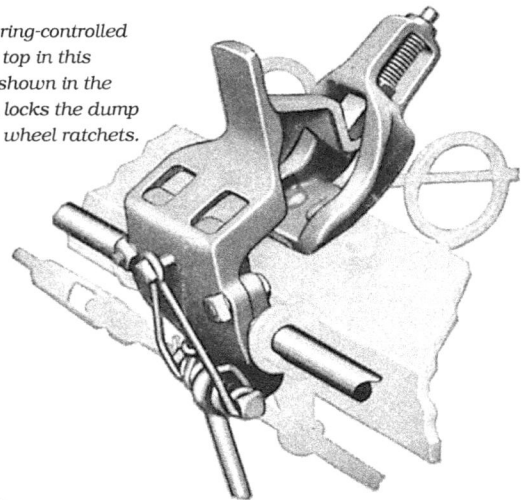

Fig. 76. The spring-controlled plunger, at the top in this illustration, is shown in the position which locks the dump rods out of the wheel ratchets.

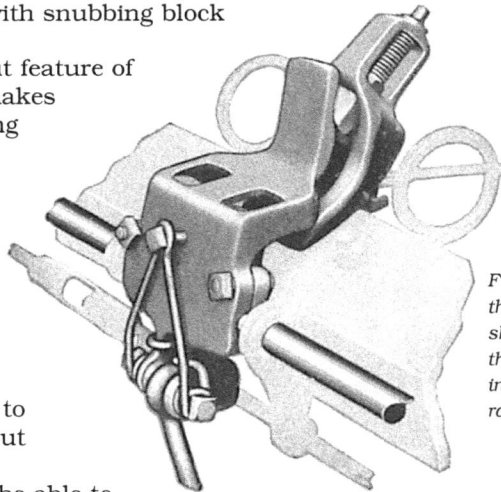

Fig. 77. Here the plunger is shown locking the dump rods into the wheel ratchets.

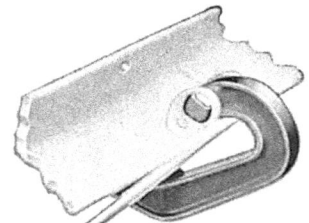

Fig. 78. The hinge pin cannot turn. Strain is taken by the entire surface of the pin.

makes riding more comfortable for the operator.

Interchangeable Parts Prolong Life of Rake

Repair costs are kept down and years are added to the life of this rake by reason of the interchangeability of important parts. For example, the wheels may be interchanged to secure double wear on ratchet teeth; the hardened steel dump rods are reversible, doubling the wearing surface.

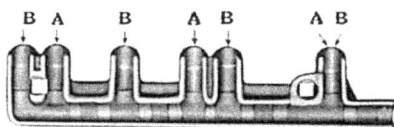

Fig. 79. Tooth Holder. Teeth placed in "A" for wide spacing, in "B" for narrow spacing. Below: Opposite side of holder, as shown above. Holders are heavily reinforced.

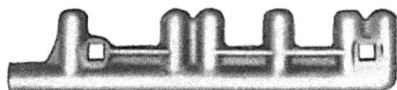

Fig. 80. Tooth holders accommodate minimum or maximum number of teeth.

Tooth holders accommodate the minimum or maximum number of teeth, thereby making two rakes in one. Only one kind of tooth holder is required for each size of tooth; no special end holders are required. (See Fig. 79).

Axles Securely Bolted

The durable steel axles, which give the wheels the proper pitch at all times, are bolted securely to rake head. Bolts are double-nutted. The easily-adjusted truss rod provides means for keeping the axles and

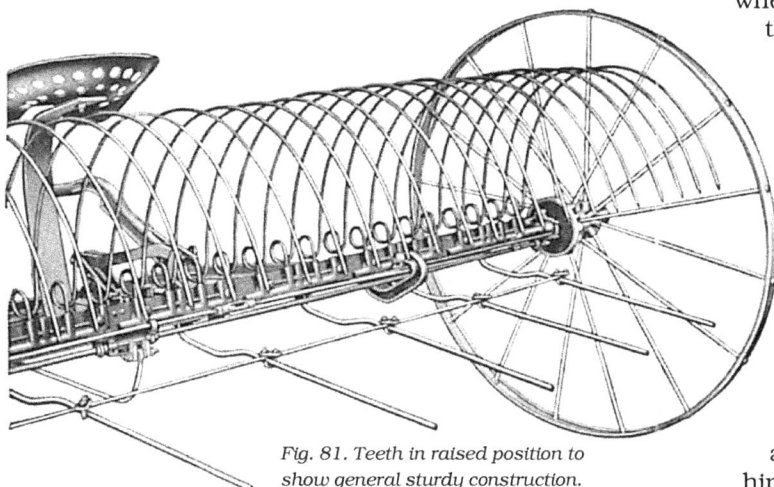

Fig. 81. Teeth in raised position to show general sturdy construction.

wheels in proper working position throughout the life of the rake.

Adjustable Truss Rod Reinforces Rake Head

On the 9-, 10-, 11-, and 12-foot sizes, the rake head is greatly reinforced by an adjustable truss rod and a heavy steel strut welded to the head angle.

Sagging of the frame and spreading of the wheels are prevented; wear on the wheel bearings is also greatly reduced. The truss rod is not needed on the 6- and 8-foot sizes.

Rake Teeth Lift Hay

The teeth are so shaped and located that the hay is lifted off the ground as the rake moves forward. The capacity of the rake is fully utilized; dragging and dusting of hay is prevented. The result is clean raking and clean hay all the time.

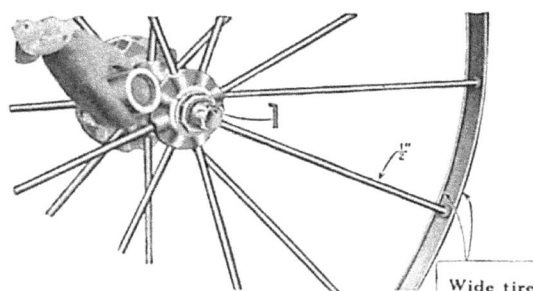

Fig. 82. All wheels have hard grease caps and extra long bearing.

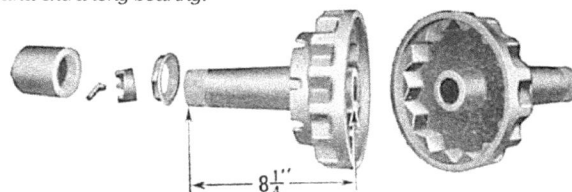

Fig. 83. Notice above the 8-1/4 inch bearing used in the sulky rake wheels, and the grease cap and assembly.

Wheels Are Unusually Strong

On the ordinary rake, the wheels are often the first parts to give trouble. On the John Deere, the wheels are built exceptionally strong and durable-- they last longer. The John Deere regular rake wheels are 54 inches high and have 16 wide, staggered spokes which are hot-riveted to the tire and cast into the strong hub.

Axles are longer than others and are thoroughly and easily lubricated with hard grease caps. (See Fig. 82).

Two types of mountain wheels may be obtained for severe conditions. The regular mountain wheel with 18 spokes and 2-inch tire; and the special heavy-duty mountain wheel with 18 spokes and 2-1/2-inch special channel tire.

Riveted Angle-Steel Frame Extra Strong

The extra-strong front and rear steel frame angles are rigidly cross-braced. Heavy malleable hinges are fastened to the frame and head angles with three rivets.

These strong malleable hinges insure a better bearing as well as more uniform hinge alignment and stronger connection between frame and rake head. Their location also aids in hastening the return of the teeth to working position. Strength is added by ribbing on the underside where the most strain comes.

The seat support bar is fastened to the rear frame angle with two rivets instead of one as ordinarily used. Foot-lever support angles are reinforced with a heavy steel plate. Large hinge pins are held from turning by special head, putting the wear on the full width of the malleable hinge bearing. (See Fig. 78).

Rake and Cleaner Teeth Made of Oil-Tempered Spring Steel

Both the rake and cleaner teeth are made to withstand severe strains. The use of oval spring steel stock for the cleaner teeth insures the extra strength

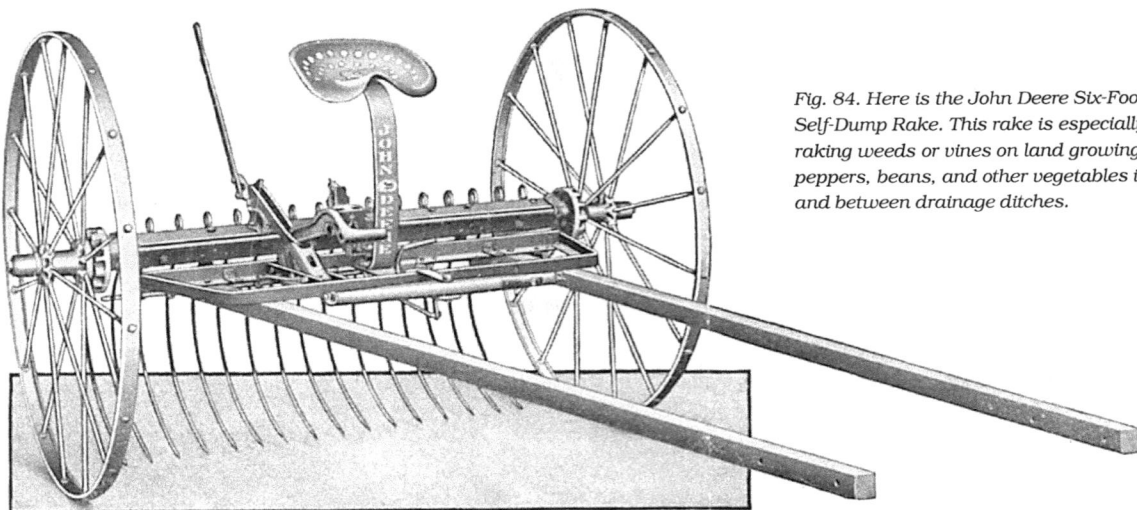

Fig. 84. Here is the John Deere Six-Foot Truck-Farm Self-Dump Rake. This rake is especially designed for raking weeds or vines on land growing such crops as peppers, beans, and other vegetables in narrow beds and between drainage ditches.

necessary in unusual conditions. Cleaner rod clamps are the U-bolt type, permitting secure clamping of fingers to the tie rod. (See Fig. 81).

Sizes and Equipment

Made in 6-, 8-, 9-, 10-, 11-, and 12-foot sizes. Relief spring is regular on 11- and 12-foot. This spring is recommended for 10-foot rake equipped with heavy teeth, or when the work is heavy.

The 6-, 8-, and 9-foot rakes are regularly equipped with combination singletree and pole extension; the 10-, 11-, and 12-foot rakes with plain pole extension.

The 11-foot cornstalk rake is regularly equipped with 27 one-half-inch, single coil teeth with pencil point

Extras

Relief spring for 9- and 10-foot sizes.
Mountain wheels with extra spokes.
Special heavy-duty mountain wheels with extra spokes, and wider tire.
Tongue truck.
Guard teeth, wheel shields, stiff pole, doubletrees

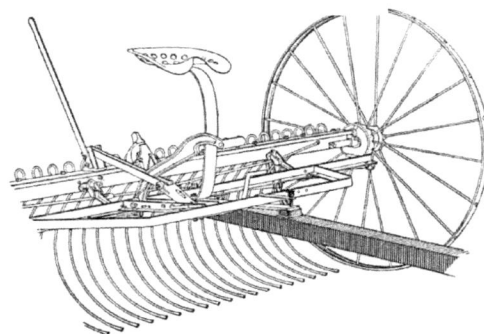

Fig. 85. Special stiff pole may be preferred in place of the two-piece pole, in rough ground or in heavy hay. Neckyoke is regular with this pole.

and neckyoke, 7/16-inch and ½-inch teeth, and overhead cleaner are also extra.

A Sturdy Tongue Truck

Those who wish to make their John Deere run even more steadily, turn still easier, and ride more comfortably, can obtain the improved Tongue Truck. (See Fig. 86).

The seat can be moved from the rake to the truck and the truck unhooked from rake, making a handy transporting truck for various light farm implements.

The wheels are equipped with replaceable chilled axle sleeves and bearing boxes--there is no wear on axle and wheel.

Fig. 86. Tongue truck can be furnished for the John Deere Sulky Rake. It prevents whipping of the tongue and promotes smooth running.

Fig. 87.

Fig. 88. This is a custom-built "scatter rake" (2 rakes welded together) in use at Hirshey Ranch in the Big Hole Country of Montana.

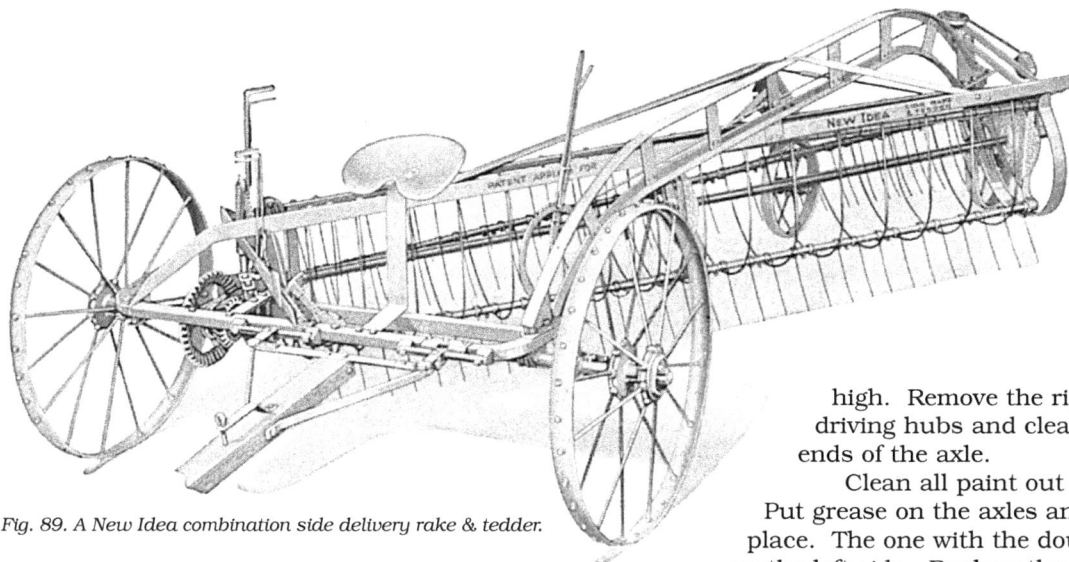

Fig. 89. A New Idea combination side delivery rake & tedder.

INSTRUCTIONS FOR SETTING UP AND OPERATING

THE NEW IDEA SIDE RAKE AND TEDDER

This information is from the New Idea manual of the period.

1. (Fig. 90) Remove all parts wired in the main frame bundle. Lay the main frame with the driving mechanism on a pair of boxes or trestles about two feet high. Remove the right and left outside driving hubs and clean all paint off both ends of the axle.

Clean all paint out of both wheel hubs. Put grease on the axles and slip the wheels in place. The one with the double ratchet hub goes on the left side. Replace the driving hubs and pins, be sure to spread the spring cotters in the cross pins.

NOTE: Do not tighten any of the bolts in the frame assembly until the entire frame has been completely assembled. Make sure that all bolts have lock washers under the nuts.

2. (Fig. 90) Bolt the front end of truss to front cross angle using the same bolts that hold the left main axle bearing in place.

3. (Fig. 90) Bolt rear reel frame angle to front main frame angle and truss. Put all bolts through from the bottom.

4. (Fig. 90) Clean all paint from caster wheel spindle and from bore of caster wheel. Lubricate wheel

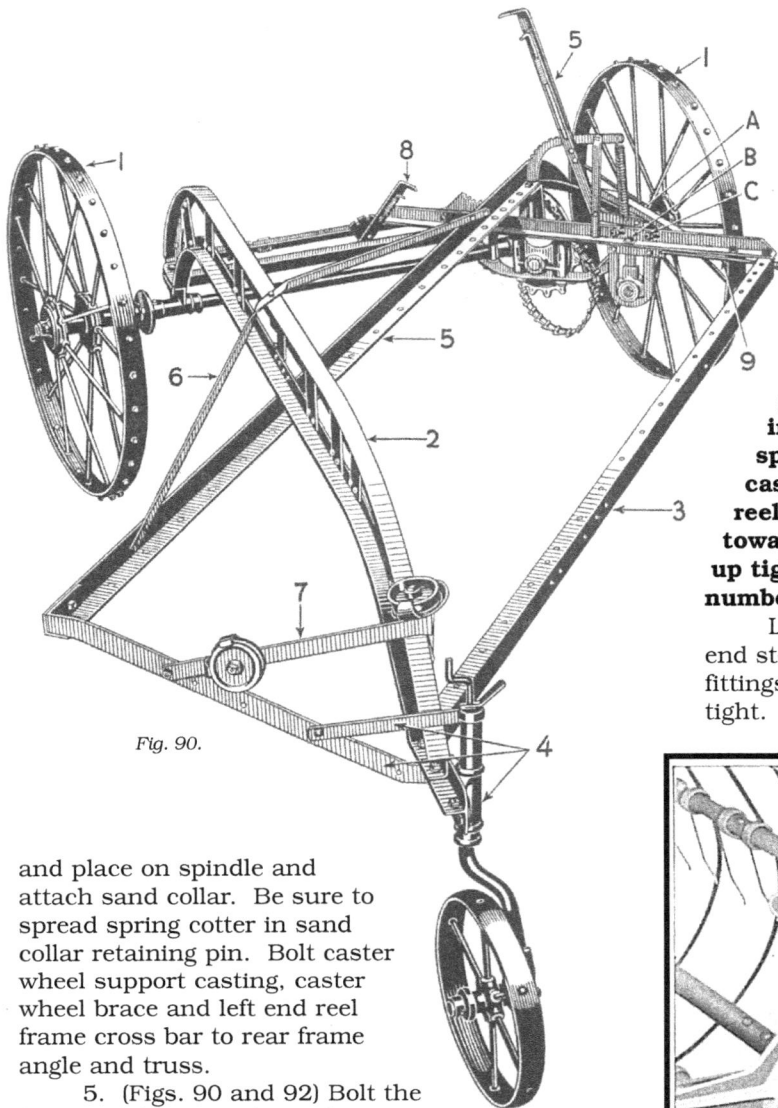

Fig. 90.

and place on spindle and attach sand collar. Be sure to spread spring cotter in sand collar retaining pin. Bolt caster wheel support casting, caster wheel brace and left end reel frame cross bar to rear frame angle and truss.

5. (Figs. 90 and 92) Bolt the front reel frame angle in place, starting with the front bolt. Then bolt the right reel controlling lever and sector in place and connect to reel bearing. The center support for the truss should be bolted to the outside of the upright angle leg of the front reel frame angle and finally bolt the angle to the left end cross bar, putting the carriage bolts in from the front.

6. (Fig. 90) Bolt truss bar to the inside front reel frame angle and truss.

7. (Fig. 90) Bolt flat sheave support to truss frame and outside of left end frame cross bar. Bolt

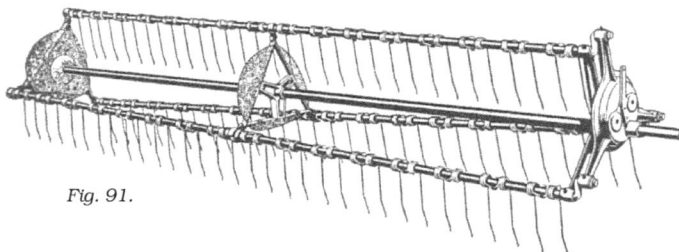

Fig. 91.

to be put through from the inside.

Now go over the entire frame and tighten all the bolts. Make sure that all bolts have lock washers under the nuts.

8. (Fig. 90) Bolt clutch lever to the sector and connect to clutch shifting rod. Shifting rod is properly adjusted when the machine leaves the factory. Do not loosen the set collar.

9. (Fig. 90) Put drive chain tightener rod in place and connect to right reel bearing hanger angle.

10. (Figs. 91 and 93) Assemble raking reel. This should be done on the floor or ground to the left of the machine. **Drive key in inner keyway of reel shaft and put right cast spider (M-310) on so that the flat side of the casting and the arrow is toward the inside of the reel and the casting number and set screw is toward the outside. Draw set screw and lock nut up tight. Lay the pipe and spider so the leg with number and arrow is in the upright position.**

Loosen supports for tooth bar bearing in the left end steel spider assembly and insert pressure type fittings in bearings, replace bolts and draw them up tight.

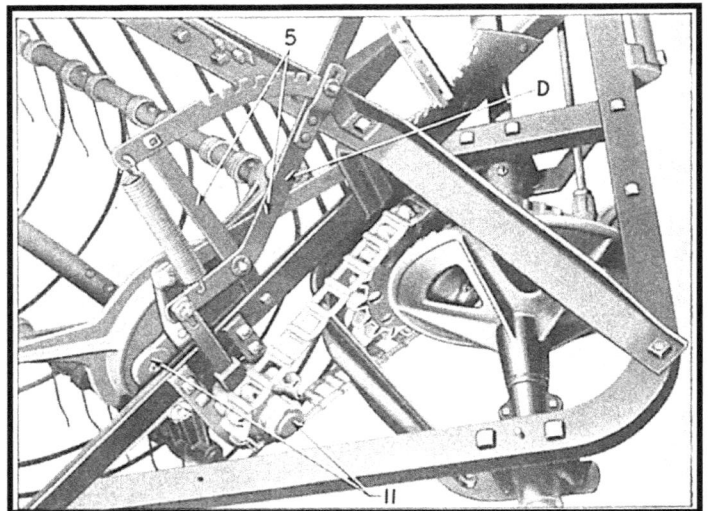

Fig. 92.

Remove the steel bearing arms from the center reel shaft cast spider (M343) and slide it into position on the center reel pipe.

Place the left steel spider assembly and large flare shield on the pipe so that when one bearing is at the top its grease fitting is toward the front, bolt it securely to the clamp and bolt already in place on the pipe and so that the U bracket is in a downward position when the leg of the right cast spider having the number and arrow is in an upward position. The above positions of the spiders are necessary to provide the proper angularity on the tooth bars so that the right or inside end of the tooth bars lead the left or outside end when raking. Bolt the large flare shield to the inside of the left

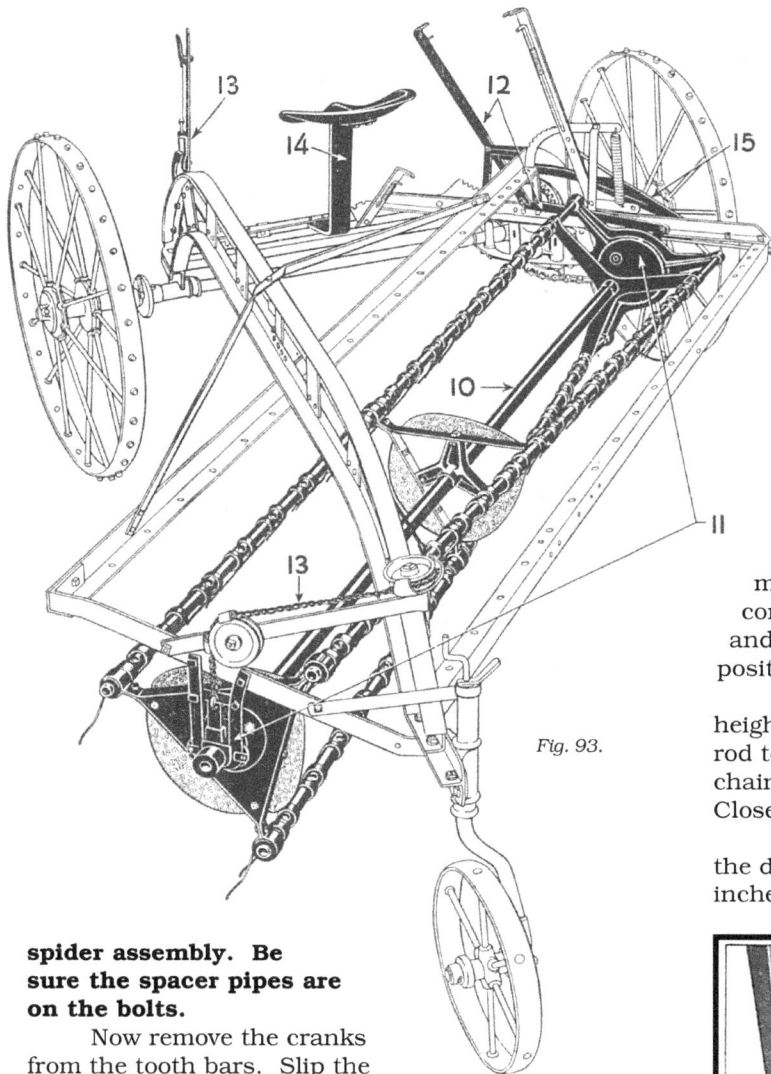

Fig. 93.

put in the machine. **Put a wearing washer on the right end of reel shaft,** slip shaft through right bearing **and slip a thin washer on the shaft,** also slip left end reel shaft bearing on reel pipe and bolt its hanger to end cross bar. Fasten drive sprocket to shaft with key and cross pin, use 8 tooth sprocket if rake is to be drawn with a tractor. Now remove the rear bolt "C" Fig. 90 from the right bearing hanger and the bolt "B" holding the hanger angle to the main frame angle. Then put reel height-controlling lever "D" Fig. 92 in the third notch from the front. The drive chain can now be easily slipped over the sprocket. Do not uncouple chain. Replace the bolts and adjust the chain so it runs with just a little slack and draw bolts up tight. Insert pressure type fittings in both right and left reel shaft end bearings.

12. (Figs. 93 and 94) Bolt tooth control mechanism lever bracket and brace in place. Then connect tooth mechanism controlling link to lever and post so that the post stands in an upright position.

13. (Fig. 93) Bolt sector and lever for controlling height of left end of reel to truss. Attach controlling rod to lever, bolt the supporting clip to truss and place chain around sheaves and connect to left reel bearing. Close connecting link on bearing and rivet over end.

Adjust clamp on height-controlling rod so that the distance between the clip and clamp is 11 3/4 inches when the lever is in its forward position.

spider assembly. Be sure the spacer pipes are on the bolts.

Now remove the cranks from the tooth bars. Slip the tooth bars in the bearings of the left steel spider, then slip the cranks through the right cast spider **and place one steel washer on each crank between the hub of the cast spider (M310) and tooth bar pipe,** slip cranks into the tooth bar and replace bolts so that the crank stands opposite the teeth. **Attach control spider to cranks of tooth bars so that its number is toward the outside, place washers on crank ends, insert spring cotters and spread them.** Insert grease fittings in center tooth bar bearings and bolt center spider steel arms to bearings and center reel support. Now space the three center reel pipe bearings evenly between the tooth bar springs and then tighten set screws and lock nuts in center reel head. Bolt the three small flare shields to the center spider so flanged portion is toward the left as illustrated. The reel should now appear as in Fig. 91.

11. (Figs. 92 and 93) Lubricate the rollers of the tooth control mechanism and make sure that they turn freely. Slip the roller assembly on the shaft and into the control spider. Insert grease fitting in roller assembly hub bearing. The cylinder is now ready to

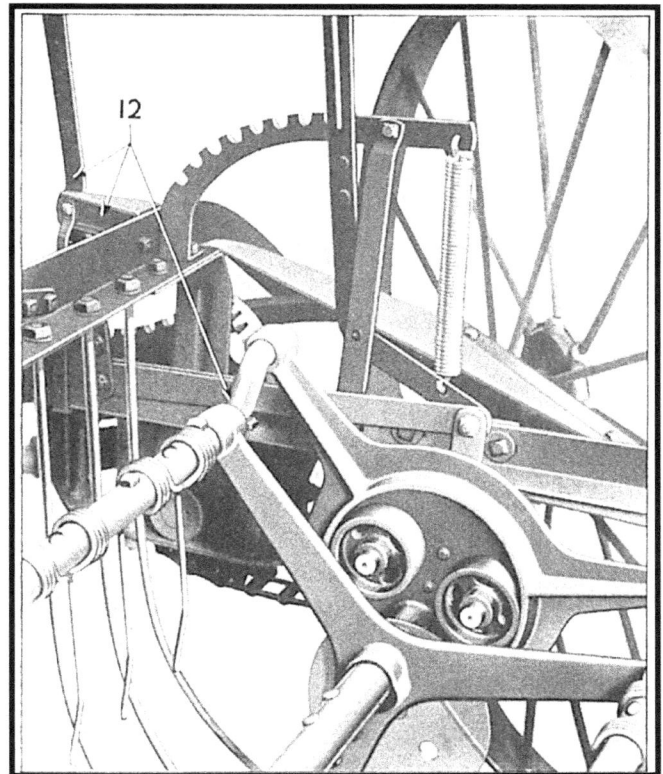

Fig. 94.

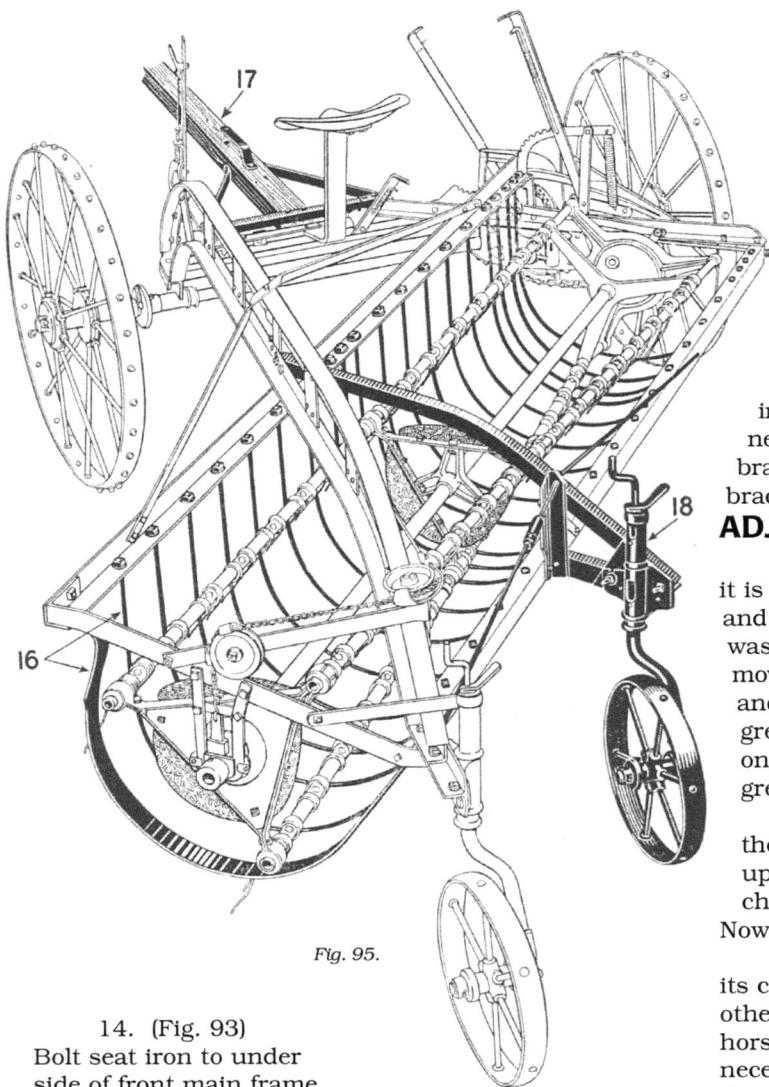

Fig. 95.

14. (Fig. 93) Bolt seat iron to under side of front main frame angle and attach seat.

15. (Fig. 93) Bolt gear shield in place over drive gears.

16. (Fig. 95) Bolt round stripper bars in place. **The shortest one is for the right end under the control spider.** Bolt the flat fender bar in place on the left end of the frame.

17. (Figs. 95 and 96) Bolt tongue irons to tongue and attach tongue to rake. Bolt neckyoke casting to

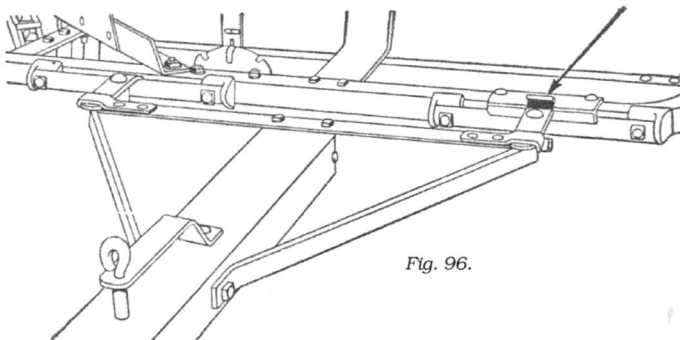

Fig. 96.

front end of pole and attach hammer strap.

18. (Fig. 95) When the machine is to be equipped with the extra caster wheel attachment bolt the bracket to the rear reel frame angle and lower angle of truss. Clean all paint from caster wheel spindle and from bore of caster wheel. Lubricate wheel and place on spindle and attach sand collar. Be sure to spread spring cotter in sand collar retaining pin. Put truss rod in place and adjust turn-buckle. Bolt caster wheel assembly to the frame bracket putting bolts through from the right side. Be sure to spread spring cotters and wrap them around the bolts.

19. Now insert grease fittings at the following points--4 in control spider, 3 in cast reel spider next to control spider, 1 in main caster wheel bracket and 1 in extra caster wheel attachment bracket if one is used.

ADJUSTING AND OPERATING

Now go over the entire machine to make sure that it is properly set up, that all spring cotters are spread and that all bolts are drawn up tight and have lock washers where needed. Give all bearings and other moving parts a generous oiling. Oil the lever latches and be sure they work freely and easily. Put a little grease in the groove of the sliding clutch casting and on the axle the clutch works on. Also put some grease on the gears and oil the drive chain.

The drive chain can be adjusted by loosening the bolts "A", "B" and "C" Fig. 90. Then by drawing up or loosening the tightening rod 9, Fig. 90, the chain can be easily adjusted, do not get it too tight. Now tighten the bolts previously loosened.

The rake is generally operated with the tongue in its center position. When raking beans and for certain other classes of work it is sometimes necessary for the horses to walk either to the right or left. If this is necessary the tongue can be easily moved by lifting up the spring controlled locking plate, indicated by the arrow in Fig. 96, and shifting the tongue to the desired position. Always use a double-tree long enough so that the horses walk in the swath board marks made by the mower.

When it is desired to turn a windrow half over or when raking beans the left ground wheel can be quickly shifted to its inner position by removing the hinged spacing collar and sliding the wheel in and turning it backwards to engage the driving dog. Then put the hinged spacing collar between the wheel and the outer driving hub. See Figs. 97, 98, and 99.

This feature is especially valuable when the hay in the windrow is rained upon, simply slip the wheel to its inner position and after the top of the windrow and the ground has sufficiently dried drive the team so that one horse walks along the side of the windrow and turn the hay half over. This then puts the wet hay on top to cure and dry out.

When the tedding feature of the NEW IDEA rake is used and if the rake is equipped with the extra caster wheel it is always desirable to swing this extra

Fig. 97.

Fig. 98.

Fig. 99.

wheel up out of the way so as not to run over the tedded hay. This is quickly accomplished by loosening the two bolts that hold the bracket casting to the frame and swinging the wheel up and laying it on top of the frame, tighten the one bolt to hold the wheel in this position. See Fig. 100. When the machine is again to be used as a rake the wheel can easily be put in its regular position.

When the machine is to be used as a rake throw the clutch lever to the right so that the clutch engages the small bevel gear on the main axle. Adjust the angularity of the teeth by shifting the tooth controlling lever so that the teeth stand in a forward position and in such a manner so that a nice fluffy windrow is obtained. Adjust the height of the reel by putting the reel controlling levers in the desired positions. Do not set the reel so low that the teeth strike the ground, this is not necessary for clean raking. For green soy bean hay or similar material that might have a tendency to wrap, use a little less tooth angularity.

For proper curing of the hay it is necessary to have the leaves on the inside of the windrow and the stems toward the outside. To accomplish this the rake should always follow the mower or in other words drive in the same direction as the mower went when the hay was cut. The NEW IDEA rake can be used wherever side raking is in practice, and furthermore can be used with great success for the various systems of hay making. The time the hay should be raked into windrows depends upon the kind of hay, atmospheric

conditions and the locality. However, the hay should always be raked before the leaves dry to such an extent that they shatter when raking. In raking make sure that all the hay is moved and turned over on clean dry stubble.

When the machine is to be used as a tedder throw the clutch lever toward the left so that the clutch engages the large internal bevel gear on the main axle. Adjust the angularity of the teeth so they stand in a rearward position. Tilt extra caster wheel up. Drive team in the swath board marks and fast enough so that the hay will be fluffed for quicker curing before windrowing. CAUTION: When used for tedding run the machine with the teeth higher off the ground than when raking.

The adjusting screws in the upper end of the caster wheel brackets M-352 and M-353 are for changing the height of the reel basket and should be turned down into the castings as far as possible and still let the cylinder teeth rake the hay. Under ordinary conditions they should be turned in about two-thirds of the length of the threads on the adjusting screw. In heavy hay, high stubble or stony ground turn the adjusting screws down so as to raise the basket as much as possible. For light crops where the teeth have to comb very close to the ground to get the hay, it may be necessary to turn them out considerably. The idea is to carry the stripper bars or basket as far off the ground as possible but close enough to get the hay. Adjustments have to be made accordingly in order to get the desired results.

Caster wheel brackets are equipped with a damping device to prevent wobble when the machine is in operation. The dampner plate should be kept free from grease. When trailing a rake on the highway be sure to fold up the extra caster wheel.

Give the machine a thorough lubrication and be sure to find all of the grease fittings.

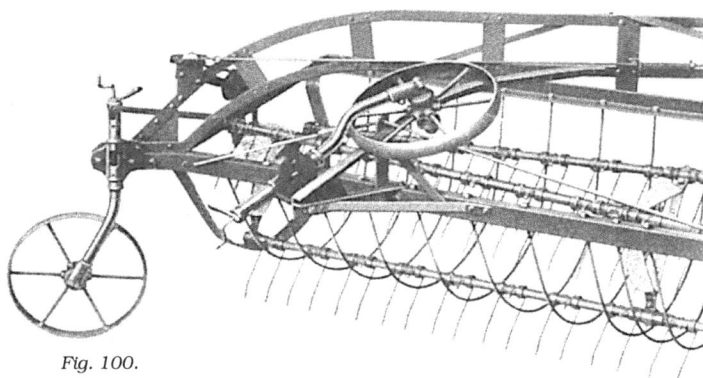

Fig. 100.

Chapter Eleven

Loading Hay on Wagons

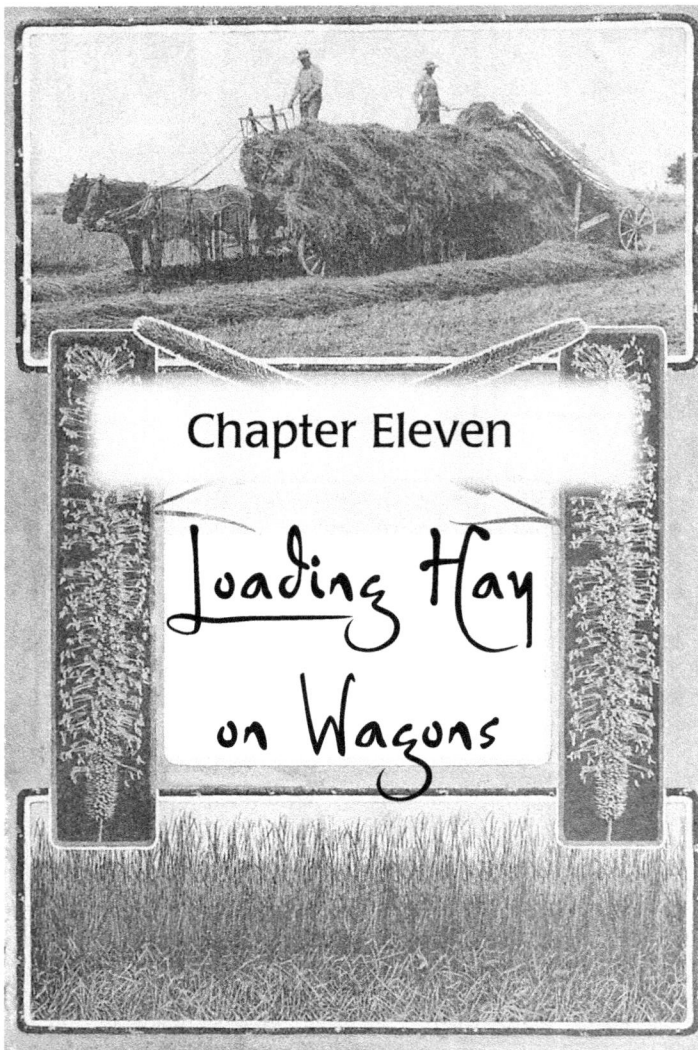

And you'll be surprised how much work smiling, singing, happy people can do.

Pitching hay up to the wagon with a pitchfork may seem like a moron's fare but *the bunch, knot and flip action* takes some thought and practise. And it's worthwhile because it helps the person actually building the load on the wagon. And building a big stable load that travels well and unloads well is definitely an art. This chapter features some information about one system design for building loads.

The other wagon loading option you have is to use a loose hay loader. This fascinating machine is towed behind the wagon and when put in gear conveyors the hay from windrow right up on to the wagonbed. A person with pitchfork then organizes the hay to build that load which takes full advantage of the trip and does not fall off before reaching its destination.

We have used our McCormick hay loader, which was a gift in 1978, for twenty-two years now. And it would appear from that experience that, when pulled at a horse's walking speed and kept well lubricated, it should last a couple of human lifetimes.

It is a very simple machine to operate. Each wheel has a lever which, when turned in towards the hub, engages the gear and

So the hay has been mown and its laying in the field cured and ready to go to storage, either in the barn or in outside stacks. How are you going to get it there? If you are going to put it on wagons or the backs of big trucks, this chapter is for you.

You have a couple of options, perhaps only one. You can get a cluster of family and friends to come over and, with pitchforks, throw that hay up onto the back of the wagon or truck. If you are using teams of horses that, in and of itself, could be reason enough for lots of people to want to help. It becomes a picturesque and fun event.

chain assembly. As the loader is pulled the revolving drum of fingers lifts the windrow up gently onto the conveyor, which may be of several different operational types. The hay moves up and drops on to the wagon bed. Most of the loaders feature an adjustable top table to assist the handlers on the wagon.

As was mentioned in chapter eight, the hayloader can cause some alarm for horses or mules. Though it's hooked way back behind the wagon there is something about the minor vibration rolling through the wagon frame. And, our experience has shown that those very first corners, when the animals are

Figure 2. Building a wagon load by hand. Ideally the hay is handed to the wagon crewmember so that he may place it where it goes.

Figure 3. Another handloading scene illustrating the physical constraints to how high a load might be built.

actually able to see that loader following them, can be a nervous challenge for our equine friends. We have taken to making a more gradual corner, even if it means skipping across some windrows, until the animals are well accustomed to the machine. We've also taken teams that were scheduled to go on the hayloader and wagon for the first time and driven them on forecart or wagon right up behind the loader, allowing them to see first hand and for quite a while that the machine wasn't going to hurt them. Then when they were transferred to the actual setup it was old hat to them.

Building the Wagonload

Your object is to get as much hay as you can on the wagon and to build the load so that it stays on until you get to the stack or barn. If you're pitching hay on by hand the height of your load will be determined by the crew's strength and ability. With a hay loader, our experience has been that the height of load is determined after a good base is built by how far down you can reach with a pitchfork to get hay off the top of the loader (in other words way up high!) But it all depends on how solid the base is. Otherwise great chunks or balls of hay will sluff off or the whole load will tip over.

A notable exception comes if you are going to

be using slings on the wagon (see chapter 14) to incorporate into the unloading. In such case judgement must be used about the aggregate weight of each sling load of hay. With heavy crops this may cause a serious strain to some barn tracks. In this case the wagon crew will want to lay on a less compacted and more evenly spread load than will be suggested in the following paragraphs.

As the illustrations on the next page suggest, we have had excellent good luck with a system which had us focus on building a solid cord-like

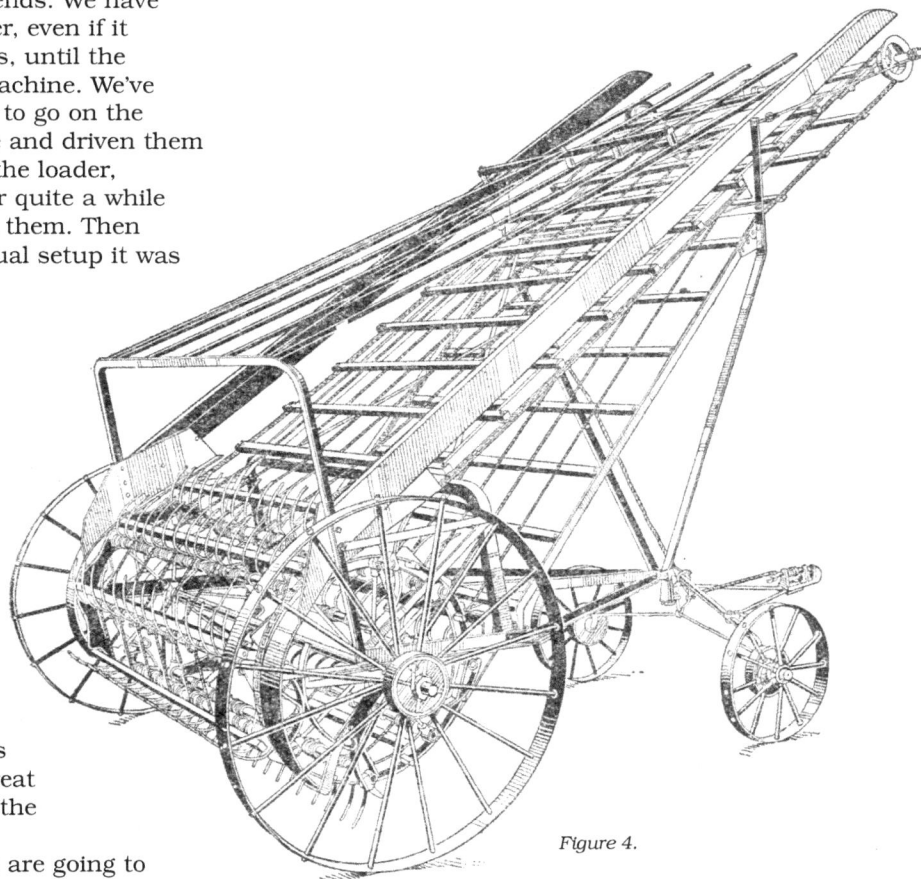

Figure 4.

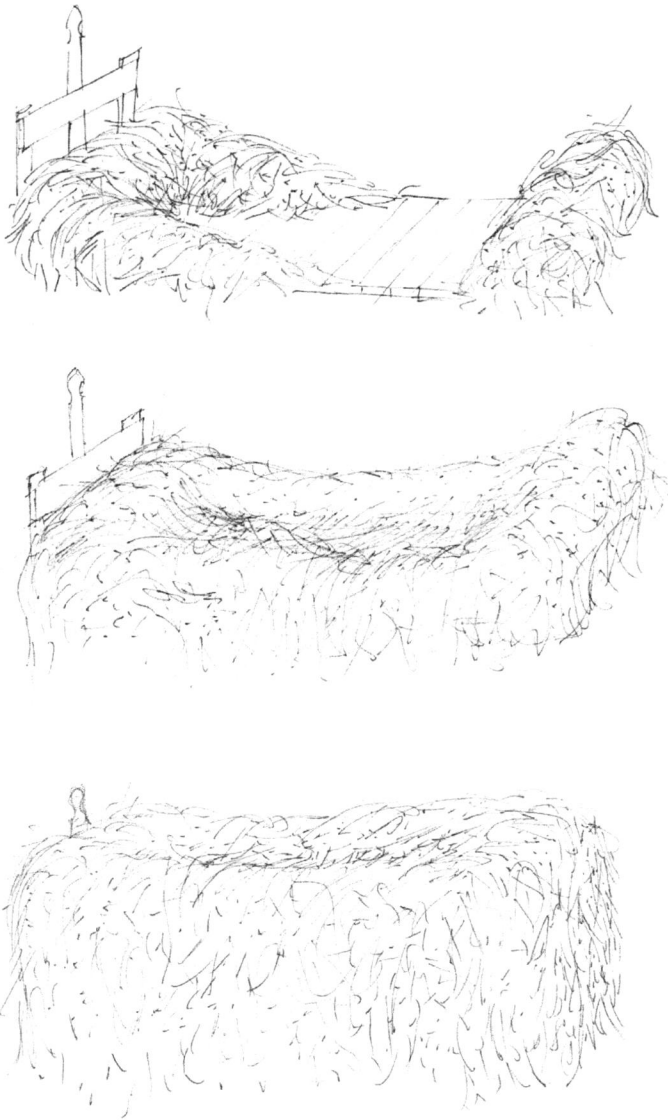

Figure 5. Building a wagon load.

stomping those as well. Gradually have the side walls connect with the back corners until a continuous cord-like wall of hay has been made. Forget about the center of the load, allow it to take care of itself. But pay extra attention to the "walls". The way to pay attention is to walk on the hay, stomping it down and making sure there are no weak spots or breaks in the wall. Keep it strong, continuous and solid enough to hold your weight as you walk on it. About one third of the way forward from the loader build a center mound for yourself to stand on. Fill the middle all around you and stomp it too but keep it generally below the wall level until you are near full height. Then when you finally fill and cap the load, the bulk and weight of that center fill will cause the walls to pull to the center tying in the whole load.

This load building system is the same procedure we have used with such success to build our outside haystacks.

It is this author's experience that the ideal working condition on the wagon is one healthy strong person (aside from the teamster) with a three or four tine pitchfork. Any more people than this and a potential for disorganization and hazard occurs. While it is certainly true that a heavy crop will keep two people working hard, we had an experience that pointed to the risk. Eleven years ago, a visiting helper on the hay wagon moved to avoid my movement and stuck his own pitchfork into his knee (under the knee cap) and had to go to the hospital. On another occasion three people working a wagon load all scrambled in unison, with pitchforks, to bale off when a rattlesnake was seen slithering out of the top of the hayloader and into the hay load. After the laughing, we realized how lucky we were that no one was skewered.

wall of hay all around the outside edge of the wagon bed. If the wagon has no tailboard, start by building up the back wall. This will make the wagon stacker's job somewhat easier because each succeeding wave of hay will tend to cascade off this wall and towards the center of the wagon. If the wagon has a tailboard and the hayloader has been hitched so that the hay falls just over the edge of that tailboard there will be somewhat less concern for getting a 'retaining' wall built early. The next area of the load to build will be across the headboard being careful to stomp and pack the hay from the very corners across that front. Allow the front corners to trail back in beginnings of side walls,

Figure 6.

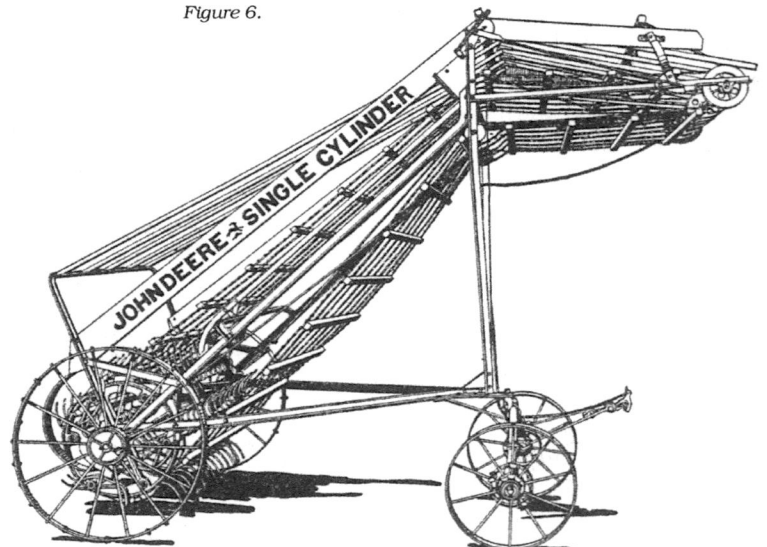

184

RAKE BAR RUNNERS

RAKE BAR TEETH

RAKE BAR

DECK

CRANKSHAFT

DECK SIDE

STRIPPERS

C

DROP GATE

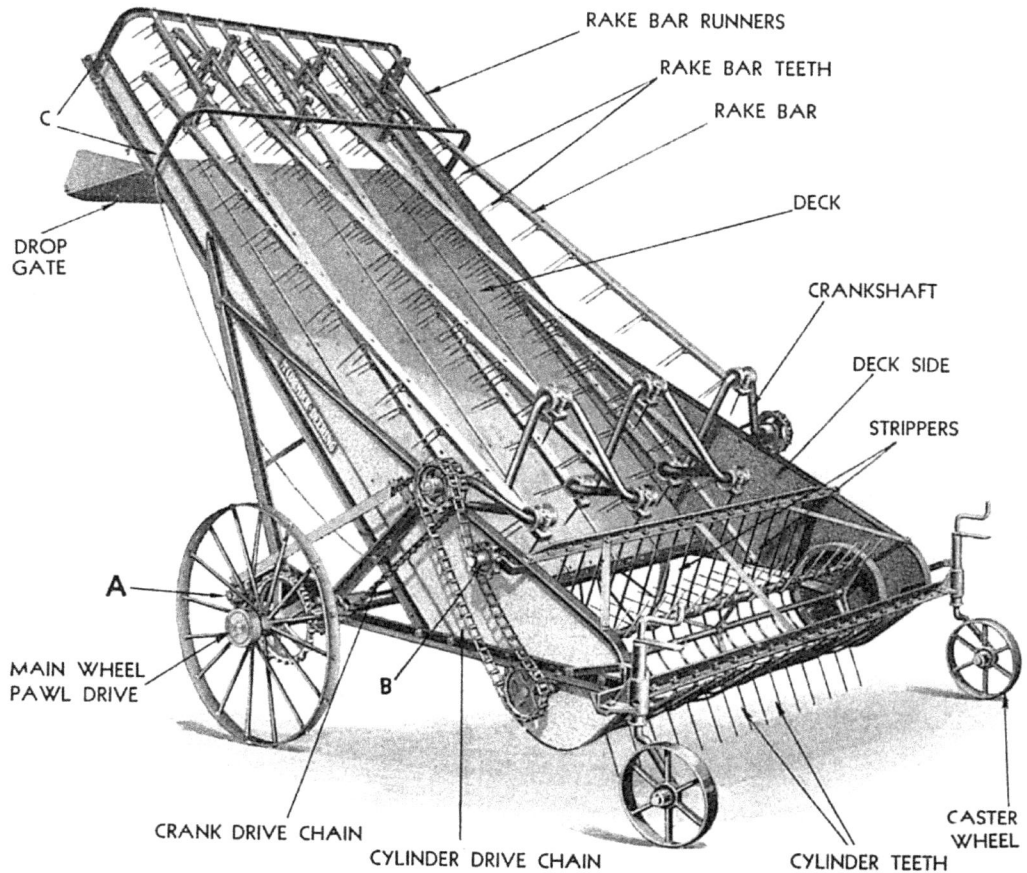

Fig. 7. Cylinder-Rake Hay Loader with various parts indentified. Letters "A", "B", and "C" indicate adjustments.

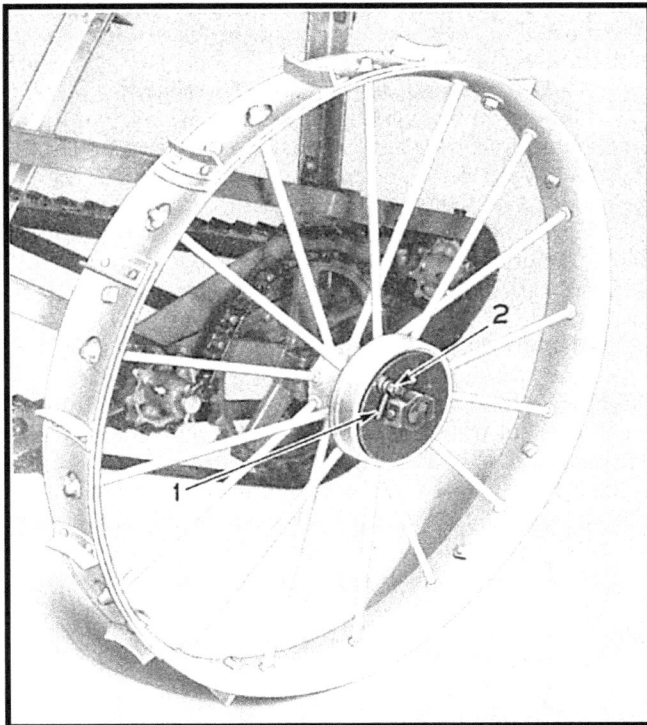

A

B

MAIN WHEEL PAWL DRIVE

CRANK DRIVE CHAIN

CYLINDER DRIVE CHAIN

CYLINDER TEETH

CASTER WHEEL

Fig. 8. Main wheel with removable-lugged sectional over-tires for model "R" green crop loader only. "1" is pawl lever; "2" horned lock nut.

HAY LOADERS

The purpose of any hay loader is to pick up all the crop from swath or windrow and deliver it high enough on to a wagon or truck. It should handle the crop gently, hold it intact in a wind, and keep the loss of edible leaves at a minimum.

Two types of McCormick-Deering hay loaders are used extensively: (1) the Cylinder-Rake Loader and (2) the Adjustable Carrier or Windrow Loader. Both of these loaders are well adapted for loading ordinary weight crops. For handling green crops such as peas and beans intended for canneries, also alfalfa, soy beans, and other heavy green crops, use either the model "R" green crop cylinder-rake loader or the model "W" green crop adjustable carrier loader. While these loaders are basically the same as the regular loaders, they are made heavier and stronger and have special features which adapt them to green crop handling.

Best loader performance can be had by running the loader in the direction in which the hay was cut or in the same direction as the mower.

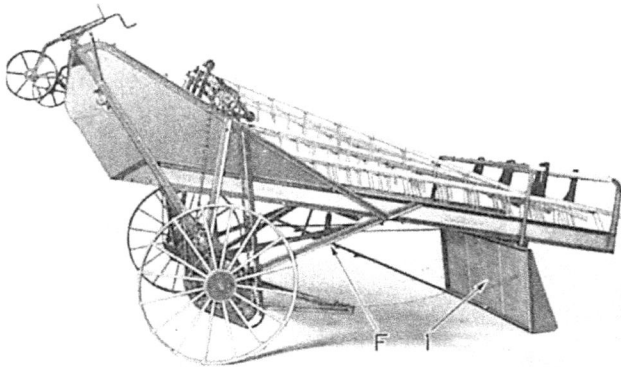

Fig. 9. Cylinder-rake loader in storage position. "1" is drop gate.

CYLINDER-RAKE LOADERS

The cylinder rake loader combines two construction features: (1) a revolving cylinder to pick up the crop from either swath or windrow and (2) rake bars to elevate the hay up to the load. The action of the rake bar keeps the hay moving upward and also has a tendency to push it away from the top of the loader.

REGULAR AND MODEL "R" GREEN CROP LOADERS

DETAILS OF OPERATION

The weight of the loader is carried principally on two main drive wheels which run freely on a main axle and transmit power through pawl mechanisms. Two sprockets, one on either side of the axle, operate the reel and rake bars by means of two drive chains. Two adjustable caster wheels at the rear of the loader control the height of raking cylinder teeth from the ground.

The siding and deck are made of heavy gauge sheet steel which reduces the loss of valuable leaves from the crop being loaded. The upper end of the loader has a drop gate which can be adjusted from the top of the load to regulate the point of delivery.

A hitch is provided which the operator can release by pulling a rope from on top of the load.

When transporting the cylinder-rake loader behind a truck, wagon, or tractor for any distance, the main wheel pawls should be locked out of gear. To accomplish this, turn the pawl lever "1" Fig. 8 to an outward or disengaged position. Remove cotter pin and tighten horned lock nut "2" so that pawl lever cannot operate; replace cotter pin. This will hold the pawl out of gear during transportation and prevent accidental engagement and possible damage. Excessive transporting speeds are destructive and should be avoided.

The rake bar teeth are held rigidly to the rake bar on the upward stroke and yield, to release from the crop, on the downward stroke.

Fig. 9 shows the position in which the cylinder-rake bar loader should be placed when in storage in a shed of limited height. It is also recommended that the loader be lowered to this position when it is left in the field over night, as heavy winds contacting the loader in the upright position might tip the loader over and cause considerable damage. The drop gate "1" may be swung up completely to allow the loader to rest more squarely on the ground if desired.

ADJUSTMENTS
A--Crank Drive Chain:

The crank shaft is chain-driven by sprockets mounted on the main axle on either side of the rake. Caution: The two crank drive chains "2" Fig. 10 must have the same number of chain links (80 links-No. 62). Crank drive chains "2" should be installed so that their link hooks face ahead in the direction of travel as shown by arrow Fig. 10 and so slotted sides are next to main drive sprocket "8". The crank drive chain is tightened by moving location of chain tightener sprocket "1". Loosen a stud nut and two lock bolts, move sprocket to take up slack in chain, then tighten in place.

B--Cylinder Drive Chain Adjustment (Timing Cylinder Tooth Bars with Rake Bars):

Care should be exercised to properly time the cylinder teeth with the lower teeth on the rake bars (see Fig. 11) The setting is accomplished by installing cylinder drive chain "6" Fig. 10 on proper teeth of drive and driver sprockets "3" and "5". Follow instructions shown on Fig. 11 for correct timing. Cylinder drive chain "6" Fig. 10 can be tightened by loosening the stud nut holding chain tightener sprocket "4"; move sprocket to tighten chain, then tighten stud nut. Cylinder drive chain has 50 links (No. 62 chain) and should be installed so its link hooks face ahead in direction of travel shown by arrow in Fig. 10 and so slotted sides face out.

C--Rake Bar (Upper Ends) Adjustment:

The upper ends of the rake bars are adjustable with relation to the deck by use of a series of holes "C" Fig. 12 in rake bar runner supports. When loading green material which has a tendency to pack, it will be necessary to move the rake bar teeth closer to the deck. Where crops are bulky and have a tendency to fluff up it may be necessary to move the rake bar teeth away from the deck.

D-Rake Bar (Lower Ends) (Model "R" Only):

The crankshaft end bearings are mounted in housings which are slotted to permit adjusting the lower ends of the rake bars up or down with relation to the deck. See "D" Fig. 13. In order to raise the crankshaft it is necessary to loosen the crank drive chain and the cylinder drive chain as covered under A-Crank Drive Chain and B-Cylinder Drive Chain Adjustments. Loosen the two bolts in each crankshaft bearing bracket, then move bracket to new position and tighten in place. Tighten chains.

E-Caster Wheel:

The height of the cylinder teeth from the ground is adjusted by raising or lowering the rear of the loader

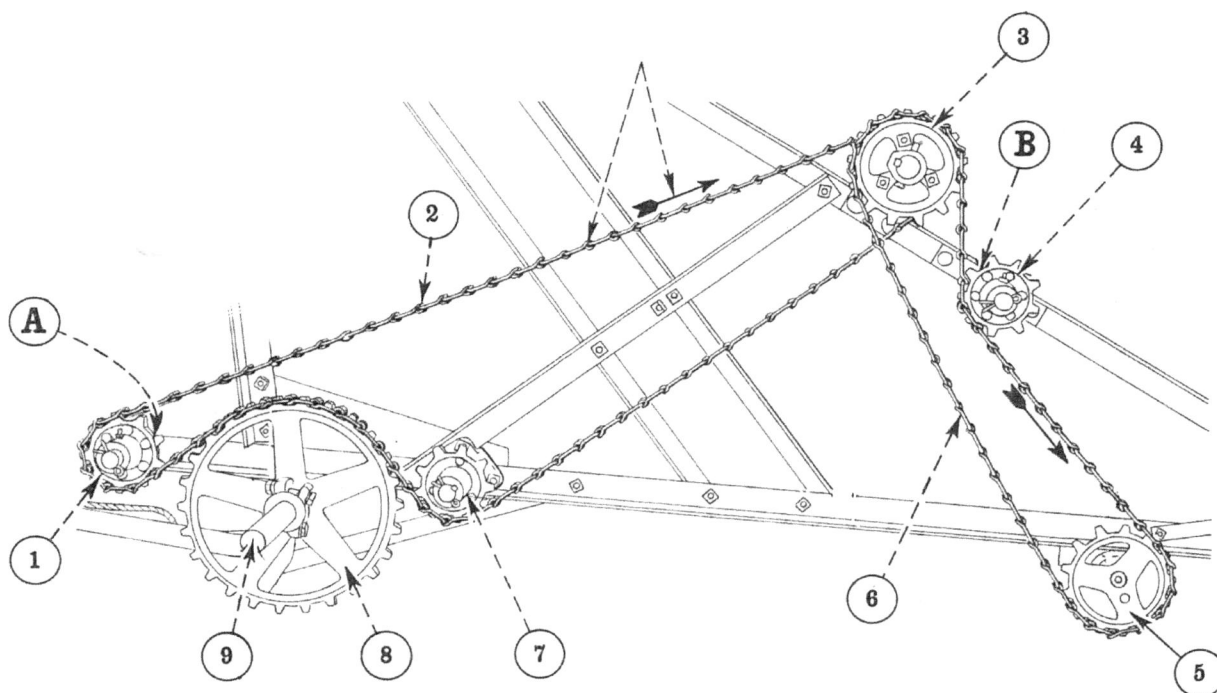

Fig. 10. Diagram of chain drives on cylinder-rake loader. "1" is chain tightener sprocket; "2" crank drive chain; "3" cylinder drive sprocket (shown) - crank drive sprocket (hidden); "4" chain tightener sprocket; "5" cylinder shaft sprocket; "6" cylinder drive chain; "7" idler sprocket; "8" main drive sprocket; "9" main axle.

For average and light crop, cylinder tooth "1" and rake bar fork "2" should be set as illustrated. For average to heavy crop, cylinder tooth "1" should be set lower than illustrated.

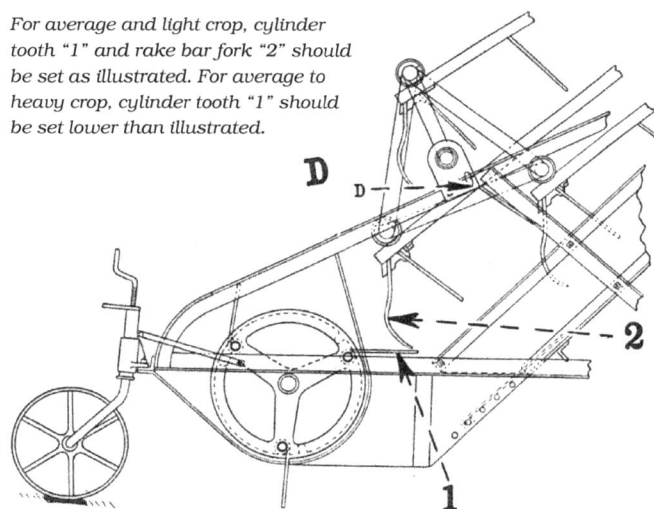

For average hay, cylinder tooth "1" and rake bar tooth "2" should be set as illustrated. For light hay, cylinder tooth "1" should be set higher than illustrated; for heavy hay, lower than illustrated.

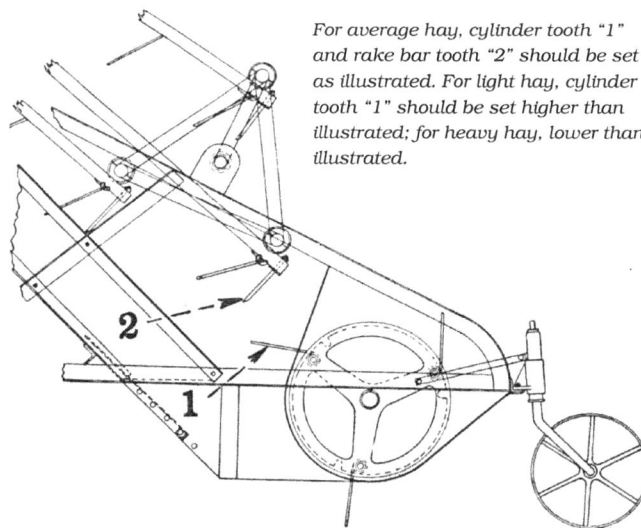

Fig. 11. Correct position of cylinder teeth with rake bar teeth for average hay conditions; left, model "R" green crop loader, right, regular cylinder-rake loader.

which is supported by two caster wheels (see Fig. 14). Adjustment is obtained by loosening caster wheel adjusting lock "1" and turning crank in the proper direction to obtain desired height; then tighten lock. Run the rake cylinder as far from the ground as possible and still gather all the crop.

F-Adjustable Drop Gate:

The upper end of the loader is equipped with a drop gate "1" Fig. 9 which can be adjusted from the top of the load to regulate the point of delivery. When

starting the load the gate can be dropped to the down position. This can be a valuable feature on windy days. As a load is filled up, the gate is raised until a maximum delivery height of 9 feet, 3 inches is reached.

G-Hitch

Three hitch adjustments are provided in the draft bracket (see Fig. 15). Under ordinary conditions on level ground use higher hitch "1". The intermediate hitch "2" should be used on rolling terrain and the lower hitch "3" should be used in hilly conditions; this

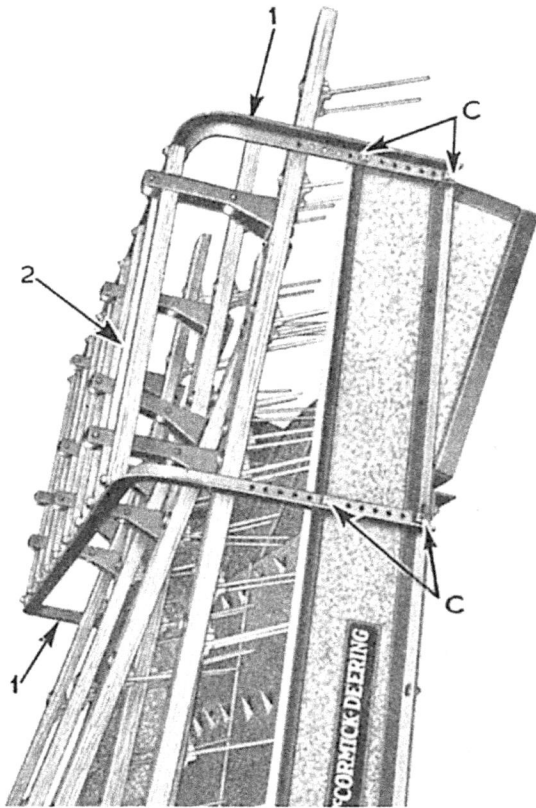

Fig. 12. Adjustment provided at upper ends of rake bars. "1" rake bar runner supports; "2" rake bar runners.

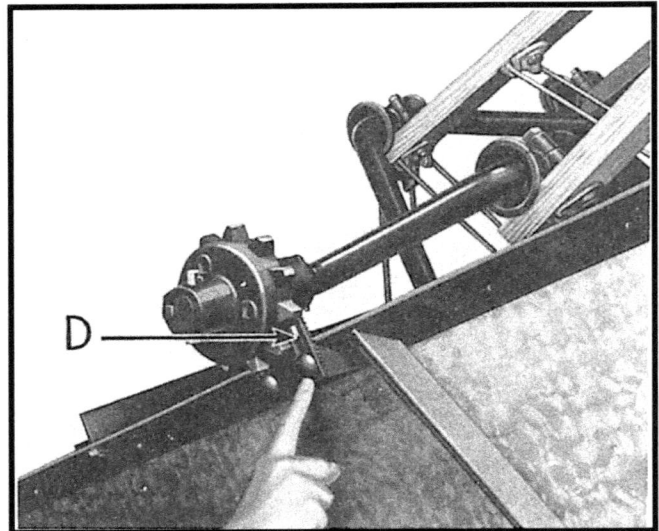

Fig. 13. Adjustment provided at lower ends of rake bars (model "R" green crop loader only).

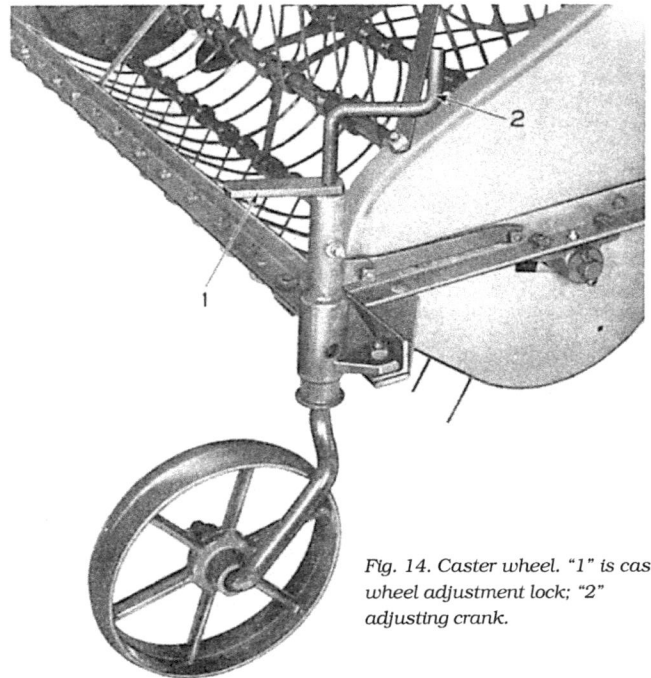

Fig. 14. Caster wheel. "1" is caster wheel adjustment lock; "2" adjusting crank.

tends to force more weight on the caster wheels and prevent the loader tipping forward.

SERVICING INFORMATION

The *cylinder tooth bar* passes through the coil spring end of the *cylinder teeth*, therefore it is necessary to remove the tooth bar from the loader to replace damaged teeth. To remove a tooth bar, first take off left cylinder shield "1" Fig. 16 and remove cylinder drive chain, cotter pin "2" and two bolts "3" holding crank "4" to tooth bar "5". Then pull crank "4" from tooth bar "5". Take out two bolts "6" fastening cylinder center arm. Slide tooth bar sideways towards its crank end until opposite end releases from its bearing in cylinder head. Broken teeth can then be removed from tooth bar.

The *caster wheel* axle is prevented from falling out of the caster wheel bracket by a grooved tapered pin. This pin is driven into a hole provided in the axle which can be lined up with cross holes in the bracket for removing or replacing.

SPECIAL EQUIPMENT

Throat shields and deck extension are furnished as special equipment on the regular cylinder-rake loader and furnished regular with the model "R" green crop loader. The purpose of the throat shield "1" Fig. 18 is to prevent short, light crops from passing out the side of the loader especially when turning corners. The deck extension "2" enables the loader to perform

more efficiently on a wider variety of crops. It is adjustable; the lower position is for light crops and the raised position for heavy fluffy crops.

Pneumatic implement tires are available in the following sizes: 4.00x30 in. 4-ply tires on main wheels and 3.00x7 in. 2-ply tires on caster wheels.

4-7/8-inch wide rim main wheels are available in lieu of regular 2-1/2-inch rim wheels for soft or stoney ground conditions.

Removable lugged sectional overtires are available for the main wheels for the model "R" green crop loader (see Fig. 8).

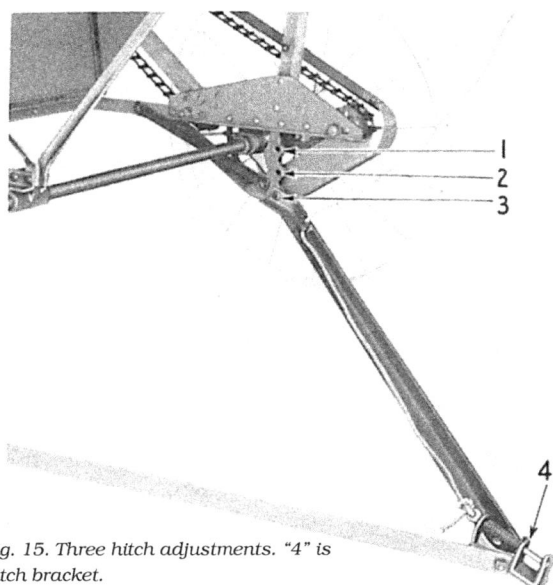

Fig. 15. Three hitch adjustments. "4" is hitch bracket.

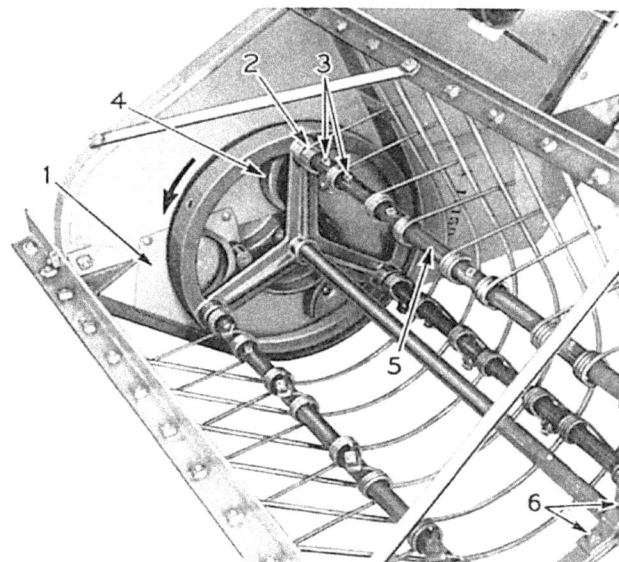

Fig. 16. Cylinder tooth actuating mechanism. "1" cylinder shield; "2" cotter pin; "3" bolts; "4" crank; "5" tooth bar; "6" bolts holding cylinder center arms.

Loading Wagons

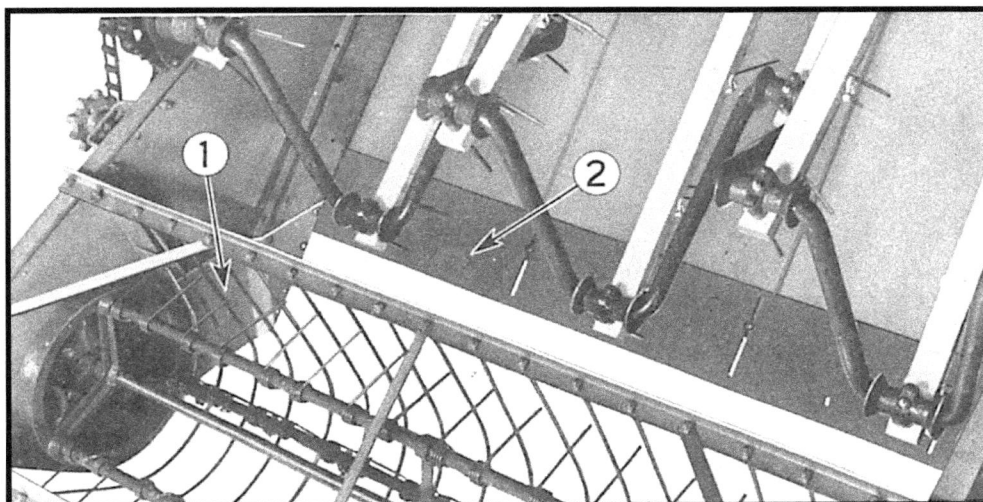

Fig. 17. Throat shield and deck extension. "1" is shield; "2" extension.

ADJUSTABLE CARRIER LOADERS

The adjustable carrier loader is primarily used for windrow loading. It consists of a raking cylinder which is mounted on the main axle and revolves in unison with main wheels. The cylinder teeth pick up the windrowed crop gently and deliver it onto a rope and slat carrier where it is conveyed up to the delivery position at the top of the load. Tooth bars are given an independent movement by means of a durable cam-and-lever arrangement which causes the teeth to enter the hay at the proper angle.

REGULAR AND MODEL "W" GREEN CROP LOADERS

DETAILS OF OPERATION

The cylinder teeth are mounted on eight tooth bars "6" Fig. 105 held in position at both outer ends by the cylinder head "7". As the cylinder rotates, the tooth bar arm "1" contacts the tooth controlling cam "2". The curved teeth are forced forward and under the windrowed crop in a "scooping" motion and the hay is carried up on the cylinder to a position illustrated by the upper cam "3".

At this point the pressure of the hay forcing back on the teeth and the tooth bar arm losing contact with the tooth controlling cam allow the teeth to swing back, releasing the hay on the carrier. The teeth are completely and surely retarded as the tooth bar arm clip "4" contacts the tooth bar arm trip and the tooth

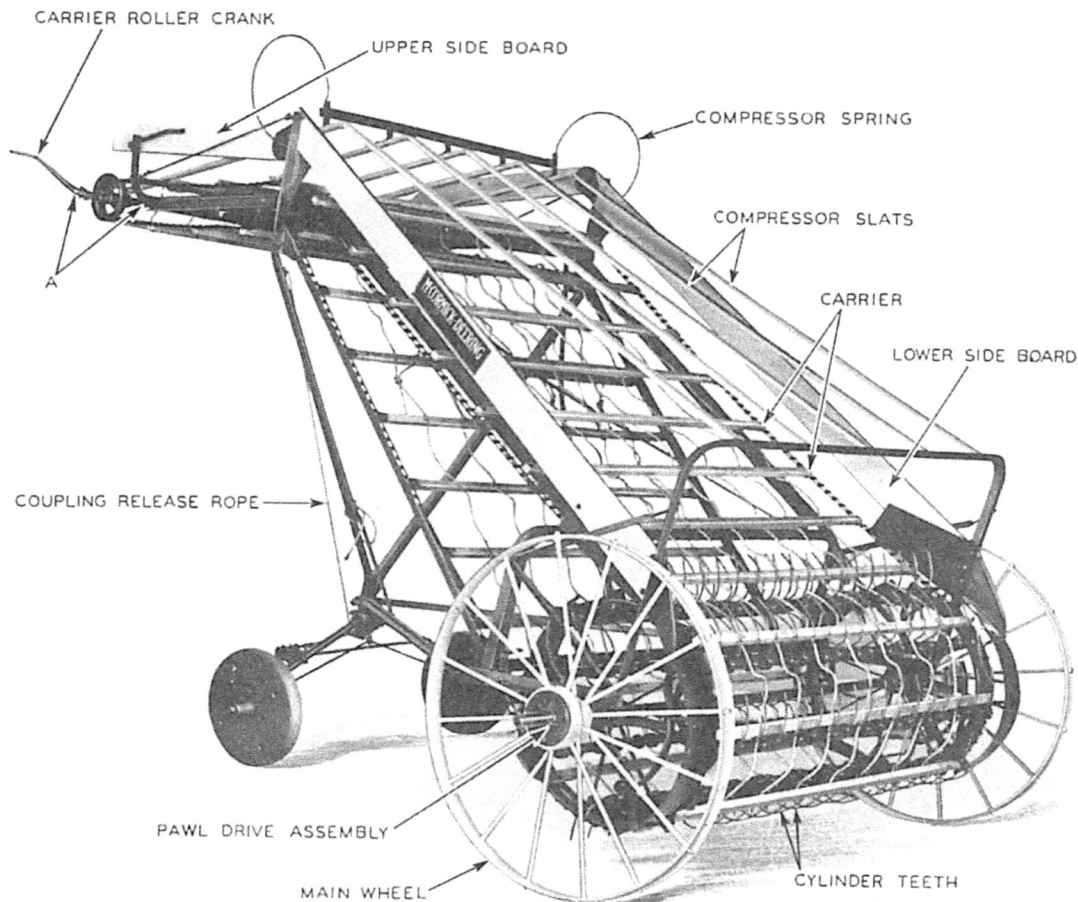

Fig. 18. *Adjustable carrier-type hay loader with various parts identified. Letter "A" indicates adjustment.*

action is again repeated. Cam and lever actions on the eight bars are controlled from both sides of the cylinder. The bar levers are mounted on alternate bars so that four rake bars are controlled from either side.

The loader can be disconnected quickly from the wagon, or truck, by pulling a release rope which extends from the top of the loader down to the coupling pin hitch.

Each wheel is equipped with a ratchet and throwout pawl. Therefore the drive for the cylinder and carrier is transmitted through the pawl assembly which is pinned to the axle. The pawl may be disengaged for transporting by manually operating pawl key lever "1" Fig. 20 to outward position. The pawl key "2" forces the pawl "3" to a retired position where it clears the teeth in the wheel ratchet "4". Tripping lever "1" to an inward position allows the pawl to engage the ratchet which places the loader in gear.

ADJUSTMENTS
A-Carrier:

The upper section of the carrier is adjustable up or down so that a man on the load can control the delivery elevation. Adjustment is made by turning carrier roller shaft crank "2" Fig. 21 which shortens or lengthens the supporting chain "3". Ratchet "4" locks it in position. Tension of the carrier is adjusted by the

two lock nuts "1" on the carrier tightener at the top of the load carrier frame. An identical adjustment is located on each side of the carrier. The tension in the carrier should be even on both sides.

SERVICING INFORMATION

To remove *web carrier* from the loader, first take the loader out of gear by disengaging pawls (see Fig. 20). Circulate the web carrier until the tied ends of the ropes are at a convenient location on the underside of the loader. Loosen the web carrier by backing off lock nuts "1" Fig. 21. Untie the ropes and disconnect the carrier chains. Web carrier may be pulled down, running over the pulleys, standing in front of the loader. Caution: When pulling web carrier from the loader, care should be taken that the last section of the web carrier leaving the loader does not fall on the serviceman.

When replacing ropes or chains on the web carrier the chains should be fastened at their full length to the floor as shown in "1" Fig. 22. Check chain linkage and number of links, which must be the same for each chain (6 links - No. 45 chain between each slat link). Ropes are stapled to slats and require a certain amount of slack. Correct slack is had when rope is drawn over a 3-inch block "2" Fig. 22 and rope is just taut.

190

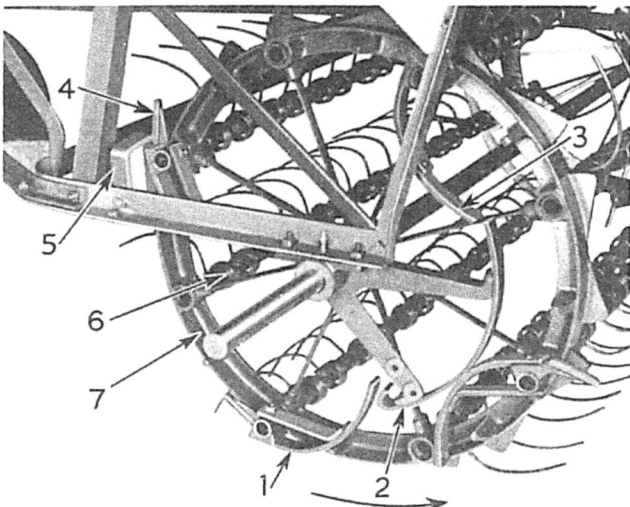

Fig. 19. Cylinder teeth operating mechanism. "1" is tooth bar arm; "2" controlling cam; "3" tooth bar arm; "4" tooth bar arm clip; "5" tooth bar arm trip; "6" tooth bar; "7" cylinder head.

Fig. 20. Wheel ratchet with pawl mechanism removed. "1" is pawl key lever; "2" pawl key; "3" pawl, and "4" wheel ratchet.

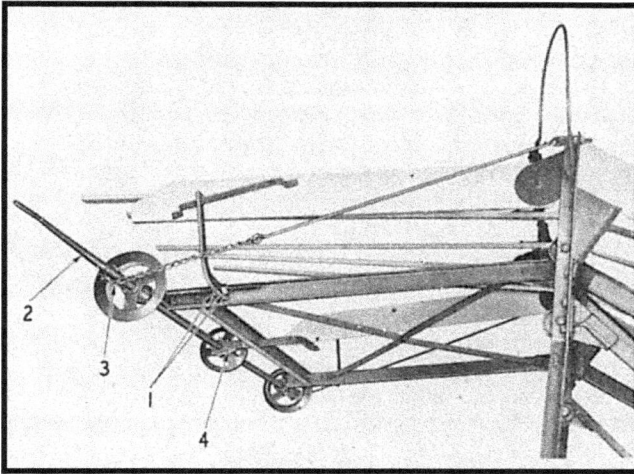

Web carrier can be replaced on the loader as follows: Stretch web carrier out in front of loader with slats facing up on top of chain and hooked ends of links facing ahead. Attach one end of a rope to center of last slat and run other end of rope up over carrier roller shaft and carrier guide cross tie and down past the cylinder. By standing in back of cylinder, pull the rope which will hoist the carrier up and onto the loader.

Pull the carrier far enough onto the loader so that it will fit around the cylinder and up a short ways below the loader. Fit the carrier onto the cylinder in its normal running position. Run the other end of the carrier down until the ends of the chains can be connected and ropes tied. Tie rope using a square knot as shown in Fig. 23; adjust carrier tension as explained in "A-Carrier," page 190.

Should *cylinder teeth* become bent out of shape, they must be immediately reshaped or replaced to

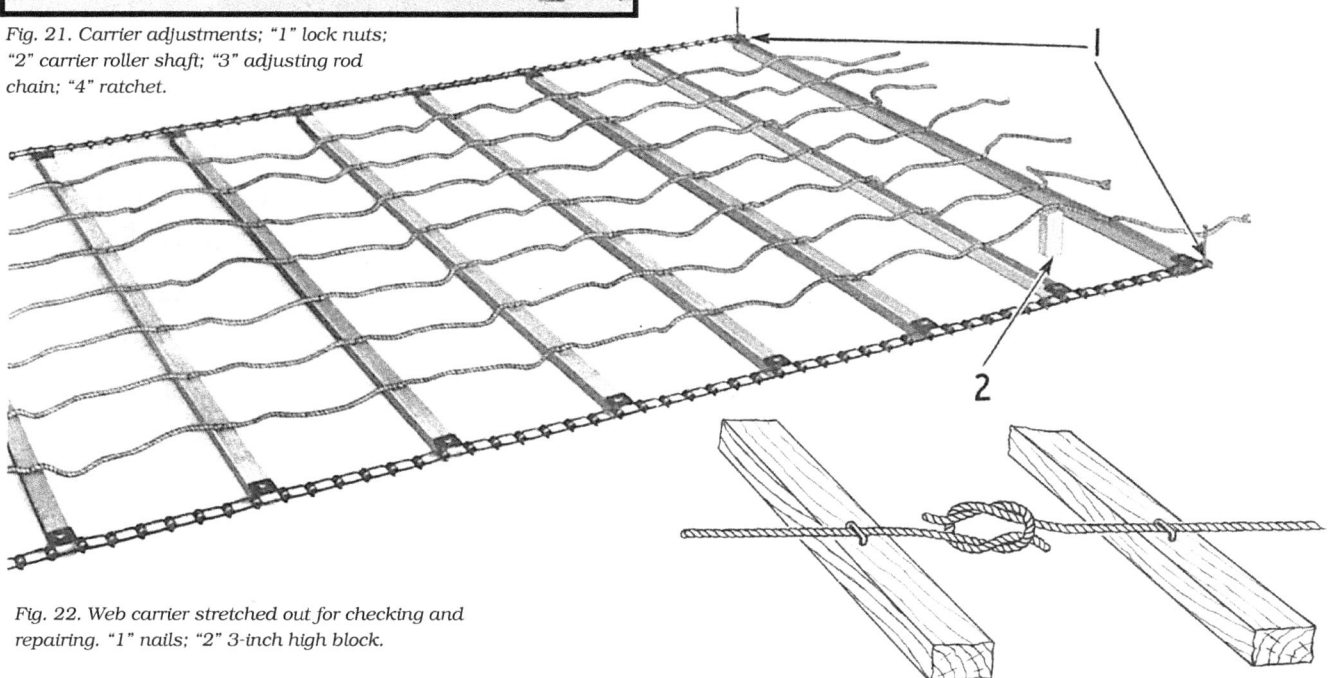

Fig. 21. Carrier adjustments; "1" lock nuts; "2" carrier roller shaft; "3" adjusting rod chain; "4" ratchet.

Fig. 22. Web carrier stretched out for checking and repairing. "1" nails; "2" 3-inch high block.

Fig. 23. Correct method of tying carrier rope ends.

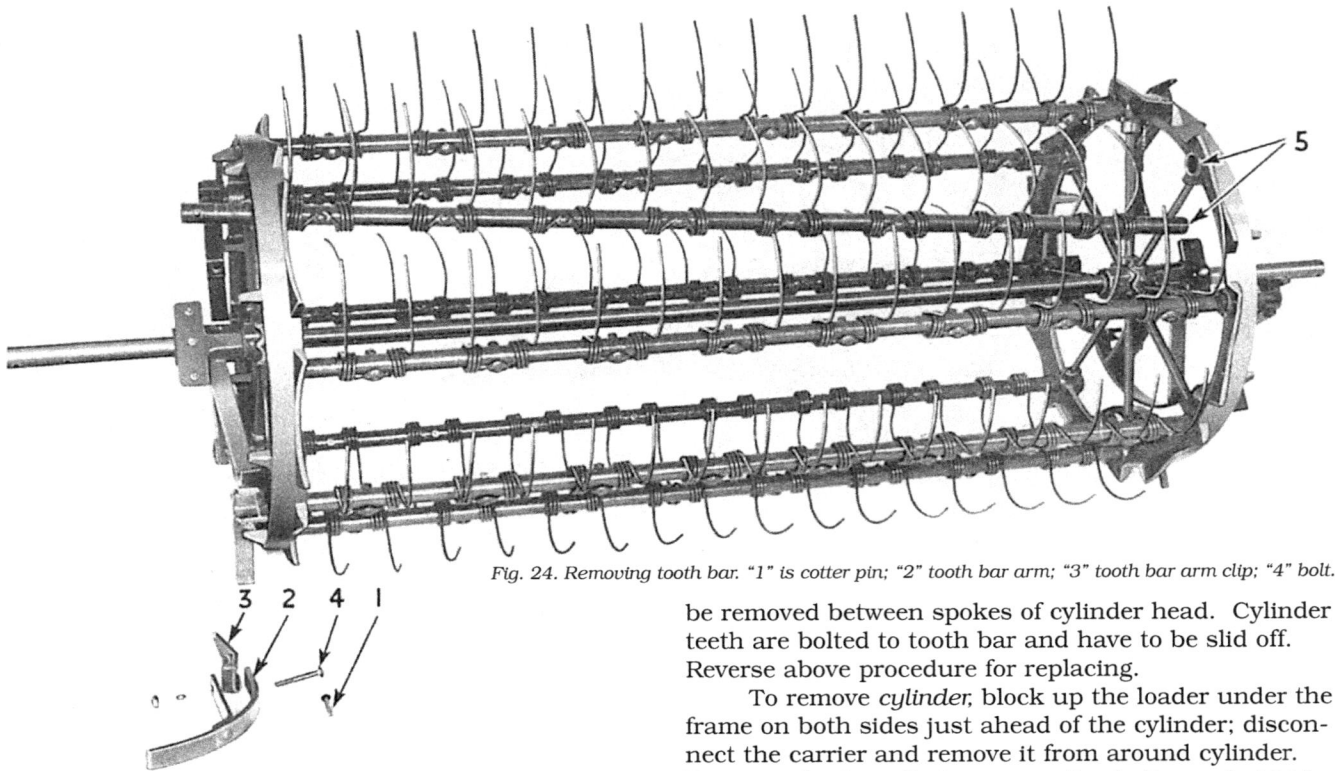

Fig. 24. Removing tooth bar. "1" is cotter pin; "2" tooth bar arm; "3" tooth bar arm clip; "4" bolt.

prevent web breakage. The teeth are formed to clear the slats on the web carrier. To replace cylinder teeth it is first necessary to remove rake bar from the cylinder. Turn the cylinder to a convenient position and remove bolt "4" Fig. 24; slip tooth bar arm "2" and clip "3" from tooth bar. Remove cotter pin "1". Slide tooth bar to one side allowing opposite end to drop free of cylinder head as shown at "5". Tooth bar can then be removed between spokes of cylinder head. Cylinder teeth are bolted to tooth bar and have to be slid off. Reverse above procedure for replacing.

To remove *cylinder*, block up the loader under the frame on both sides just ahead of the cylinder; disconnect the carrier and remove it from around cylinder. Remove wheels, cylinder supporting bolts, and shields. Cylinder can then be removed from loader.

SPECIAL EQUIPMENT

Gleaning cylinder attachment can be supplies as special equipment. The gleaning cylinder makes a double cylinder loader, valuable in short hay. 4.00x36-in. 4-ply pneumatic tires for main wheels and 4.00x12-in. 4-ply pneumatic tires for forecarriage wheels are also available.

Fig. 25.

Instructions
For Setting Up

New Idea Easy-Way Hay Loader

This material is a reprint from the New Idea Operators manual and once again features information on how to assemble a new machine, information which is useful when having to deal with the restoration of an older machine.

(These instructions also cover the Heavy Duty Loader)

1. Bolt angles "A", "B" and "C" Fig. 27 in place between right and left main side frames. Also bolt corner braces "D" in place, and be sure to get them on as shown in the illustration. Don't forget the lock washers.

2. Insert bearings in main axle bearing plates, bolt center main axle bearing in position and slide axle through bearings.

3. Drive woodruff keys in main axle for sprockets-- place sprocket wheels, ground wheels and driving hubs on main axle, Fig. 27. Sprocket wheels should be put on so that the long end of the hub is toward the inside. Be sure to spread the spring cotters in driving hub pins.

3A. Bolt the cylinder cam shoe O-282 in place on lower left frame angle as shown in Fig. 27.

4. Remove all parts wired to raking cylinder. Put cylinder bars in position and then bolt left double spring teeth and stationary teeth in place. Parts will be found on each bar. Place cylinder in position as shown in Fig. 28 and bolt cylinder bearing plates to lower side frame angles.

5. Bolt caster wheels in place, Fig. 28.

6. In the bundle of round stripper bars will be found a separate bundle of three special bars; the one is for the center position directly under the crank shaft bearing arch which should be bolted in place at the same time, Fig. 28. The other two are for the extreme end positions and fasten through the horizontal leg of the rear frame angle. The one on the left also holds in place the flat rear corner brace.

Fasten the remaining round cylinder stripper bars in place, Fig. 28. Don't forget the lock washers.

7. Bolt right and left deck side panels to frame, Fig. 29. (Be sure spacer bushings are used when bolting to frame angles.)

8. Bolt arch to upper end of side

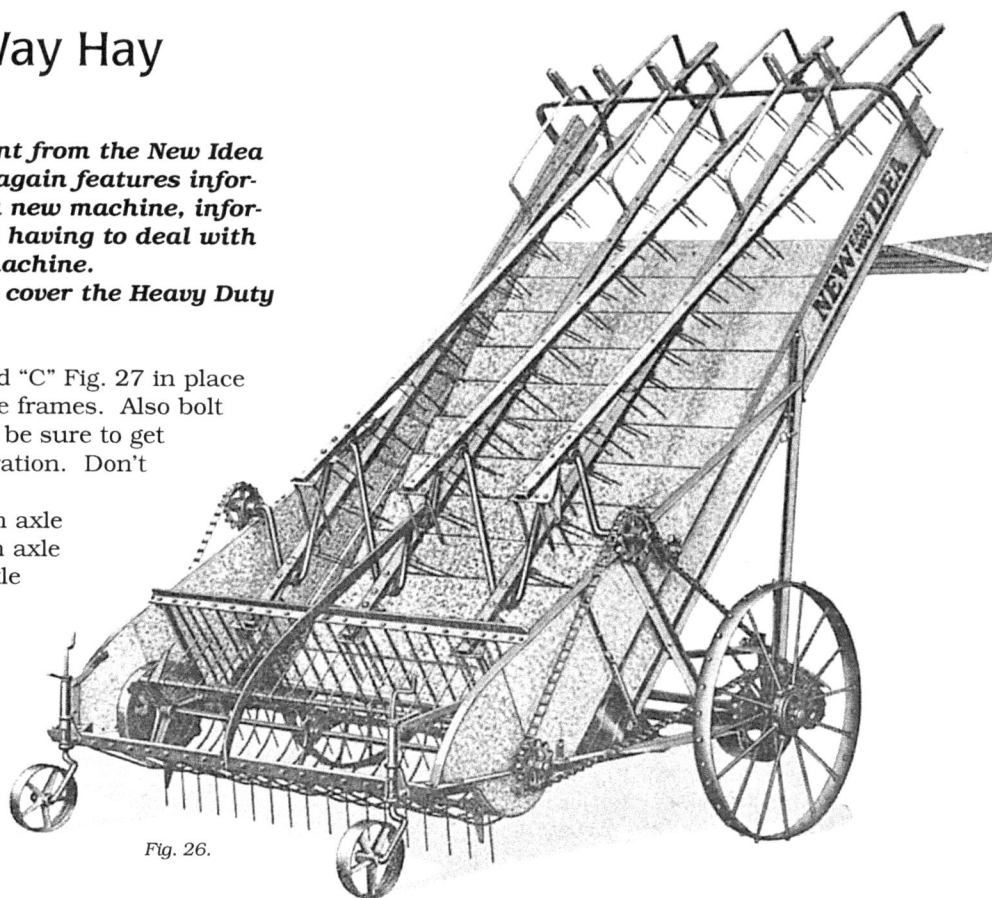

Fig. 26.

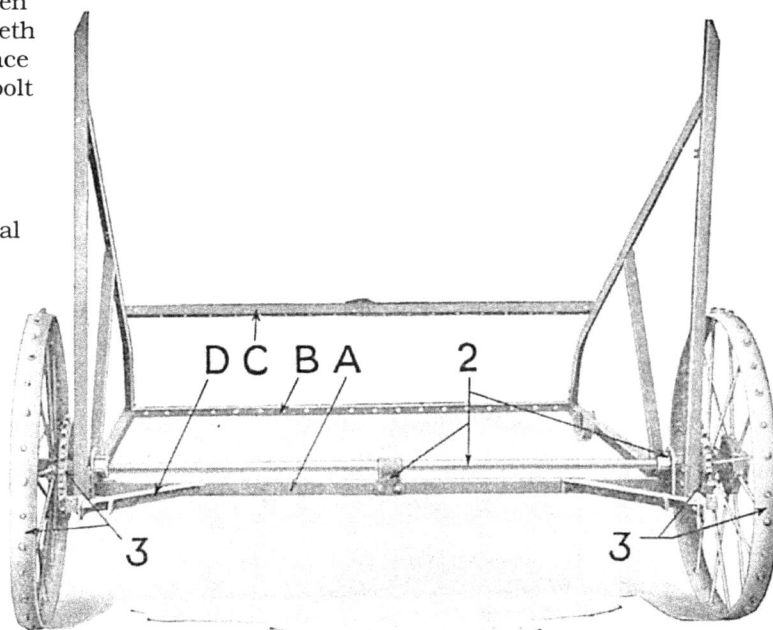

Fig. 27.

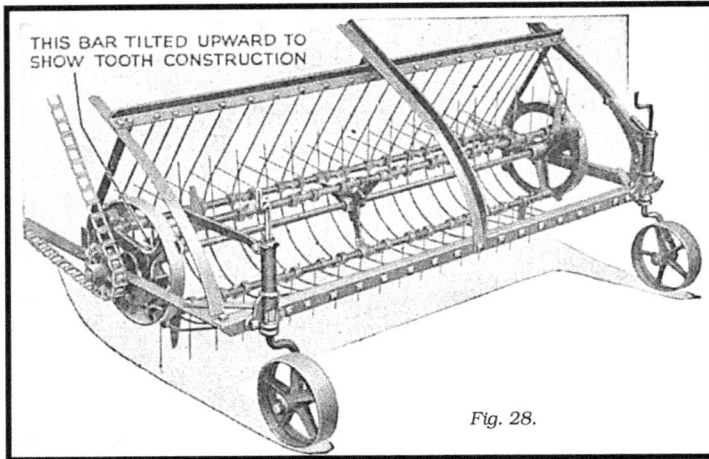

THIS BAR TILTED UPWARD TO SHOW TOOTH CONSTRUCTION

Fig. 28.

the top end up to deck side angles and bolt in position. The leather washers are in the bag with pressure fittings in bundle of sides.

10. Bolt right and left galvanized side enclosing sheets along crank and the right and left upper and lower galvanized cylinder side pieces to frame, Fig. 30. Also fasten upper caster wheel braces to side pieces and frame.

10A. Attach the right and left flat cylinder head fenders to the upper cross angle, bolt angle bracket to lower main frame angle and attach fenders, Fig. 30. The bottom of the fenders should set close to the cylinder heads. Attach oil can holder to the right flat fender bracket and the crank side sheet.

panels, Fig. 29. (Be sure the spacer bushings are used.)

8A. Bolt tool box to the right deck side panel. Be sure to put bolts through from the inside.

9. Place stationary deck section (shortest section) between side panels and bolt to angles as shown in Fig. 29. Place large floating deck section in position as shown in Fig. 29. Bring

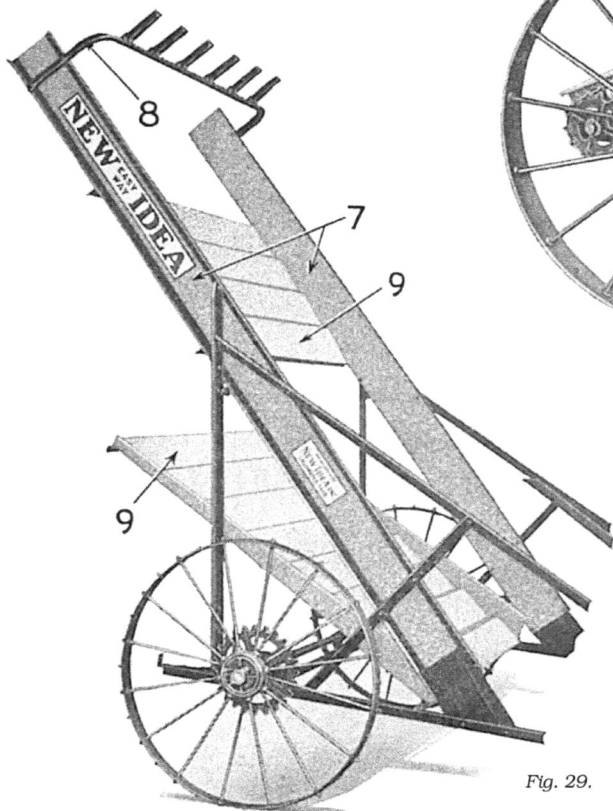

Fig. 30.

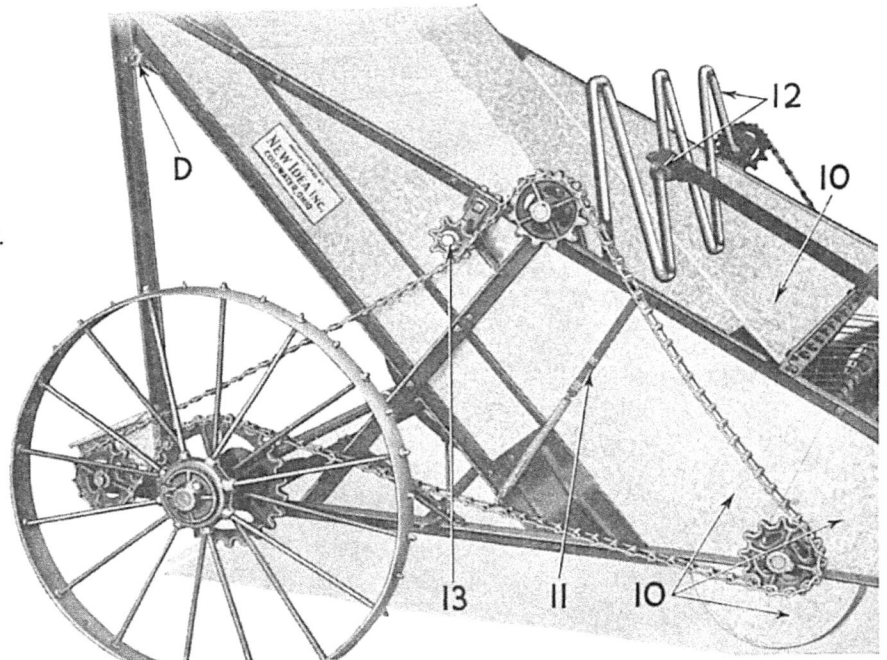

Fig. 29.

11. Bolt right and left yielding deck spring supports in place and connect springs to yielding deck. Fig. 30.

12. Place crank shaft in position as shown in Fig. 30 and bolt right and left bearing plates to top of main frame angles, also fasten center bearing.

THE CORRECT WAY TO PUT SPROCKET CHAINS ON NEW IDEA HAY LOADERS

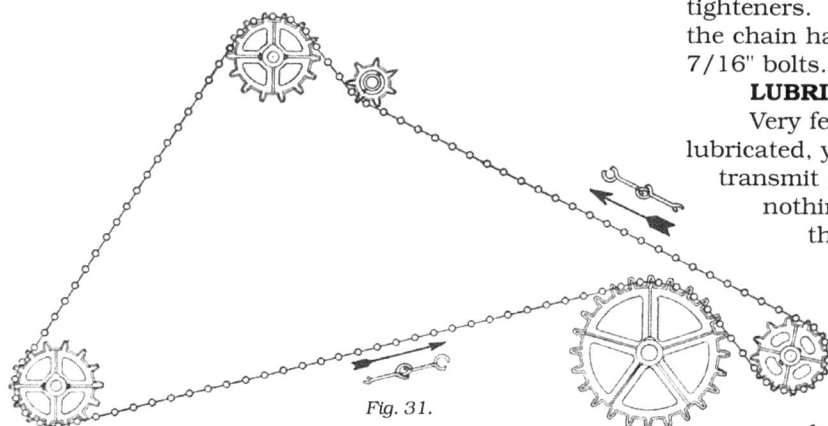

Fig. 31.

--AND THE REASON WHY.

13. There is a great difference of opinion among implement men as to **How the chains should be put on.** In the majority of instances chains are run without any regard to rules or good engineering practice and the result is that chains and sprocket wheels wear unduly, chains break, power is wasted, and the user, failing to check the chains with the instructions, blames the manufacturer of the particular implement.

A variety of conditions come up in sprocket chain drives which require careful analysis and individual treatment; but in the average chain drives, especially where the driven wheel is small in diameter and the driver comparatively large, the conventional rule is: "Run drive chains with hook forward and slot to the outside." This rule applies to the chains on the NEW IDEA Hay Loader. See Figs. 30 and 31. On some other implements where the drive sprocket is smaller than the driven sprocket, the chain should be run with the bar of the link forward and the slot to the outside.

Now what is meant by saying: "Run drive chains with hooks forward and slot to the outside." It means that where chains are used to transmit power from a large driver to a small driven wheel, the hook or barrel end of each link should travel forward in the direction of travel of the chain. In this way the chain when it disengages the small wheel, which it does under strain **does not** cause any rubbing with consequent wear on the links and teeth of the driven wheel; however, if the chain in this same drive is run with the **bar forward** it would have to leave the tooth of the driven wheel under full strain in rubbing contact, causing excessive wear on the chain links and also on the teeth of the driven sprocket. Thus in the chains driving the cylinder and crank shaft make sure that the links run **Hook Forward** in the direction of travel and slot to the outside.

Bolt the right and left spring chain tighteners in

place as shown in Fig. 30, and then adjustable chain tighteners as shown in Fig. 37. To adjust the tension of the chains open the 7/16" bolts clamping the tighteners. Move tightener with long 3/8" bolt until the chain has the correct tension and then tighten the 7/16" bolts. Both chains should be adjusted evenly.

LUBRICATING CHAINS

Very few chains running on implements are lubricated, yet when it is considered that these chains transmit all the power and since these chains are nothing but a series of joints bending around the different small sprocket wheels under strain, it would seem strange that so much care is put on lubricating bearings and none of the chains. A properly lubricated chain transmits the power with much less loss than a dry running chain and wears considerably longer. Complaints are sometimes made that chains running in the open are susceptible to dust and grit if lubricated. This is not near as bad as generally imagined, however, it is better in such cases to use a lubricant that is not so susceptible to collect dust and grit, than to run chains dry. Put chains on correctly and then keep them properly lubricated with a suitable lubricant.

13A. Attach drive chain shields as shown in Fig. 32.

13B. Bolt flare shields above crank shaft in place as illustrated Fig. 35. (Used on heavy duty loader only.)

14. Tilt the loader downward and bolt the elevating bars with bearings to crank shaft and bolt the slides to the upper end of the bars, Fig. 33. The two bars having the steel plates at the lower end are for the outside positions, the other four are interchangeable. Make sure that the elevating bars are bolted on top of the bearings as shown in Fig. 34. Then again set the loader in its natural position.

Turn the long spring teeth on the lower end of each of the elevating bars so they will stand in operating position as shown in Fig. 34.

Bolt the right and left plates having the three teeth and the long threaded single tooth to the lower end of the outside elevating bars and the plates having the double teeth to the four center bars.

15. Bolt channel corner braces and cross angle "A" in place and fasten round cross braces underneath deck, Figs. 36 and 37. Rod "B" must be next to deck. Now adjust brace rods at "D" Fig. 30, so that yielding deck will be central in the loader and work freely. In order to get the desired result it may be necessary to loosen one rod and tighten the other. Fasten brace rods together at center with clip "C", Fig. 36.

16. Fasten the galvanized corner shields over drive chain tighteners to main frame upright angles, Fig. 36.

17. Fasten assembled hitch to clips on front corner braces, Figs. 36 and 37. Be sure to spread the spring cotters in hitch pins. Fasten the trip rope to

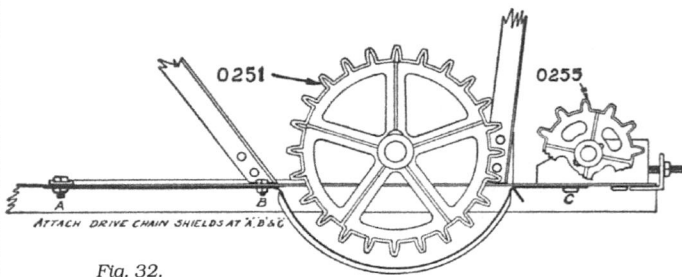
Fig. 32.

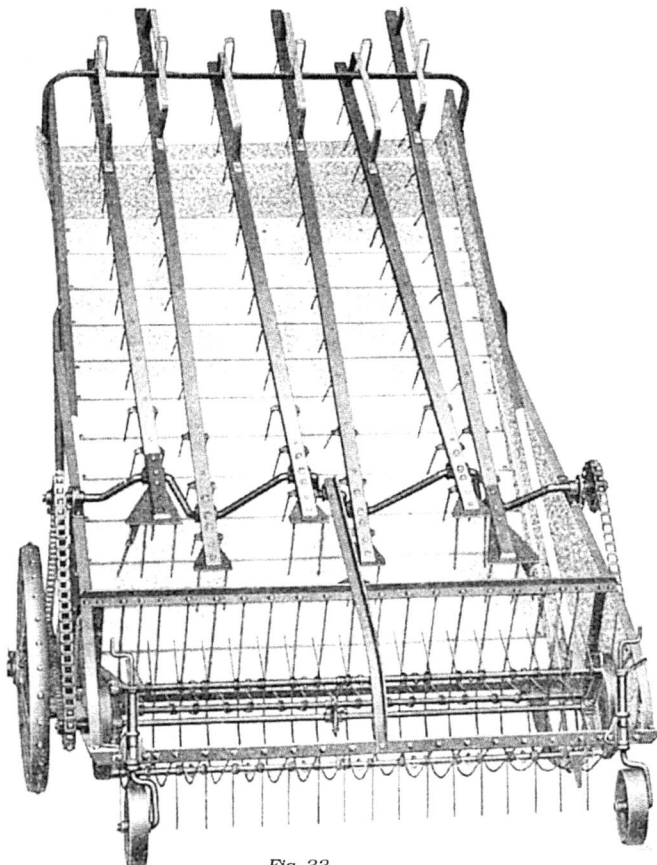
Fig. 34.

hitch eye pin and fasten one pulley to the center of the hitch and the other pulley can be fastened to either the right or left upright angle of the main frame as desired, Fig. 36.

18. Fasten the steel apron and its braces and supports in place, Fig. 38. Be sure to spread the spring cotters in the hinge pins.

Now insert Alemite oilers at the following points: 3 main axle bearings; 2 cylinder shaft bearings; 3 crank shaft bearings; 6 elevating bar bearings and 1 in left upper side shield along cylinder for oiling cam shoe.

Give the bearings and all working parts a generous oiling before the machine is started in the field.

Attach draft ring to wagon bed so wagon hitch is no more than 27" above ground when connected.

DECK EXTENSION AND DECK HOOKS

When the heavy duty loader is to be used to load heavy green material the Deck Extension and Deck Hooks should be used. These can be secured for a slight additional cost. Instructions for attaching are furnished with the attachment.

ADJUSTMENTS

The raking cylinder can be raised or lowered by releasing the lock handle, and turning the crank screws at the upper end of the caster wheel brackets, until the cylinder teeth are at the proper distance from the ground for clean raking. Then again tighten the locking handle.

The drive chains can be adjusted by loosening the nuts on the 2-7/16" bolts that hold the chain tighteners in place "13" Fig. 37 and by drawing up or loosening the adjusting screw until the desired results are obtained. Both chains should be kept at about the same tension. Be sure to always tighten the nuts on the 7/16" bolts after adjustments have been made.

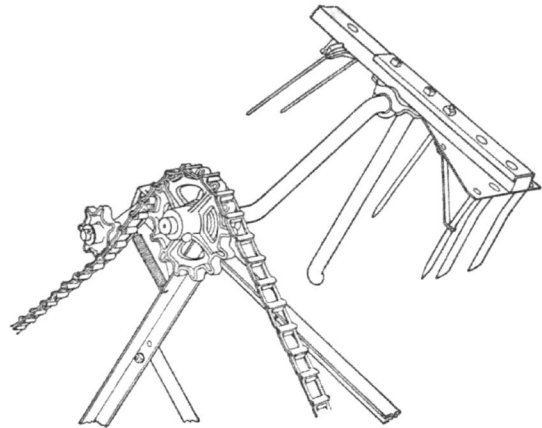
Fig. 33.

Fig. 35.

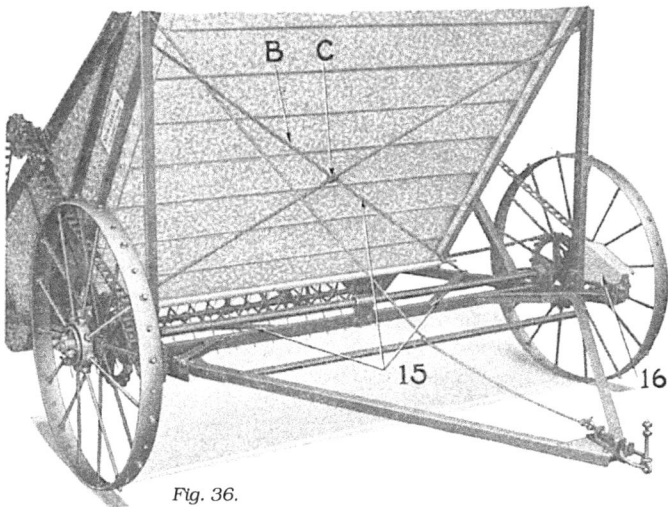

Fig. 36.

The yielding deck should always work freely in the machine. If it does not work freely adjustments can be made at "D", Fig. 30. To get the desired results it may be necessary to tighten one rod and loosen the other. Always keep brace rods tight.

How to Put the Machine In and Out of Gear

The hay loader is put in gear by turning the small levers on both wheel ratchet hubs toward the center as shown in Fig. 39 and out of gear by turning both levers away from the center against the small lug on the outside rim of ratchet hub as indicated by dotted lines. Make sure that the levers on both ratchet hubs always stand in the same position.

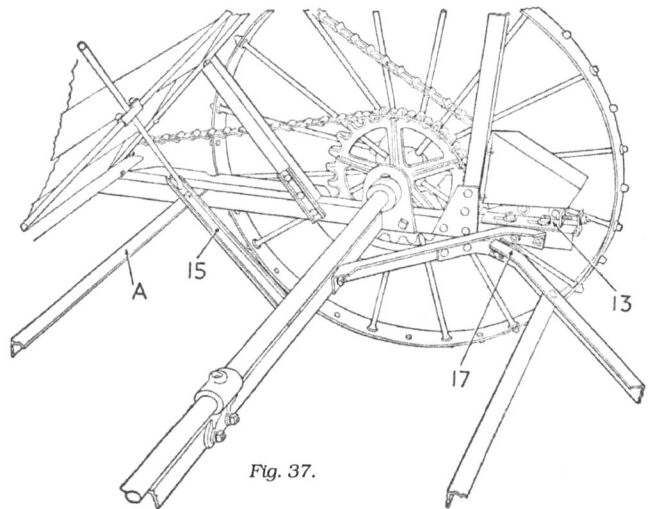

Fig. 38.

EXTENSION RIMS

Where the soil is very loose such as the sandy and muck soil of the mint growing sections and in the irrigated sections where it is necessary to straddle ditches it is desirable to have an extra wide wheel. Where such conditions exist extension rims can be furnished (at a slight additional cost) which can be bolted to the regular wheels. See Fig. 40. The use of these extension rims will increase the traction considerably.

Use machine oil on the rollers at upper end of push bars and other moving parts. Also occasionally put some oil on the drive chains.

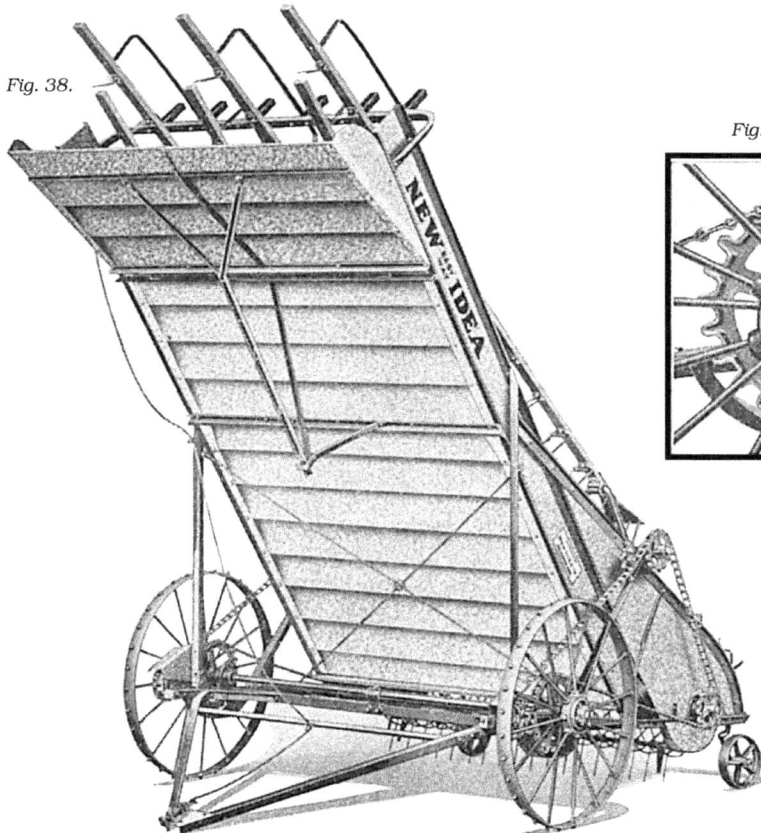

Fig. 37.

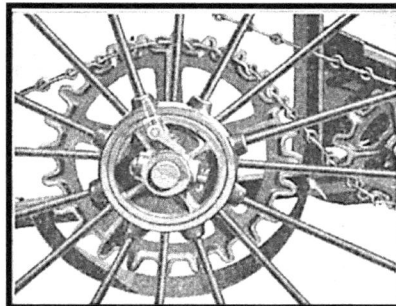

Fig. 39.

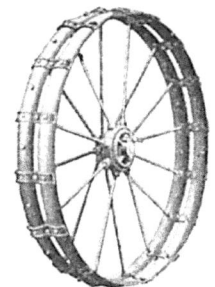

Fig. 40.

Hay Loader Review Questions

What advantages are there in the use of a hay loader?

Wherever a farmer has from eight to ten acres of hay, a hay loader is almost indispensable, not only because it is a great labor-saver, but because haying comes at a time when the farmer is in the busiest part of his season. It enables the farmer to get his hay up much quicker than he would if he pitched it by hand. Speed is very essential because it often happens that a few hours' saving in getting the hay in the barn, after it is cured, will prevent it being rained on. After hay is cured and rained on, it does not have the high quality that it has when put up without rain.

What general types of loaders are in use?

The raker-bar type (many times called the "fork" type); the single-cylinder, and the double-cylinder types. The latter two types are, at times designated as "carrier" loaders.

What type of loader should we use?

Double-cylinder type loader. It has, as it is often spoken of, a "double pick-up" which controls the swath of windrow hay from both front and rear as it is lifted from the ground. At the front is the elevating cylinder, mounted on the main axle, while the rear is the gathering cylinder, which is spring-floated.

How does the floating cylinder do its work?

The work of raking is done mainly by the spring-floated gathering cylinder which is thickly studded with flexible steel fingers. Since there are five rows of gathering fingers, these rows of fingers come to the ground often. In this way a continuous lift of hay is maintained. Because the gathering cylinder is spring-floated, it constantly follows the changing surface of the field.

Explain the action of the hay as it passes between the cylinders.

As the hay is gently lifted and tipped forward by the gathering fingers, it is received by the elevating cylinder fingers and thus, being held between the rising fingers of the two cylinders, is carried into the elevator.

What happens to the hay as it leaves the cylinders?

In the elevator mouth the hay is guided by the strippers. The concave part of these strippers hold the hay against the axle cylinder, they hold it against the elevating fingers in their travel with the cylinder, and under the lower ends of the compressor slats. Thus the hay moves in a stream upward and forward on the endless carrier.

What is the purpose of the compressor slats?

The compressor slats act to prevent wind interference in elevating the hay on the carrier. The slats also press the hay upon the cross slats so as to provide positive elevation and delivery.

How are the side boards arranged?

The distance between the side boards at their top is less than the distance between the boards lower down.

This means the stream of hay is narrowed to five feet or less at the delivery point. This puts the hay in the middle of the load.

Explain the action of the gathering fingers in passing over rough ground.

Because the gathering cylinder is flexibly spring-mounted, the gathering fingers spring back, forming a "sled-runner" contact on touching any surface elevation. Thus the cylinder is lifted over the obstruction, whereupon the fingers straighten out, and the cylinder is lowered again for level ground work.

How is the tension on the elevator chains adjusted?

The carrier tension can be regulated by turning the tension screws at the top of the elevator.

What is the arrangement of the wheels with reference to the frame?

The wheels are located inside the frame. Outer ends of the wheel hub are extended to carry the frame.

Explain the transmission of power from the drive wheels.

The drive for the axle cylinder and elevator is direct from the axle, while the floating cylinder is chain-driven. The chain in turn derives its motion from a sprocket that is direct-connected to a pinion that meshes with the drive gear on the left hand axle extension.

How are the gathering fingers set?

No two of the five rows of fingers in the gathering cylinder stand at the same angle to the cylinder center at the same time. They are constantly rocked back and forth by the eccentric control in the cylinder head. The teeth are so set for work that the row passing, nearest the ground, point straight down in raking position. The row at the top point straight ahead so as to back out of the hay, in stripping.

What is the general construction of the wheels?

The wheels are made of steel, with concave tires, and removable hubs. The hubs are, therefore, easily replaced in case of wear. Traction lugs are formed on the extended ends of the spokes, being seated inside the concave of the tire.

How is the loader coupled to the wagon?

There are two angle irons clamped to the rear axle with four holes bored through them, and to these is attached a malleable clevis, the rear end of which is in a concave shape. In backing the wagon up to the loader, the clevis hits the spring jaw on front of the loader tongue and automatically adjusts to the wagon hitch. By the use of these four holes, this can be adjusted up or down so as to be adapted for either a high-or low-wheeled wagon. Underneath this hitch is a tongue support that is also adjustable for a high or low wagon.

Can the hitch be uncoupled from the top of the load?

Yes. Pulling the trip rope uncouples the loader.

What distance behind the rack should the loader be placed?

In adjusting the hitch parts, keep the loader at such a distance that the rack cannot catch in the carrier or strike the frame back of the loader in turning.

At what position should the gathering cylinder be set?

It should be set in the highest position in which good raking will be done; usually not lower than to touch the stubble.

How much hard work in loading can be saved?

A lot of hard work in loading can be saved if a few simple facts are taken into consideration.

The hay rack should have a tail board. Then, by allowing the hay to build up at the rear against the tail board, and by keeping the hay at the rear of the load higher than in front, a load can be put on with little effort.

Why should you use a tongue truck on a double-cylinder loader?

By using a tongue truck on the double-cylinder loader insures the rear gathering cylinder being the same distance from the ground all the time; also by the use of the tongue truck, there are no adjustments necessary for either a low- or high-wheeled wagon. A farmer can use a low-wheeled truck and a high-wheeled wagon on his farm without any change in adjustments.

Fig. 41. New Idea Hay Loader loading sequence.

Fig. 42. Very unusual and elaborate Wood Bros. Hay Loader.

Homemade HAY RACKS

The information in this segment comes from a 1932 Cornell Extension Bulletin by L. M. Roehl and was prepared to aid farmers who desire to construct their own hay racks from material obtainable at the local lumber yard, hardware store, and blacksmith shop.

The plans for two racks are herein presented. One for a combination wagon bed and rack, or sectional rack, and the other for a flat rack. Each is assembled in sections, thus making it possible for one man to place it on or off the wagon.

The combination bed and rack is used by some farmers in preference to a flat rack, as it is better suited to a hilly country. The flat rack makes the better rig for hauling ensilage and crates.

COMBINATION WAGON BED AND HAY RACK, OR SECTIONAL HAY RACK

The hay rack with removable side racks has several advantages: Since it may be removed from the wagon in three parts, one man can place the rack on and off the wagon more easily than if it were assembled in one piece. The sectional rack also can be stored in less space than can the flat rack, for the side racks may be hung on brackets in the implement shed or at some other place that is otherwise not used. The bed without the side racks may be used when potato crates, bags of grain, or baled hay are to be hauled. The front and rear standards may be removed by taking out the bolts at the lower ends. These standards may be stored until needed for haying. By the removal of the four bolts at the lower end of the front posts, the

Fig. 43. Combination wagon bed and hay rack empty.

bed may be cleared for the minimum of space for storage.

Bill of materials
Lumber:
2 pieces, 2" x 8" x 16'
2 pieces, 2" x 2" x 16'
11 pieces, 2" x 4" x 12'
1 piece, 2" x 6" x 3'6"
19 pieces, 1" x 6" x 16'
1 piece, 1" x 4" x 12'

List of pieces:
2 pieces, 2" x 8" x 16' for stringers
3 pieces, 2" x 4" x 3'6" for sills
1 piece, 2" x 4" x 3' 1 3/4" for sill (front)
1 piece, 2" x 6" x 3'6" for sill (rear)
8 pieces, 2" x 4" x 5'7" for arms
2 pieces, 2" x 2" x 15' 5 1/4" for cleats inside of stringers
14 pieces, 1" x 6" x 16' for top boards and floor
2 pieces, 2" x 4" x 2'3" for bottom posts of front standard
2 pieces, 2" x 4" x 6' for front standard
1 piece, 1" x 4" x 2'10 1/4" for cross bar of front posts
3 pieces, 1" x 4" x 2'6" for cross bars of front standard
2 pieces, 2" x 4" x 4' for back standard
2 pieces, 1" x 6" x 6' for cross bars of back standard
8 pieces, 2" x 4" x 2'6" for posts of basket or crib rack
2 pieces, 2" x 4" x 8'4" for cross bars of basket rack
4 pieces, 1" x 6" x 16' for sides of basket rack

Hardware required:
10--hay-rack clamps, figure 45
58--1/4"x5" carriage bolts to fasten boards to arms and standards
20--3/8"x4 1/2" carriage bolts to fasten front and rear standards at bottom, cross ties of basket rack, and 2"x2" pieces to stringers
4--3/8"x5" carriage bolts to fasten cross bar of front posts
4--1/4"x4 1/2" carriage bolts for bottom ends of standards
16--3/8"x3" machine bolts to hold iron clevises to ends of arms for basket-rack posts

Fig. 44. Getting the load on.

200

8 pieces, 1/4" x 1 ½" x 16" iron for
clevises
5 dozen, 1 ½-inch No. 10 flat-head
wood screws for floor boards
62--1/4-inch washers
24--3/8-inch washers
16--3/8-inch lock washers
20--½-inch lock washers
½ pound, 6-penny common nails
2 quarts, priming paint (white lead
and linseed oil)
2 quarts, outside paint or
1 gallon, outside paint or
1 gallon, creosote oil

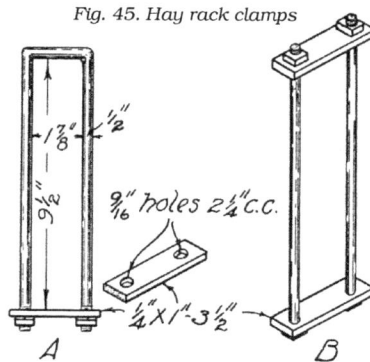
Fig. 45. Hay rack clamps

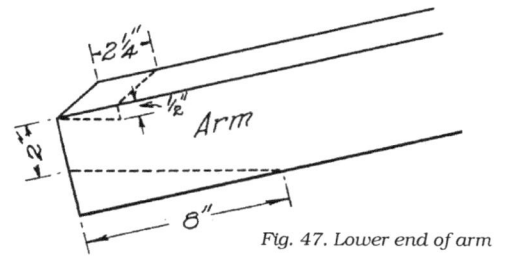
Fig. 47. Lower end of arm

Directions

In assembling the sills and stringers with the hay-rack clamps, the workman needs to be careful and make the outside spacing 37 3/4 inches. The distance between wagon stakes is usually 38 inches, and to make the bed of that width would make too tight a fit. Two styles of hay-rack clamps are shown in figure 45. The one shown at A is made by heating the rod and

A 2- by 6-inch piece is used for the rear sill to give added strength for the attachment of the hay loader.

The floor may be fastened to the sills with nails, wagon-box rivets, or screws. Nails soon allow the boards to loosen from the sills and are not recommended. In placing the screws, the heads are countersunk slightly below the surface, to leave the floor smooth.

As shown in figure 46, the front posts are bolted to the stringers at a point 3 inches from the front end with two 3/8-by 4 ½-inch carriage bolts. Lock washers are placed on all bolts used on the rack. Two 3/8- by 5-inch carriage bolts are used to hold the cross bar to the posts at the top.

The lower ends of the arms are laid out and cut to the dimensions shown in figure 47, and they nose under the 2- by 2-inch pieces which are bolted to the inside of the stringers, as shown in figure 5, at 2 inches from the floor. A 3/8 by 4 ½-inch carriage bolt is placed through the 2- by 2-inch piece from the inside of the bed right above each arm.

The 6-inch boards are fastened to the arms with 1/4- by 5-inch carriage bolts (figure 49). The construction may be further strengthened at this point by placing a screw or nail at each side of each bolt. The screws will hold the boards down tight on the arms at the edges.

The length to cut the 2- by 4-inch pieces for the rear standard (figure 50) is partly determined by the hay loader at hand. The pivotal bolts at the lower

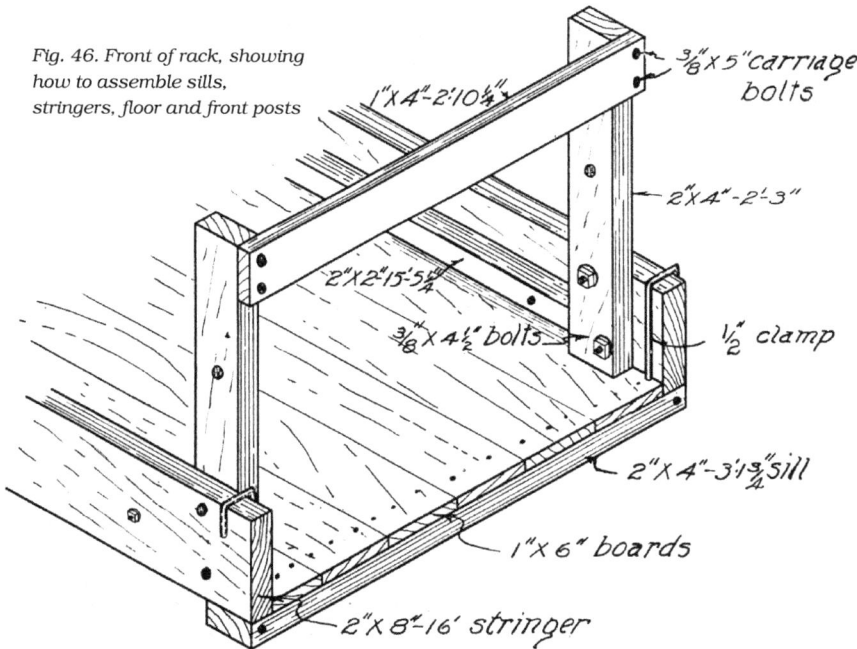
Fig. 46. Front of rack, showing how to assemble sills, stringers, floor and front posts

bending it into shape. The one shown at B is made by threading 11-inch rods at both ends. If this clamp is used, the upper ends of the rods project through the nuts on the upper-plate washer only 1/16 inch and are riveted.

The front sill is only as long as the width of the bed, which is 37 3/4 inches. One bar of the front clamp is cut off as shown in figure 46. If the sill projected as the others do, it would catch the spokes of the front wheel when turning.

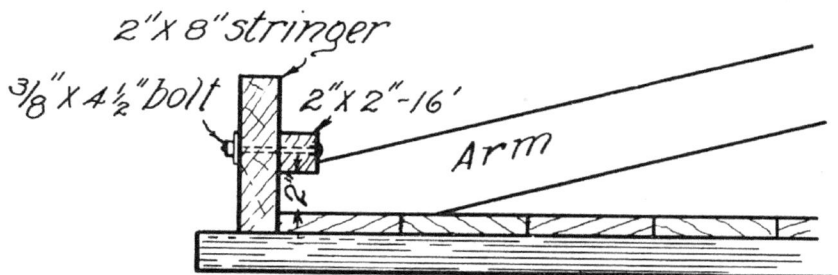
Fig. 48. The arm fits under the 2 by 4 inch piece which is bolted to the stringer.

ends are placed 12 inches from the ends and 2 inches from the upper edges of the stringers. A piece 2 by 2 by 10 inches is fastened to the underside of the left rear standard to make it even with the other as they rest on the arms.

The lower ends of the front standard are strengthened by placing 1/4- by 4 ½-inch carriage bolts as shown in figure 51. A similar bolt is placed at the lower ends of the rear standards just above the pivotal holes. The front standard is fastened to the front posts with 3/8- by 4 ½-inch bolts which are placed 12 inches from the top of

iron. The size indicated in figure 53 is merely suggestive. It is fastened to the arms with two 3/8- by 3-inch machine bolts.

The sides of the basket rack are held together at front and rear with pieces (2 inches by 4 inches by 8 feet 4 inches) which are bolted to the front and rear posts, as shown in figure 54. The front 2- by 4-inch cross piece may be bolted to the front standard to aid in keeping the basket rack vertical.

The life of the rack can be increased by painting it.

The first, or priming, coat may consist of white lead and linseed oil and the second coat of any outside paint, such as is used for buildings.

If white lead and linseed oil is not available, the rig may be finished with two coats of outside paint or with one coat of creosote oil.

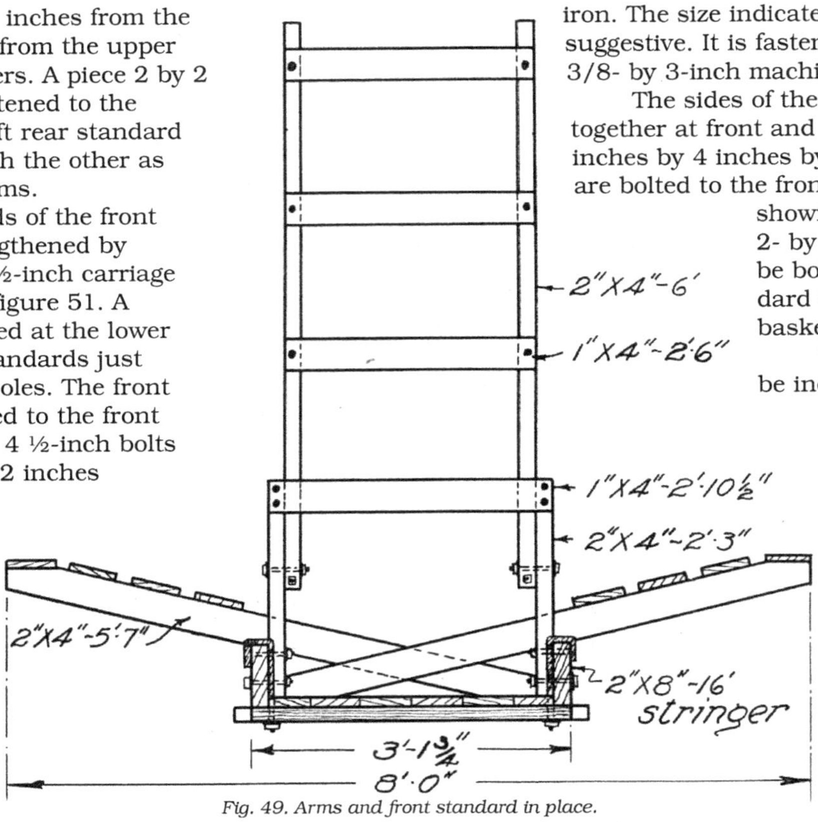

Fig. 49. Arms and front standard in place.

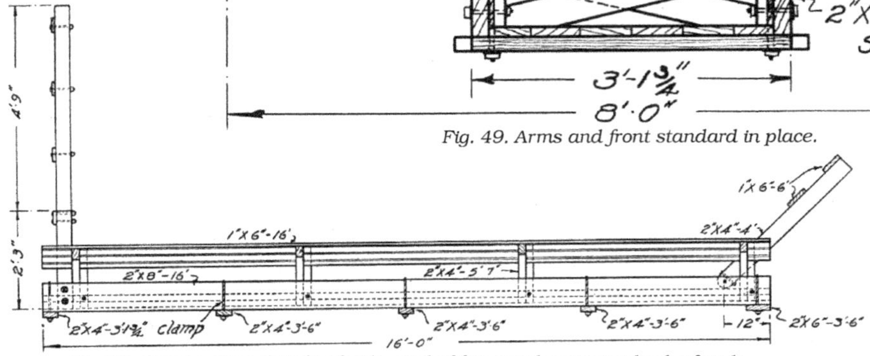

Fig. 50. Construction details of side, and of front and rear standard, of rack.

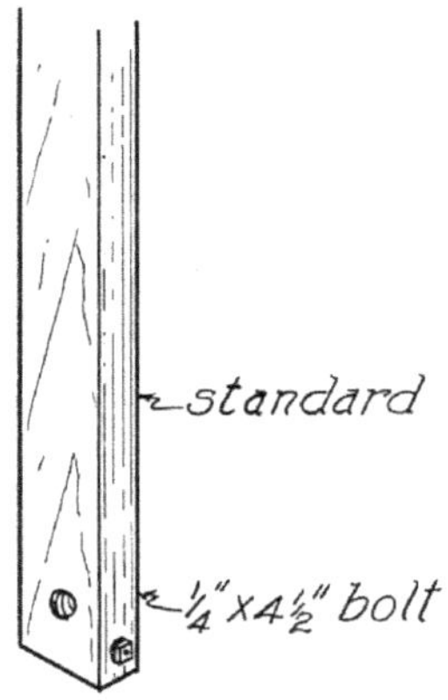

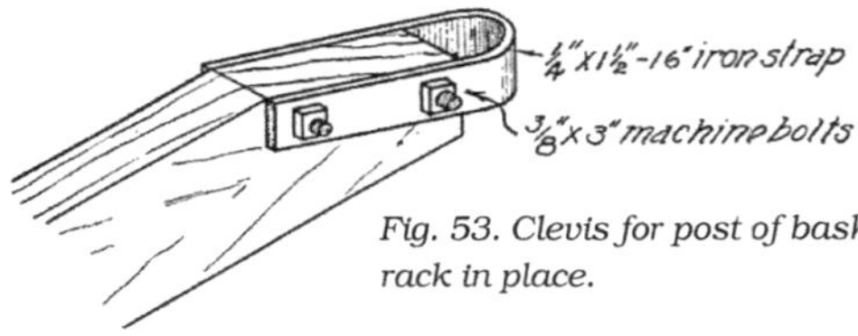

Fig. 53. Clevis for post of basket rack in place.

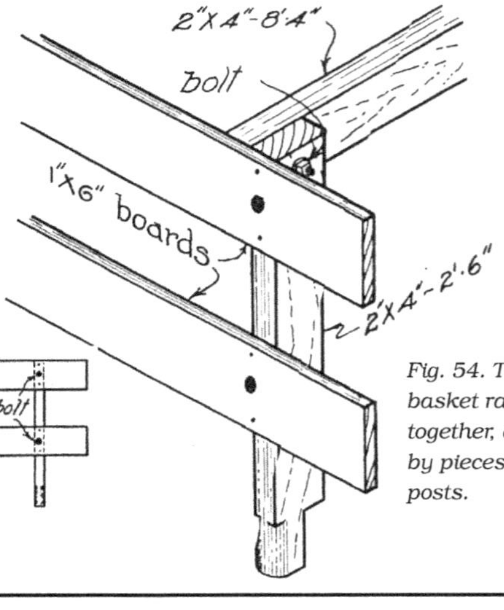

Fig. 51. Lower end of front standard.

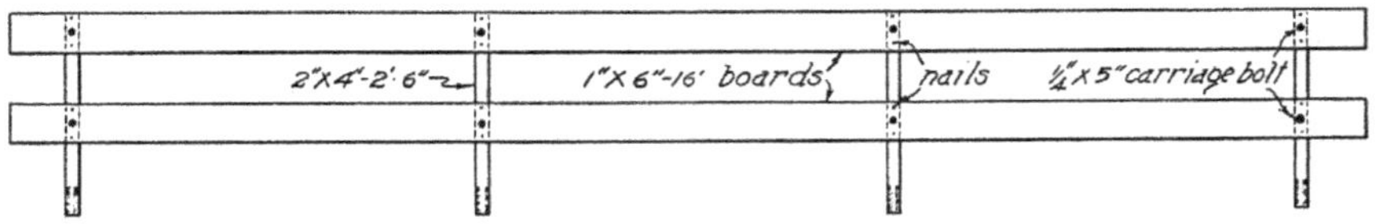

the posts (figure 49). This construction allows the standard to tip back on the rack when the rack is empty.

On many farms the basket, or crib, rack (figure 52) may find sufficient use to warrant its addition to the rack. It offers considerable advantage where one man needs to load hay or grain.

The clevises for the ends of the arms to hold the posts of the basket rack may be made of a variety of sizes of

Fig. 54. The sides of the basket rack are held together, at front and rear, by pieces bolted to the posts.

Fig. 52. Construction details of basket, or crib, rack.

Fig. 55. The combination wagon bed and hay rack in use.

FLAT HAY RACK

The flat hay rack, in addition to service as a rack, is convenient to use with a low truck wagon for hauling green corn for ensilage, potato crates, and baled hay.

Bill of Materials
Lumber:

2 pieces, 2" x 8" x 16' for stringers
4 pieces, 2" x 4" x 8' for arms
2 pieces, 2" x 4" x 16' for sides
1 piece, 2" x 4" x 14' for sills
1 piece, 2" x 4" x 14' for front standard
1 piece, 2" x 4" x 14' for back standard and pockets of standards
26 pieces, 1" x 4" x 16' for edge-grain matched flooring or
18 pieces, 1" x 6" x 16' for platform or floor
4 pieces, 1" x 4" x 8' for straps for floor
1 piece, 1" x 4" x 8' for cross bars of front standard
1 piece, 1" x 6" x 8' for cross bars of rear standard

Hardware:

8--hay-rack clamps, figure 57
8--3/8"x6" carriage bolts to fasten side pieces to arms
8--3/8"x4" carriage bolts to make pockets for standards
1 pound 6-penny common nails
2--1/4"x4 ½" carriage bolts for lower end of rear standard
14--1/4"x5" carriage bolts to fasten cross pieces to posts
16--½-inch lock washers for clamps.

Directions

The framework of the rack is shown in figure 56.

The stringers are placed 24-inches apart (outside) at the front end and 3 feet 6 inches at the rear, to make possible shorter turns with the wagon.

Four 2-inch sills are used. The front one is 2 feet 4 inches long, the rear one is 4 feet long, and others as shown in the drawing. The sills project 2 inches past the stringers at the sides to allow space for the hay-rack clamps.

The hay-rack clamps are made as shown in figure 57. They need to be made 3 inches

wide to clamp over the 2- by 4-inch arms diagonally.

In assembling the sills, stringers, and arms, it is convenient to temporarily nail a piece of board to the front and rear ends of the stringers, to hold them at the proper space for measuring. The upper edges of the arms may be notched to let the clamps down flush, or grooves may be cut on the underside of the floor to allow the floor to rest evenly on the arms.

A piece 2 inches by 4 inches by 16 feet is bolted to the end of the arms with a 3/8- by 6-inch carriage bolt at each joint. The spaces between the arms are equal. (Figure 56.)

A false bolster, as shown in figure 58, needs to be placed on the front bolster to hold the rack in place at the front end.

The construction of the front end of the frame is shown in figure 62. The cross arm is clamped to the stringer at 1 inch from the end. Two pieces of 2- by 2- by 8-inch material are bolted to the insides of the stringers. One immediately under the cross arm and the other far enough back on the stringer to make a pocket for the lower end of the front standard. The detail drawing

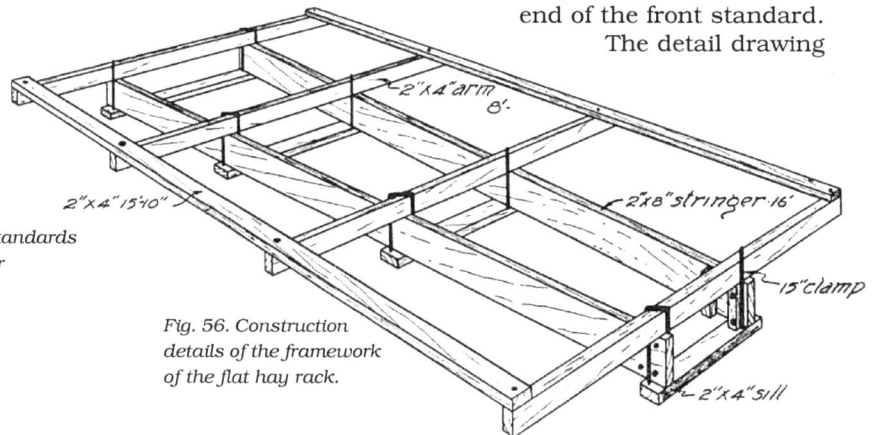

Fig. 56. Construction details of the framework of the flat hay rack.

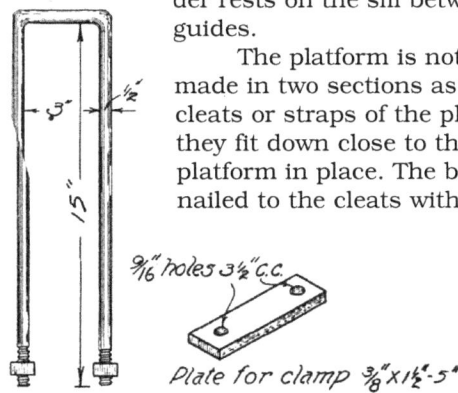

Fig. 57. Hay-rack clamp.

at A (figure 62) shows how a 1- by 2-inch shoulder is cut at the lower end of the standard. The shoulder rests on the sill between the 2- by 2-inch guides.

The platform is not fastened to the arms. It is made in two sections as shown in figure 59. The cleats or straps of the platform are so placed that they fit down close to the arms, thus holding the platform in place. The boards of the platform are nailed to the cleats with two 6-penny common nails at each joint. The nails are clinched on the cleats. The edges of the boards are painted when the floor is assembled. This tends to keep the edges dry and adds to the length of service.

37 3/4"

27"

2"X6"-7 1/2"

2"X6"-3'4"

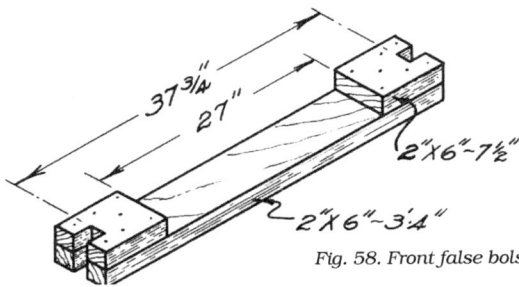

Fig. 58. Front false bolster.

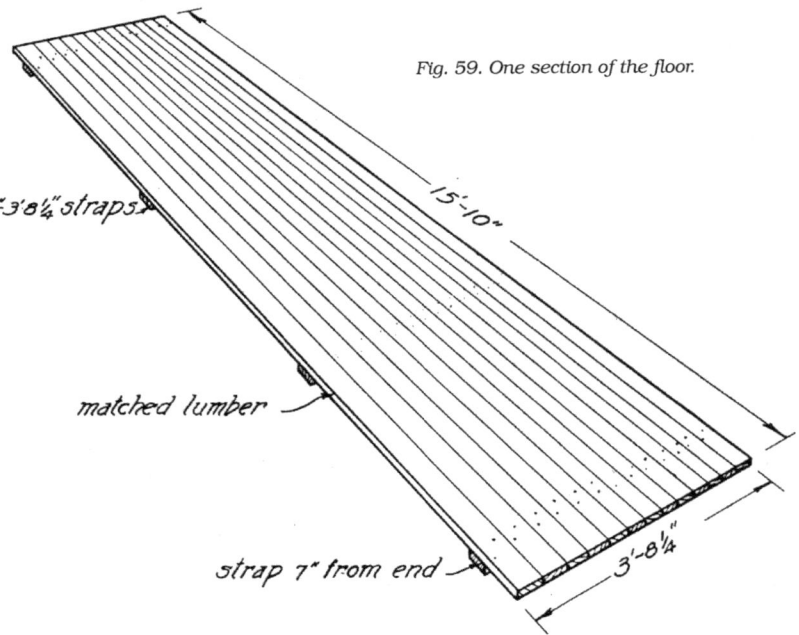

Fig. 59. One section of the floor.

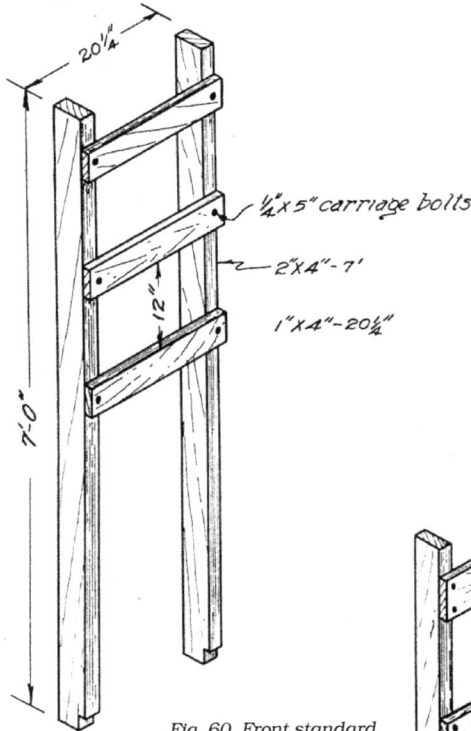

1"X4"-3'8 1/4" straps

15'-10"

matched lumber

strap 7" from end

3'-8 1/4"

20 1/4"

1/4"x5" carriage bolts

2"X4"-7'

1"X4"-20 1/4"

12"

7'-0"

Fig. 60. Front standard.

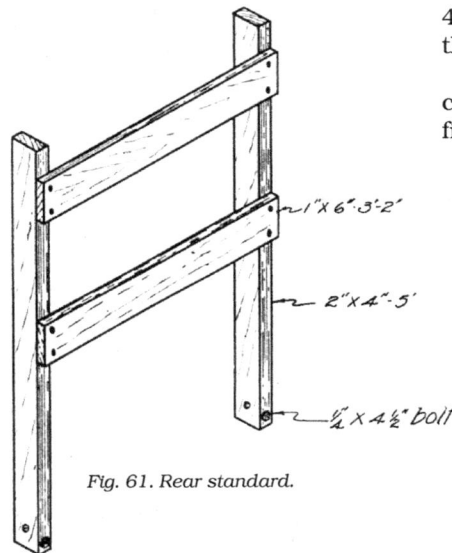

1"X6"-3'-2'

2"X4"-5'

1/2" X 4 1/2" bolt

Fig. 61. Rear standard.

4 ½-inch bolts through edgewise below the pivot holes as shown.

If it is desired to use a basket, or crib, rack, the scheme suggested in figures 52, 53, and 54 may be employed.

Paint the rack as suggested for the combination bed and rack.

1"

2"

lower end of front standard

A

If the platform is made in two sections and is not fastened to the arms, the rack, when not in use, can be removed and stored much easier than if the platform were in one piece.

The front standard is shown in figure 60. Holes are cut in the platform or floor of the rack immediately above the pockets, and the standard is dropped in place when its service is desired.

The rear standard is made to the dimensions shown in figure 61. Instead of standing upright like the one at the front it is bolted to the stringers at points on the stringers 2 inches from the top edges and 1 foot from the ends. It rests on the rear cross arms. The length of the 2- by 4-inch pieces is regulated by the hay loader in use. The lower ends of the 2- by 4-inch pieces are strengthened by placing 1/4- by

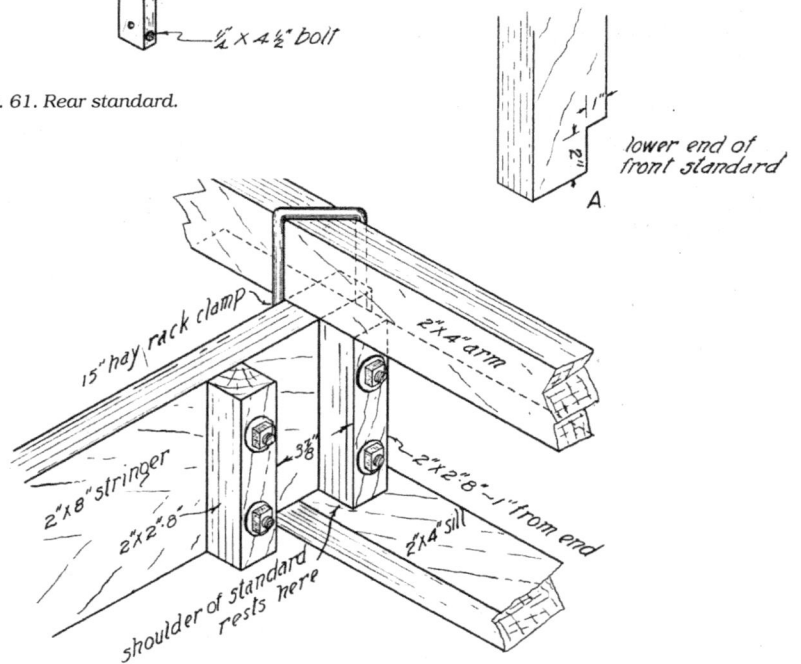

15" hay rack clamp

2"X4" arm

2"X8" stringer

2"X2"-8"

3"

2"X2"-8"-1" from end

2"X4" sill

shoulder of standard rests here

Fig. 62. Construction details of front end of the frame.

Buckraking Loose Hay

Chapter Twelve

The common methods of bringing the hay from the windrow or cock (bunch) to the stack are by means of buck- or sweep rakes or by loading the hay on slips or wagons. The slip loading is done by pitchfork or cabling as demonstrated in the feeding chapter. Wagon loading was covered in the previous chapter.

The buckrake, as the drawings and photos show, is an implement with a basket of long teeth which is driven into waiting wind-rows or bunches of hay. When a load has been accumulated on the basket it is driven (pushed) to the waiting stacker.

For hauls of a half mile or less, hay may be moved more rapidly and cheaply and with less labor with a buckrake than by any other means. For hauls longer than a half mile wagons are usually just as economical as buckrakes.

The sweep rake or hay buck, as it is called in some localities, may be of the side-hitch type, with the horses hitched one on each side of the rake that carries the hay, or of the newer push type, with the horses hitched to the rear of the rake. The side-hitch type was considered by some in the past to be more

Figure 1. (top) A scene from the Hirshey Ranch in the Big Hole of Montana. Figure 2. (above) The author bringing in a jag of hay with a John Deere Power Lift Buckrake at Singing Horse Ranch in Central Oregon.

Haying with Horses

Figure 3. Heavy duty homemade buckrake.

Figure 4. A factory buckrake with pushbar.

Figure 5. A four horse buckrake with very large jag.

Figure 6. A hay slip which has been loaded by pitchfork.

satisfactory where the hay ground is extremely rough. The push type has the advantage of working up closer to fences, and the work of the driver is less strenuous than when using the side-hitch type. In some sections where extremely large loads of hay are hauled, homemade sweep rakes of a heavier pattern are better than those usually built by hay-equipment manufacturers. Figure 3 shows a homemade sweep rake and Figure 5 shows a four-horse sweep rake used on ranches in northern Nebraska in the 1900's.

On level ground two-horse buckrakes can push anywhere from 600 to 1200 lbs. of hay (weight varies more due to crop and to cure and less to bulk). If the travel distances and windrow layout are the same, a buckrake can move the same amount of hay as a wagon and hayloader in 1/2 to 1/3 the time.

We have used buckrakes for ten years with tremendous success. Both of ours are John Deere Power Lifts. The first one came as a pile of hardware, no wood, which was purchased for $25. The photos on pages 219-221 show what it looked like right after all new wood (done by following the old manufacturer's drawings). The second one came a few years later and was purchased complete from an auction. The information we share as to hitching and operation is specific to this model and our experience with it. Though our procedures with this implement have always been perfectly satisfactory it needs to be noted, for the sake of thoroughness, that several 'old-timers' have taken issue with us. There is a humorous short tale to illustrate this:

After three successful years of buckrake operation we were visited by an old timer who watched as I hitched up. He shook his head and said I had it all wrong, that the horses should be tied ahead and the lines needed to be setup as modified team lines with long cross-checks. I explained that it was working this way and that I appreciated his information but no thanks. Several days later he showed up again, this time with two other older gentlemen all in a white Cadillac. He was anxious that they should see that the *world famous L. Miller(?)* didn't have a clue what he was doing. So we all went out to the buckrake and I once again explained how I hitched. They were all in agreement that I was doing it wrong. AND none of them could agree on what was the right way! They could, however, agree that I was doing it wrong. I said, "okay, I'll go get Tuck and Barney and one

206

Fig. 7. John Deere Side-hitch Buckrake

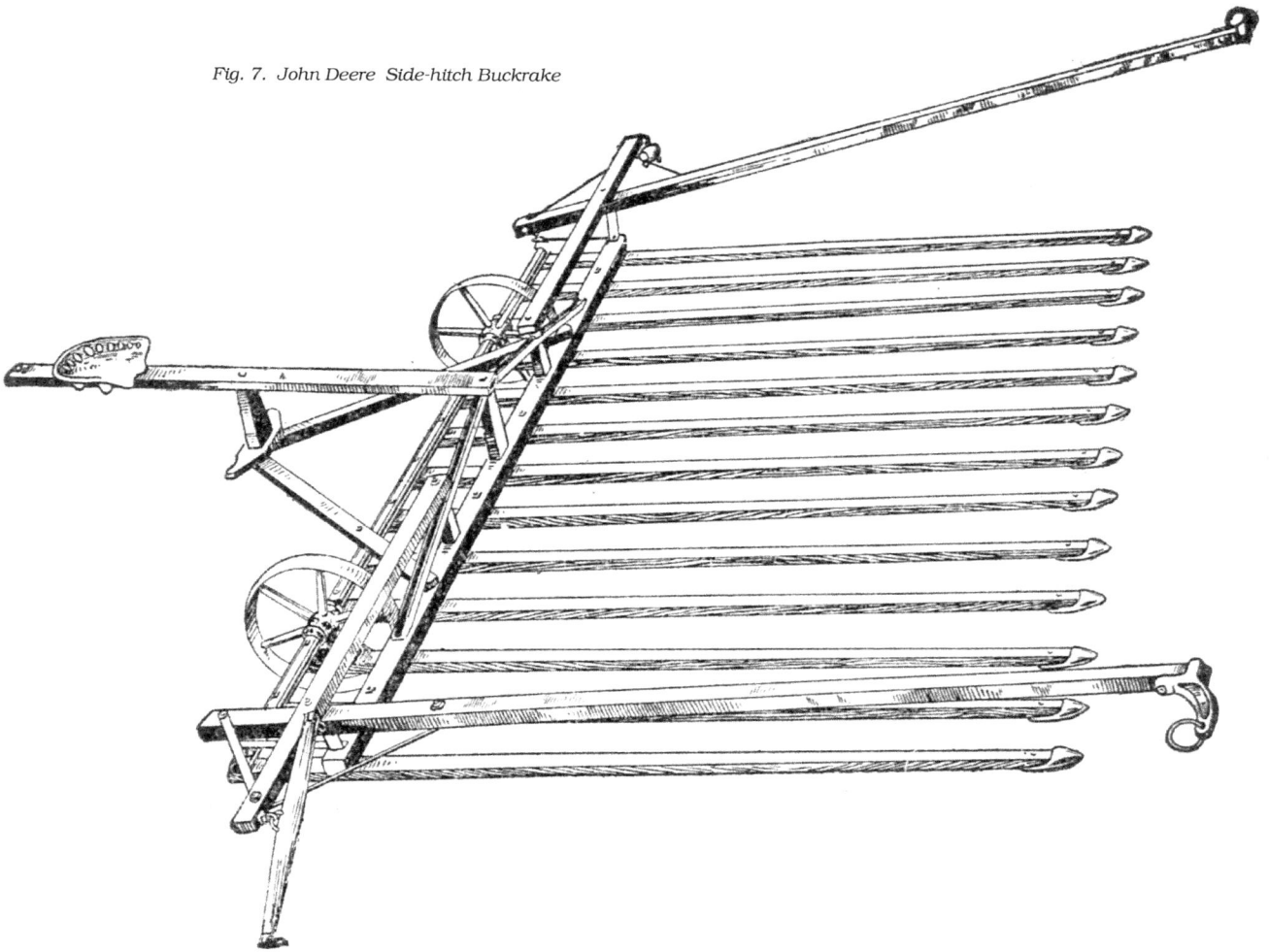

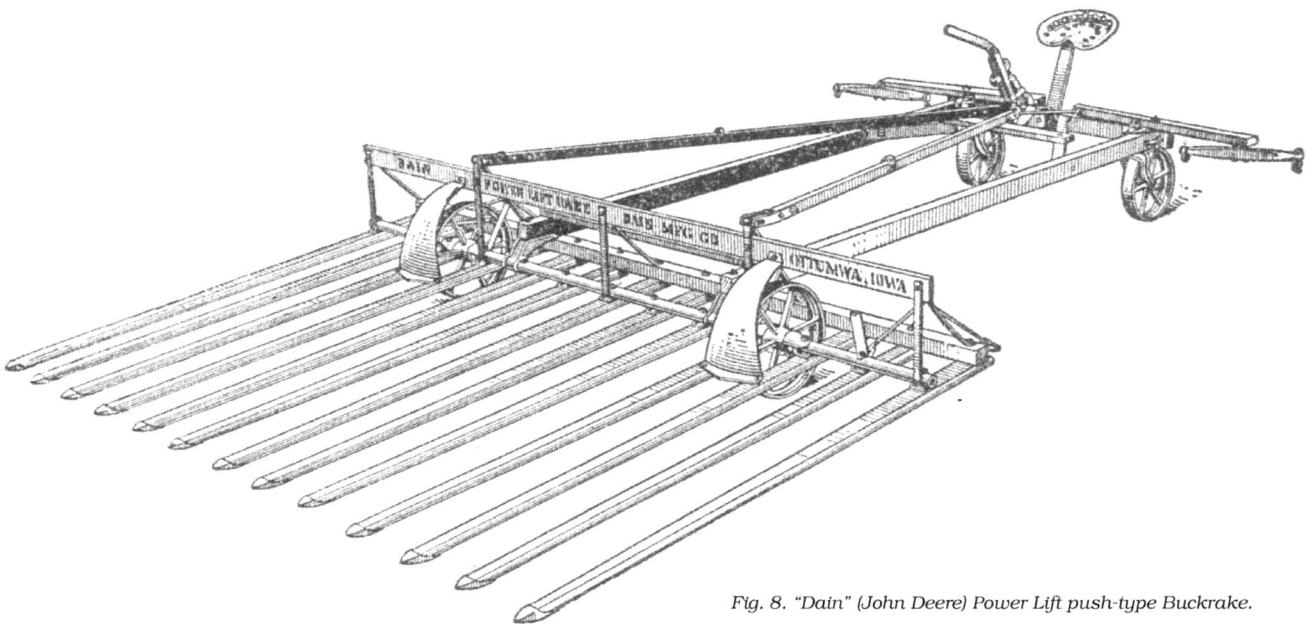

Fig. 8. "Dain" (John Deere) Power Lift push-type Buckrake.

of you guys can hook them your way and show me how to do it right. Decide which one of you it'll be. I'll be right back with the horses." I didn't even make it to the barn. One of them said, "I'm not dressed for this work." another said, "They're not **my** horses!" and the third one just scowled and shook his head. They left and we're still using the same system and happy as clams. In fact I have my serious doubts as to whether any of those gentlemen had ever buckraked hay.

The story illustrates two important points. Don't assume because the man walking up to you is old and talks bold that he must be an authority on working horses. And, also, it is altogether possible that you may come up with a completely different way to make any of this hay equipment work for you and work well.

Fig. 9 Breast strap extenders made of hame straps and rings.

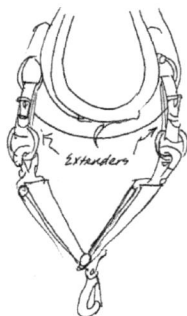

The Buckrake Hitch

This author has never experienced any application of animal-power so geometrically unique and adventuresome as the buckrake. As illustrated in Figure 10 the horses are hitched quite a ways apart from one another and, in our setup, rigged with a single (inverted) team line to each horse. The low-hung pole strap is attached to a backing rod attached to the front axle bracket (We use snap-on breast strap assemblies and attach breast strap extenders to the bottom hame rings. These are hame straps with rings.) The traces hook to singletrees which scissor off the basket-lifting mechanism. In our setup the horses are wearing halters under their bridles for this job and lead ropes are tied sideways and in to the frame.

The horses are driven as two singles. The line pressure slows or stops the horse which is feeling the pressure. Slowing or stopping one horse as the other one continues makes the buckrake turn, sometimes quite fast. Pulling back and giving the command to back, with well trained horses should result in the buckrake backing up. This procedure is the one thing which bothers new animals the most because as they back up they can see the basket follow them. It is, however, amazing how quickly good animals pick up on the process.

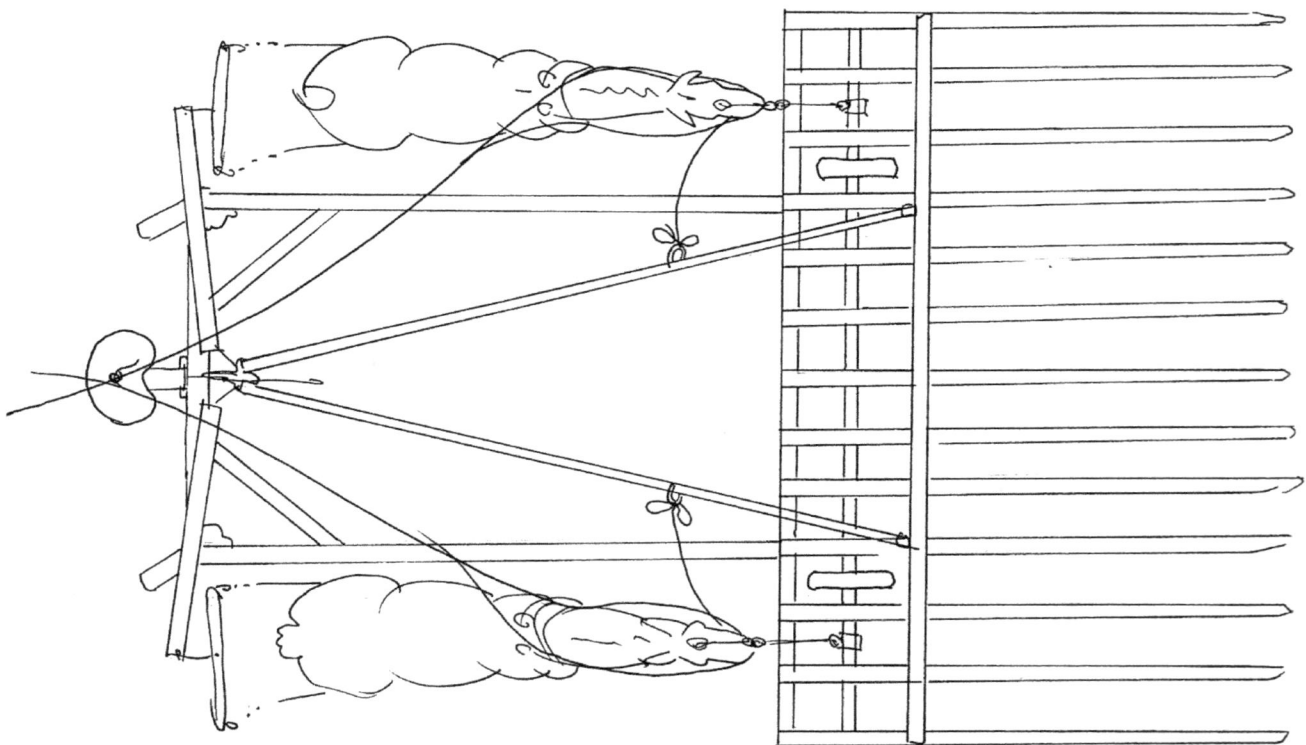

Fig. 10. Showing the push type buckrake hitch with team lines which have been inverted (or switched over).

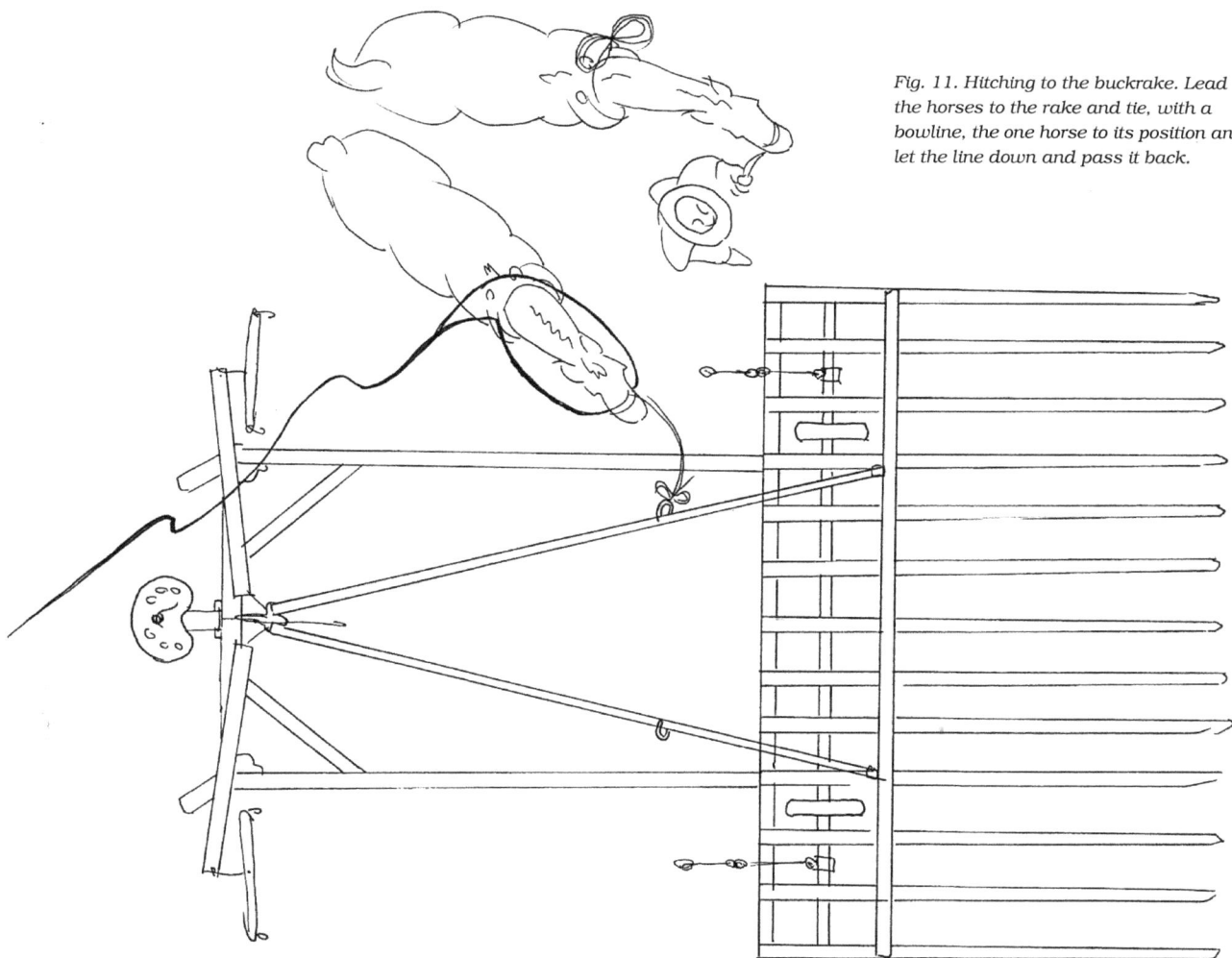

Fig. 11. Hitching to the buckrake. Lead the horses to the rake and tie, with a bowline, the one horse to its position and let the line down and pass it back.

Hitching to the Buckrake

Leave the halters on under the bridles. Make sure the breast strap extenders are in place and that the team lines have been inverted (or flipped) over so that the solid line goes to the inside and the cross check to the outside. Until you are VERY comfortable with the procedure take along a helper. At the buckrake, tie each horse to the frame ring (see Fig. 11) and pass the lines back to the seat area. Fasten the pole strap assembly of one horse to the backing rod and then hook traces taut to the single tree. Go to other side and repeat procedure. Be careful if it's the first time, as when you pull the single tree and trace to hook up, the buckrake may move some and startle unsuspecting animals.

If you are hitching for the first time it might be helpful to have two people available, one to each side, to help with any fine tunings such as lead rope lengths or adjustments to trace chains. When you first head out to the hay field with a new team DO NOT AT-TEMPT to push a load of hay into any waiting stacker UNTIL you have backed the buckrake up repeatedly. Otherwise you may find yourself having to unhitch a balky team and pull the buckrake out by hand. If you try to force a team to back up when telescoped into a waiting stacker's basket, in all likelihood, one horse will back and the other will not which will cause the buckrake to turn and it may break teeth. Play it safe and practise out in the field until everything is working fine. And remember that the buckrake is NOT the time to teach a horse to back, he should know the command and procedure well before being hitched.

Buckraking Procedures

Field size, shape, distance to stackyards, nature of hay crop and other variables may all come to play in whatever procedural system a person employs with buckrakes. The goal is simple. You want to "push" the cured hay to the stacker where it will be hoisted up on to the stack. Anything you can do to save travel time and work steps will figure as a bonus. By way of simple example, two healthy horses and a well-built buckrake can push MORE hay in one jag to all styles of stackers than those stackers can lift (the single exception being larger models of Beaverslides, i.e. Big

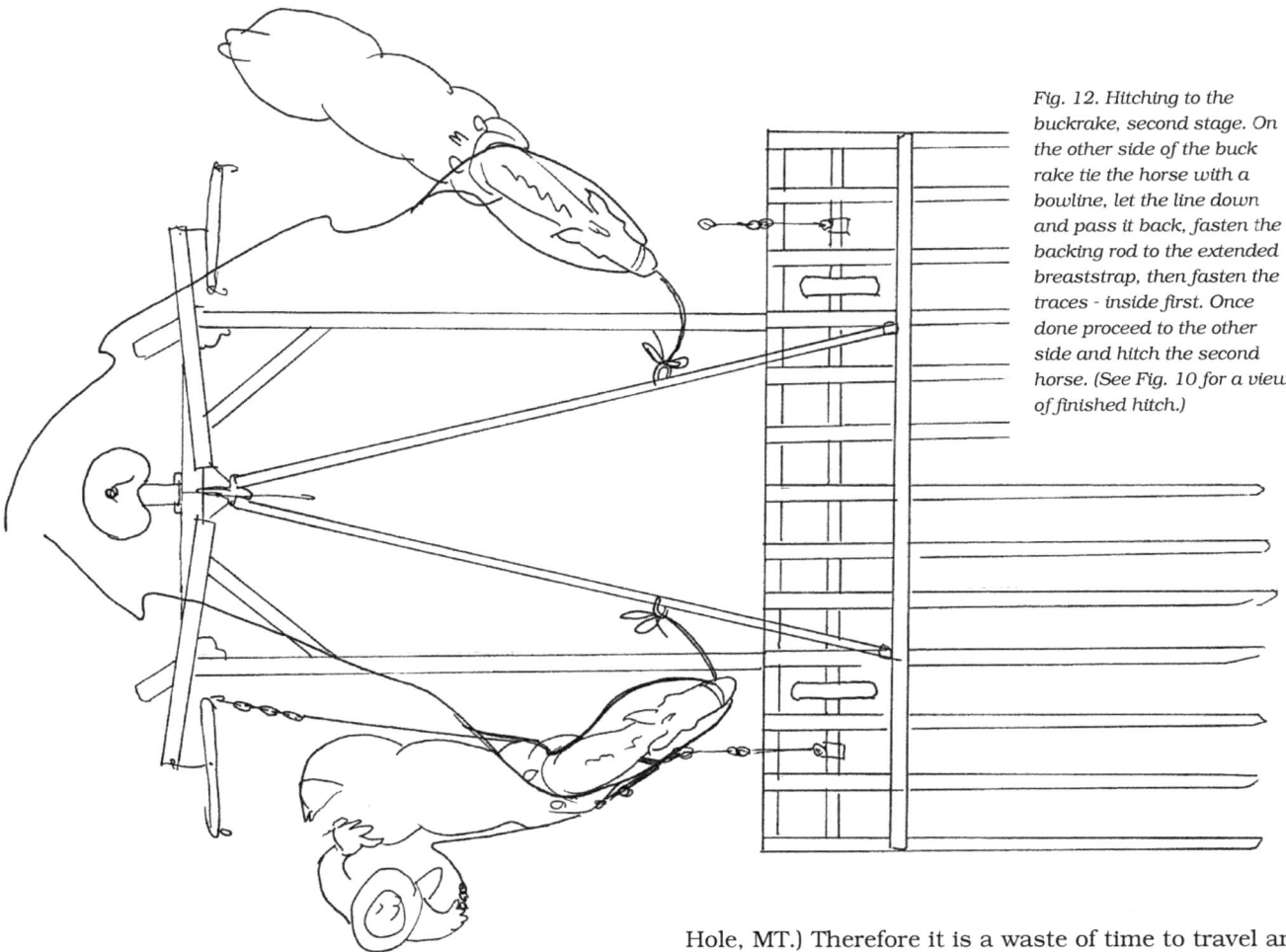

Fig. 12. Hitching to the buckrake, second stage. On the other side of the buck rake tie the horse with a bowline, let the line down and pass it back, fasten the backing rod to the extended breaststrap, then fasten the traces - inside first. Once done proceed to the other side and hitch the second horse. (See Fig. 10 for a view of finished hitch.)

Hole, MT.) Therefore it is a waste of time to travel any distance to a stacker with less than a full jag.

This author, when leaving the stacker with team and buckrake, chooses to bunch hay (as in Fig. 13 and Fig. 14) and then turn to gather the bunches into one jag. (By bunching, and then gathering bunches into jags, it is usually possible to put more hay on the buckrake than with a straight shot of gathering.) The full jag is then taken to the waiting stacker basket. The horses push it on and then back up and go after another load. Since our field is 1/4 mile long, some jags get to the stacker quicker than others. For this reason, if we should return before the load has been hoisted, we'll drop off full jags short of the stacker. This works well as a working stockpile because, as the horses tire, it is possible to rest at the stacker and push up one of the stockpiled jags when needed to keep the stack crew busy.

This system works well in part because our operation is small. Larger systems may deserve some additional planning. One system which addresses this is that which was used on Nebraska's Triangle Ranch. In his own words, Pete Lorenzo shares the information on the next pages.

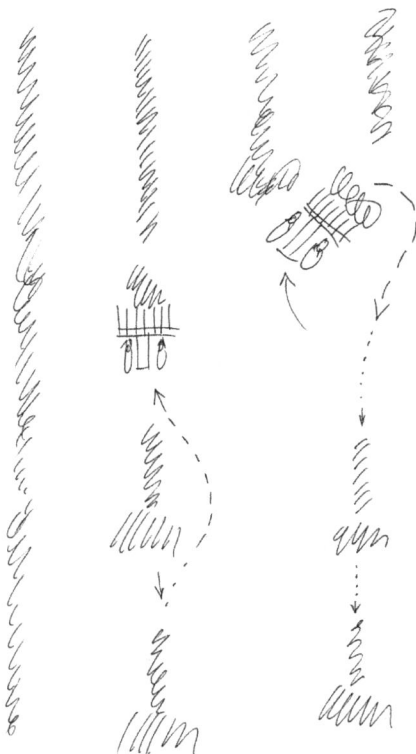

Fig. 13. Going away from the stacker we bunch hay, unconcerned about windrow tails, turning on third bunch to gather two more and head for the stacker.

Fig. 14. At Singing Horse Ranch Jess Ross heads away from the stacker bunching with his two Percheron mares.

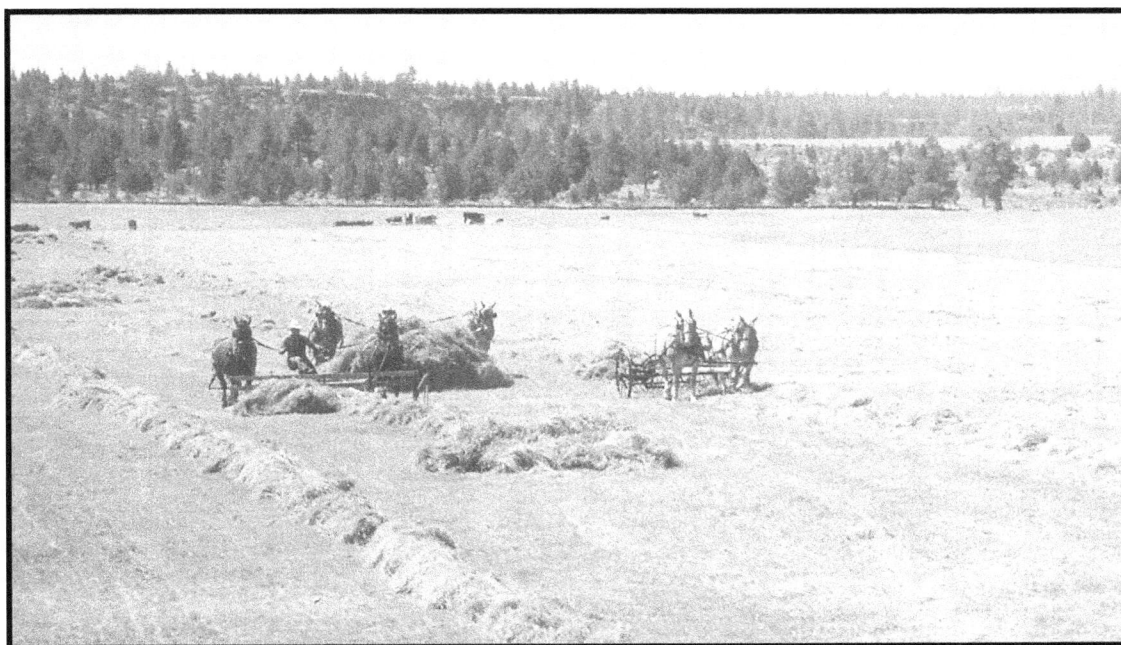

Fig. 15. In this picture, taken from atop the stack in progress, Jess is returning to gather bunches. Behind him is the author with a full jag and to the right is Justin Miller raking additional hay.

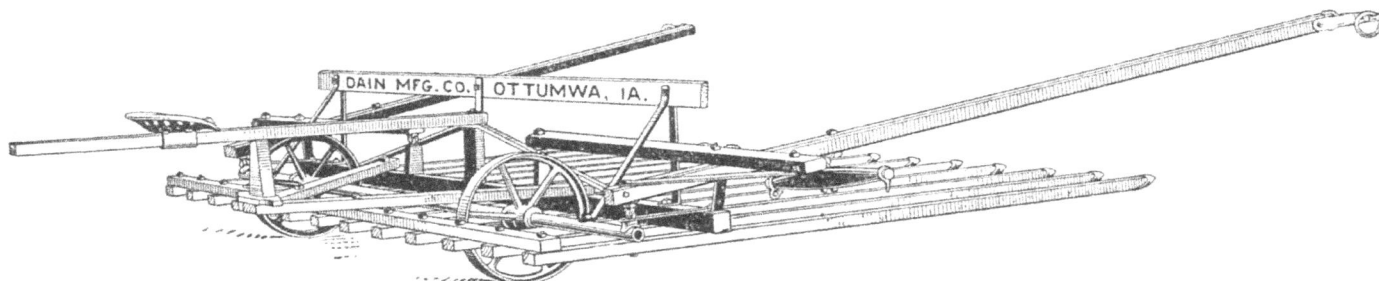

Fig. 16. *Showing Jess continuing to gather bunches.*

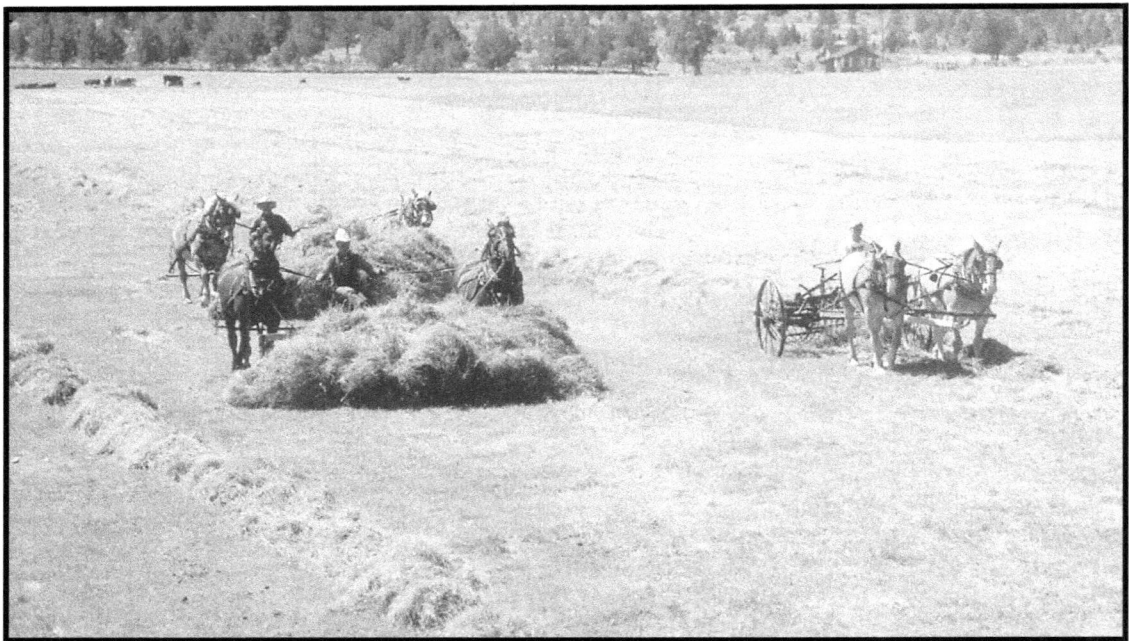

Fig. 17. *Jess with a full load. A clear view of how quickly large amounts of hay can be gathered with buckrakes.*

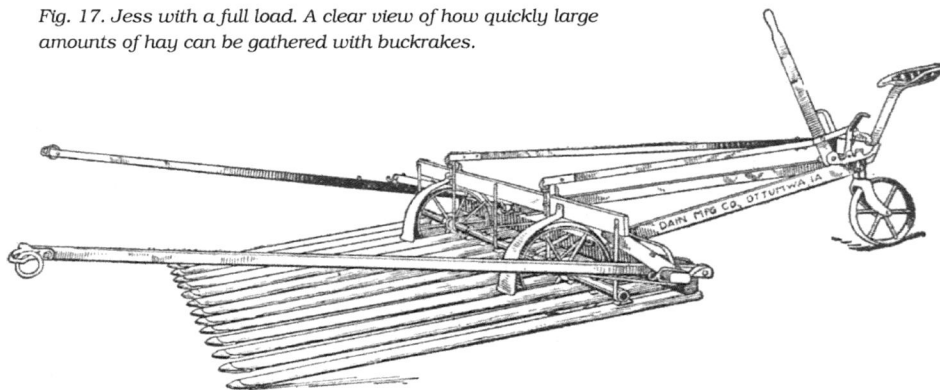

Triangle Ranch Bucking System

by Pete Lorenzo

I'll tell you a little about the size of the Triangle Ranch near Antler, Nebraska so you'll know how much hay we put up. They ran about 11-1200 cows there, yearlings plus bulls, replacement heifers, milk cows for ranch and camps. They had about 40 saddle brood mares and I think 14-16 draft mares, while I was there. Three saddle studs and 2 draft. The old draft stud was cut the last year I was there. We used him in the haying string that year. I'd jingle in the string every morning and that old guy would have every mare cut out into his own bunch every day. And each cow hand would have 4-5 saddle horses in his string. There was about 3 of us who would have 7-8 with the colts we broke out. In the winter there was at least six 6-horse teams hooked every day to feed. We would put up from 300 to about 350 stacks of hay at 7 ton. We always started just after July 4 and hoped to be done by the end of August. We loaded up usually 20 head of horses a day. With a total of about 50 head of draft horses in the hay string. The extra horses came from another ranch they owned, about 15-18 miles away. I usually got to trail them to the ranch for haying. It was a good job. Well I guess I'll start.

Most of the sand hill meadows are good sized. A lot would average 3/8 to ½ mile wide and 1-1/2 to 1-3/4 long. Some smaller some bigger. We had three 9' tractor mowers ahead of us. They were contracted to mow. One power sweep that the foreman would use to sweep the stacker, you could use horses but there's a lot of backing. And the straight rake we used was a 30' Rause rake pulled by a Ford tractor, about 30 hp. A smaller haying could be done with horses alright using the rakes we used, not dump rakes, unless you have enough teamsters and equipment. The guy we had on the straight rake could fly and he was good. The straight raker had to judge his hay and know how to lay out a land.

Most stackers in this country are slide stackers (see Figs. 21 & 23) or a few overshots. They are moved to the hay instead of sweeping to stacker a long ways.

Our machinery for horses was two 4-horse sweeps (Figs. 18 & 19) with 12' heads I think and two-four horse rakes. Ground drive hydraulic dump and stacker 4 head. We also had 2 side delivery rakes hooked to an evener to turn the windrows with horses if rained on. Maybe you have seen these straight rakes in your travels? They are really good to use. I don't remember their store bought name.... The sweeps were made at the ranch on pickup frames. They are light, fast and strong. One had a lever cable lift. The one I used had a boat winch cable lift. You can sweep to the stacker head with them real easy. The two rakes we used most had 16' length. We also had one 18' but needed 6 head on it and didn't use it a lot.

I can't say if any slide stackers are factory made. The ones I've been around all had something a little different about them. I was on a couple of ranches that used tractors to stack with. The one on the

Triangle was made so the slide arms would drop to go under power lines. As you can see they would be a high cost item to build. All pulleys have bearings and grease zerks. We had a Dodge power wagon to carry the cage and hydrofork to stack with. But I worked stacking by hand in the cage on one ranch (got paid extra). If you don't fight the hay its not that bad.

An average day without much down time we could put up 25 to 28 stacks a day. The best I was ever on was 33 in one day. These are 7 ton stacks or better. Actual stack building time was 12-15 minutes. Its a lot different on my 80 acre farm now with just a team of mares, hay rack and New Idea loader. I and my family did build one stack of hay that got wet a few times.

Fig.18. Roy Stevens on sweep.

Fig. 19. Author on 4 horse sweep, note pull back bar.

Fig. 20. An average size bunch.

Fig. 21. Roy Stevens driving stacker team and Brady Demsey stacking.

Fig. 23. Roy Stevens on lines, Brady Demsey stacking.

Fig. 24. Load going up.

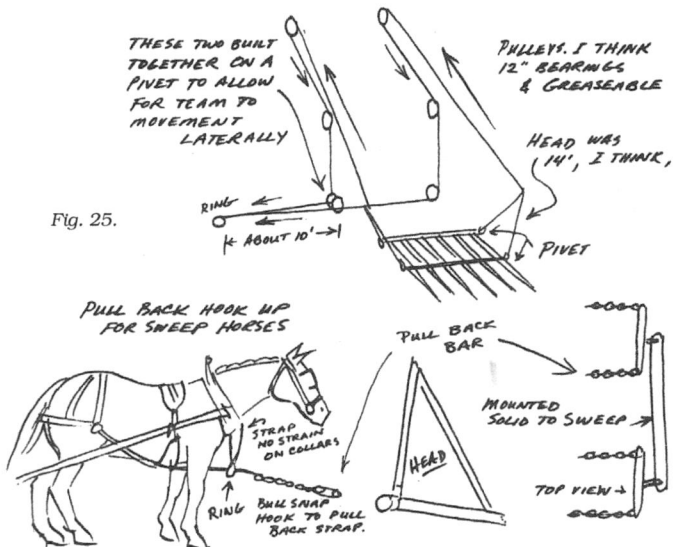

Fig. 22. Cal Long building a stack with a hydra fork on a Dodge Power wagon.

THE CABLE HOOKUP OF THE STACKER

THESE TWO BUILT TOGETHER ON A PIVET TO ALLOW FOR TEAM TO MOVEMENT LATERALLY

PULLEYS. I THINK 12" BEARINGS & GREASEABLE

HEAD WAS 14', I THINK,

RING

|← ABOUT 10' →|

PIVET

Fig. 25.

PULL BACK HOOK UP FOR SWEEP HORSES

PULL BACK BAR

HEAD

STRAP NO STRAIN ON COLLARS

RING   BULL SNAP HOOK TO PULL BACK STRAP.

MOUNTED SOLID TO SWEEP

TOP VIEW →

I will show how we hayed a land or one small meadow that doesn't have to be cut in more than one land.

Hope I keep it simple enough to understand.

A few thoughts as I think of them about ranch haying. We always put our faster horses on the sweeps. I would sweep most of the time, but I did scatter rake and drove the stacker team. The sweep drivers always have one extra team to rotate with and colts to work most times. The stacker team would never change unless there was one lame or sore.

The stacker team sure figures out the end of the cable fast. But you can get them to pull easy to drop hay in front. Or pull hard to throw in back.

Sweeping you figure how to build your load. And if you hit a rat run or soft spot and bury a tooth or two, you might pick the back of the sweep up maybe 3' off the ground. That sure wakes you up. I've drove a lot of big loads where the back wheels were off the ground a foot, and those horses just chugging right along. Haying like this is real enjoyable when you have a good crew and working horses.

Hope this might help you.

Pete Lorenzo, Excello, MO

① Straight rake comes in to rake one full rake pass makes two bunches. It don't matter were the gate is.

①

Not to Scale

RAKE · DUMP · RAKE · DUMP · RAKE — ALL THE WAY DOWN FENCE

DUMP

③ RAKE AROUND ⟶

④ RAKE OUT ENOUGH ROWS TO GET THE LAND LAYED OUT THEN GO TO START AND TURN IN ENDS SO SWEEPS CAN TURN IN,

DUMP   DUMP

MAKING CIRCLES ALL THE WAY DOWN. THEN FINISH RAKING AND TURN IN OTHER END.

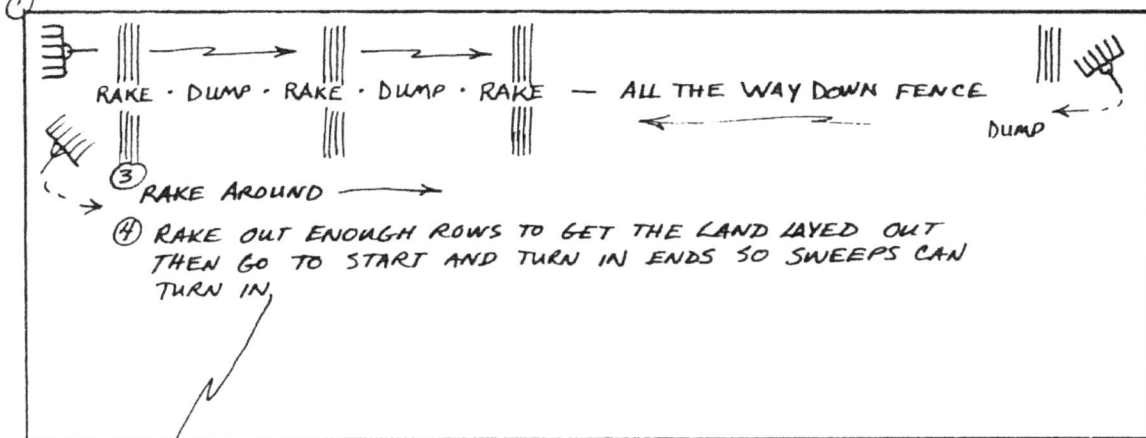

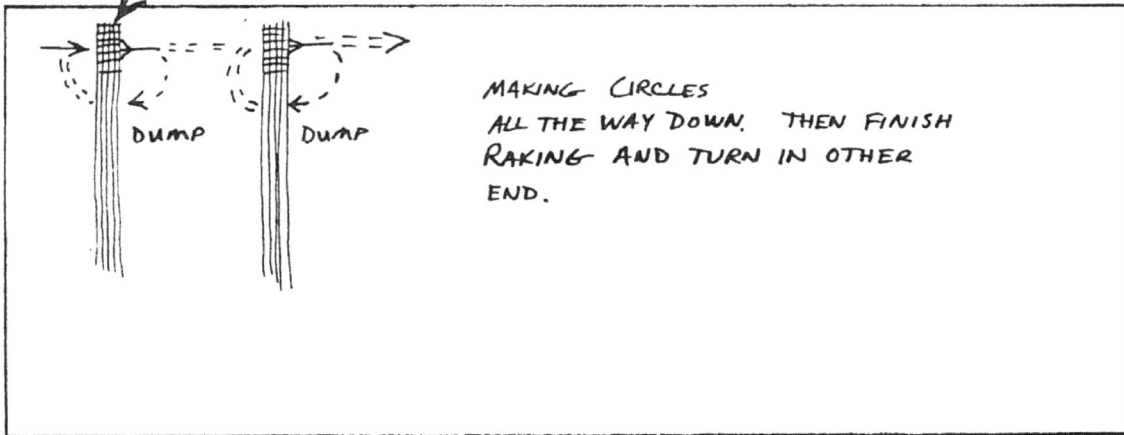

Sweeps start as soon as the straight rake gets ahead enough

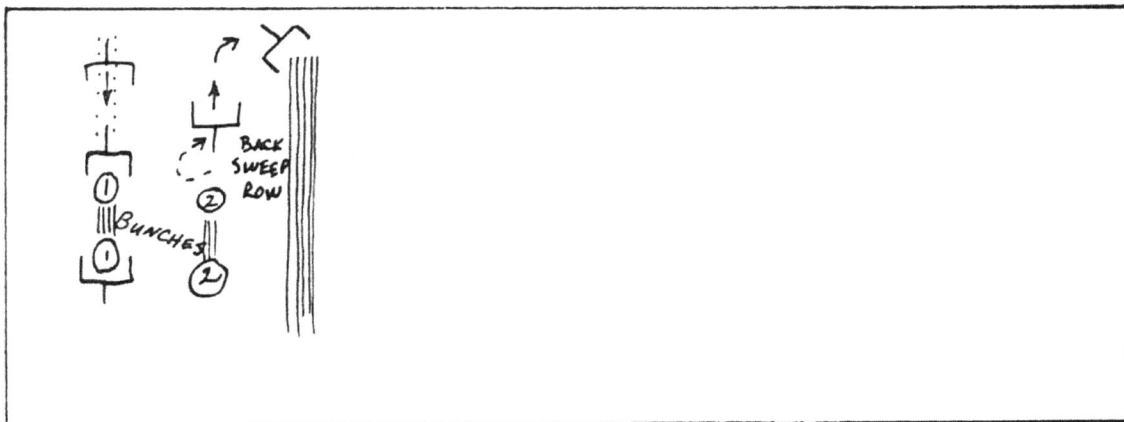

① ① Bunches

② ② BACK SWEEP ROW

THE HAY LEFT BETWEEN BUNCHES IS PICKED UP BY THE SWEEP PUSHING TO THE STACKER. AS THE SWEEPS WORK THE ONE SCATTER RAKE WORKS AFTER THEM AND PICKS UP SCATTERS AND DROPS ON WINDROW AHEAD, OR BEHIND BUNCH FOR STACKER SWEEP TO PICK UP. ONE RAKE CAN KEEP WITH 2 SWEEPS ALLRIGHT.

Fig. 26.

Buckraking

215

THE STACKER COMES ON WHEN THERE IS 2-3 STACKS READY IN BUNCHES

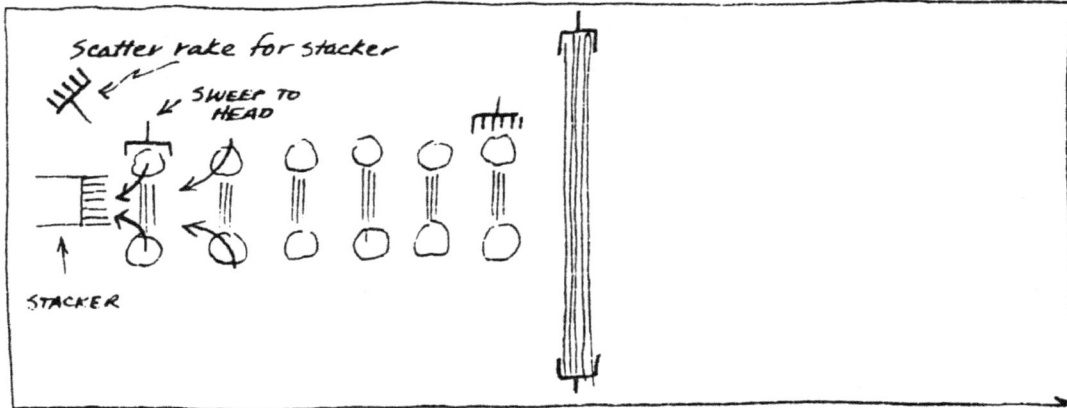

Scatter rake for stacker
SWEEP TO HEAD
STACKER

THE SCATTER RAKE FOR THE STACKER CLEANS UP AROUND
EVERY STACK. PULLS ALL HAY IT GATHERS TO FRONT OF
THE STACKER HEAD TO BE LOADED AND CLEANS AROUND LOADED
BUNCHES. BUT THEY <u>STAY CLEAR</u> OF THE SWEEP PUSHING TO THE STACK.

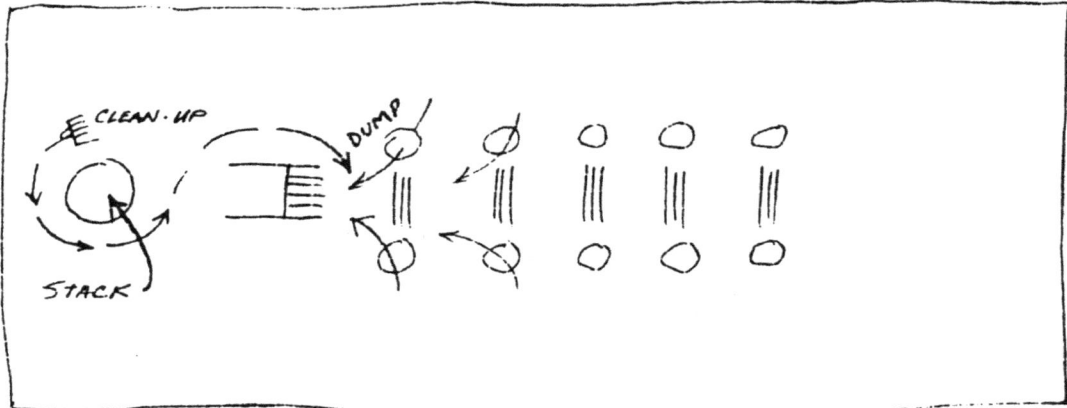

CLEAN·UP
DUMP
STACK

THE HAY IS ALWAYS AHEAD OF THE STACKER, NO BACK TRACKING
AND NO WASTED HAY OR MOTION. This way of haying can
be scaled down to one sweep and one straight rake and
one side delivery rake with a stacker.

We pulled two high-wheeled wagons with covers full of grain
and two low box wagons rigged with feed boxes with rings
bolted through and boxes bolted down to wagons. Worked
horses fed three times a day unless changed at noon. Sure
was nice to see all the horses harnessed and tied eating
at noon. Or all the harness and collars hanging on posts
at the end of the day. There were always windmills near
so horses got watered at noon.

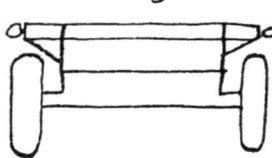

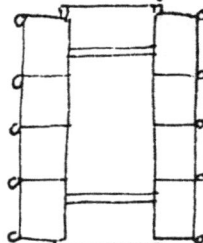

IF ONE SPOOKS
THEY ALL SPOOK
THEY ALL PULL
EACH OTHER.
NO TROUBLE

Fig. 27.

DIRECTIONS for SETTING UP and OPERATING
THE DAIN THREE WHEEL FOLDING SWEEP RAKE

TO SET UP

1. Place large wheels on axle pipe D-5079, between third and fourth holes from each end, with bearing ends of D-14-E washers in each end of the wheel hubs.

2. Bolt two of the three-hole teeth under the axle pipe at the fourth holes from the ends of the pipe, with riser irons (bundle No. 11) in place at A and B, as shown in Fig. 29. Put the bolts in from below through the second holes from the rear of the teeth, the flat end of the D-14-E washers, the hooked ends of the riser irons, the axle pipe, the steel caps over the wheel hubs and the ends of the diagonal riser iron braces, in the order named.

3. Bolt the other two three-hole teeth under the axle at the third holes from each end in the same way except that there are no riser irons.

4. Bolt other teeth under the axle, with steel washers (bundle No. 1) between the teeth and the axle, using ½ x 4 3/4-inch bolts. Hook ends of shorter riser irons should be bolted between washers and pipe at the center and end teeth, as shown at C, D and E (Fig. 29).

5. Place head bar D-5084 (bundle No. 1) on top of teeth at rear ends with steel shoes D-5077 (Fig. 30), on the rear side of the bar. Bolt braces on riser irons A and B (Fig. 29) on top of head bar, putting long end of double eyebolt, found in riser irons, through brace, head bar and fourth tooth from each end, in the order named. Bolt brace on end riser irons C and D on top of head bar at end teeth, using ½ x 5 ½-inch machine bolts. Bolt head bar to the other nine teeth with ½ x 5-inch bolts, put in from below. Flat braces found in place on ends of the pipe should be bolted on top of the head bar at the second tooth and on top of the pipe at end holes.

6. Bolt slat D-5088 (bundle No. 1) between riser irons and braces at A, B, C and D, and to center riser iron at E (Fig. 29).

7. Bolt shields (bundle No. 6) to slat and teeth, as shown at F (Fig. 29), using bolts found in shields and slat.

8. Bolt push bars D-5086 (bundle No. 2) to R-54-E casting (bundle No. 10) with hinge rod on ends of bars away from casting and pointing outward.

9. Put hinge rods on ends of push bars, through eyes in rear ends of riser iron braces A and B, and fasten in place with cotters. Bolt push bar race D-5089 (bundle No. 2) on top of push bars, using bolts through hinge rod and push bar.

10. Put shank of caster wheel fork (bundle No. 7) up through R-54-E casting in rear of frame and fasten in place with cotter and washer.

11. Bolt wood hand lever (bundle No. 2) to R-54-E at F (Fig. 29) and bolt loop to lever at G.

12. Attach plain loop ends of pull poles D-5085 (bundle No. 2) to loop in front of hand lever and attach other ends of pull poles to riser irons at A and B (Fig. 29) by inserting eye pins through eyes on riser irons and hooking loops on pull poles over the castings in the top of the riser irons.

13. Bolt seat spring in socket in R-54-E, using bolt found in place.

14. Untie right tongue (bundle No. 3) and bolt draft bar D-5083 to tongue D-5104, using bolt found in tongue. (See Fig. 30) Attach brace rod D-5100 to tongue from underside.

15. Pass free end of draft bar D-5083 under end riser iron brace and attach to eyebolt already in place at A (Fig. 30). Attach end of tongue to eye on end of riser brace at B, using bolts and R-52-E washer

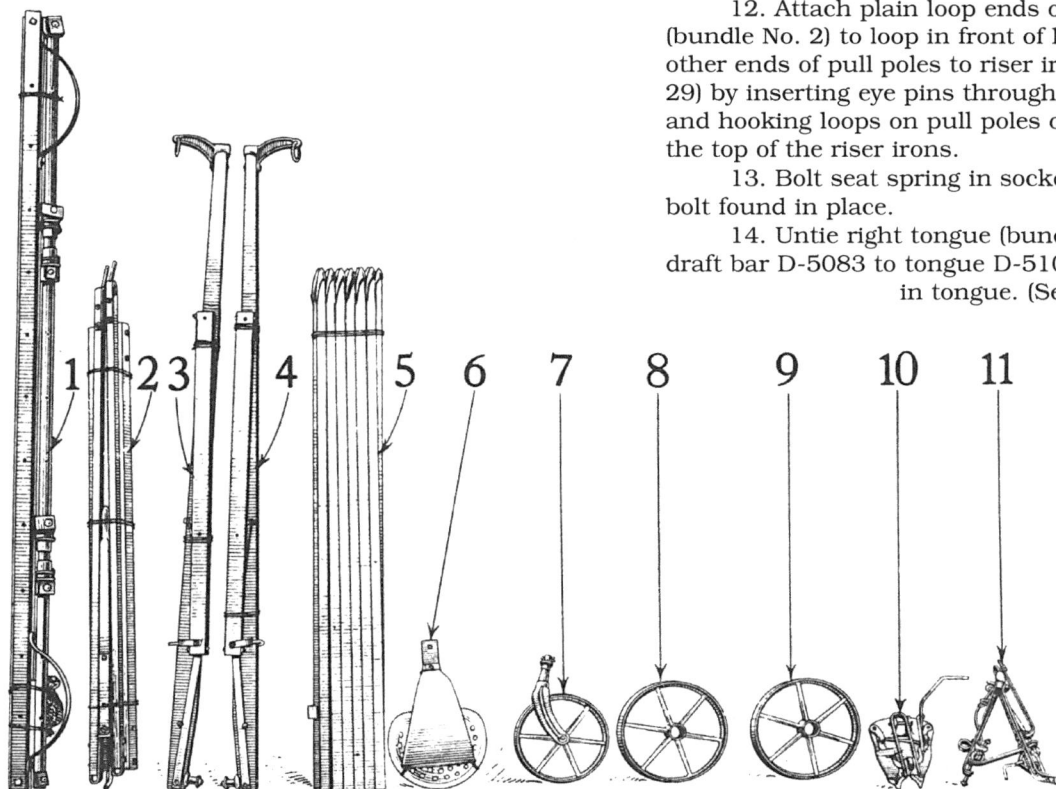

Fig. 28. Packed for shipment, the Dain Folding Rake consists of eleven bundles.

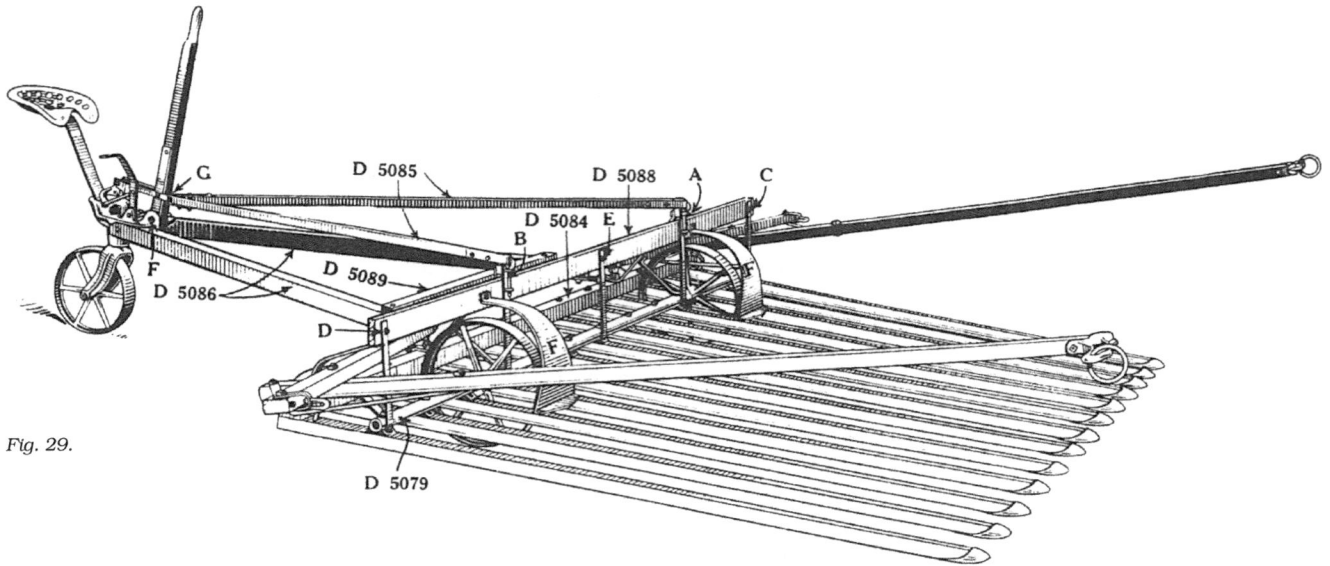

Fig. 29.

found on tongue.

16. Attach left tongue and draft bar (bundle No. 4) in the same manner.

17. Tighten all bolts and oil wearing parts. See that the automatic lever lock works freely, will lock easily and trip by pushing on foot lever.

TO OPERATE

Hitch the horses in place, taking care to have tugs the proper length to allow the horses full room to work, and fasten breast straps to rings in castings on the ends of the tongue.

Use separate line to each horse, by shortening the inside rein on each line and fastening both reins of each line to the same horse. With this arrangement,

the rake can be turned by holding one horse and allowing the other to go forward.

Foot pressure on the wood lever will force the teeth to the ground for gathering hay that has been beaten into the stubble. To raise the teeth, push forward on the foot lever and pull back on the hand lever until lock can be pushed down so it will hold the teeth off the ground.

To fold rake, disconnect pull poles at head and lift teeth to vertical position. Insert eye pins at ends of pull poles in holes in axle at C and D (Fig. 30). Hold tongues in vertical position by fastening hook E on the end riser iron over the draft bar. Machine can then be pulled endwise by hitching to the end of the draft bar at S, as shown in Fig. 30.

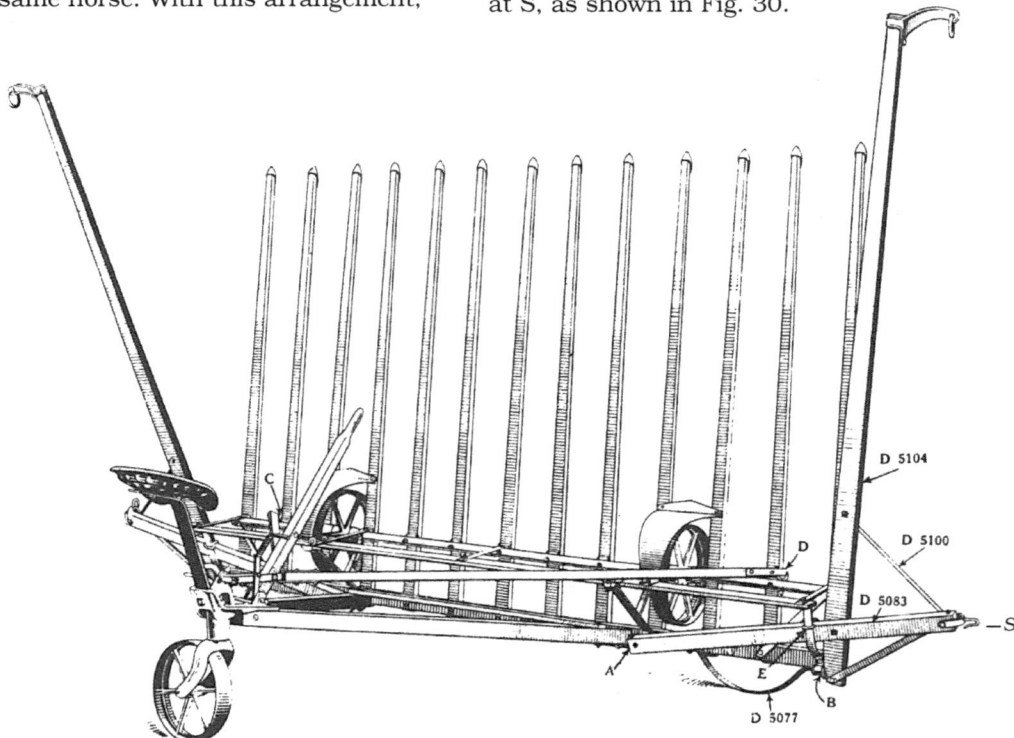

Fig. 30

Building A Buckrake

In 1989 we purchased, for 25 dollars, the hardware from a JD Power Lift Buckrake. With peeled poles for teeth and dimension lumber for the balance it took 2 weeks, working from engineer's drawings with no dimensions, to completely rebuild this implement.

Fig. 31. The finished beauty.

No changes were made, even after field testing and, now, 11 years of use. Throughout the West the fence rows are still littered with the remnants of old Buckrakes which, we hope, will begin to see new use after rebuilding.

Fig. 32. (left) A rear view showing the turning wheels.

Fig. 34. (Below) Showing a closeup of the basket, axle assembly and forged backing rod.

Fig. 33. (Below) The important cast housing for the wheel yoke.

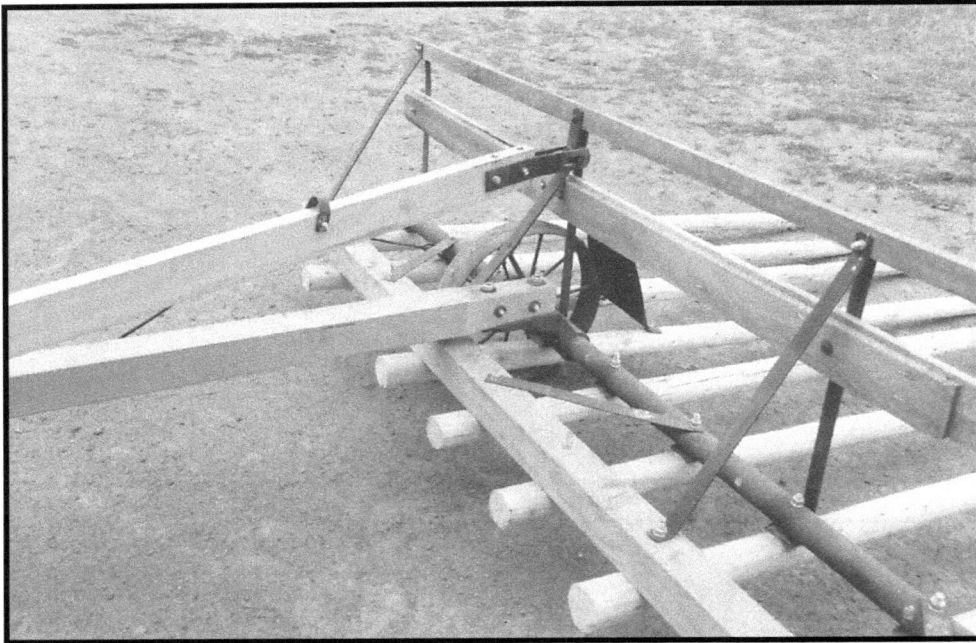

Fig. 35. (Left) In this closeup you can see how the mainframe timers, with cast brackets, receive the axle. The upper cross-member is one half of the basket lift. Note the small clevis on the arm where the horse is tied.

Fig. 36. (Below) Arm and pedal foreward... Fig. 37. (bottom) means teeth down.

Fig. 38. (Left) Arm and pedal up.... Fig. 39. (above) teeth up.

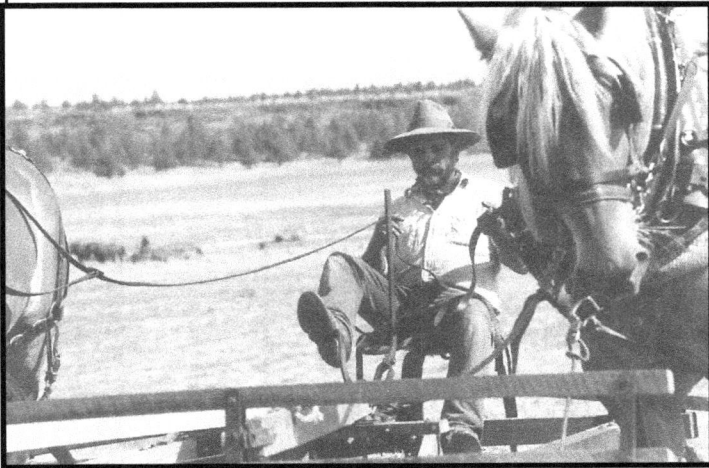

The eighteenth century parent to the modern buckrake.

Fig. 40. To turn, pull back on one horse and tell him to whoa while giving the other horse a slack line and command to go.

Fig. 41. To stop or back up, pull back both lines and give appropriate command.

Fig. 42. One horse held back, one free, and the buckrake turns.

Fig. 43. Buckrake teeth down and in gather position.

Baling Hay

*Figure 1. Old John Deere
Stationary Baler*

Chapter Thirteen

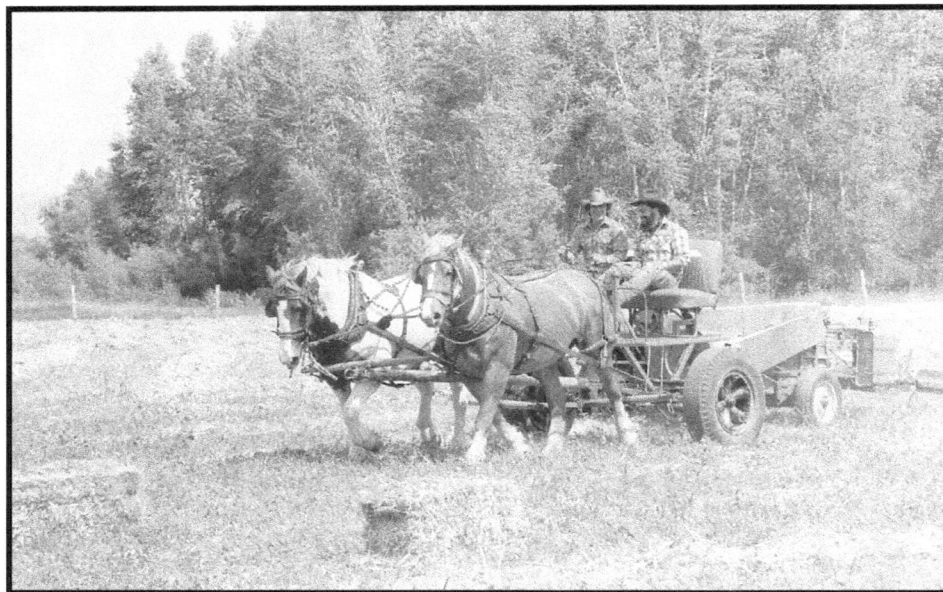

Figure 2 & 3. Robert Clark of Whitehall, Montana baling hay with a forecart and New Holland baler in 1983

Figure 4. A three horse Amish team in Ohio with a New Holland baler.

adequate hydraulics and *power take off* to power any baler, including the big round models. As the photos in chapter six showed, if you wish to use lighter horses or ponies for baling you'll be adding animals to the hitch.

While most of the time brakes are superfluous on forecarts (any team can drag any forecart that has its brakes set) the clear exception is with balers. As the plunger rams the hay in the chamber the baler tends to roll back and forth and swing a little to the side, this creates a back and forth pressure on the horses. While the motorized baler is being pulled into a windrow this slamming action is minimized but should a heavy windrow or temprorary problem cause the teamster to stop the horses all of that movement is applied to the forecart and on to the horses. Unless, of course, the forecart has brakes and the teamster sets those brakes whenever the forward motion is stopped. The brakes won't get rid of all the offending motion but they do tone it down considerably. Some people have gone so far as to build spring loaded hitches for the baler that absorb some of that action.

This is a short chapter. There is not enough room in this book to even begin coverage of all the mechanical vagaries of balers and baling. What is important for the subject of this book is to clearly state again that **Yes,** you can pull a motorized baler with a team of horses. And, if you wish to spend the money, motorized forecarts are available which provide more than

Figure 5. The Indiana modified ground-drive John Deere baler as seen in 1998 at Ohio Horse Progress Days. Note tongue truck and seat mounted on top of baler.

Figure 6. *These four Belgian geldings are actually pulling the fully operational ground-drive JD baler.*

Figure 8. *A view of what sort of bales are being spit out of the Indiana GD JD baler.*

Figure 7. *A view straight down, while moving ahead and baling, of the modified 'bull' or drive wheel*

Ground Drive Baler News

At the '98 Ohio Horse Progress Days a modified John Deere Baler was demonstrated which this author feels may be the beginning of a significant technological step for animal powered farming. An Amish shop from Indiana, Zehr's Repair Shop, converted this JD baler to ground drive by making the leftside wheel into a power source. As the photo shows the wheel was rebuilt with expanded metal traction surface and gear sprocket. The drive chain was routed to the plunger flywheel assembly. Without having the exact gear ratios to offer we cannot give enough information for just any amateur farm shop mechanic to build his own unit. But, that said, it is easy to see that such a setup could be custom engineered by most of the better farm shop mechanics.

The MOST important information is this - it works! Most of us would have argued, and did argue, that the concept of a ground drive baler was dead on arrival. If forward motion is necessary in order to "build up" speed and torque then the concept of a GD baler just didn't seem possible because MOST of the time the baler would be in a windrow feed mode when it began to roll. That, theoretically, should mean that there would be insufficient operational speed and torque to avoid the baler jamming up as it started to move. In other words the baler, it was thought, had to be up to operating speed BEFORE it hit any ready hay. Then along came this beauty. As the photos will attest this author actually rode atop the machine as it worked. It was amazing, and it seemed impossible but the baler was up to full torque BEFORE those four Belgian geldings hit their second step! And with an ungainly rained-on, heavy windrow there was NO (ZERO) problem. It worked slick. Though the horses did have to all lean in together to get it started, once it was going their walking work (or the draft) seemed to the author's eyes to be equal to pulling a plow or less. We're certain that the actual size of the drive wheel gearing is critical (within tolerances). When that baler moves at 2 to 2.5 miles per hour the power to the system needs to be equivalent to 540 rpm PTO torque and it obviously was. While this baler was seen working at Horse Progress Days, a college professor approached this author and stated, "That's incredible, do you realize that what we're witnessing here is impossible by any ag-engineering measure!"

This is exciting engineering news because the dynamic at work here can be applied to many different field procedures and implements.

Getting Animals Accustomed to Balers

With mowers, rakes, hayloaders and buckrakes we've made an issue of how to get the work animals used to the machine's noises and action before actually hooking up. While the same sort of common sense approach is applicable to balers it is perhaps time to qualify all those cautions.

After thirty years of working with these magnificent draft horses I've come to realize that their adaptability makes we humans look poor indeed. The experienced teamster[1] with horses or mules that trust him or her can go to any new piece of equipment or procedure or environment with no concern, only excitement for the new adventure.

While in Manhattan (NYC) in July of 1999 this author marvelled at the patience and acceptance of the carriage and police horses which had to deal almost at any moment with some new sight, noise, movement,

threat or calamity. There were plenty of folks crippled or panicked by the same stimuli.

Here's an ideal introductory routine: If you're green or the animals are green, ground drive them up to an operating baler and allow them to see first hand what the noise and action is all about. Keep them there until it is old hat. Shortly after that, hook them to the baler and go to work. Until things are working smoothly, plan on having someone knowledgeable handle the team AT ALL TIMES and someone knowledgeable handle the baler (setting tension, watching for broken knots or ties, etc.).

Over twenty years ago this author visited the Cheatum Bros. Dairy in Arizona where haying was being done with horses - at night. Seven hundred acres of Alfalfa and horses were being used because it was more economical. The work was done at night because of the heat and humidity in the summer. Yes, the midday sun was bad on the men and the horses but more important for the dairy, it was harder yet on the hay. The cool and humidity of the nighttime helped the baling process. This author spoke to the gentleman who was responsible for the baling teams and he told a wonderful story. Part of it explained how the baling teams were so well accustomed to the work and the night that their teamsters tied up the lines and walked along with the baler, occasionally speaking to the horses as they walked along. The teams were always careful to keep the windrow feeding the machine. Only when they got near the corners did the teamsters take up the lines because the horses were always wanting to turn a little too soon. This author asked if the equipment was rigged with lights for the work and he laughed, "nope, even when we couldn't see the horses could see, never missed with the mowing, raking or baling." They started at dusk with four mowers. A couple of hours later, mowers still going, a four abreast with a double-wing 20 foot side-delivery went out to rake up the hay that had just been cut. A couple hours later two balers went out and started baling. By then it had usually cooled down to 80 or 90 degrees. This routine went on pretty much all summer long as when they had finished the first cutting it wasn't far from the second cutting being ready where they had originally started mowing.

Yep, you can bale hay with horses. This author would encourage you to think about putting it up loose, at least part of one season, and compare the work and the thrill of the two different approaches.

Doug Hammill baling with two Clydesdales, a New Holland Baler and homemade forecart in 1978, Montana.

---

1. *Let's use this criteria for <u>experienced</u>; six successful years of depending on animals in harness as a power source.*

Unloading Wagons

Chapter Fourteen

If you had to pitchfork hay off a well-built wagon load and into a barn or onto a stack just once you might ask yourself "Why am I fooling with loose hay?" It can be a grueling task.

But if you were able to hook on to three slings and in three jags effortlessly clean off nearly a ton and a half of hay from a wagon--you'd be smiling. Or if you positioned the four prongs of a loose grapple fork and marveled as several hundred pounds of hay went skyward to loft or stack--you'd feel fortunate. These are such affordable tools to use and wonderful toil easers. Up until World War II there were millions and millions of hay fork and sling systems in use across North America. The forks and slings are still littering the landscape, junk stores and barns. Many years ago, when we thought we needed to be diligently collecting the old hardware, we bought (for pennies) any trolley, track hardware, and grapple forks we could find. Long ago when the pile got big, we quit collecting. When you learn what to look for and where to look there is just no shortage.

Figure 1. A nice load of hay from the turn of the century. Pitching this off by hand will take quite a long time whereas lifting this hay off with a double harpoon or grapple fork setup would take three or four effortless jags at five minutes per.

Figure 2. Lifting hay off the wagon and in to the barn loft. A sling is being used and, as can be seen, is doing a clean job of taking the hay up. Notice that beside the wagon is a grapple fork available should something go wrong with the sling. The old motor and drum replaces the haul back horse. This is not a necessary upgrade since a single horse and competent youngster can do a keen job of lifting and moving the hay.

Figure 3. Grapple forks in use with a cable stacker.

Unloading

Different crops hold together differently on loaded wagons and in fork loads. Various crops weigh substantially different amounts from one to another. A green clover crop, well packed, will weigh far more and hold itself together better than a well packed oat hay crop. The person experienced with unloading loose hay off wagons will know to be careful not to take up too big a load of clover or similar crops. Whereas with oats every effort is made to maximize the amount taken up.

Of course unloading procedures may be affected by where the hay is going (we cover stacks in the next chapter and barns after that) but some principles apply across the board. For instance, it can be a real nuisance if a piece of a load is lifted off only to have the balance fall off the wagon. How uniformly the wagon was loaded can matter (unless of course slings are used).

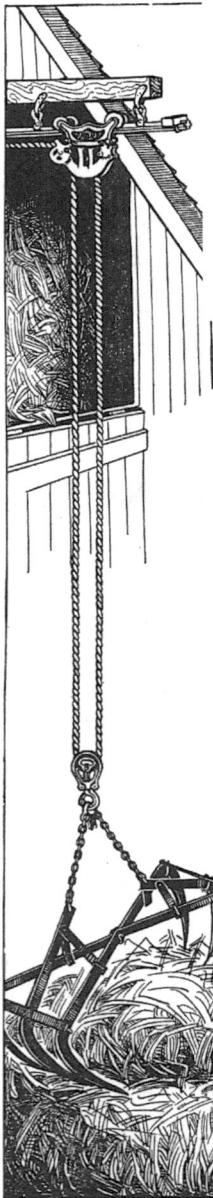

Figure 4. Grapple forks biting into a waiting wagon load.

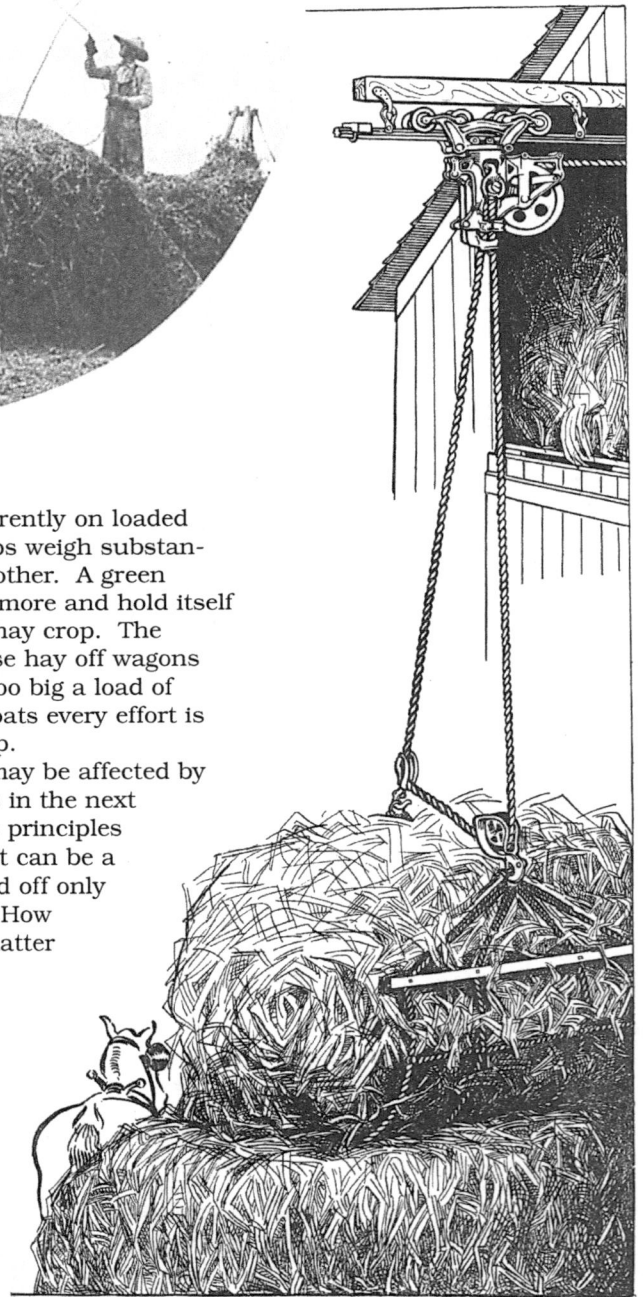

Figure 5. Showing one style of sling lift employing two pulleys as a width extension to hookups.

There are four different systems for unloading wagons that we will discuss. 1.) Pitchfork 2.) Parbuckle 3.) Hay forks 4.) Slings.

1.) Pitchforking doesn't require much explaining except to suggest that in some instances--when the hay doesn't have to be thrown up it can be "pulled" off easier using heavy-duty bent tine potato forks. For any pitchforking, stay with three or four tine forks. Five tine large forks are for 'serious' folks.

2.) One year without a stacker, we constructed a sling with nylon rope and aluminum pipe cross bars. Locating the stack up near a tree, we anchored a block (or pulley) up high and used a team to literally roll the load off the wagon and up on to the loaf stack. As the photos on page 77 show, such a system may be used to load or unload into a barn. Also see Fig. 8 for another parbuckling idea.

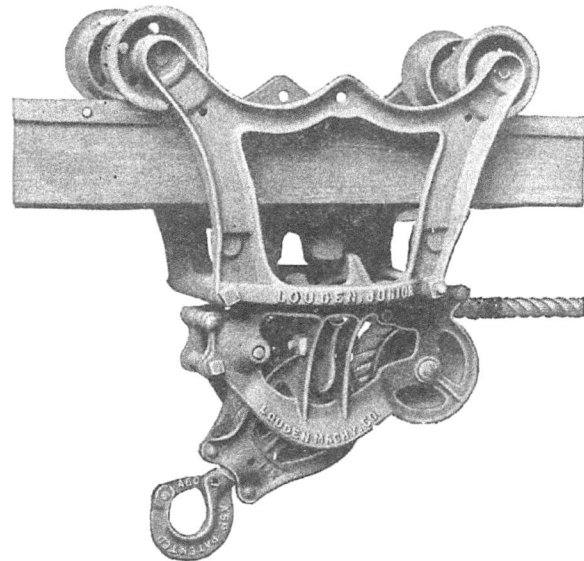

Figure 6. A wood track trolley for barn loading.

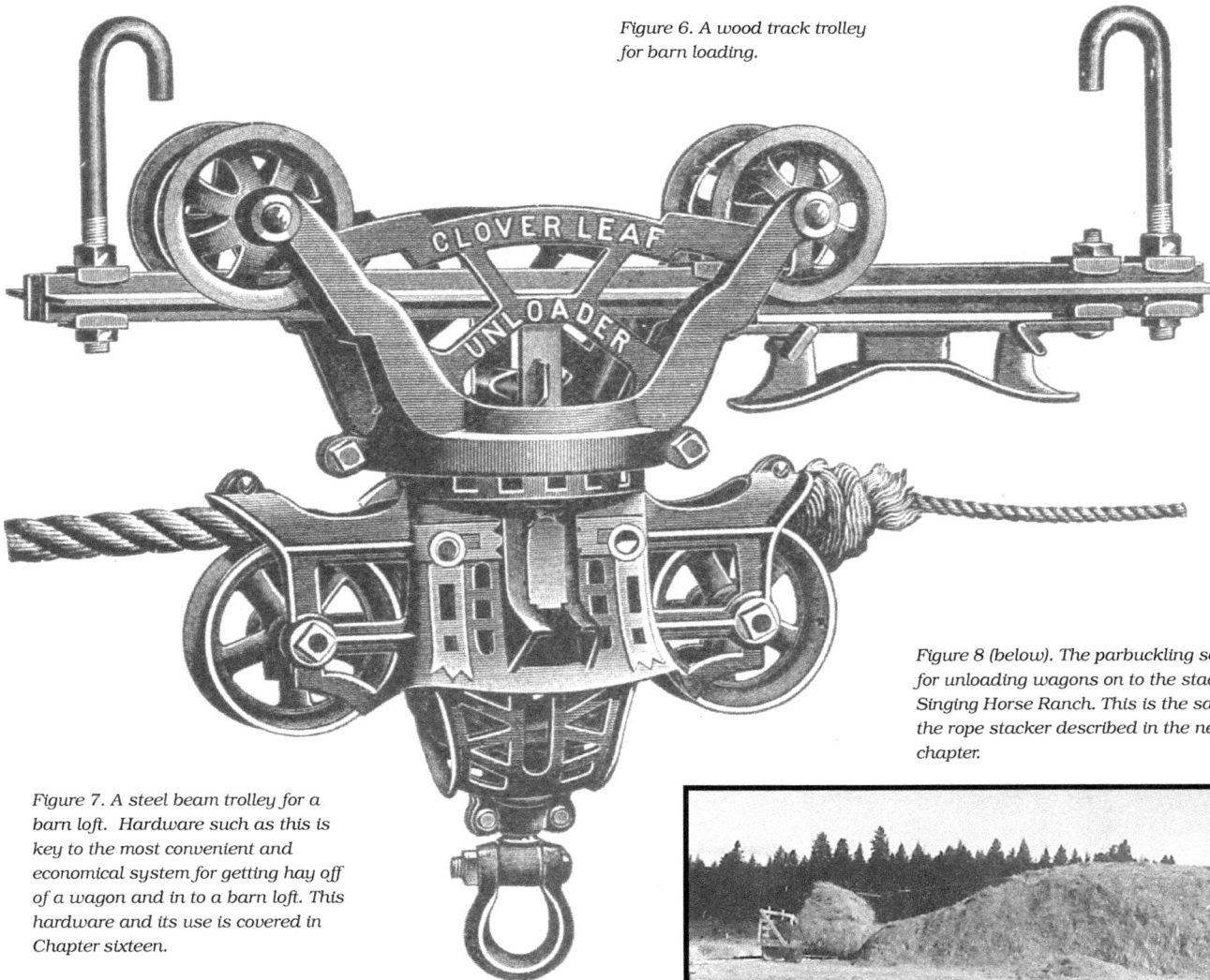

Figure 7. A steel beam trolley for a barn loft. Hardware such as this is key to the most convenient and economical system for getting hay off of a wagon and in to a barn loft. This hardware and its use is covered in Chapter sixteen.

Figure 8 (below). The parbuckling setup for unloading wagons on to the stack at Singing Horse Ranch. This is the same as the rope stacker described in the next chapter.

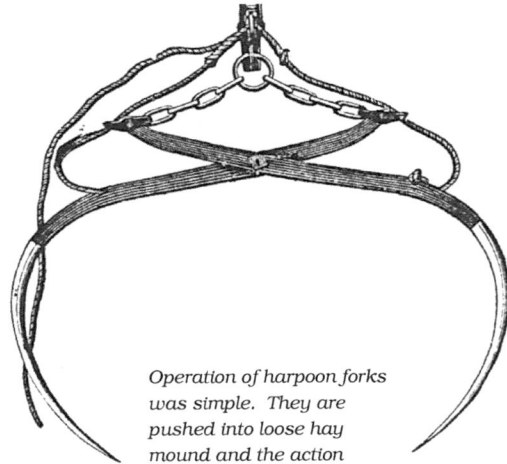

Figure 9. This "Beardsley" brand hay tong setup was sold circa 1900 and proved less effective and popular than the double harpoon styles which followed.

Operation of harpoon forks was simple. They are pushed into loose hay mound and the action engages the teeth. When the load is where it is wanted the trip rope is yanked and the teeth go straight to release the load.

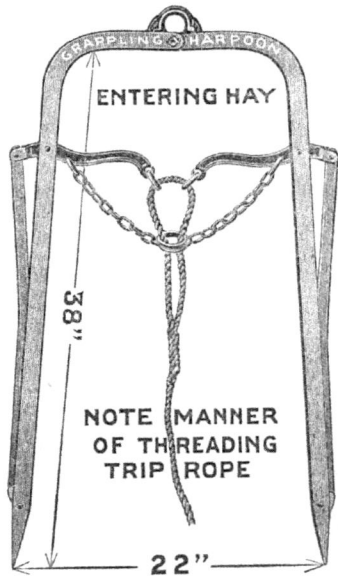

GRAPPLING HARPOON

ENTERING HAY

38"

NOTE MANNER OF THREADING TRIP ROPE

22"

Figure 10. Double harpoon fork set to drive in to a wagon load.

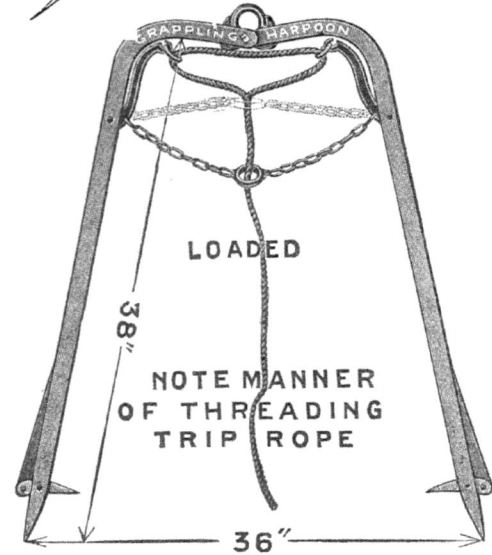

GRAPPLING HARPOON

LOADED

38"

NOTE MANNER OF THREADING TRIP ROPE

36"

Figure 11. Same double harpoon fork in load with harpoons spread and teeth engaged.

Figure 12. Single harpoon fork.

Figure 13 and 14. Two double harpoon forks.

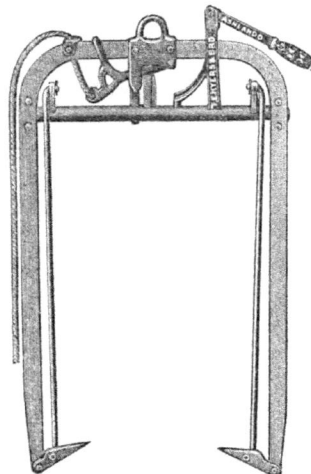

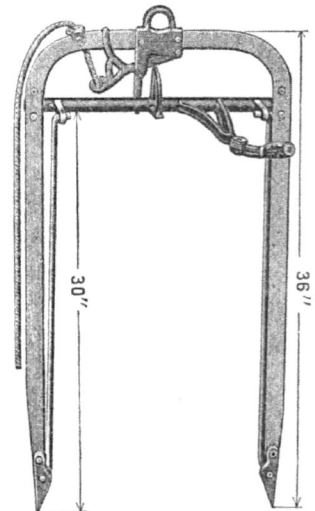

NELLIS

30"

36"

3.) The use of hay forks is and was perhaps the most common approach. Though this author feels they are inferior, the double harpoon fork is the tool used by more farmers. The Jackson Derrick fork would handle larger jags but could hang up on a barn so it was used exclusively in outside stacking operations. The grapple forks, coming in a few different configurations, are this author's vote for superior tool. Especially the loose prong grapple. This setup allows the smart farmer great leeway in selecting how much and what to grab.

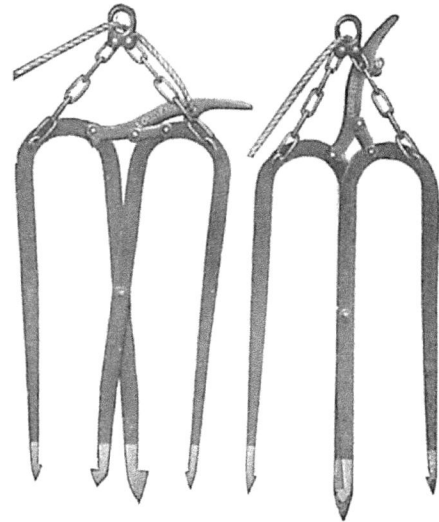

Figure 15. A rocker-bar harpoon fork which saw limited favor. On the right, the straight forks were pushed into the hay. The top lever was then pulled and locked down causing the action on the left which grabbed the load. When the cord was yanked, the forks returned to straight and the load slid off.

Figures 16, 17. Closed and open illustrations of grapple forks

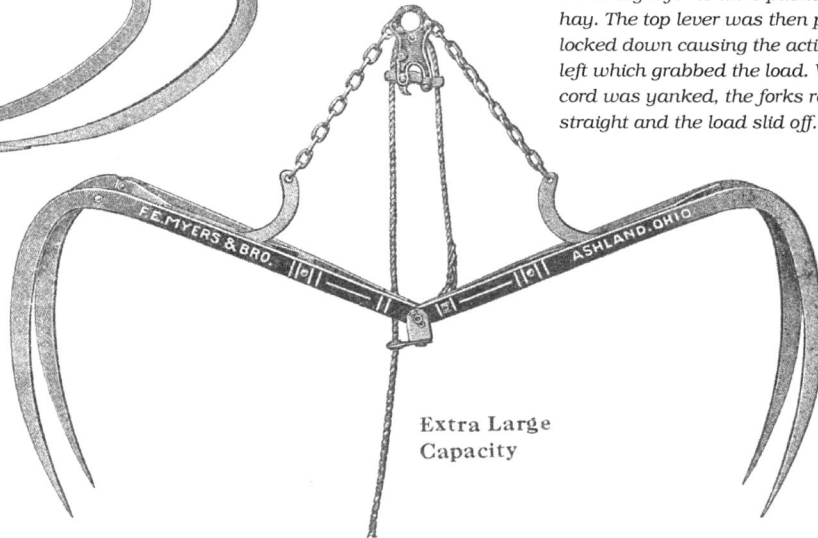

Extra Large
Capacity

Figures 18, 19. The Jackson Derrick Fork. On the right the fork is lowered and pushed into the load. The hand is then employed to rock the teeth into the load and lock the top of the frame into the metal triangle. When the cord is tripped, the fork easily swings to dump position.

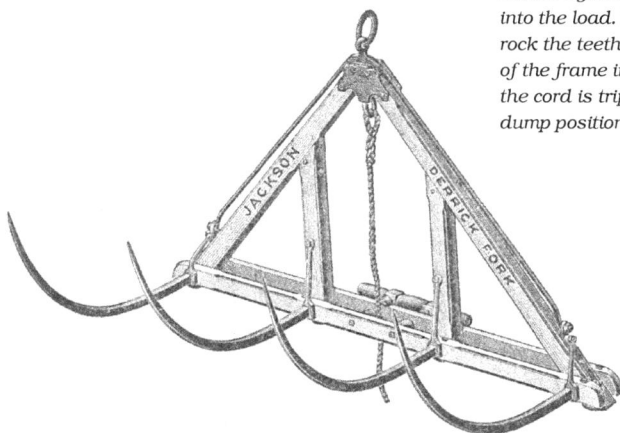

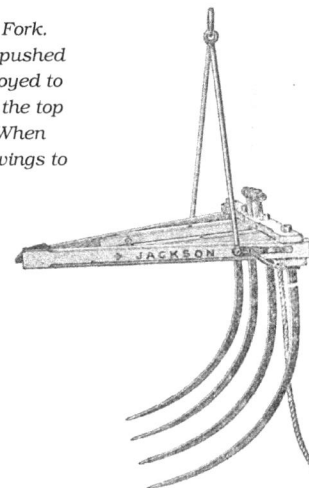

4.) Slings are, in theory and by all appearances, the best way to unload wagons, but this author found them cumbersome and time consuming. Slings need to be draped, first over the empty wagon bed, then over the first layers of hay, and the next layers. They need to be clearly marked so that the wagon person doesn't accidently hook the top sling on one end of the wagon and the middle or bottom sling on the other. This would lift no hay and possibly upset the load. The slings need to be kept untangled and hauled to the field in some out of the way place. A standby fork of some sort needs to be kept in case a sling trips before the load is lifted. When they are working right, they make clean work of unloading. The challenge is with the farmer to keep

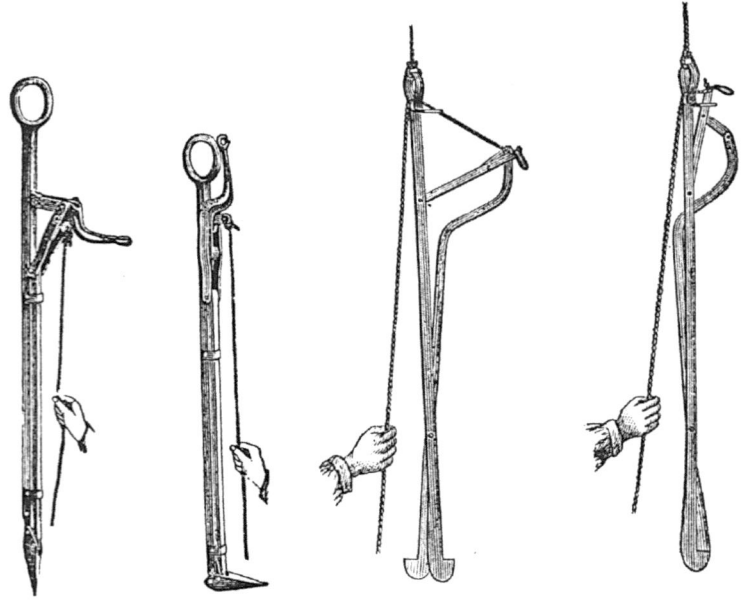

Figure 20. Two styles of harpoon forks showing hook and trip mechanisms.

Figure 21. Showing the clean ease of sling unloading.

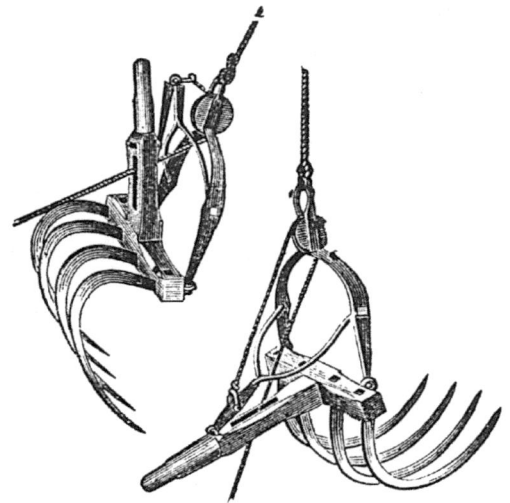

Figure 23. The predecessor to the Jackson Derrick Fork, the "Wheeler" fork.

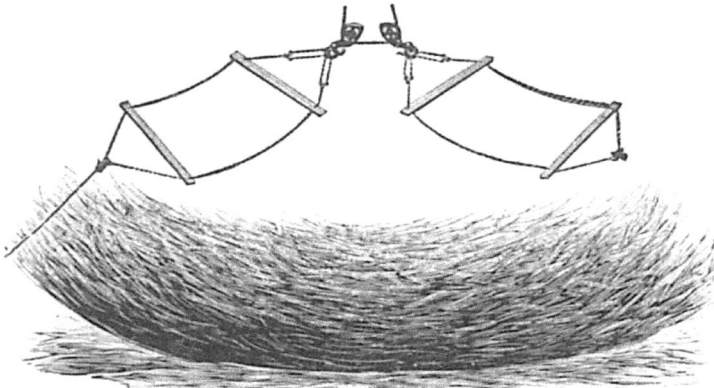

Figure 22. The action when the sling is tripped and load dropped.

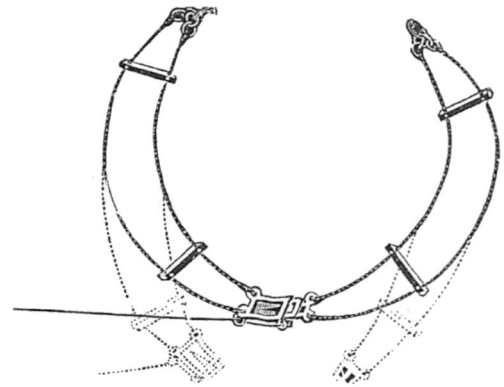

Figure 24. Demonstrating one sling setup hooked and then released.

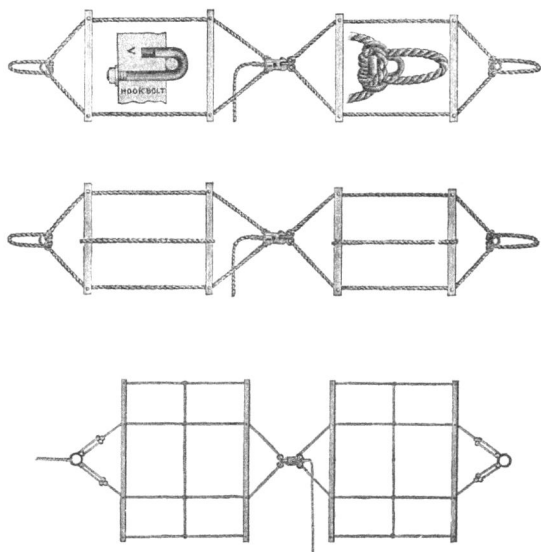

Figure 25. Three different styles of slings.

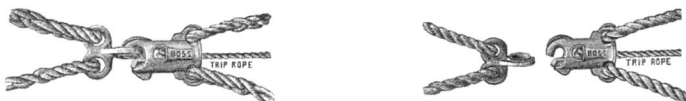

everything neatly organized.

When unloading wagons up into barns or stacks care needs to be taken that children, animals and the infirm are clear out of the way. 700 pounds of hay falling from the sky can do quite a bit of damage.

Figure 27. (right) A sling pulley.

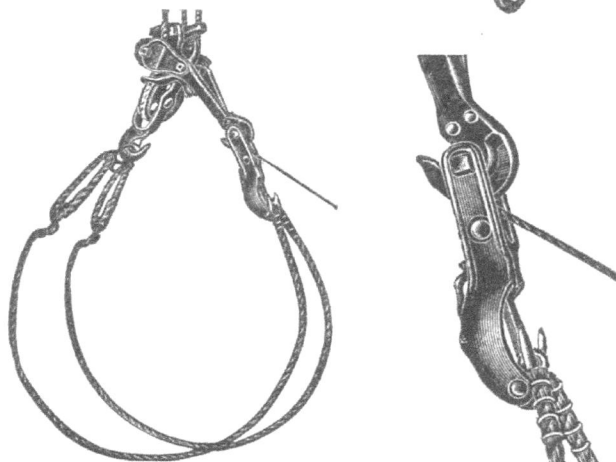

Figure 26. A closeup on one type of sling release clasp.

Figure 28. A cable sling setup with catch close up.

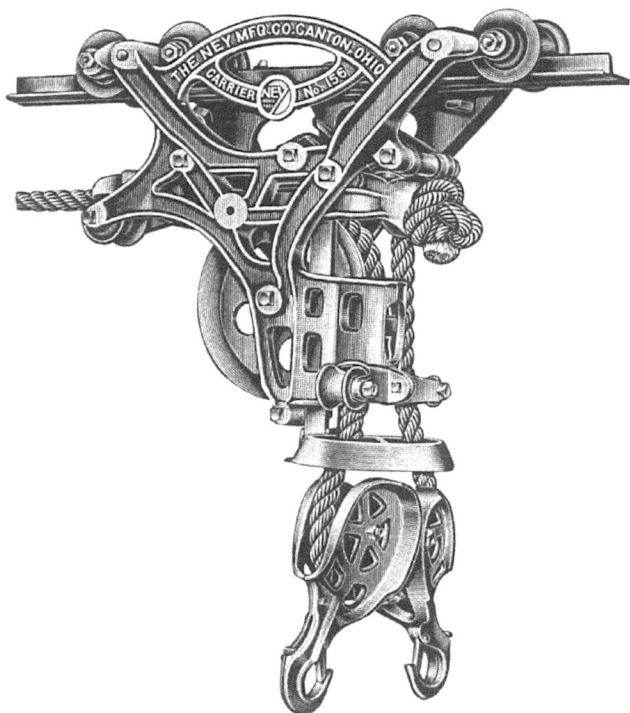

Figure 29. A barn trolley set up to receive a sling.

Figure 30. A sling trolley with an adapter to receive grapple forks.

Figure 31. (right) A close up showing a rope sling attached with a full load.

Figure 32. (left) Showing how a sling pulley can be used with a double harpoon fork.

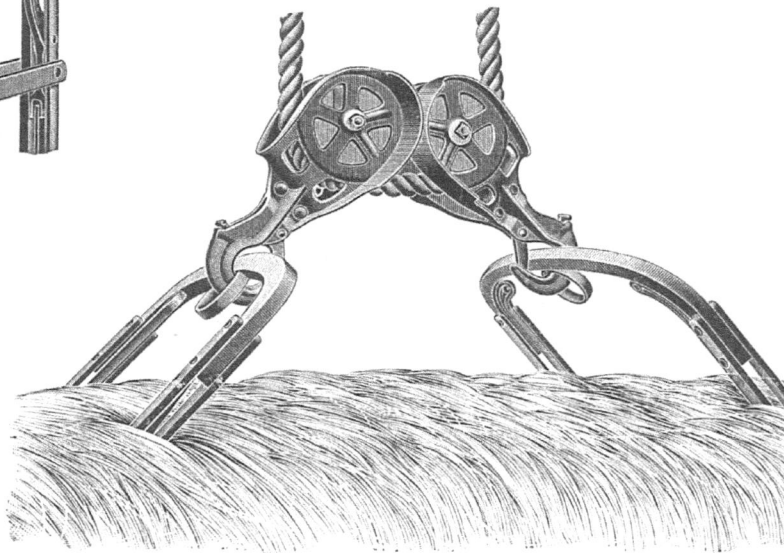

Figure 33. Two double harpoon forks set at angle to a sling pulley.

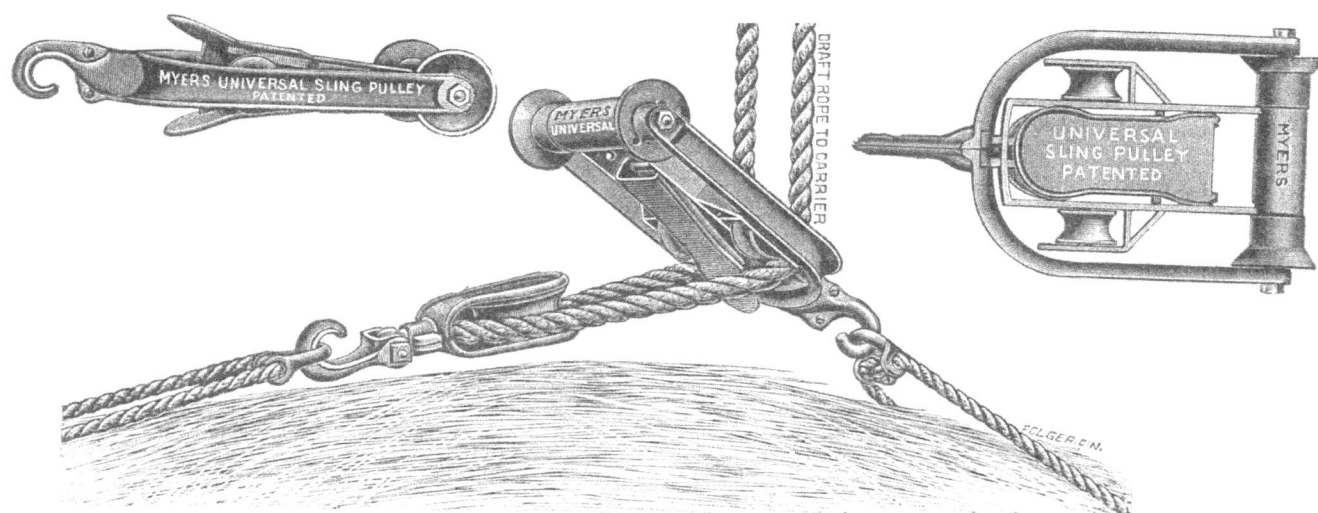

Figure 34. A nifty apparatus for converting a double block fork pulley to a rope sling.

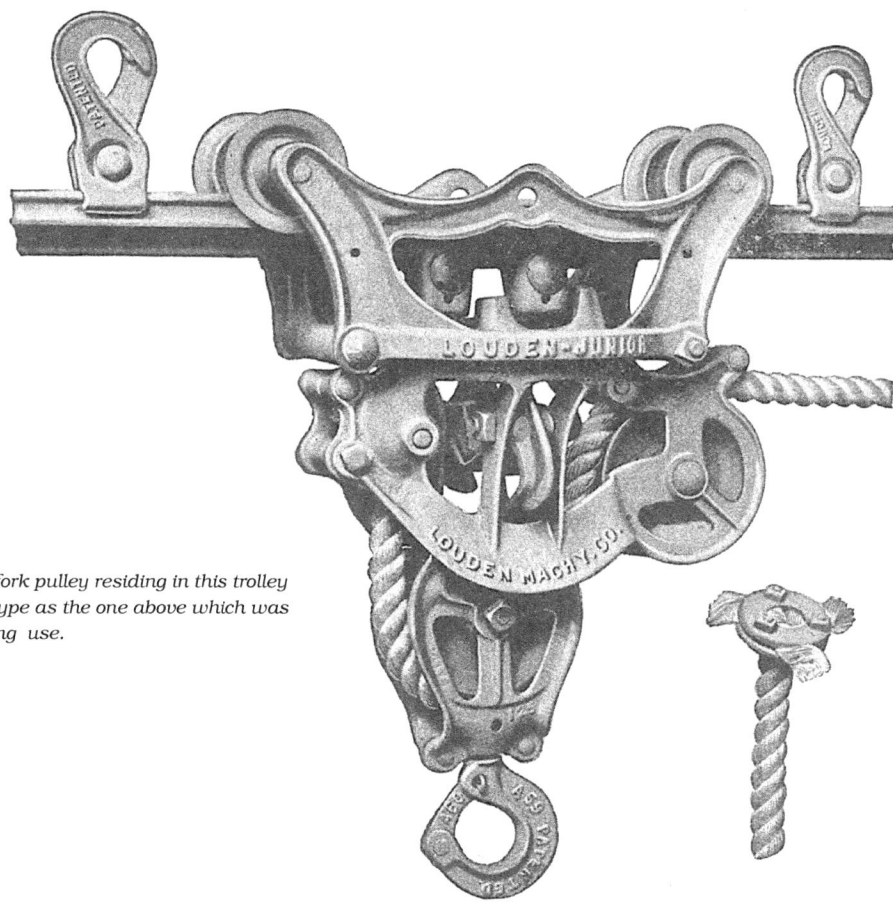

Figure 35. The fork pulley residing in this trolley is of the same type as the one above which was converted to sling use.

Figure 36. An exposed view of how one barn system might be set up, for central unloading, with wagons. In chapter sixteen a wide assortment of set ups are presented.

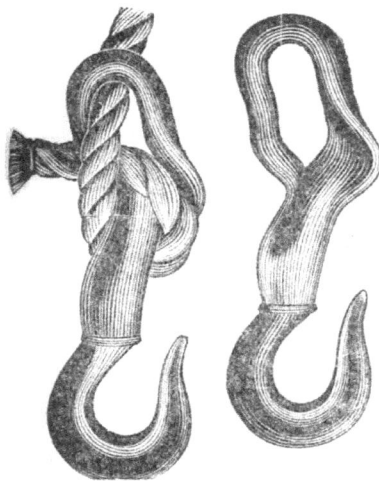

Figure 37. There are many different inventions for quick hookup of hay ropes either in-line or on the ends. This is one such setup specifically designed for use on the end of the haul back rope.

Figure 38. A haul back wiffletree setup, as seen below, to ride like a brichen and allow quick and easy detachment of the haul back rope and, more important, backing up without tangling in a singletree and trace chains.

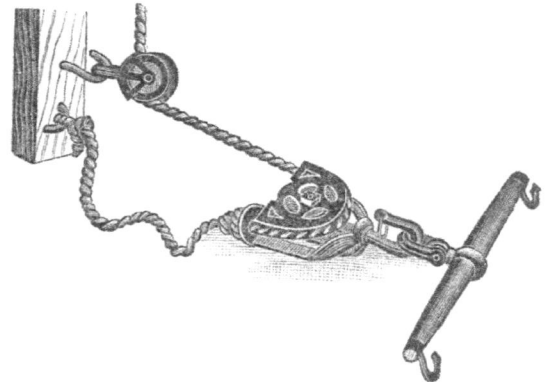

Figure 39. Showing how a block setup might function with a haul back rope.

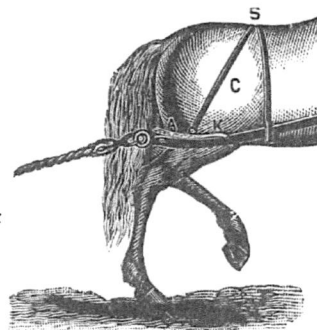

Figure 40. The haul back whiffletree in place

Haystacks,
How To Build Them

Chapter Fifteen

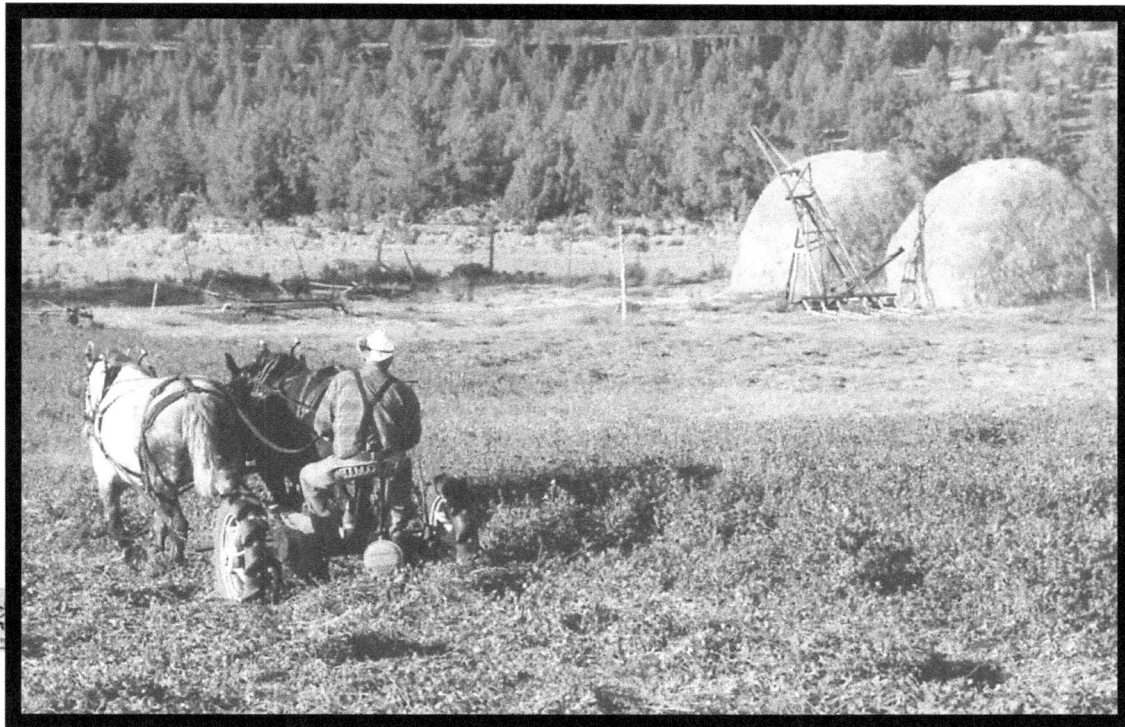

Figure 1. Bud Evers and his Percherons mowing at Singing Horse Ranch with first cutting stacks in the background. Stackmaster, Tony Miller.

There are two ways to build stacks out of doors: by hand or with a stacking device. There are two kinds of hay stacks: ones which are built to withstand weather with a minimum of loss, and those built with no concern for loss (perhaps because such a concern was too much of a constraint to rapid stacking).

For hundreds of years hay has been stacked outside out of necessity, either the barn or shed storage was full or it was unaffordable.

The most common question we hear is

Figure 3. Loading and/or unloading wagons by hand can be done but it is hard work and will limit the amount of hay to be put up.

Figure 2. A homemade derrick stacker and long loaf stack.

Figure 4. Classic conical stacks on which livestock have been eating.

Figure 5. A small beaver slide stacker is moved to various positions to build a long loaf stack.

Figure 6. A Swinging derrick stacker.

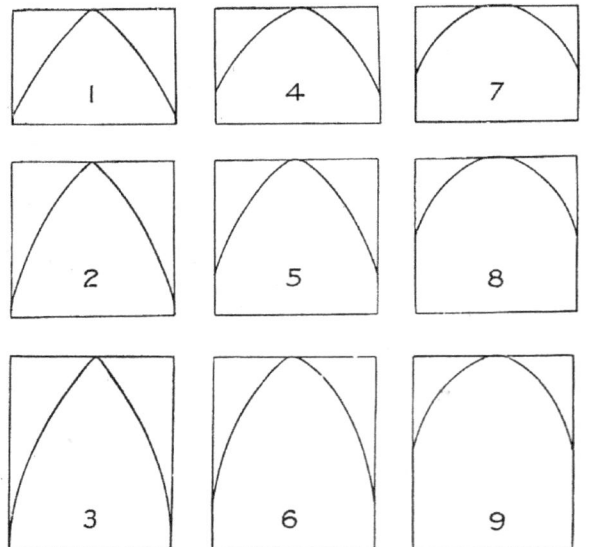

Figure 7. Our work at Singing Horse Ranch over a dozen years of stack building have concluded, for us, that certain stack shapes are decidedly superior as regards the quality of the hay come feeding time. 1,2, & 3 are the worst shapes resulting in the most spoiled hay while 7, 8, & 9 are the best shapes.

Figure 8. This 1920 stack was graded as excellent but we would not agree as the wide flat side and top gather too much moisture.

Figure 9. A photo of a 1920 haystack. We would grade this as very poor.

Figure 10. Tony Miller, Singing Horse Ranch stackmaster, stands before a stack we would grade as very good. See Figure 1 for two stacks Tony built which we grade as excellent.

from folks who wonder if stacks would work in their climate.

Haystacks were a matter of course from Scandinavia to Spain--from the British Isles to the Balkans--a land mass covering many climate extremes. In drier climes, bare-headed simple stacks were more the norm. Whereas in wetter climes (i.e. England) efforts were made to build superior stacks and even cover them with thatching.

We've been building stacks and experimenting with the process, for over a decade at Singing Horse Ranch. We've come up with a set of observations and preferences. In our work with **Small Farmer's Journal** and these horse farming books, we've unearthed quite a bit of literature on the subject from 1860 to 1925 and have always been surprised to see that far more emphasis has always been placed on building as big a "lump" of hay as possible with little or no concern for shape, cap or finish. What the "experts" of the past called good stacks we, in our dry climate experience, would consider a waste of our time.

To MEASURE HAY in the Stack or Mow

Experts cannot agree as to an average number of cubic feet per ton of hay. The numbers vary from 485 to 620 cubic feet. Below is one simple formula using a low average. If you have a mixed legume grass hay or a clover or alfalfa hay you should figure 575 to 620 c.f.

The quantity of hay in a mow or stack can be only approximately ascertained by measurement. The weight of a given bulk depends upon its compactness. If hay, either in the stack or mow, is thoroughly settled, 512 cubic feet will make a ton. To illustrate:

If a mow measures 12 feet wide, 20 feet long, and 12 feet deep, multiply 12x20x12, which will give a product of 2880. Now divide 2880 by 512 (the number of cubic feet in a ton), which will give 5.62 tons of hay in the mow.

Stack Building as a Craft

One might think that there is nothing much to this stack building business. Push the hay around into a pile and move on. Wrong. The sheer volume of hay which can go into a stack of a given size may be doubled by how well the stack is packed. And the way it is constructed allows for the height and finally the shape. The height may add to the compacting as, obviously, the weight and mass of the hay itself will pack the material. And the height, relative to the diameter, or in the case of a loaf--the width, most definitely influences the angles and shape of the 'cap'. It is true that hay can be simply piled, as in the case of frame-built beaver slide stacks--with little or no concern for what's best. Indeed, on big outfits such is usually the case. If there's hundreds of acres to get in to the stack, 'finesse' may go to the bottom of the list of priorities.

In our time with stack building no one has taken the task more seriously or done a better job than the author's brother, Tony Miller. Perpetually experimenting with small subtle variables, he was always in search of the perfect shape and the best cap building technique. Though he retired a couple of years ago, no one has yet lived up to his standard. He was a true craftsman. In the world of his new vocation he may only ever be a 'good' golfer. In the world of farming, if this book bears any weight, he'll be known as one of the all-time great stackmasters.

How to Build a Stack

What follows is a description of the process we use, the process perfected by Tony Miller. It is, for the most part, identical to that which was described in building a wagon load. With our swinging stackers vertical reach of 20 to 24 feet we start with a base that is at least 24 x 24 feet. As we push hay into place, in the beginning, with the buckrakes we drive steel posts in for the corners. This allows the teamster to see where the edges are. As the work proceeds and a height of four feet is reached the posts are removed. Marking a 24 x 24 stack base results, with settling, into a 30 x 30 base.

From the very beginning the walls of the stack are critically important. They need to be kept straight, free of any breaks, and extremely well packed. From the beginning, the walls should be at least three feet thick. And they need, for at least ten feet of height, to be perfectly vertical. This is the focus of the stack construction--the walls. Little or no concern is taken for the middle because the stacker drops its load there. The drop tends to help pack the middle. And the stack crew takes that center drop hay and moves it to the walls.

Right from the beginning there are concerns. It may be helpful to speak of those things to watch for and avoid. This may help to give a clearer picture of what is sought. It is important to make sure the hay gets all the way out on the walls. If the hay for the walls creeps in the vertical is lost. It is important, as the stackers walk along the wall to pack it, that they find no breaks where they sink down. It is important that no big central mound is formed separate from the walls. It is important to keep moving the ladder to avoid a 'channel' break forming in the wall.

At ten or twelve feet the sensation is that it takes more hay to gain each foot of height. And it's true because the shear bulk is constantly compressing the stack. And at ten or twelve feet the walls should, very gradually, move in to begin the angle that leads to the cap. Since the object is for symmetry (because symmetry means each cap side will be equal and less likely to dish and gather water) and since most stackers will drop the hay in the center of the stacker, you may need to begin at 14 to 16 feet to make the wall away from the stacker higher.

As the walls come in and the stack top surface gets smaller it becomes important to pack all of it equal. The photos illustrate that the very top is tricky as the jags of hay either cannot or should not be

dumped because much of it will be lost off the sides. So the last jags are hand pulled off the stacker. Not all machines will allow this. Many of them feature hay in motion and it's impossible to have it hang up there for a casual cap building. Our swinging stacker and combination buckrake stacker, however, do permit this luxury.

All of these stacks do settle so it is important to make the cap a bit exaggerated in its peak. After about a month it will settle down a few feet.

As can be seen in the photos our stacks look 'neat.' This is not an accident and it certainly is not a given. You have to work at it. After the walls are five feet high we 'comb' the sides with our pitchforks, every so often. Besides helping us to define the walls and find flaws early on, it also removes those divots, dents, clumps, holes, and irregularities which reach out for or collect moisture. The "combed" surface of the stack (always vertical) serves like a fine thatching, helping to shed rain.

A last caution: make sure your ladder is notched in to the ground. This author once spent a couple of hours atop a 22 foot hay stack waiting to be missed. Seems that the ladder fell over. Two hours, by the way, is either not long enough or too long for a stranded stack maker to lay in the sun. It was discovered that the human sense of humor evaporates after an hour of hundred degree sunlight. It's a good idea to always have a second person in attendance when working atop a haystack.

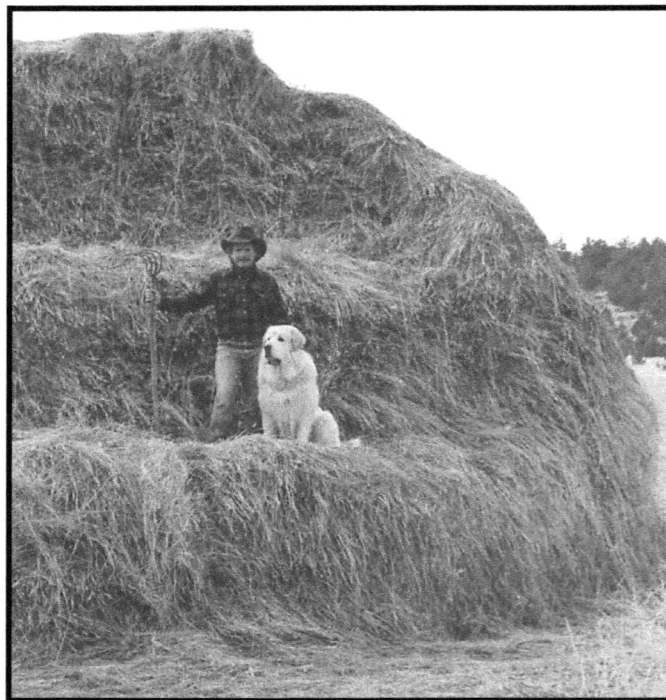

Figure 11. Tony and Ziggy stand on a shelf of one of Tony's stacks in the spring of the following year. As can be seen, this outside stack weathered a winter and, as the cutaway shows, the quality of the feed remains excellent.

Building stacks by hand

By this we mean without the use of a mechanical stacker. The size of any stack built by hand is limited to how high the hay might be pitched. As Fig. 4 on page 67 shows, some height can be added by staging the pitch to some sort of intermediate platform or staging (i.e. a third person standing on a wagon and relaying the pitch to the top of the stack.) In order to arrive at a shape which has the best chance of withstanding weather the height of the freshly built stack should be at least 75 to 80 percent of the base diameter. In other words, a square or round stack with a 12 foot base should finish out at least 9 feet tall (or approximately 2.5 tons).

The small hand-built stacks may be easier to cover with tarps but, for the labor involved, any crude stacking system allowing a larger stack will quickly pay for itself.

Hay Stackers

Hay stackers can all be classified under one of the following types: Overshot, swinging, combination, slide, derrick, cable, and rope.

The overshot, swinging, combination, and slide stackers are better adapted for use with buckrakes. The derrick, cable, and rope stackers which use slings or forks for elevating the hay are preferable when the hay is brought to the stacks on wagons or slips.

Figure 12. An overshot stacker mid stroke.

Overshot Stackers

Overshot stackers were very popular and affordable throughout the west. There is currently, as with all of the other styles of stackers, no one manufacturing this implement. That's unfortunate. This author believes a small farm shop could make some extra money building a couple of stackers each year to sell.

This style of stacker features a basket setup in catapult fashion without spring-loading. Like an overhanded toss, after buckrakes deliver the jag of hay by pulling on a cable, the arm arches and tosses the hay on top of the stack. There is no control over where the hay will land outside of moving the entire stacker. Few of the patented factory-made type of stackers were built strong enough to stand up where heavy loading was or is the practice. If one were built today reinforcement would be relatively easy.

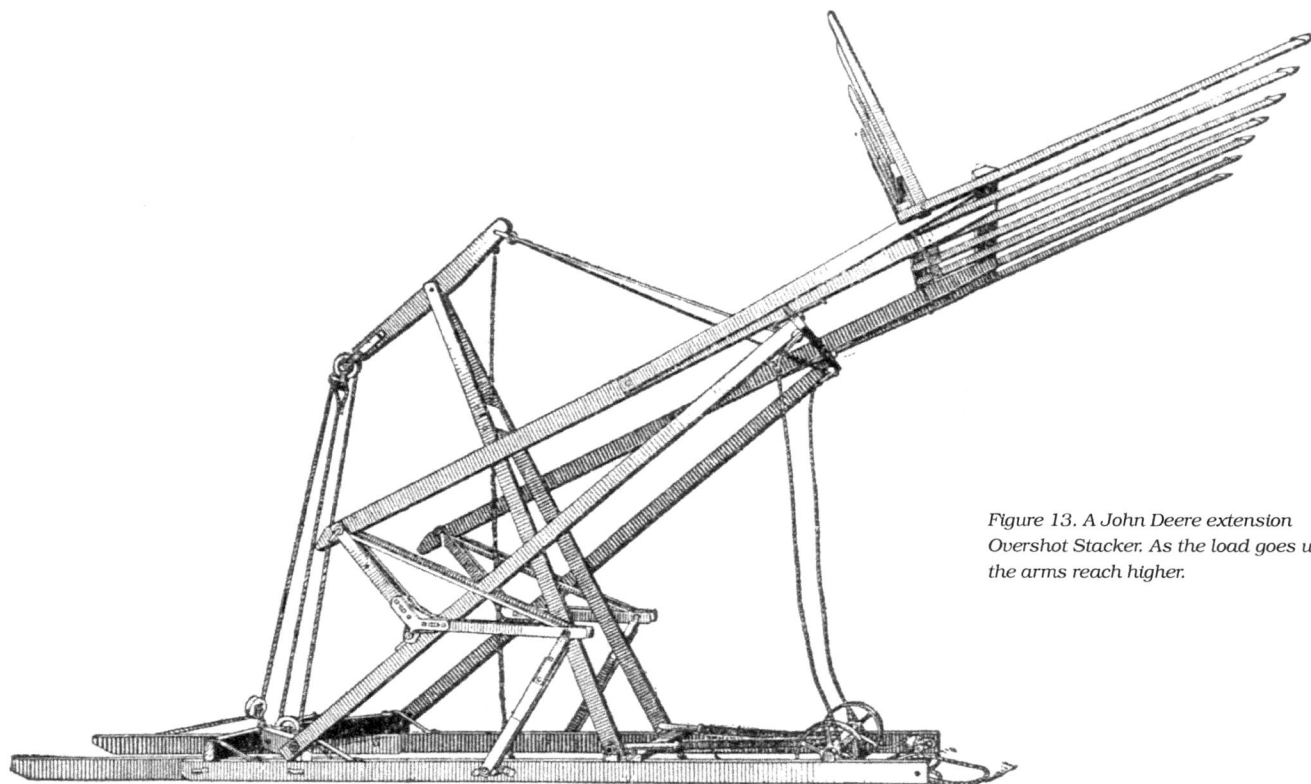

Figure 13. A John Deere extension Overshot Stacker. As the load goes up the arms reach higher.

Swinging Stackers

The swinging stacker is similar to the overshot type except that the basket carrying the hay swings sidewise instead of vertically, the hay sliding off the ends of the rake teeth when being tripped to be deposited on the stack. The advantage of this type of stacker is that the hay can be deposited at almost any point on the stack, and with some styles of swing stackers it is possible to load from either side at the will of the operator so that the sweep rakes do not interfere with each other as much as when using other types. As many as six and eight sweep rakes are sometimes used with this stacker. The success of this type of stacker depends largely upon the man operating it. In the hands of a good operator it has greater capacity than any other type of stacker in use, but in the hands of a poor operator it is one of the most unsatisfactory types. The patented swing stackers have the same disadvantages as the overshot type - that is, too light construction for heavy loading.

We own and use a "Mammoth" Swinging Stacker and are extremely well pleased with its operation. It would be possible, out west, to purchase a decrepit swinging stacker for very little money. With care and intelligence, the wooden parts could be replaced. Hopefully there is enough information in the pictures on these pages to help someone rebuild a swinging stacker.

Figure 14. A Mormon derrick. The boom pole is attached solidly to the upright which is pivoted at the base of the stacker.

Figure 15. Another style of swinging stacker. The weighted box on the lower end of the swinging arm is used to counterbalance the weight of the head.

Figure 16. A swinging stacker of the same type as we use.

Figure 17. A "Fleming" swinging stacker.

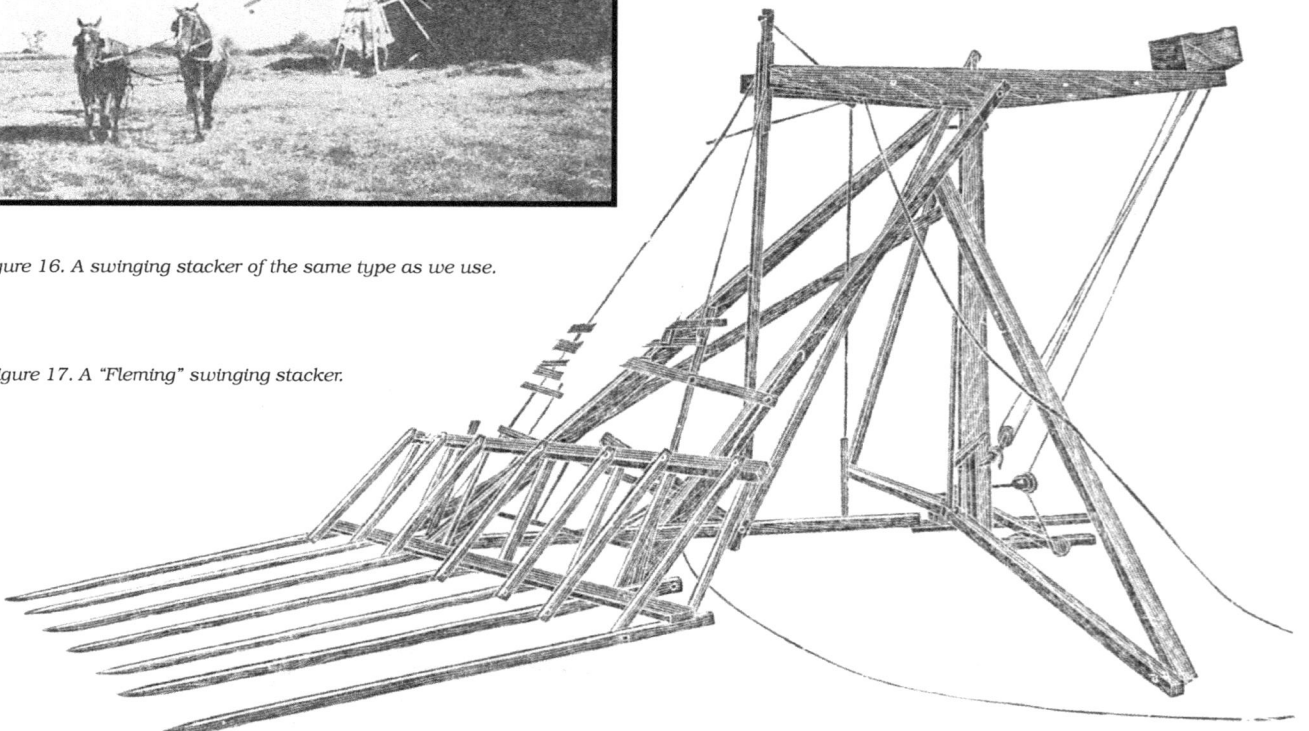

Combination Buckrake/Stackers

Combination stackers are less frequently used than either of the two types just described. This type of stacker is basically a buckrake with articulated 'grasshopper-like' arms which are cable-driven to lift hay. They are mounted on wheels and fully mobile. Hay can be deposited on the stacker basket by means of a another conventional buckrake or it can be collected directly from the windrow with the combo/stacker and brought to the stack. The elevating mechanism is engaged when the stacker is about 80 to 100 feet from the stack, and the basket with its load is then lifted as the stacker advances until it is sufficiently high to be deposited upon the stack. The advantages of this type are that no time is lost in setting up when starting a new stack, the hay can be easily placed at any point desired on the stack, and the position of the stacker in relation to the hay stack can be easily changed in case of a sudden change of the direction of the wind. The disadvantages are high initial cost or difficulty in finding a unit, rather low load capacity, unsatisfactory work on rough or uneven ground, and excessive weight for the operating team, especially if the hay is brought in directly from the windrow.

We have two *Jayhawk* brand combination buckrake stackers (one for parts). We don't recommend their use to go out and gather hay from the windrows although the machine does work for this. If building stacks with fairly level approaches they excel in getting hay up and to where ever it is needed.

Figure 18. A "Jayhawk" brand combination buckrake stacker.

Figure 19. A homemade combination buckrake stacker

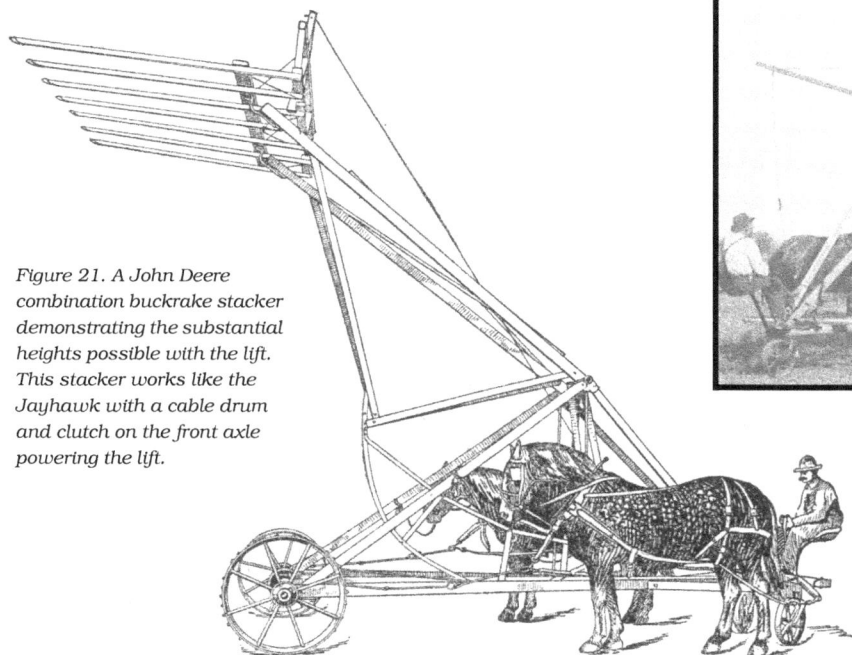

Figure 21. A John Deere combination buckrake stacker demonstrating the substantial heights possible with the lift. This stacker works like the Jayhawk with a cable drum and clutch on the front axle powering the lift.

Figure 20. An early model Jayhawk stacker does a neat job of delivering a jag to the rising stack.

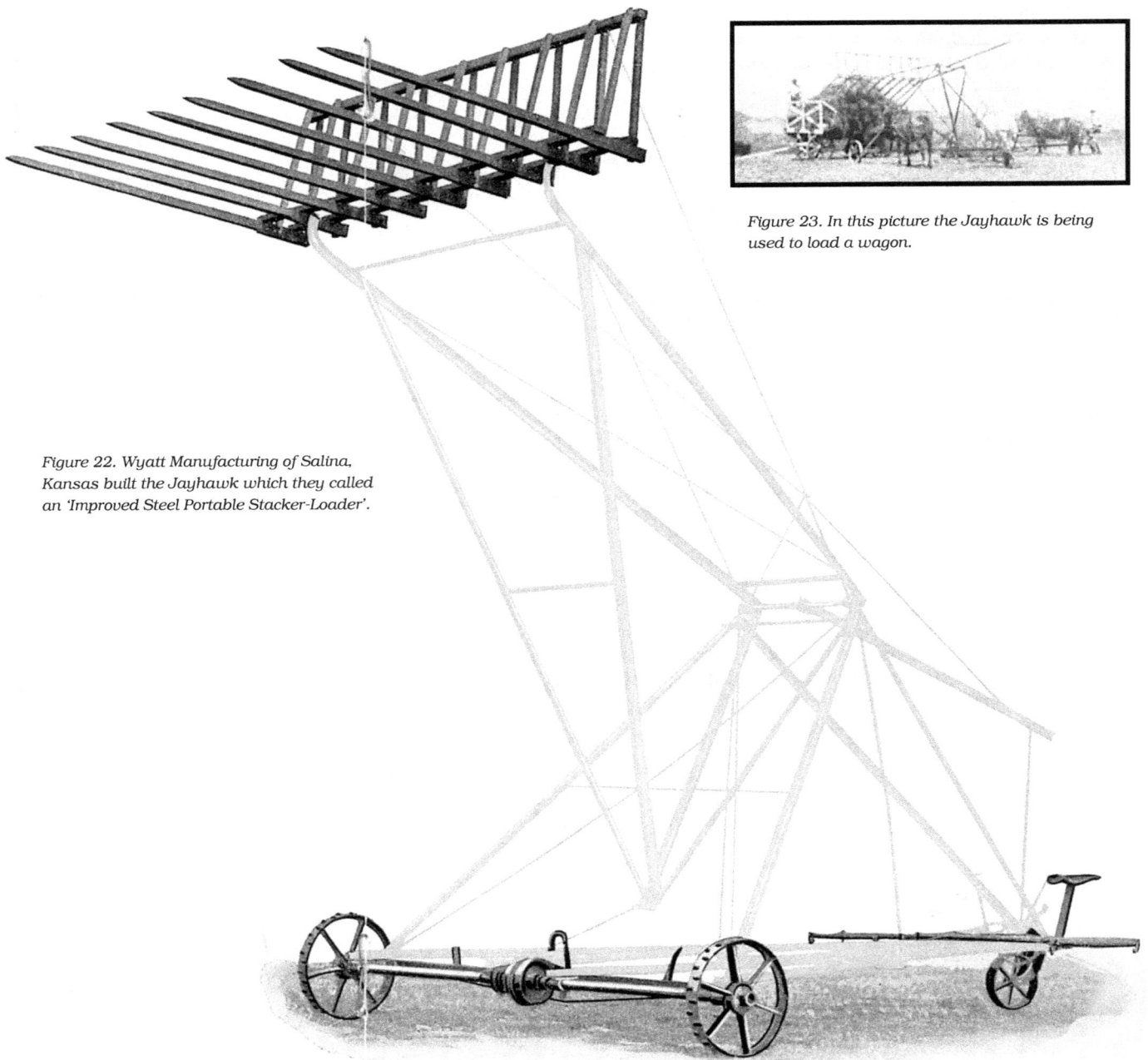

Figure 23. In this picture the Jayhawk is being used to load a wagon.

Figure 22. Wyatt Manufacturing of Salina, Kansas built the Jayhawk which they called an 'Improved Steel Portable Stacker-Loader'.

Figure 24. A good demonstration of the limited reach of the extended Jayhawk.

Figure 25. With the Jayhawk, hay can be deposited all the way around a stack.

Figure 26. Another view of a Jayhawk at work.

Figure 27. Loading a wagon. The Jayhawk makes short work of this job.

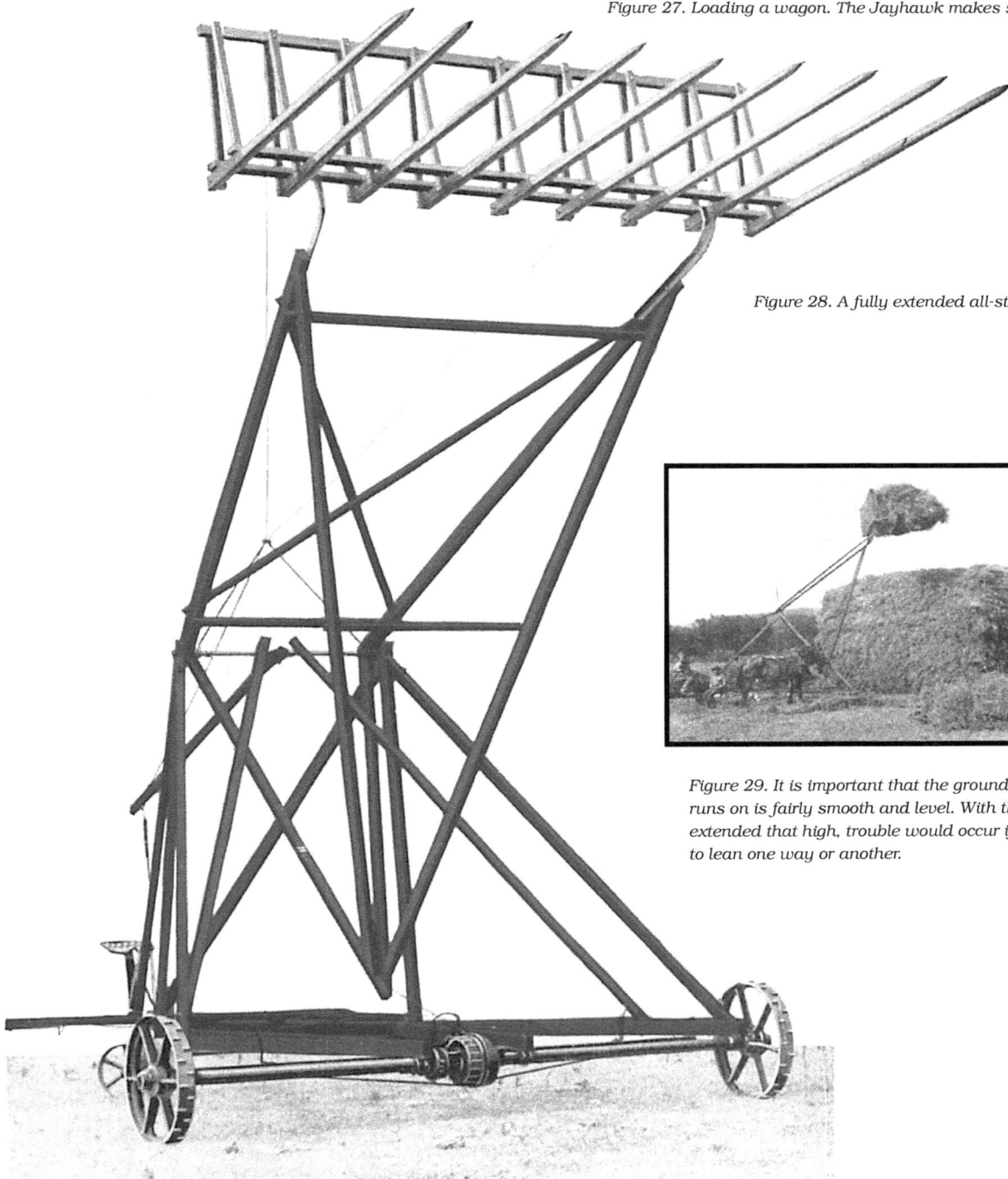

Figure 28. A fully extended all-steel Jayhawk.

Figure 29. It is important that the ground the Jayhawk runs on is fairly smooth and level. With this much weight extended that high, trouble would occur if the load was to lean one way or another.

Slide Stackers

There are two distinctly different slide stacker designs. The plunger style and the beaver slide. The plunger style is limited to the capacity of the team on the plunger. Whereas the beaver slide, with cables and pulleys set up to 'run' the basket up the slide, are only limited by the physical width of the slide.

This type was developed for stacking wild hay where particularly large tonnages are handled daily by each stacking crew. The stacker is a form of slide built of poles and mounted on skids.

Two styles of plunger slide stackers were in general use. One was wider at the top than at the bottom, and the other was the same width throughout. The stackers were and are constructed of poles from which the bark and all rough spots have been removed.

The lower six feet of the slide is usually constructed of movable sections of six or eight teeth each. When the stacker is moved to a new set these sections are lifted and shoved back into the body of the stacker. The poles on the slide are continuous from the movable sections to the top. On each side of the slide is a guard rail which prevents the plunger from running off when the hay is being pushed up.

The plunger, used to push the hay up the slide and onto the stack, is made of a pole about 35 feet long, on the front of which is a heavy frame 12 feet wide, 12 to 16 feet long, and about 3 feet high. To reduce friction, the frame runs on two wooden rollers. A doubletree is attached on a swivel so that the team can be turned around when the plunger is pulled back from the slide. A pair of small wheels 8 to 10 inches in diameter are sometimes attached to the rear end of the plunger to prevent its digging into the ground when being pulled back.

Of the two styles, the fan-shaped slide is the better where large stacks are built, as the hay can be thrown to one side or the other of the stack with the plunger and the stackers saving a great amount of work in pulling the hay to the ends of the stack.

The beaver slide style stacker is equipped with cables running through pulleys to the top and down one corner through another pulley and to the lifting power source.

In stacking with these types of stackers the hay is brought to the base of the slide by the buckrake, as in Figure 30, and left as it slides off the rake teeth. The plunger (if that is the style), which is in position as in Figure 31, is then pushed forward by the

Figure 30. Plunger-style Slide Stacker.

Figure 31. Plunger at work.

Figure 32. A Montana Beaver Slide Stacker and cage. This entire apparatus is drug to the stack sight.

team. The frame on the front pushes the hay ahead
and up the slide. The speed with which the team is
driven determines whether the hay will fall at the front
or at the rear of the stack. By handling the team
properly the driver can also, with the fan-shaped slide,
dump the hay toward either end or in the middle of the
stack. Two or three men are usually required on the
stack, as this outfit handles hay very rapidly and in
large quantities. A heavy team is required to operate
the plunger, which, together with the hay, makes up a
considerable weight. With the beaver slide setup, after
the hay is pushed on to the slide basket, a team is
used to pull on the cable and raise the load.

Men who have worked with all types of stackers
say they prefer the slide for wild hay because the hay
then comes over loose and when it drops it falls apart,
making the work of stacking easier.

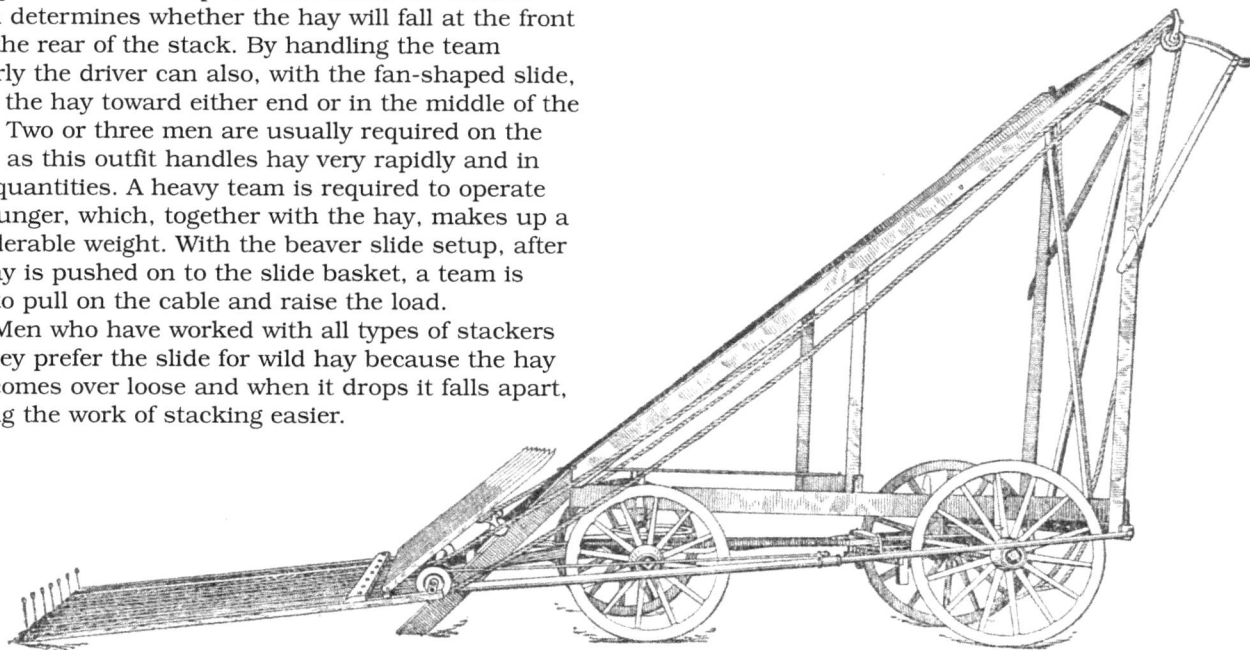

Figures 33 & 35 (bottom) Though sometimes referred to as a portable or mobile overshot these John Deere stackers actually function like small beaver slides. The basket conveyors up the slide when the pulley and cable mechanism is used.

Figure 34. In this Nebraska hay scene, horses pull up the beaver slide and a tractor is used to buck a huge pile on. This style of stacker was favored in the Mountain west because large amounts of hay could be piled fast. In the back a man sits on a hydra fork which is being used to spread the hay around. It is not possible to use much finesse with this style of stacking.

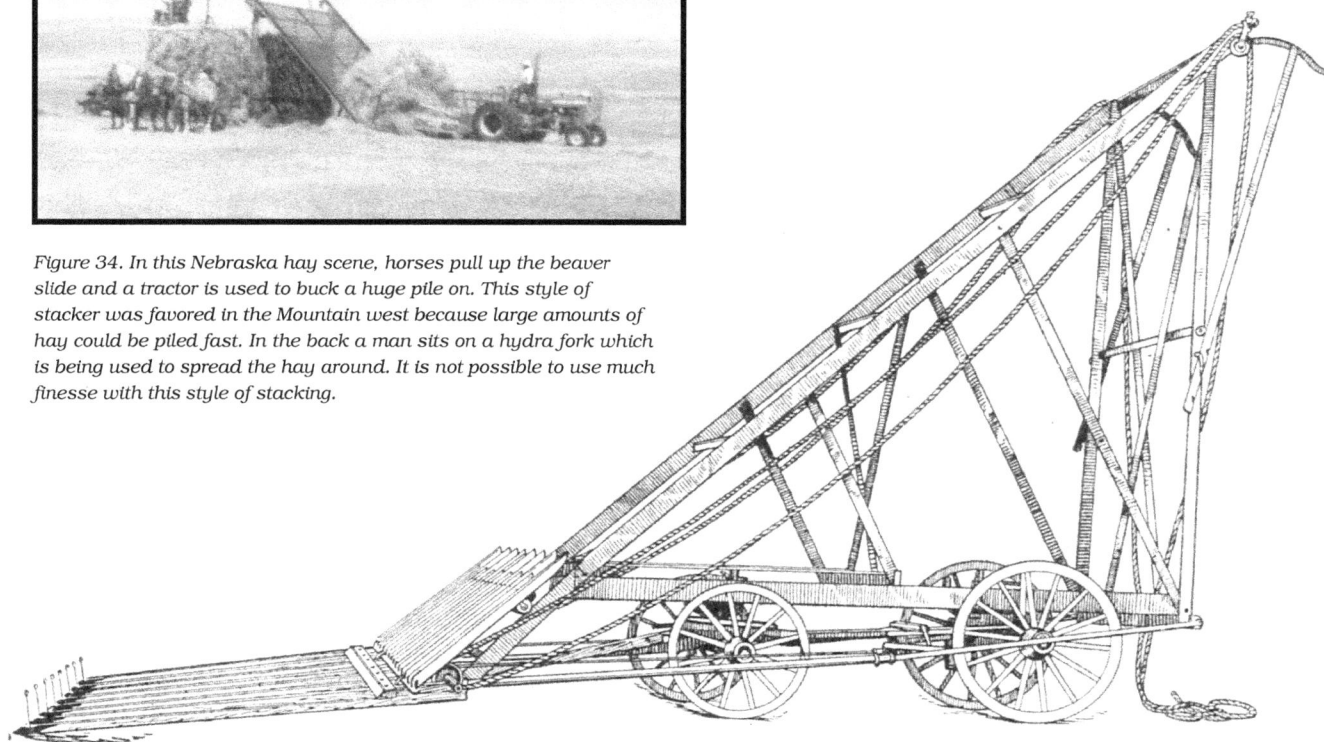

Figure 36. A large heavily constructed homemade derrick mounted on wheels. This stacker can handle very large forkfuls of hay and can be moved quickly to a new stack site.

Figure 38. A simple homemade pole stacker capable of making a very tall stack. As the hay leaves the wagon, the stacker turns and the hay is swung over the stack and dropped at the point desired. Stackers of this type must be held by heavy guy wires or cables.

Figure 39. A style of Wilson derrick with one adjustment of the boom pole. The lower end is solidly attached to the upright pole.

Figure 37. The Wilson derrick beside a finished stack. (One we would rate as poor.) This derrick is held in place by three wire cables attached to the top of the upright pole. The boom pole is attached to the upright by means of a steel collar which permits two adjustments, the elevation and the extension of the boom pole.

Derrick Stackers

Derrick stackers are especially adapted to alfalfa-hay sections where the hay is cured in the cock, but are seldom used in sections where wild and mixed hay is made.

For handling alfalfa with a minimum loss of leaves from shattering and a saving of labor on the stack, the derrick stacker when used with slips, excels all other types.

All stackers of this type are homemade and therefore are usually built to conform to the ideas of individual builders. The *Mormon derrick*, the style in most general use and perhaps the best, varies slightly in construction in different sections.

Two variations of this style of derrick stacker are shown in Figures 14 & 39. The principal difference in the two stackers is in the method of bracing and attaching the boom pole to the upright pole or mast.

The *Wilson* style of derrick stacker is of lighter construction than the Mormon, but because of the cables it takes more men and time to move it. Other variations of this style are shown in Figure 37.

In stacking with a derrick, the hay is usually transported from the field to stack yard on wagons or slips. Slips are more desirable for short hauls and wagons for long hauls. With the slip, a sling, or two chains (one on each side), are first placed on the bed of the outfit and the hay is then pitched on by one or more men. About 1,000 pounds of hay is handled at a load. When the slip reaches the stack, the sling or chains are attached to the pulley and the entire load is elevated to the stack and dumped where the stackers want it.

It is also possible when using a derrick stacker to use a sweep or push rake in delivering hay to the stack. Where this method is employed, the slings are sunk into the ground at the stack until they are flush with the surface. The rake is then driven onto the sling, where the load is left to be carried to the top of the stack. This is not a very desirable method for handling alfalfa, because more leaves are lost and more dust sometimes gets into the hay than when a slip or wagon is used.

Figure 40. A tripod stacker which can be easily made on the farm. It is self-supporting and can be taken down and moved easily but not so quickly as those mounted on wheels. Stacks larger than this are seldom built by a small crew using a tripod stacker.

Cable Stackers

The cable stacker is used when the rancher wishes to build unusually large stacks in a permanent location each year. It can be used for any kind of hay but is not so desirable for alfalfa if the hay is to be sold, as it may be necessary to mix cuttings to complete the stack, a practice which usually lowers the value of hay.

If the rancher wishes to stack hay in a stack yard close to the feed lots each year, this type of stacker is very satisfactory. The cable stacker can also be used where the stack yard is permanently located in the fields. The important constructional features of the cable stacker are shown in Figure 54.

Figure 41. Stacking hay with a homemade cable stacker.

Figure 42. A cable stacker with which a sling or fork is used to unload the wagon at the end of the stack. After the hay is elevated, the load is tripped at any point on the stack indicated by the stackers. Stacks up to 200 tons may be made this way.

Figure 43. A substantial and handsome cable stacker best left in the same site.

Rope Stackers (or Parbuckling)

Rope stackers are used for alfalfa or mixed hay in conjunction with buckrakes or wagons. Cheapness and lightness are perhaps the only advantages of this type of stacker.

In construction of this stacker, three ropes about twice the length of the ordinary stack are fastened to an iron ring. This ring is put over the head of a stake driven into the ground at one end of the stack (far enough away to allow a sweep rake or wagon to drive between) and the three ropes are passed over the stack. A buckrake delivers a load of hay on the ropes and is then backed away. A fourth rope, placed over the top of the stack and the load of hay, is hooked in the ring, which is now lifted from the stake. A team is hitched to the other end of this rope, and when driven ahead the hay rolls up the stack. The other ends of the three ropes are held by the man on the stack, two in one hand and one in the other. If one side of the load rolls ahead of the other the rope on that side is slackened and the rope on the other side is tightened. When the hay is in the desired place on the stack the ropes are released and the hay remains on the stack as the ropes are pulled from underneath. The man driving the team throws the ropes back to the man on the stack, who arranges them for the next load. If a wagon is used instead of a buckrake, the hay is rolled off the rack by means of ropes.

Stacks made with rope stackers are built low at the end where the load is rolled up, and high at the opposite end. At best this is a poor method of stacking, compared with other methods.

Backboard and frontboard

Backboards and frontboards are extensively used with good results by ranchers who put up wild hay with overshot stackers. The frontboard is attached to the stacker, whereas the backboard is built on wheels or shoes so that it can be easily moved about the hayfields. By using these devices, the front and rear of the stack can be built without much work on the part of the stackman. All that is necessary is to walk along the sides of the stack and force the hay down so it will shed water. The ends, being open, are made in the usual way. This method of stacking requires only one man on the stack to handle the hay which ordinarily requires two men.

Frames for the backboard and frontboard should be 10 to 12 feet high and about 18 feet long and made of 2 by 6 inch lumber. Smooth, heavy wire is stretched up and down the frame at about 1-foot intervals. Boards may be used in place of the wire but they increase the weight, and breakage is more frequent.

A somewhat similar device, but one that is more clumsy to move, is a wooden box or rack about 16 feet wide, 20 to 24 feet long, and 10 feet high in which the

Figure 44. A large portable sliding overshot stacker. The basket is pulled up and racks over the top to dump its load similar to a beaver slide.

Figure 45. This portable backboard regulates the width of the stack. A stake is driven in the ground and to which the caster wheel in front is attached by a chain preventing the back board from moving.

Figure 46. A combination frontboard backboard used with an overshot stacker.

stack is built. The sides and ends of the box are built separately, or a side and an end may be hinged together. When the stack is finished the rack is taken down and folded flat on the ground for moving.

Figure 47. A Swinging stacker at work.

Figure 48. A swinging stacker takes up the final jag to cap a nice old stack.

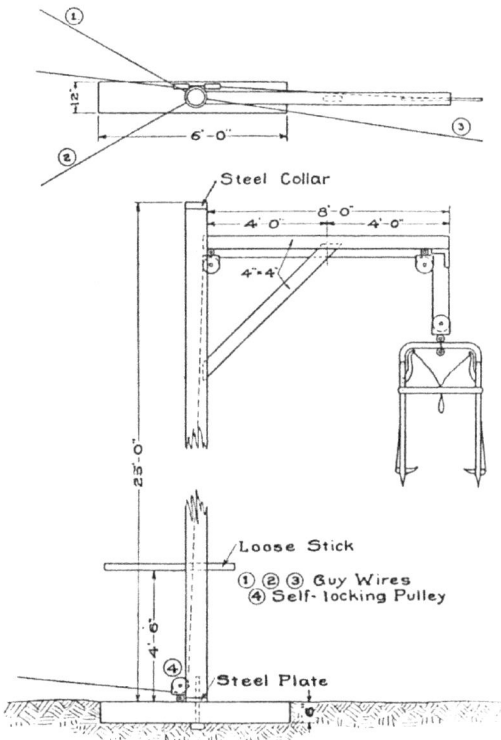

Figure 50. A homemade cable stacker.

Figure 51. A swinging or gin pole stacker.

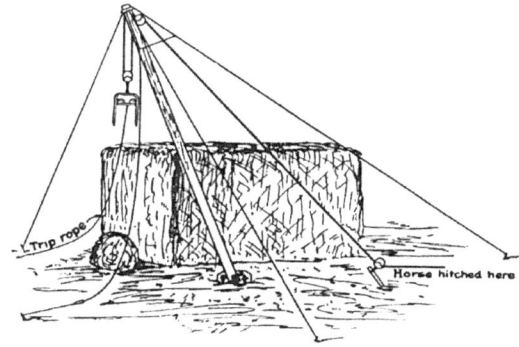

Figure 52. An A-frame stacker which pivots to shift the load over the stack.

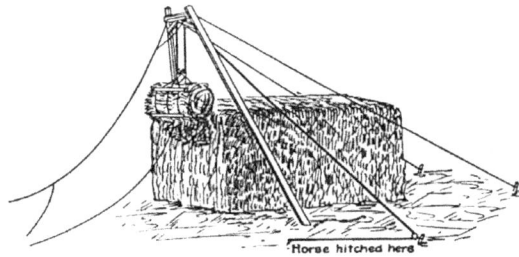

Figure 49. A design for a homemade pole stacker. Note the relative position of the bottom pulley to the top (pole) pulley. As the load is drawn up this half wrap of the rope or cable is what turns the pole to put the load over the stack.

Figure 53. A "Palmer" derrick with a Jackson fork.

Figure 54. A cable stacking system.

Figure 55. A phenomenal scene from 1910. A modified A-Frame pivot stacker. Three buckrakes wait their turns to deposit loads at the base of the stack where forks are set and the load pulled straight up. When at the top, the left side support cable is loosened and the frame leans over the stack where the load is tripped. In this scene two men ride the load up, using the stacker to get up and down, a dangerous practise! Those poles are anchored with pivots at the ground, usually connected to the end of a long pole running alongside the stack and anchored to a buried "deadman". Erecting these poles at the site involved first setting the anchors, then poling up the frame to at least 20 degrees where the support cable had sufficient angle to pull it up the rest of the way. Care had to be taken to have the opposing support cable in place so that the frame didn't come on over and crash.

Figure 56. A small true Overshot stacker by John Deere. If some small farm shop were to build a few units per year of this or a similar stacker, they'd be surprised by the nice little extra income it might bring in.

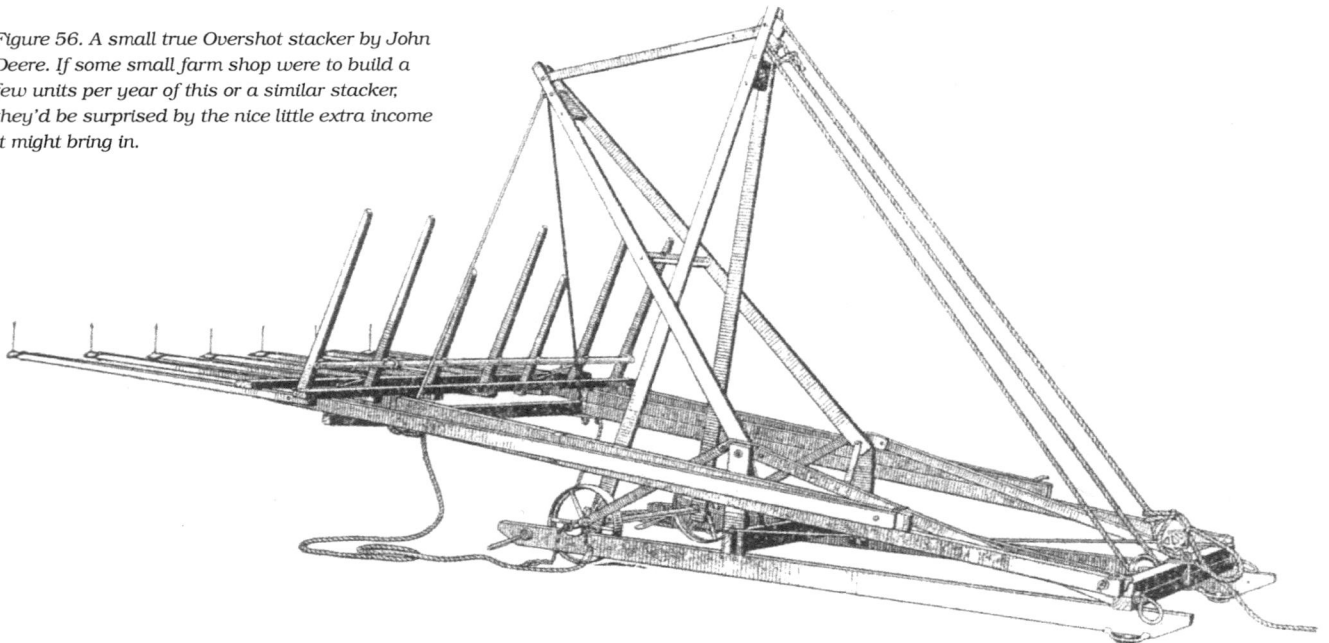

Stacking Crew Thoughts

The introduction to this book offers a bunch of information about crew organization which, when combined with what follows, we hope won't seem too daunting.

In stacking hay with buckrake and stacker, if not enough men are used on the stack, they will be overworked and the buckrakes will be idle a part of the time, while if too many are used, the buckrakes will not be able to keep them busy. In either case labor will be wasted.

The smallest possible crew for this method is composed of one person on the stack, one to drive the stacker team, and one with push rake. This crew will put 12 ½ tons in the stack a day, which is at the rate of 6 1/4 tons per adult per day.

On many farms one person on the stack will handle the hay brought by two push rakes, which is at the rate of 8.33 tons per adult per day.

It is seldom that one person stacking is able to keep up with three buckrakes. This can be done, however, if the stacker is experienced, or if the yield is light, say one-half ton to the acre, and the push rakes bring the hay to the stack at regular intervals so as to give the stacker time enough to stack the hay properly. Beginners in the use of the stacker, especially if the labor obtainable is not of the best quality, will probably get the most satisfactory results by using more people on the stack, at least until the workers become accustomed to the work.

When four buckrakes are used, and the yield is one ton per acre or better, it is customary to use three people and sometimes four on the stack. When four to seven push rakes are used, it will be necessary to use four on the stack, as the stacker is kept fairly busy bringing up the hay to the stack, and this necessitates a person for each corner of the stack.

It is seldom that more than six or seven push rakes are used. In order to use this number efficiently the stacker should be mounted on wheels so that it can be moved quickly, since it may be necessary to move it once or twice a day. When a new stack is started, some of the push rakes should begin to bring in hay from distant parts of the field, while others work nearer the stack; thus they will be able to keep the hay coming in steadily. This number of push rakes is too large to be used except on farms where hay is grown on a large scale.

A crew using three or four buckrakes will be found about the right

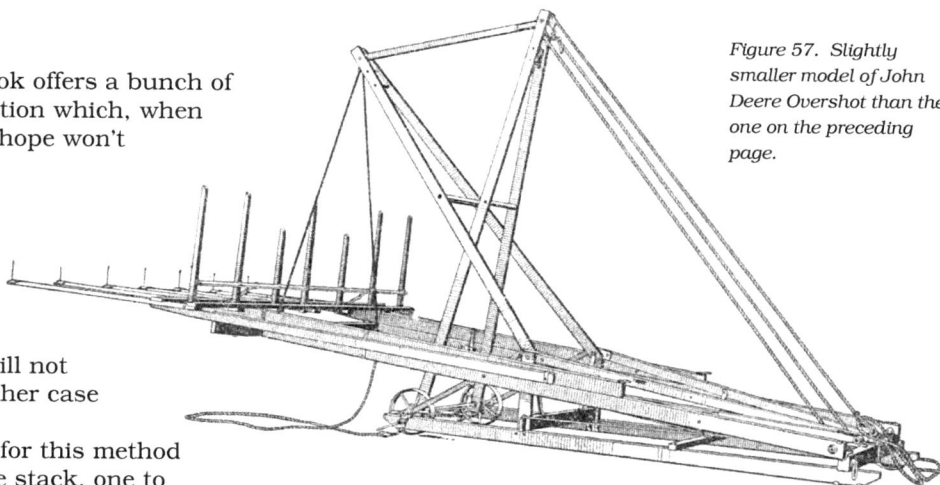

Figure 57. Slightly smaller model of John Deere Overshot than the one on the preceding page.

size for farms growing from 100 to 200 acres of hay.

In some instances it will be found necessary to have one or more extra people in addition to those who are engaged in operating buckrakes, or working on the stack. With some types the stacker requires the services of a man to operate it. Again, when several buckrakes are used, especially by "green" workers or teams, it will be necessary to use one person to clean up the hay dropped by the buckrake in unloading around the front of the stacker.

When several buckrakes are used and they are working fast, more or less hay will be left in front of the stacker. Keeping this hay cleaned up is a very light job, which can be done by a youngster.

SIZE OF STACK

The amount of hay put into the stack has an important bearing upon the efficiency of the crew. The

Figure 58. Used corrugated roofing being used to protect a hay stack. Concrete weights are hung from the sheets to keep them in place. A covering like this will last for years and properly installed will keep the hay free from rain.

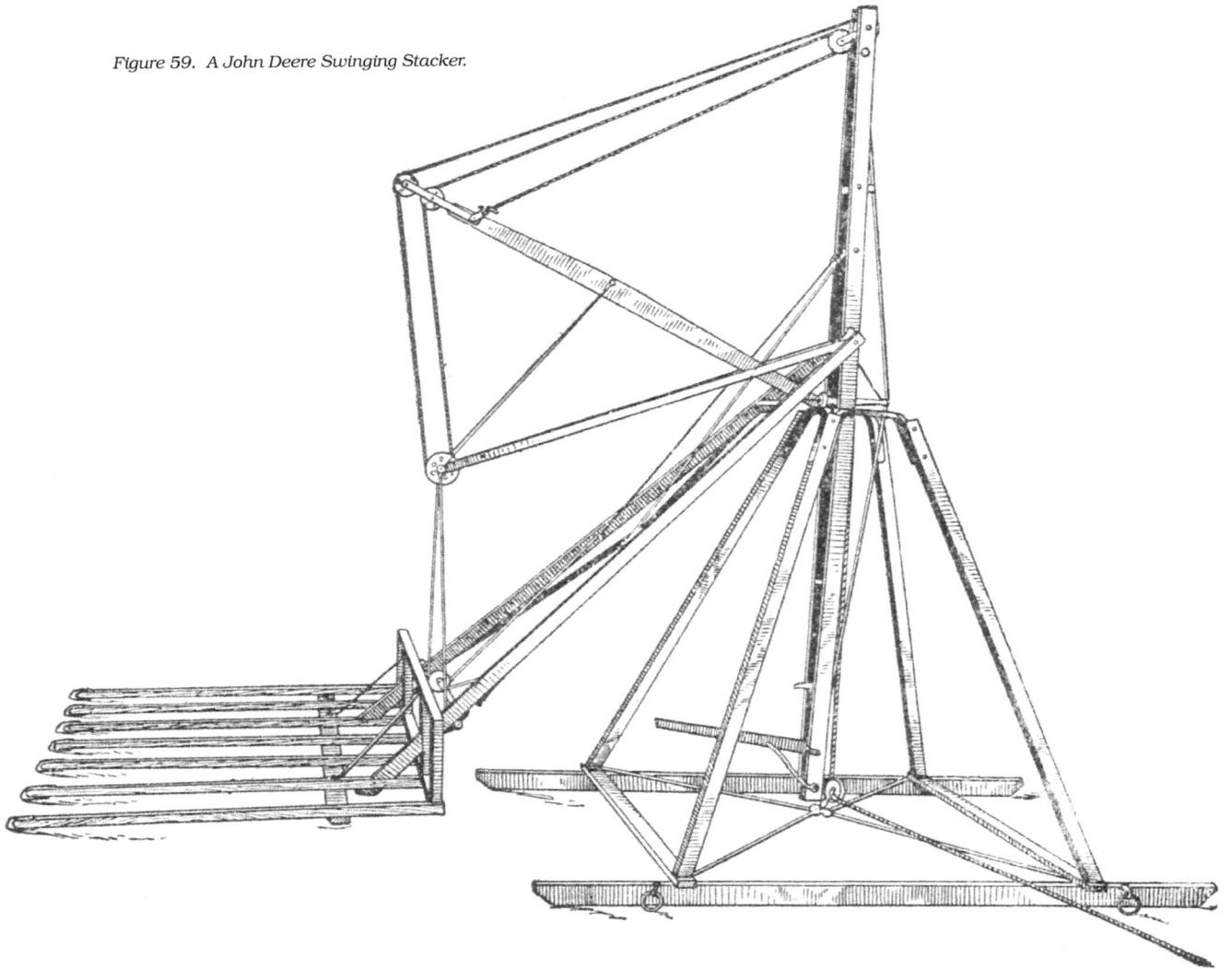

Figure 59. A John Deere Swinging Stacker.

size of stack depends upon type of stacker used, yield, number of people working on the stack, and amount of hay brought to the stack per hour. It is scarcely economical to build anything smaller than a 10-ton stack when using a stacker.

STACKS MADE WITH CABLE STACKER

To move and set up a cable stacking outfit requires considerable time, and fairly large stacks ordinarily should be made in order to prevent a loss of time by the crew while waiting for the stacker to be moved. The ordinary cable stacker is capable of making ricks up to 40 feet in length, although longer ricks can be made when an extra long cable is used. The width of the rick varies from 12 to 20 feet, and the height from 20 to 30 feet. The contents of the stacks may be anywhere from 10 to 30 tons or more. When the amount of hay stacked per hour or per day is known, the size of the stack may be regulated so that it will be finished about noontime or by evening. If this is done, the crew will lose little time, since the stacker

can be taken down and moved in the morning, while the crew is waiting for the dew to dry off, or at noon, or even late in the evening after the day's work is done.

STACKS MADE WITH DERRICK STACKERS

The size of stacks made by the different types of derrick stackers varies considerably. Small stackers of this type are often too low to permit the building of tall stacks, but some of the larger stackers used in the West build stacks higher than the cable stacker. Ricks made by derrick stackers are in general, not as long as those made by the cable stacker, because of the large amount of labor required in moving the hay to the ends of the rick. Long ricks sometimes are made by making a rick of moderate length, say 30 feet, and then moving the stacker so that the next stack can be joined to the first one. This practice is not to be recommended, unless it is necessary to move the hay to a high point to avoid injury by overflow water, or to place the hay where it will be handy for feeding, as is done in the West by cattle or sheep feeders.

STACKS MADE WITH THE OVERSHOT AND SWINGING STACKERS

The overshot stacker is perhaps better adapted to building stacks of different sizes than any other type. It is much more popular with the average hay grower. It can be used with a small crew or a large one, and, when mounted on wheels, can be moved to a new stack site in a few minutes, with hardly any loss of time on the part of the crew. With this outfit, stacks ranging from 5 to 30 tons or more can be built.

With the swinging stacker, stacks the same size and larger can be made as with the overshot.

STACKS MADE WITH COMBINATION BUCKRAKE/ STACKER

Ricks of any desired length can be made with the combination stacker without necessitating any extra labor on the stack, because this type of stacker places the hay on any desired part of the stack, and from either side or end. It can be used to build very small stacks as easily as long ricks. The extreme height of stacks made with the combination stacker is from 20 to 23 feet. Many users of this type claim that when it is worked to its capacity, one or two less men are needed on the stack than when stacking with the overshot stacker.

EFFECT OF YIELD ON SIZE OF STACK

Size of stack is often dependent, more or less, upon the yield of hay, especially when hay is brought to a stack with push rakes, but not when hay is hauled on wagons. Given a fixed number of push rakes, it will be found that larger stacks are, or should be, built when the yield is heavy than when it is light, since building large stacks with a small yield entails a considerable waste of labor.

If the yield is 1-½ tons per acre, if three people are working on the stack, and if three push rakes are used,

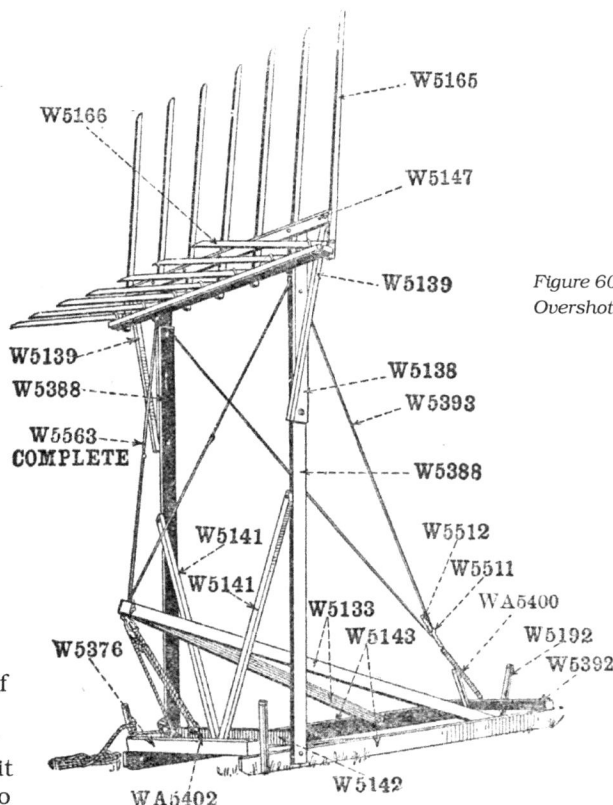

Figure 60. McCormick Deering Overshot stacker.

McCormick-Deering Overshot Hay Stacker

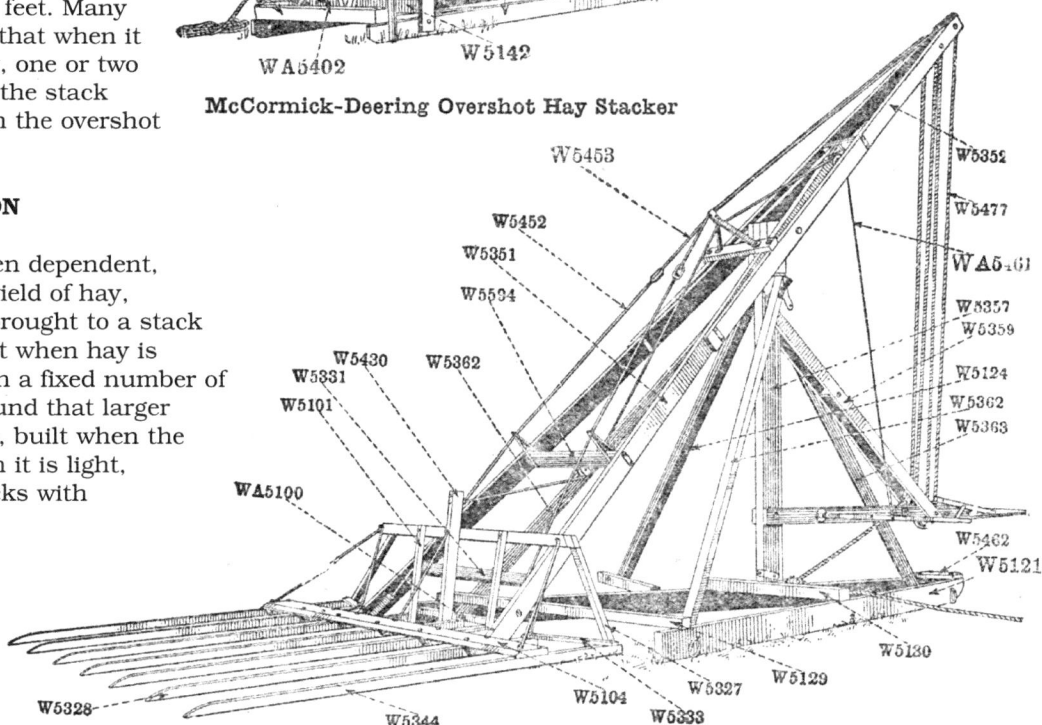

Figure 61. McCormick Deering Swinging Stacker.

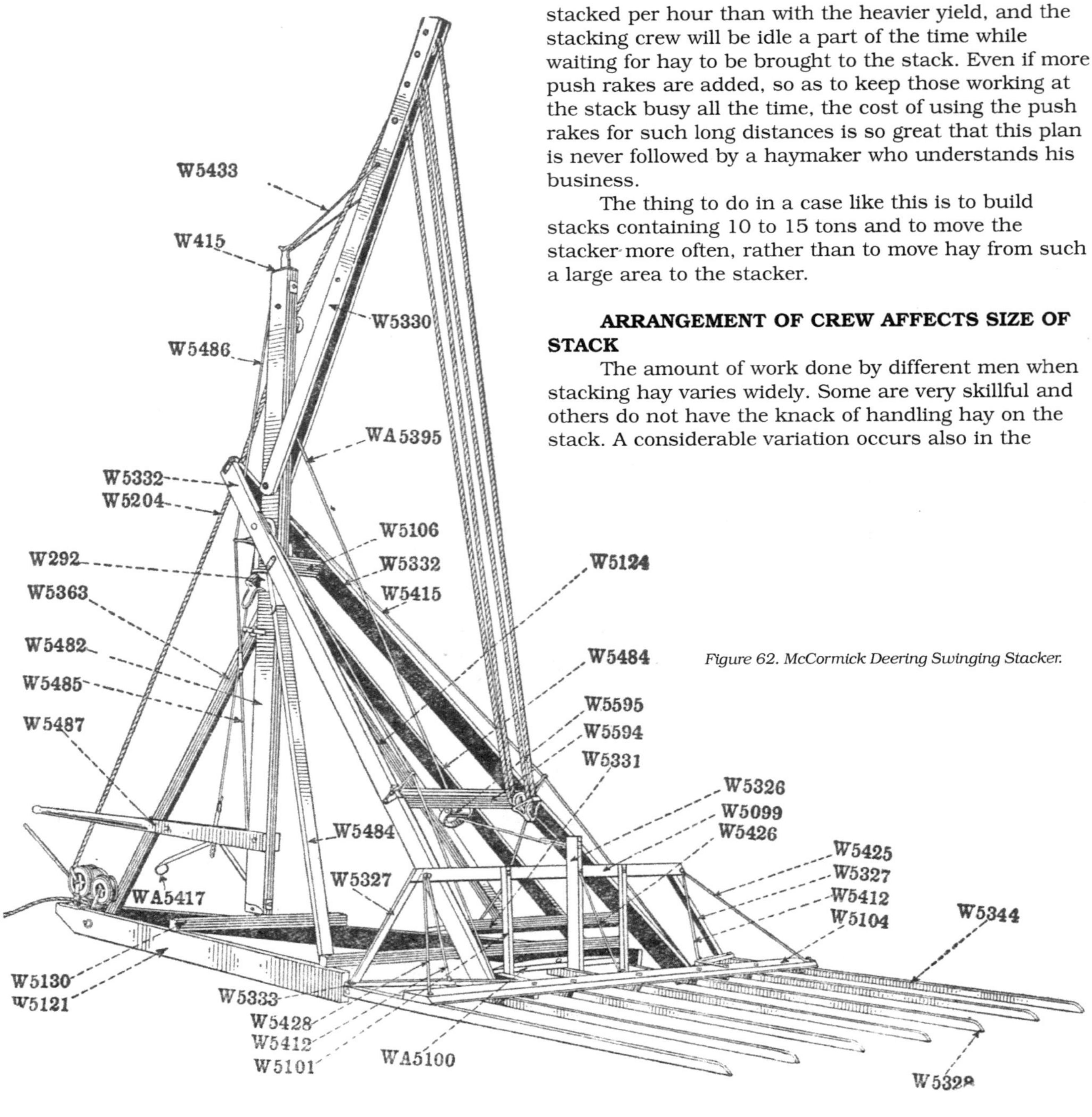

stacked per hour than with the heavier yield, and the stacking crew will be idle a part of the time while waiting for hay to be brought to the stack. Even if more push rakes are added, so as to keep those working at the stack busy all the time, the cost of using the push rakes for such long distances is so great that this plan is never followed by a haymaker who understands his business.

The thing to do in a case like this is to build stacks containing 10 to 15 tons and to move the stacker more often, rather than to move hay from such a large area to the stacker.

ARRANGEMENT OF CREW AFFECTS SIZE OF STACK

The amount of work done by different men when stacking hay varies widely. Some are very skillful and others do not have the knack of handling hay on the stack. A considerable variation occurs also in the

W5433
W415
W5330
W5486
WA5395
W5332
W5204
W5106
W5332
W5415
W292
W5363
W5124
W5482
W5485
W5484
W5487
W5595
W5594
W5331
W5326
W5099
W5426
W5484
W5425
W5327
W5327
W5412
W5104
WA5417
W5344
W5130
W5121
W5333
W5428
W5412
W5101
WA5100
W5329

Figure 62. McCormick Deering Swinging Stacker.

the three push rakes will easily handle 10 tons each per day, making a stack containing 30 tons of the hay from 20 acres. Under conditions just mentioned, all the men in the crew will be kept busy all the time. If the yield of the next cutting (Johnson grass and alfalfa mixed, for example) happens to drop down to half a ton per acre, it will require the hay from 60 acres to make a 30-ton stack. Each push rake will have to cover 20 acres instead of 6 2/3 acres, and, consequently, can not bring the hay to the stack as fast as when the yield was three times as much. To build a stack of the same size will, in this instance, result in less hay being

amount of hay a man and team will bring to the stack. Owing to these facts, those using stackers for the first time cannot hope to start right out and build a given sized stack in a given length of time. After the crew has become accustomed to the work, and to working together, and when everything begins to go smoothly, the hay grower or manager by taking careful note of the increase or decrease in the amount of hay made, can begin to make adjustments until the crew is doing the greatest amount of work possible without over-working either man or horse.

Figure 63. John Deere Overshot.

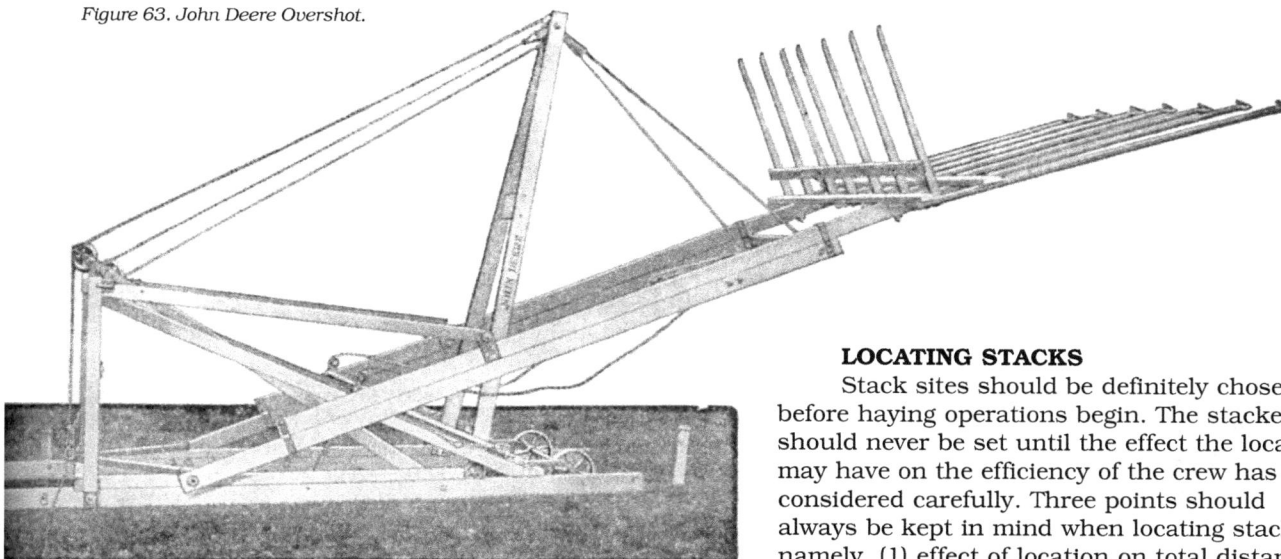

Figure 64 and 65. Two close ups of the John Deere Overshot stacker.

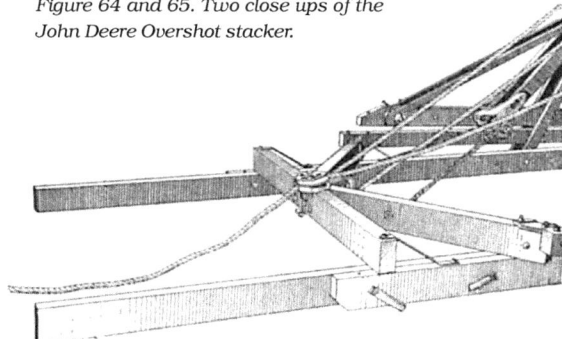

LOCATING STACKS

Stack sites should be definitely chosen before haying operations begin. The stacker should never be set until the effect the location may have on the efficiency of the crew has been considered carefully. Three points should always be kept in mind when locating stacks, namely, (1) effect of location on total distance traveled in bringing the hay to the stack, (2) lay of the land and its influence on ease of hauling the hay to the stack, and (3) danger of loss of hay in the bottom of the stack due to surface water or ground water.

AVOID LONG HAULS

Long hauls should be avoided whenever possible, especially when hauling with buckrakes. It is very common, on farms where hay is hauled with a wagon and stacked by hand, to see stacks of hay at the end or corner of the field nearest the farm buildings, the site being chosen in order to have the hay handy for feeding.

With buckrakes, this practice will seriously affect the size and efficiency of the crew. The total distance traveled in bringing hay to one side of a square field is approximately 50 percent greater than in hauling to the center. If the stack is located at one corner of the field, the total distance traveled is 100 per cent greater. It is necessary to travel about 30 per cent farther in bringing hay to the corner of a square field than to bring it to a side of the field. If four push rakes are kept busy in bringing hay to the center of a square field, six will be required to bring the same amount of hay to the end of the field, and eight to bring the same amount to a stack located in the corner, if it is to be stacked in the same length of time as when stacking in the center of the field.

TAKE ADVANTAGE OF THE LAY OF THE LAND

To haul hay up hill reduces the amount handled per day, as a well-loaded push rake, especially of the three- or four-wheel type, is usually a load for an

ordinary team on level land. In order to have a down-hill haul, it is sometimes a good practice to locate the stack away from the center of the field, thereby decreasing the labor requirement, though increasing the total distance the hay is hauled.

AVOID STACKING IN LOW PLACES

Stacks should not be located in low or marshy places. Stacks near creek beds are often subject to considerable damage by freshets. To avoid loss of this kind, stacks should always be built on dry, well-drained ground, even though the hay must be hauled a little farther than otherwise would be necessary.

Building the Stack

THE STACK BOTTOM

To minimize damage by moisture from the ground, care should be taken to keep the hay in the bottom of the stack from coming in contact with the soil. Usually more or less spoiled hay is found in the bottom of a stack that has stood for a considerable time. Such loss may be greatly reduced or altogether avoided by placing old rails, poles, boards, or even straw on the ground so as to keep the bottom of the stack dry.

SETTING THE STACKER

In setting stackers, the direction of the prevailing wind should be considered. The setting of a cable stacker or a derrick stacker can not be changed easily, because the "dead men" to which the guy wire are attached are usually buried deep in the ground, and to move them requires considerable time. Stackers are usually set to stack with the wind. The combination stacker can be used at either side of the stack, depending on the direction of the wind.

Derrick or pole stackers should be set so that the upright part is perpendicular. The bottom should be level. It is especially necessary that the overshot stacker mounted on wheels be set level, because it will not work to best advantage otherwise and, if much out of level, may be blown over when taking up a full load.

SHAPING THE STACK

To build a good, symmetrical stack when using a hay stacker requires practice. Hay pitched from a wagon is put on the stack a forkful at a time, which allows plenty of time for the man on the stack to place each forkful and tramp it down. The stacker, however, drops upon the stack larger bunches of hay, which must be quickly spread out or rolled into position without being

tramped very much. The weight of the falling hay tends to settle the hay in the middle of the stack, which is a help to those building the stack.

The aim of all good stack builders is to make a stack that will not "take water" that is, one built so as to shed water away from the center and toward the sides or edges of the stack. (Our system of building haystacks, which is different from the old theories is outlined in Chapter 15.)

The shape of the stack has a direct bearing upon the tendency of the stack to take water, and the percentage of hay liable to be damaged. Nine end views representing the ordinary shapes of stacks or ricks are shown in Figure 7, page 238. Stack No. 1 is perhaps the poorest shape. Stack No. 9 represents a shape in which the loss due to the weather will be least, provided it is built properly. The other seven shapes represent various intermediate types.

PROTECT THE TOP OF THE STACK

The stack should be well topped out and the sides should be raked down with a garden rake or pitchfork to prevent rain from entering the sides. In sections subject to high winds the top should be held in place until the stack is settled. This is done with weights attached to wire, thrown at intervals across the top of the stack. Old fence posts, tile, or rocks may be used for weights.

Canvas stack covers are used by some good haymakers, practically eliminating loss from rain. A more desirable protection of this kind consists of sheets of corrugated galvanized iron roofing, which can be bought properly curved for this purpose. These are easy to put on, will keep all the rain off the stack, and will last for years.

Figure 66. At Singing Horse Ranch, Tony Miller watches, disgusted, as a gust of wind prepares to take a swinging stacker jag and blow it into the neighboring garden. Bob Oaster is tripping the stacker.

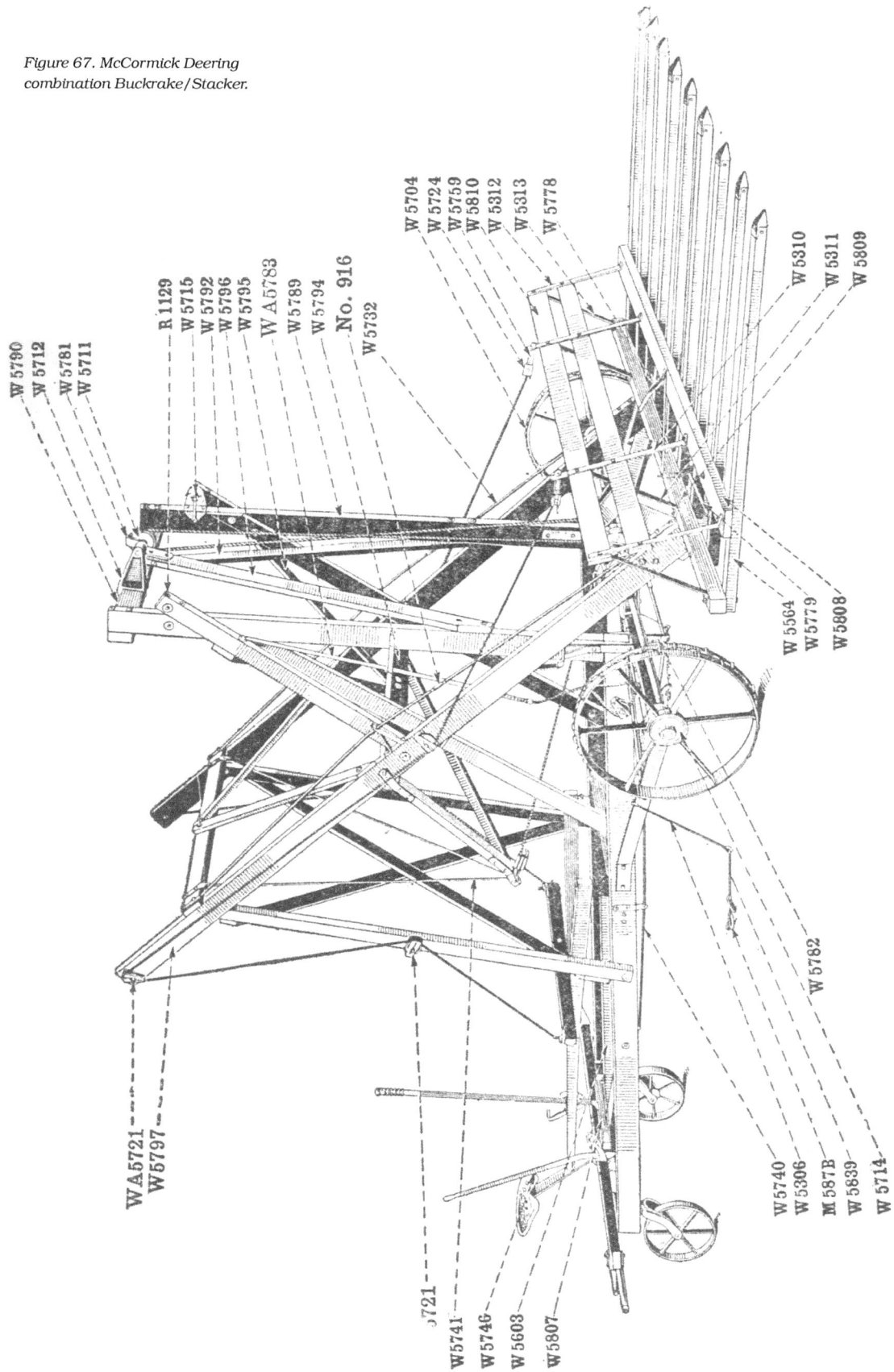

Figure 67. McCormick Deering combination Buckrake/Stacker.

259

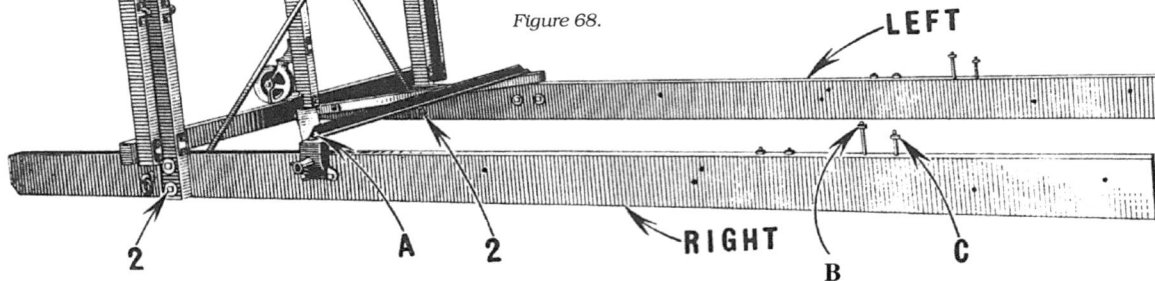

This information is typical of that which John Deere provided when they shipped their disassembled Overshot stackers. These details and this information may be helpful to anyone ambitious enough to try to build their own Overshot. LRM

Figure 68.

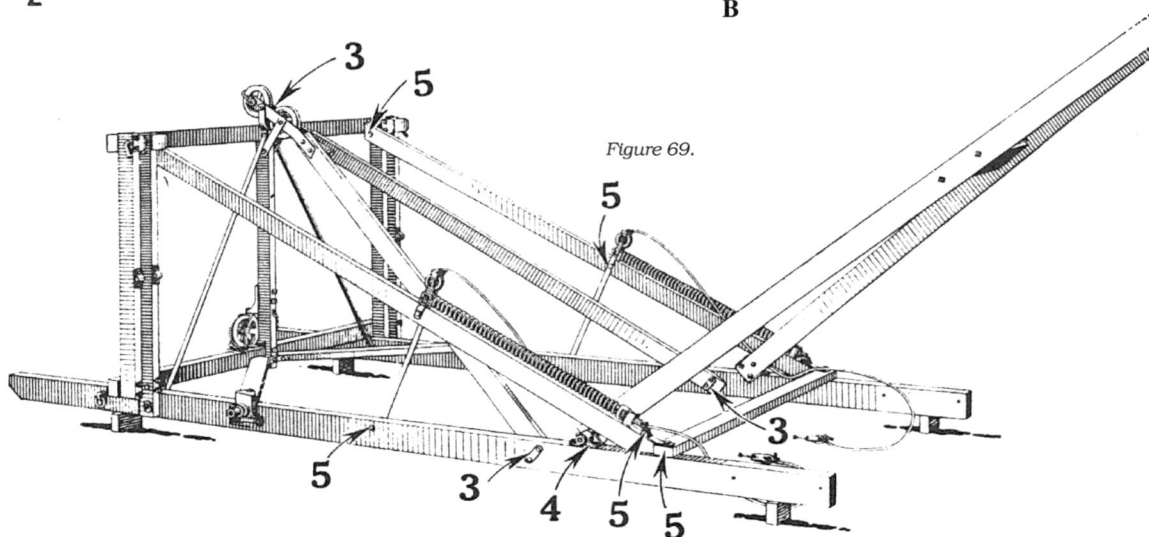

Figure 69.

DIRECTIONS for SETTING UP and OPERATING

JOHN DEERE No. 1, No. 2 AND No. 2-A STACKERS

In reading these directions, the part described in the paragraph can easily be found by looking at the arrow with the number corresponding to the number of the paragraph.

TO SET UP THE STACKER

The words RIGHT and LEFT used in these instructions mean the right- and left-hand sides, respectively, when standing behind and facing the machine.

These directions cover the No. 1, No. 2 and No.2-A Stackers. The No. 2 and No. 2-A Stackers stack higher and have higher back frames.

1. Lay ground sills parallel, with small rings at rear end on outside. Insert three carriage bolts in each sill as follows;

For No. 1 Stacker.
½" x 8 ½" Carriage Bolt at "A"
½" x 10" Carriage Bolt at "B"
½" x 8" Carriage Bolt at "C"

For No. 2 and 2-A Stackers
½" x 10 ½" Carriage Bolt at "A"
½" x 12" Carriage Bolt at "B"
½" x 10 ½" Carriage Bolt at "C"

2. Fig. 68 -- Assemble back frame, as shown.

For No. 1 Stacker bolt back frame to outside of ground sills with four ½" x 5 1/4" machine bolts and washers. Attach long steel braces to same bolts.

For No. 2 and No. 2-A Stackers bolt back frame to outside of ground sills using top holes with two ½" x 6" machine bolts and washers. Bolt long straps at bottom hole in ground sill using ½" x 6 ½" machine bolt and washer.

3. Figs. 69--69-A. Assemble stationary A-frame on back frame, as shown. The No. 1 Stacker uses only one J 19 E washer and one ½" x 5 ½" carriage bolt on each side.

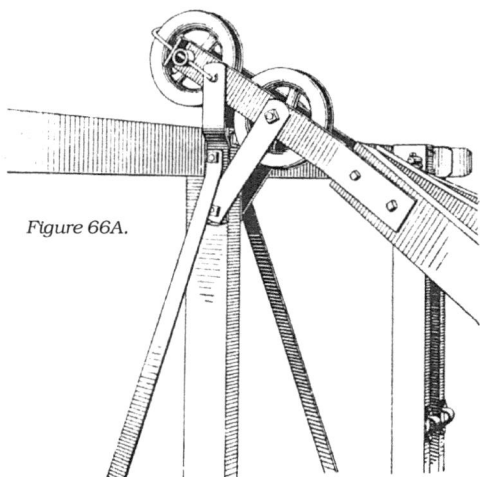

Figure 66A.

The No. 2 and No. 2-A Stackers use two J 19 E washers, two ½" x 5-½" carriage bolts, and a steel strap on each side, as shown. Washers, bolts and straps are found in bag.

4. Fig. 69 and Fig. 70. Assemble hinged A-frame and hoisting rods, as shown.

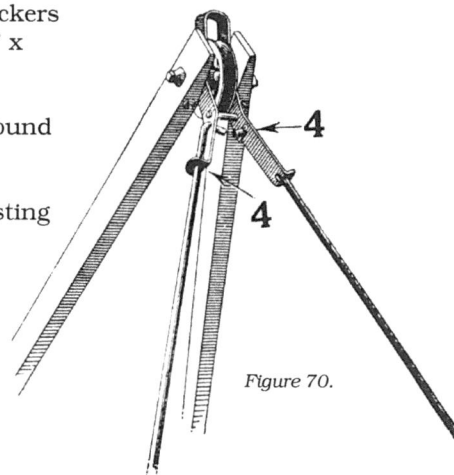

Figure 70.

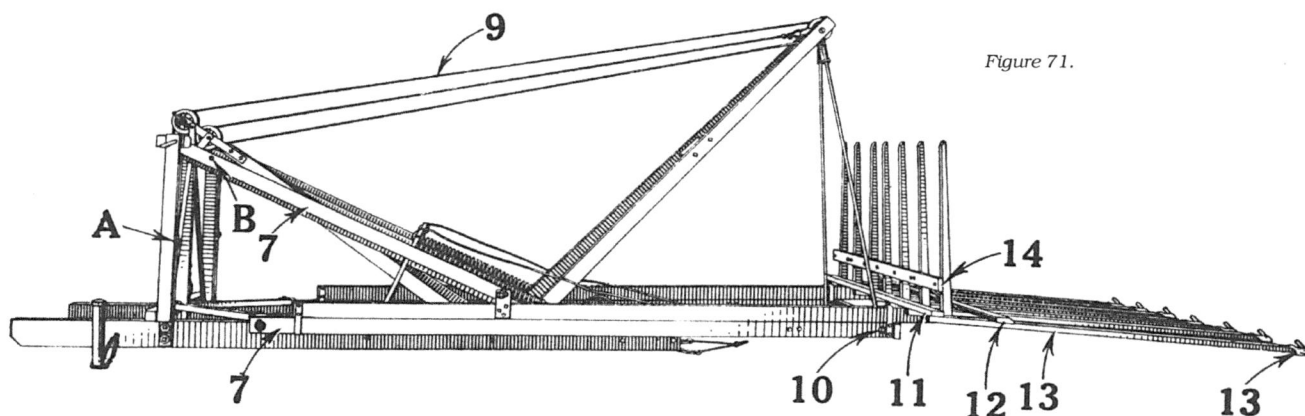

Figure 71.

5. Fig. 69. Assemble recoil spring assemblies and center sill, as shown.

6. Fig. 73. Assemble transport truck and stake chain, as shown.

7. Fig. 71. Place rear ends of lower hoisting arms on pivot castings, take off push bars and attach to hoisting arms and back frame, as shown. Hoisting arms are telescoped as far as possible for shipping. It will be necessary to slide upper hoisting arm out in

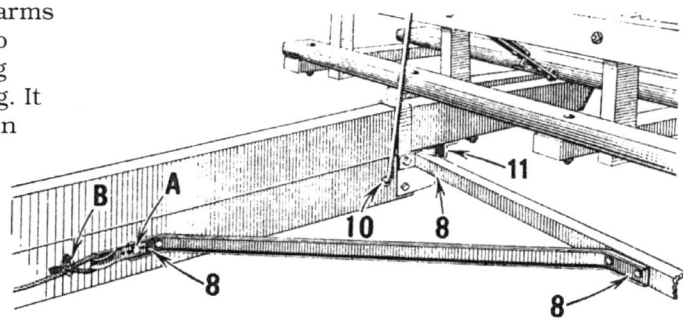

Figure 71A.

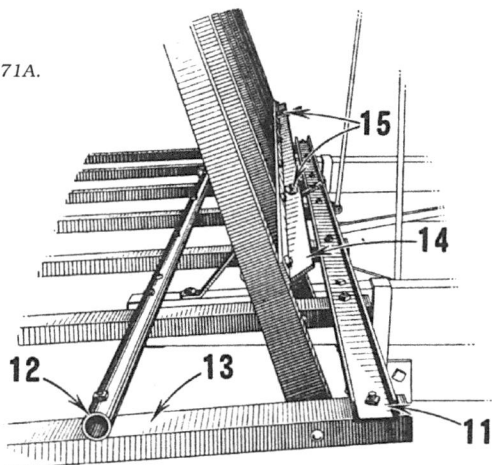

Figure 72.

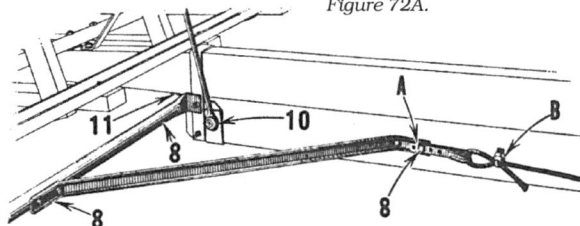

Figure 72A.

order to attach push bar to back frame.

8. Fig. 72 and Fig. 72-A. Bolt cross tie and channel brace to lower hoisting arm, as shown in Fig. 72 for No.1 and No. 2 stacker and Fig. 72-A for No. 2-A stacker. Be sure clip on end of stop cable is fastened, as shown.

9. Fig. 71 Thread cable through pulleys and attach, as shown.

10. Fig. 72 and 72-A. Insert lower end of hoisting rods in hoisting arms, as shown Fig. 72 for No. 1 and No. 2 stackers, and bolt lower end of hoisting rods to hoisting arms as shown Fig. 72-A for No. 2-A stacker. Raise hoisting arms by means of the cable until two or three feet above the ground, and hold in place with brake on swivel pulley. See Fig. 75.

11. Fig. 71 and 71-A. Bolt carrier bar to third hole from front end of upper hoisting arms with carriage bolts found in bag with stop angles on them. Put stop angles on the bolts below hoisting arms as shown Fig. 72 and Fig. 72-A.

12. Fig. 71 and Fig. 71-A. Bolt bevel bar on top of upper hoisting arms with bevel edge on top and toward front as shown Fig. 71 for No. 1 and No. 2 stackers, and bolt tooth pipe on upper hoisting arms as shown Fig. 71-A for No. 2-A stacker. Bolts found in hoisting arms.

13. Fig. 71 and Fig. 71-A. Bolt stacker teeth to bevel and carrier bars, with ½" x 4" carriage bolts for No. 1 and No. 2 stackers and ½" x 2 3/4" carriage bolts for No. 2-A stackers. Use washers between bolt head and tooth. Bolts and washers found in bag. Slab side of teeth must be up. Bolt tooth points to outer end of teeth.

14. Fig. 71 and Fig. 71-A. Bolt pitcher teeth to pitcher tooth bar with 3/8" x 4 3/4" carriage bolts for No. 1 and No. 2 stackers and 3/8" x 3 ½" carriage bolts for No. 2-A stackers. Bolts found in bag.

15. Fig. 74 and Fig. 71-A. Bolt adjusting straps to hoisting arms by means of eyebolts. Bolt adjusting rod eyebolt to pitcher tooth bar and connect strap and hook, as shown Fig. 74 for No.1 and No. 2 stackers, and bolt adjusting strap to hoisting arms and pitcher tooth angle as shown Fig. 71-A for No. 2-A stacker.

TO OPERATE

Set machine on level ground.

Use plenty of oil and grease on all working parts.

Under ordinary conditions, only one stake is required. For best results, this stake should be placed as shown in Fig. 71, by inserting a chain through small ring on ground sill. Provision is made for an additional stake at the front end. If front end stake is used, allow plenty of slack in stake chain, as shown in Fig. 73.

For high lift, set push bar as shown in Fig. 68.

For medium lift, insert pin through loop "A" on back frame and hole "B" on push bar. (See Fig. 71)

For low lift, detach push bar from stacker.

Pitcher tooth adjustments can be made at 15, Fig. 71-A and Fig. 74.

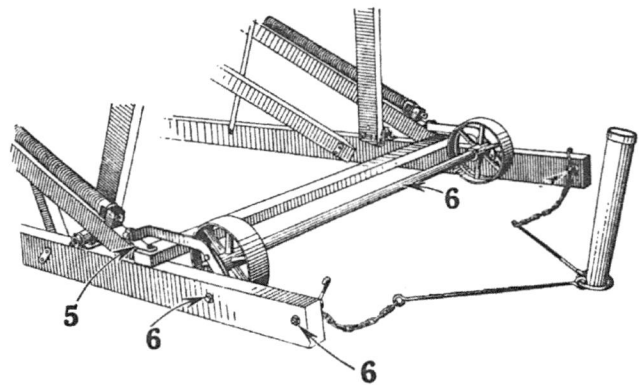

Figure 73.

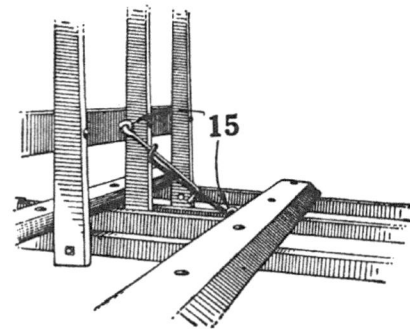

Figure 74.

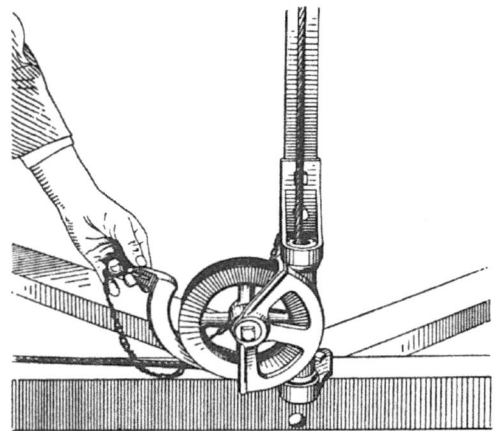

Figure 75.

The head may be held at any height for transporting stacker by inserting cable brake between pulley and cable, as shown in Fig. 75.

Transport trucks should be placed as shown in Fig. 73 when stacking hay. Swing foot levers over and attach to hooks on stake chain when transporting stacker.

Building a Singing Horse Ranch Stack

These photos were taken different seasons and different stackyards.

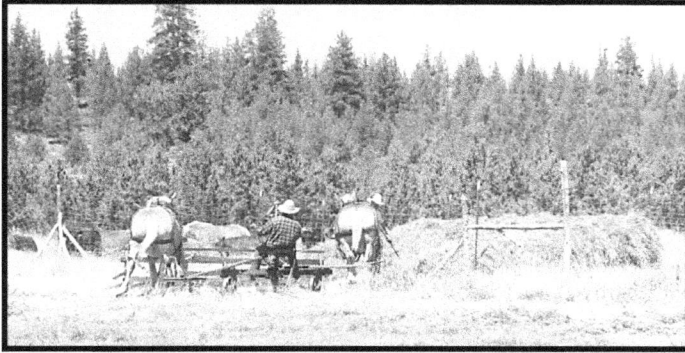

Figure 76. The author taking a buckrake load in to a new stack base..

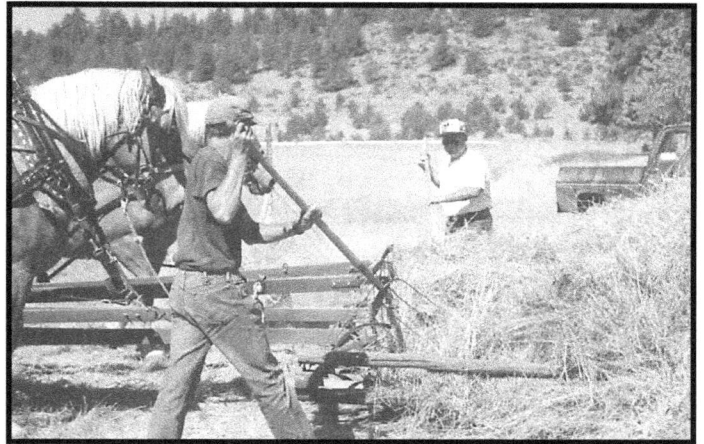

Figure 77. Justin Miller and Bill Gilman hold back a jag as the buckrake backs up.

Figure 78. Left to right Carl Leonhardy, Bill Gilman, and Tony Miller building the base walls.

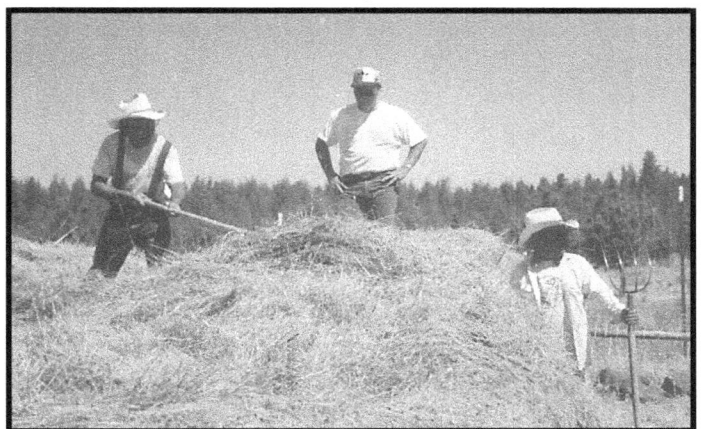

Figure 79. Same crew members from head on. The wall needs to be packed solid enough to hold up the stack crew's weight.

Figure 80. Stack begins to climb.

Figure 81. Buckrake has backed away after depositing a load on the swinging stacker. Notice that the stack walls are being kept vertical.

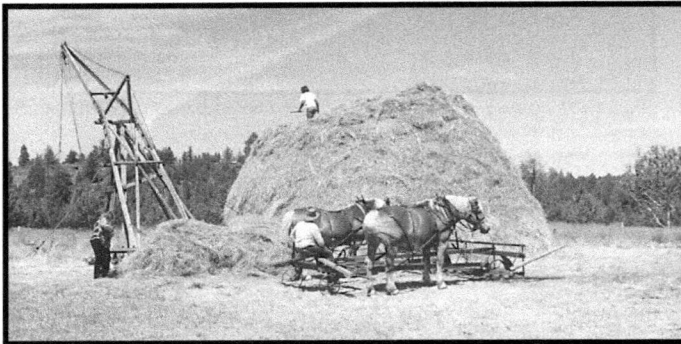

Figure 82. At this stage it takes more hay for every foot of height as the stack becomes more and more packed.

Figure 83. In this stack the walls have been kept straight up and down for an excellent height. This will translate to more hay kept clean than in the stack in figure 82.

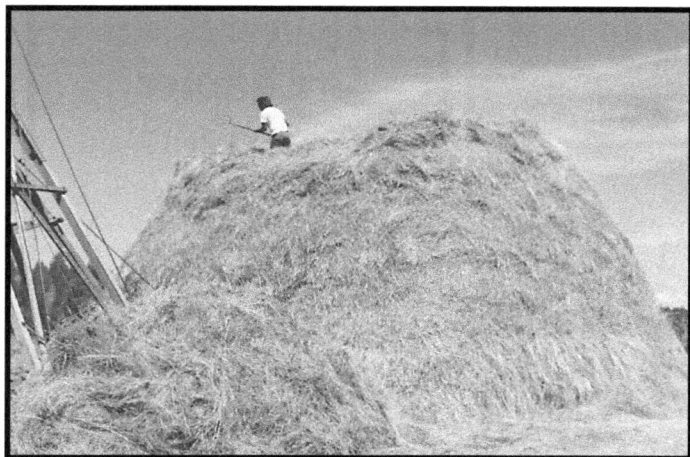

Figure 84. As the stack nears the top, it becomes important to pack the middle as well as the walls.

Figure 85. (top right) The team backs the buckrake away.

Figure 86. (right) This jag had insufficient weight to naturally trip and dump. Tony is pulling the teeth down with his fork.

Figure 87. (Below) At the top, as the cap loads are sent up, the hay has to be pulled off by fork as there is no longer height enough to dump from.

Figure 88. (bottom right) The finished haystack always looks like a part of the landscape, almost as if it was always there. We never tire of their beauty.

Chapter
Sixteen

Putting
Hay
in the
Barn

To fill a big old barn
plumb full of gorgeous
loose hay is to feel like
farming belongs to you.

There are two approaches to filling barns with loose hay which we wish to discuss in this chapter. One is to push it in with combination buckrake stackers, an admittedly rare approach. The other is to use track and trolley systems. A very common and still viable approach.

As can be seen on this page, the 'pushing in' of hay is very straightforward (pun intended). Whether the barn is open in the front (preferably with enclosed back to the weather) or with high big doors across one long side, the stacker takes the hay to the waiting gap or hole and folks on pitchforks mow it back in. Getting the hay to the barn is a relatively easy task, working inside on the forks can be very hard work.

The track and trolley systems involve either lifting hay up into a gable end door or driving the loaded wagon into the barn and directly under the ridge track.

In chapter fourteen we covered the hardware and procedural variables for getting hay off of wagons. This most notably included fork and sling design variables. That information is important to fill out the understanding of these barn filling schemes.

If the reader is considering the construction of a barn we encourage you to give more than passing thought to allowing the structure of the gable to be open enough to accomodate the hanging of a trolley track. It is difficult or impossible to retrofit a truss-built barn, which may have many supports crisscrossing the inside gable, to receive hay jags. At least allowing for the option in a new construction design will leave the option for loose hay systems in the future.

Figure 1. Open front (or side) barn. The 'Jayhawk' stacker takes the jag right up to the hole.

Figure 2. High door open. The teamster very carefully guides the Jayhawk in and dumps the load.

Figure 3. Another open side barn. Regular buckrakes bring hay in for the combination stacker (Jayhawk) to pick up. When the clutch for the cable drum is engaged the forward motion of the Jayhawk winds in the cable causing the stacker to rise. A certain number of feet of forward motion translates to a given height. When the desired height is reached the teamster can lock the drum to hold the load while he proceeeds forward. If the lifting is started too late the load has to be backed up with the drum locked to allow for more 'wind'.

Figure 4. A shed roof barn especially constructed, with high side doors, to be filled by Jayhawk stacker.

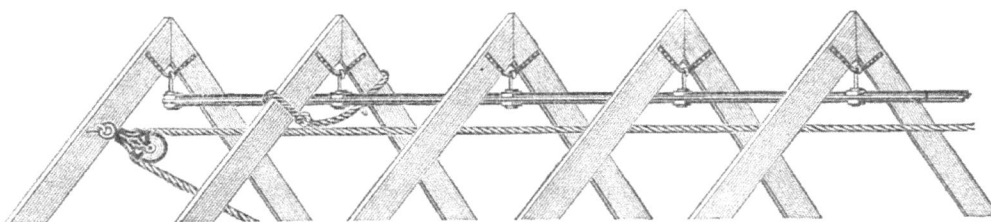

Figure 5. This diagram shows the barn rafters, as they would appear under the roofing, with a trolley track installed. This is a steel track hung from metal ties (seen below). Below the track and running parallel is the haulback rope and pulley. On the second rafter set from the left there is a rope tied in to function as a brake for the returning trolley.

Figure 6. A track hanging strap or metal rafter tie.

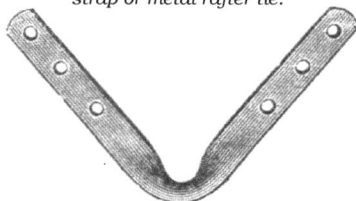

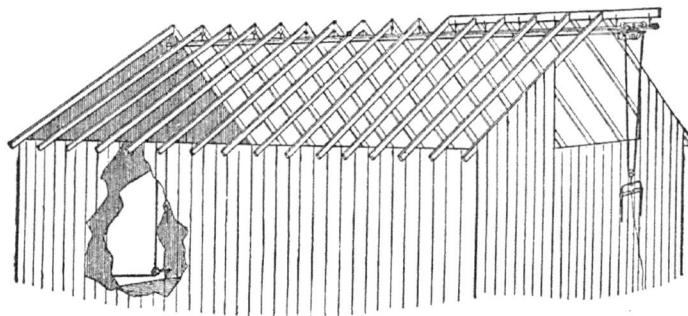

Figure 7. This cutaway of a simple barn shows the path of the haulback or hay rope. Also evident is the ridge extension to carry the trolley on outside the barn and over the waiting wagons.

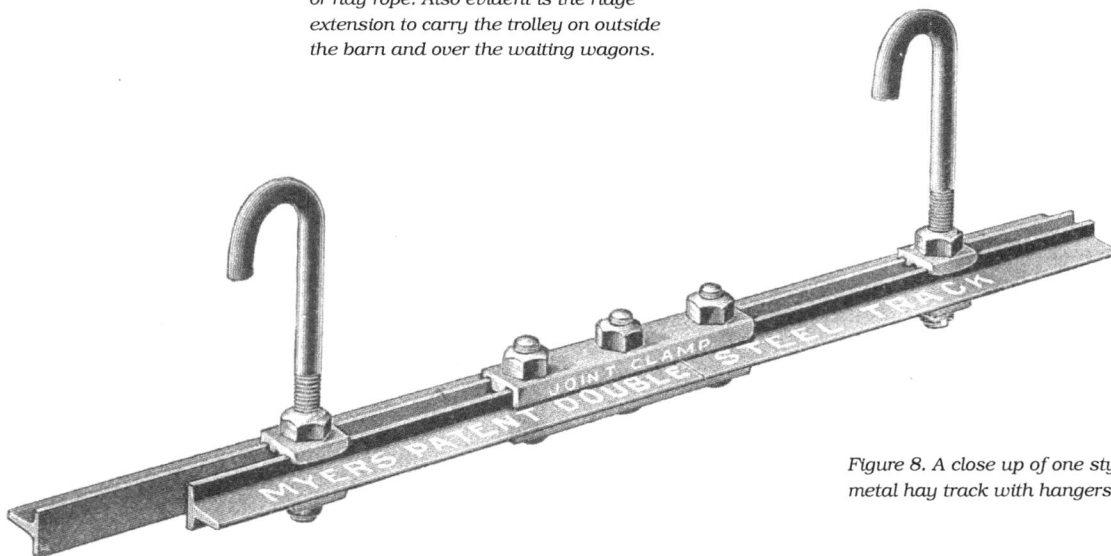

Figure 8. A close up of one style of metal hay track with hangers.

In the track and trolley systems, a steel or wood beam track is hung, as diagramed, up under the ridge of the barn. The notable exception to the placement of the track, parallel to the ridgeline, as in round barns as shown on page 278 . Although metal ties or track hangers can be used to tie between rafters and carry the track, most barns feature 1" x 4" boards nailed in place (as high rafter ties) and featuring drilled holes to receive the hooked bolt hangers. Though the simplest and most direct design features hay mow door(s) high up in the gable end(s) which required wagons to pull up along side the end of the barn, some barns featured either center alleyways or ramped wagon entries inside the barn.

Figure 9. A narrow-style barn with an extra big hay mow door receives a sling load of hay off the parked wagon.

Figure 10. This barn cutaway shows the hay rope travel as the haul back horse pulls up a double harpoon fork load off the attended wagon. The man on the wagon is holding a light trip rope. When the trolley is where the barn man wants the load dumped he'll holler out and the man on the wagon will yank the rope, tripping the load. In some cases the trip rope is used to pull the emptied trolley back out to the wagon.

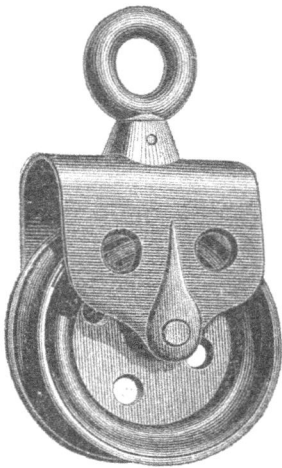

Figure 11. A simple, low-end, metal hay rope pulley.

Figure 12. Another barn cutaway, this time showing an elaborate rope and trolley system, showing that hay can go in either direction.

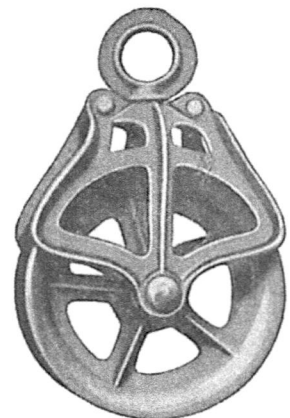

Figure 13. A cast iron hay rope pulley with swivel eye.

Figure 14. Again showing the rigging for a center alley hay trolley. To go to the other half of the barn the pulleys would be moved to the right side and the rope threaded in that direction.

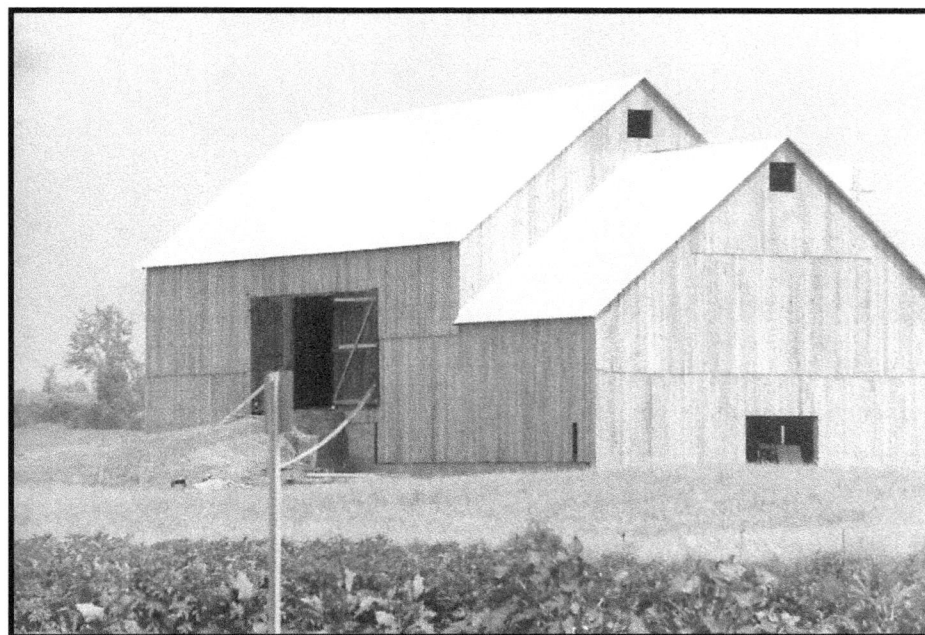

Figure 15. A simple wood hay rope pulley.

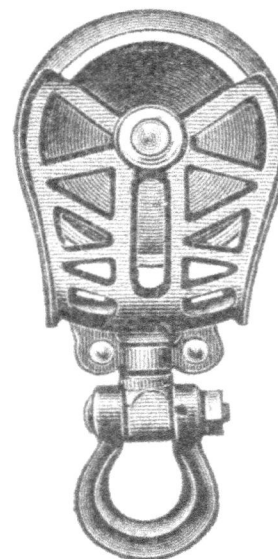

Figure 16. A hay rope snatch block with double swivel allowing opening to receive a loop rather than threaded rope.

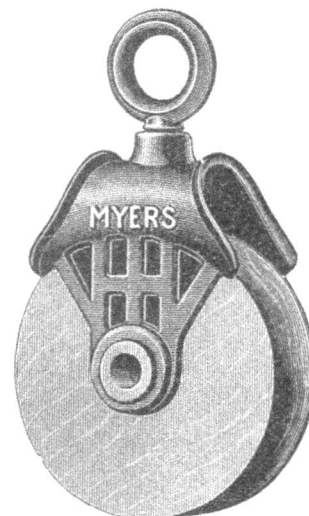

Figure 17. A new Amish barn in Heuvelton, New York featuring a ramped side entrance for the hay wagons.

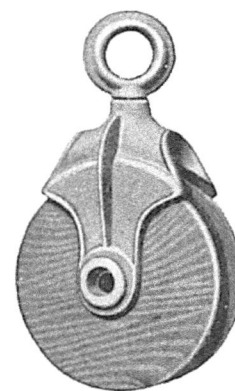

Figure 18. A Myers brand wooden hay rope pulley.

Figure 19. This cut and figure 20 below show the same basic trolley
and track systems but one is steel and one wooden beam.

Figure 20.

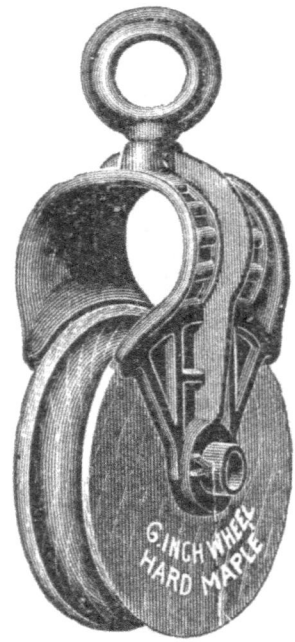

Figure 21. A wide throat
large wood hay pulley.

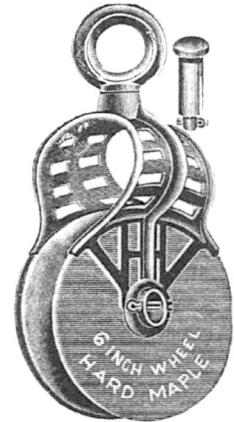

Figure 22. A wide throat large
wood hay rope pulley with a
quick release pin.

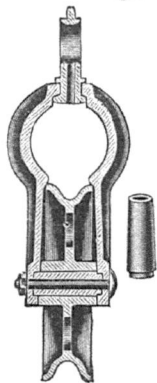

Figure 24. A unique style of
heavy duty track hanger.

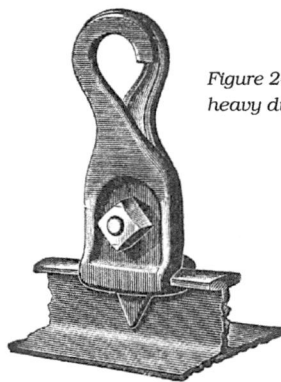

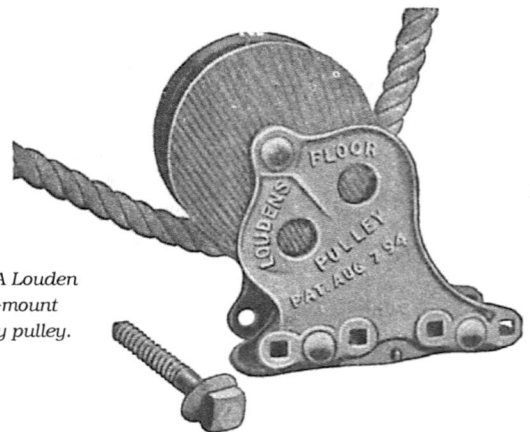

Figure 25. A Louden
brand floor-mount
wooden hay pulley.

Figure 23. A cutaway of a cast
iron hay rope pulley.

Figure 26. Yet another barn cutaway, this time with track installed in a timber frame.

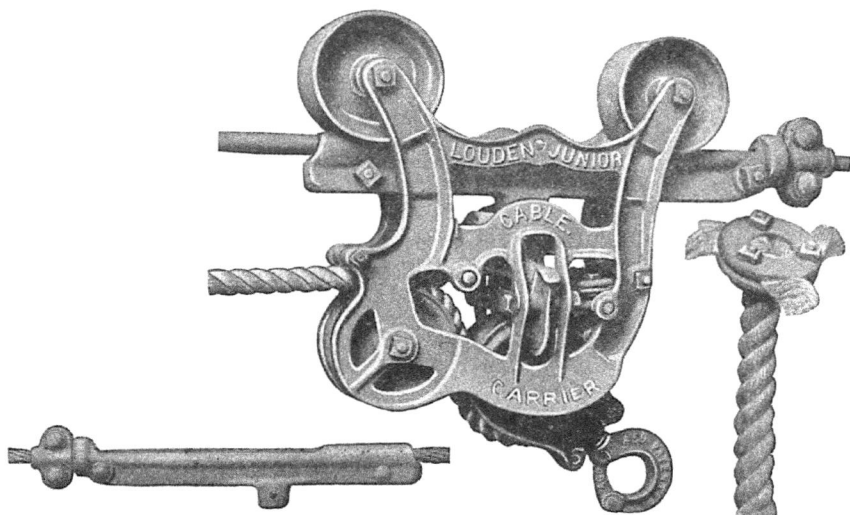

Figure 27. A Louden "Junior" cable trolley setup as used in outside cable stacking systems. Note the difference in design between this and those on pages 274, 276, & 277.

Figure 28. A slightly more modern gambrel-roofed barn with center alley, set up with track and trolley like the older styles on previous pages.

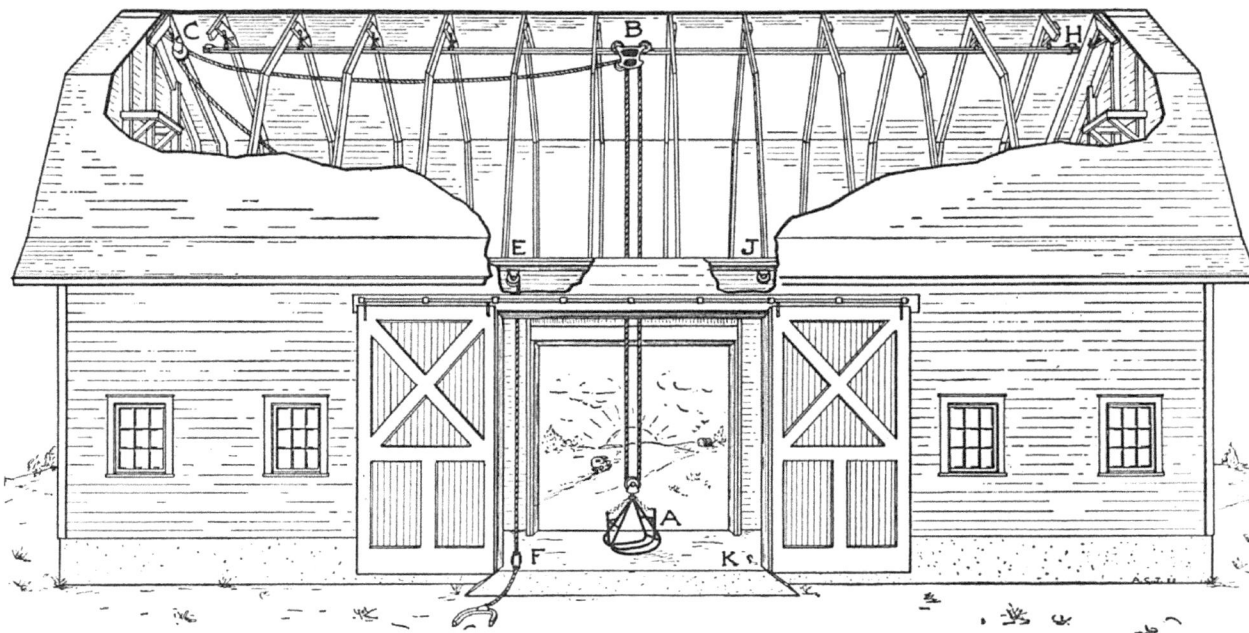

Hay in the Barn

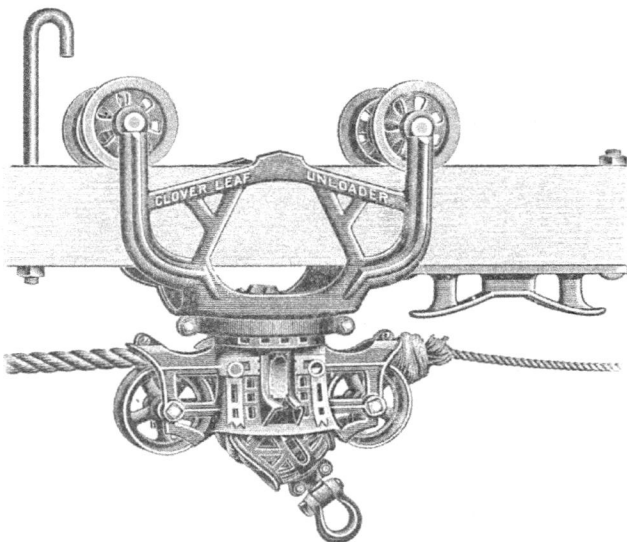

Figure 29. A small one story barn with a center alley and grapple fork setup.

Figure 30. A Louden Junior Hay trolley on a wood beam track.

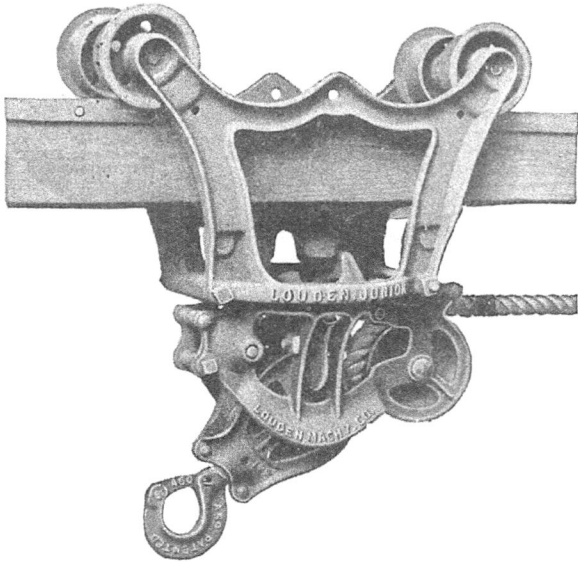

Figure 32. Another style of wood beam track trolley. Literally millions of trolleys were manufactured by dozens of companies during the period between 1910 and 1925

Figure 31. A barn set up with both gable end and center alley systems.

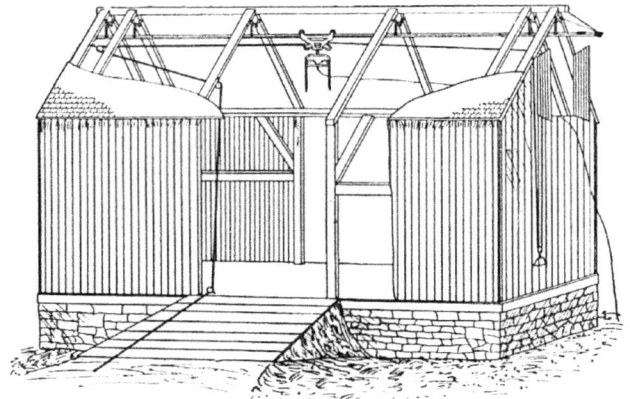

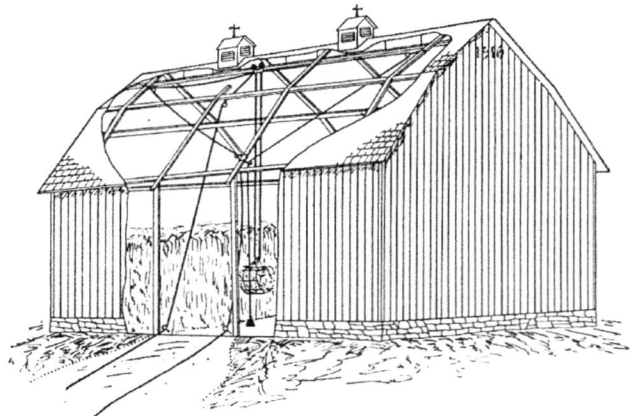

Figure 33. In this system weights on secondary ropes provide the power to take the loaded forks to the side of choice.

Figure 34. The basic track and trolley system in place on a classic midwestern dairy barn.

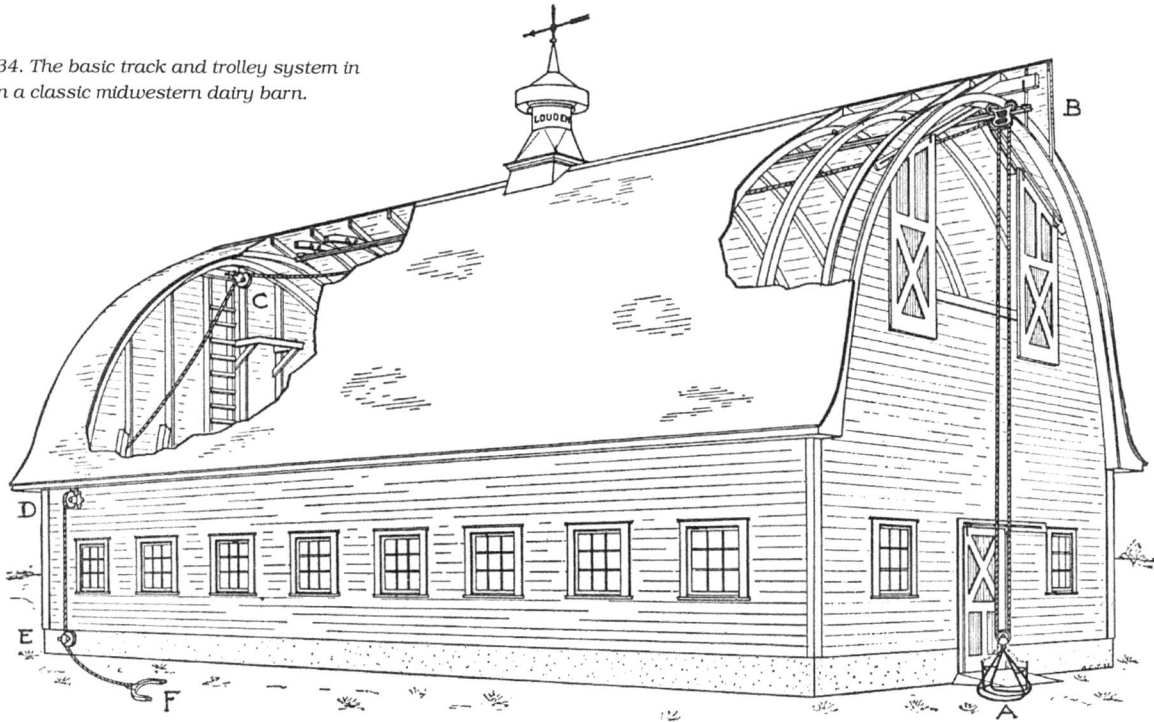

Figure 35. A Myers steel track and hanger system

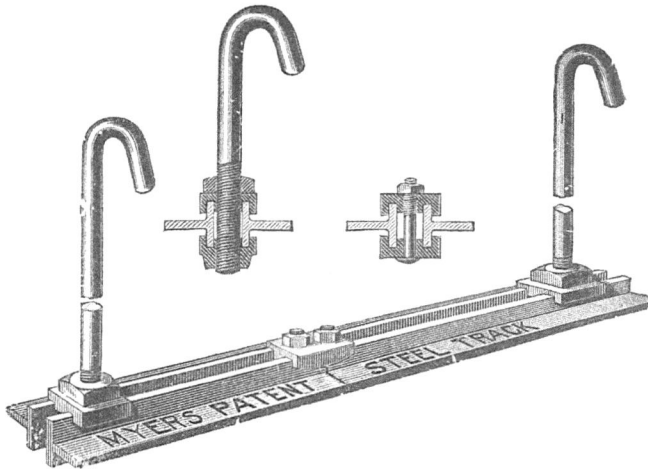

Figure 36. In this barn set up, hay is brought in from both gable ends. The haul back rope is set at center.

Figure 37. *Showing various ways to mount hay rope pulleys in barn corners. Pulleys should always be set so they will stand straight with the line of draft, as shown by the dotted lines. When the pull is crossways it will bend the hook. From the left the first, third and fifth are right, second, fourth and sixth are wrong.*

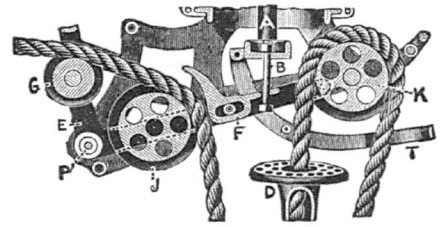

Figure 38. *A sectional view showing the locking mechanism in the Louden built Iowa Sling carrier. Also shows threading of rope.*

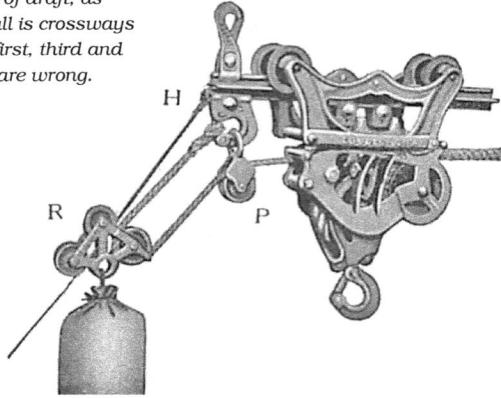

Figure 39. *A weight return for returning the carrier to the trip block after the load has been deposited in the mow.*

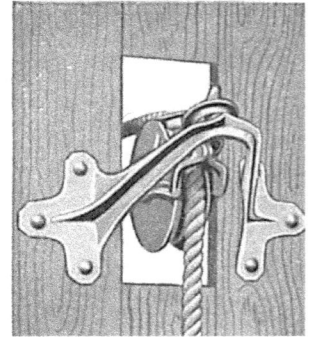

Figure 40. *A nifty Louden Bracket Pulley Holder.*

Figure 41. *The Louden built Iowa Sling carrier.*

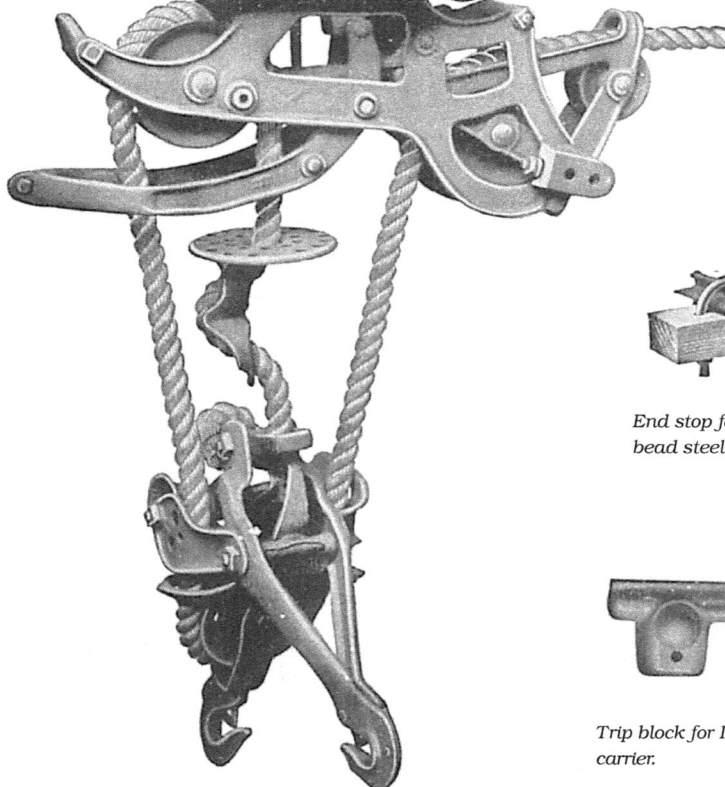

Adjustable trip.

End stop for double bead steel track.

Trip block for Iowa sling carrier.

Figure 43. An economy model steel beam trolley.

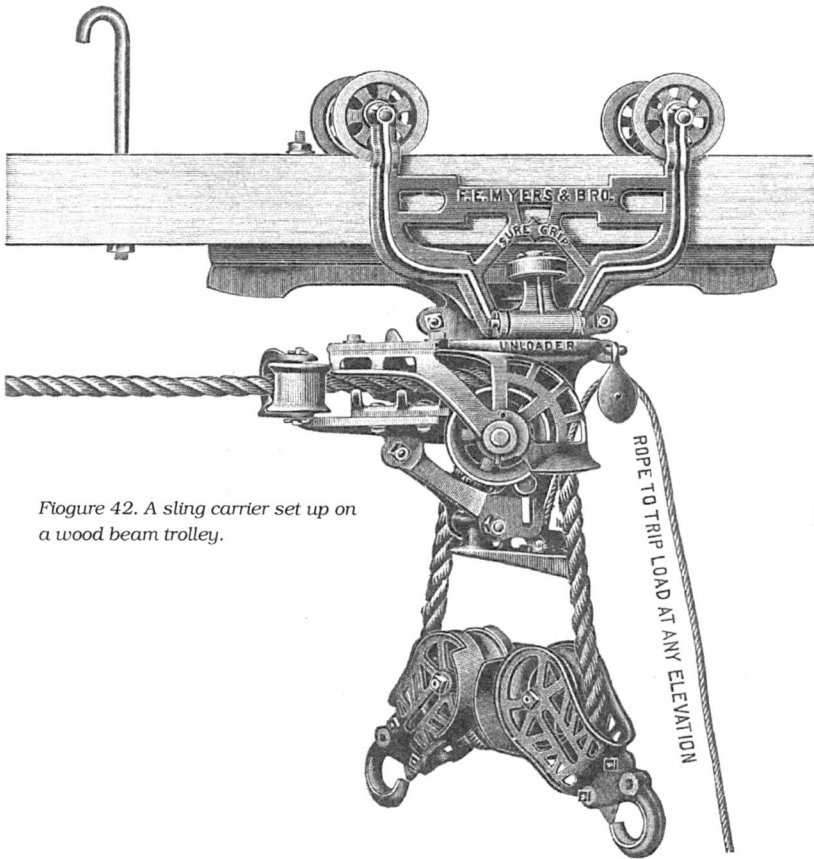

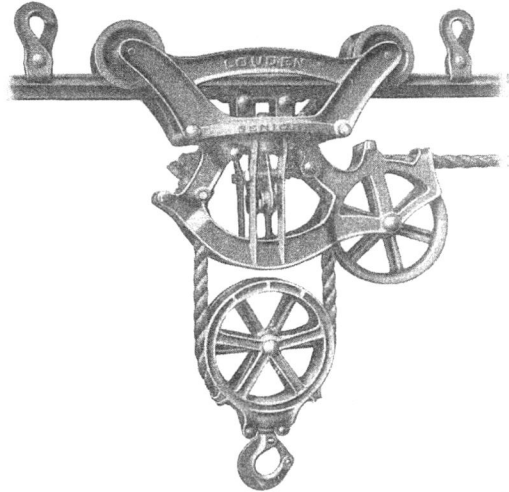

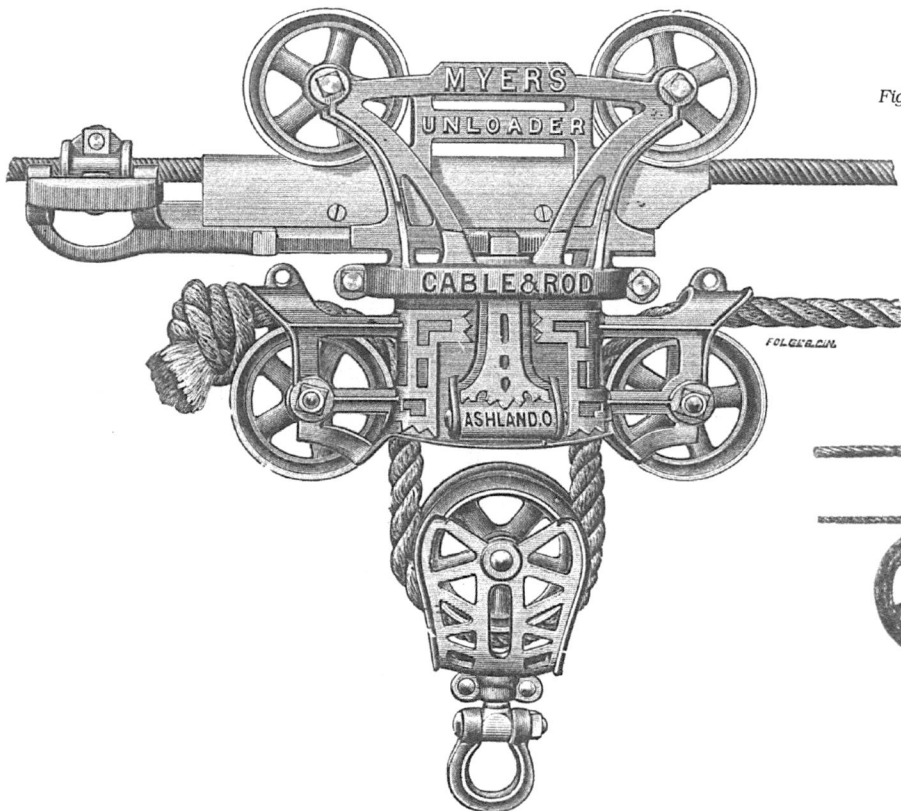

Fiogure 42. A sling carrier set up on a wood beam trolley.

ROPE TO TRIP LOAD AT ANY ELEVATION

Figure 44. A Myers cable and rod trolley.

Figure 45. A smaller, lighter weight Myers cable and rod trolley for outside cable stacking systems.

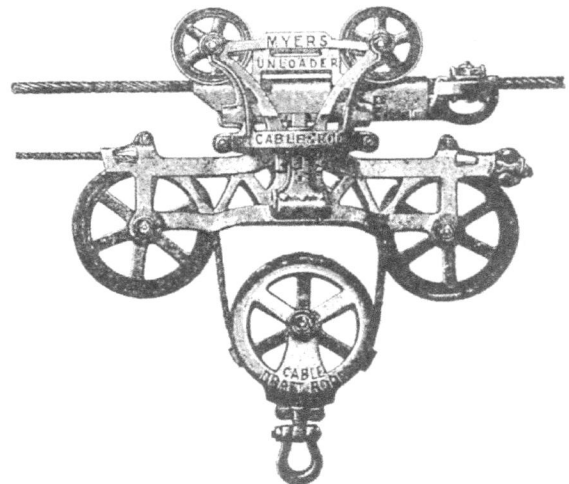

277

Figure 46. A round barn trolley set up.

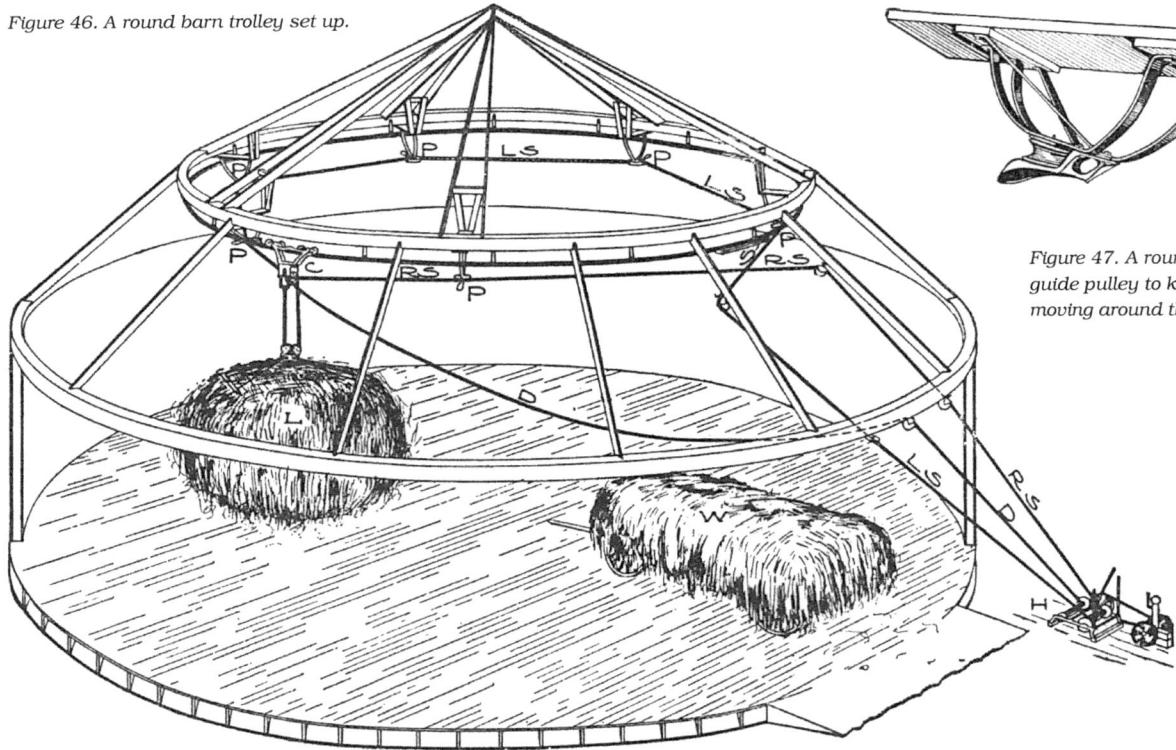

Figure 47. A round barn guide pulley to keep rope moving around the circle.

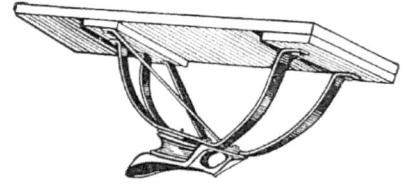

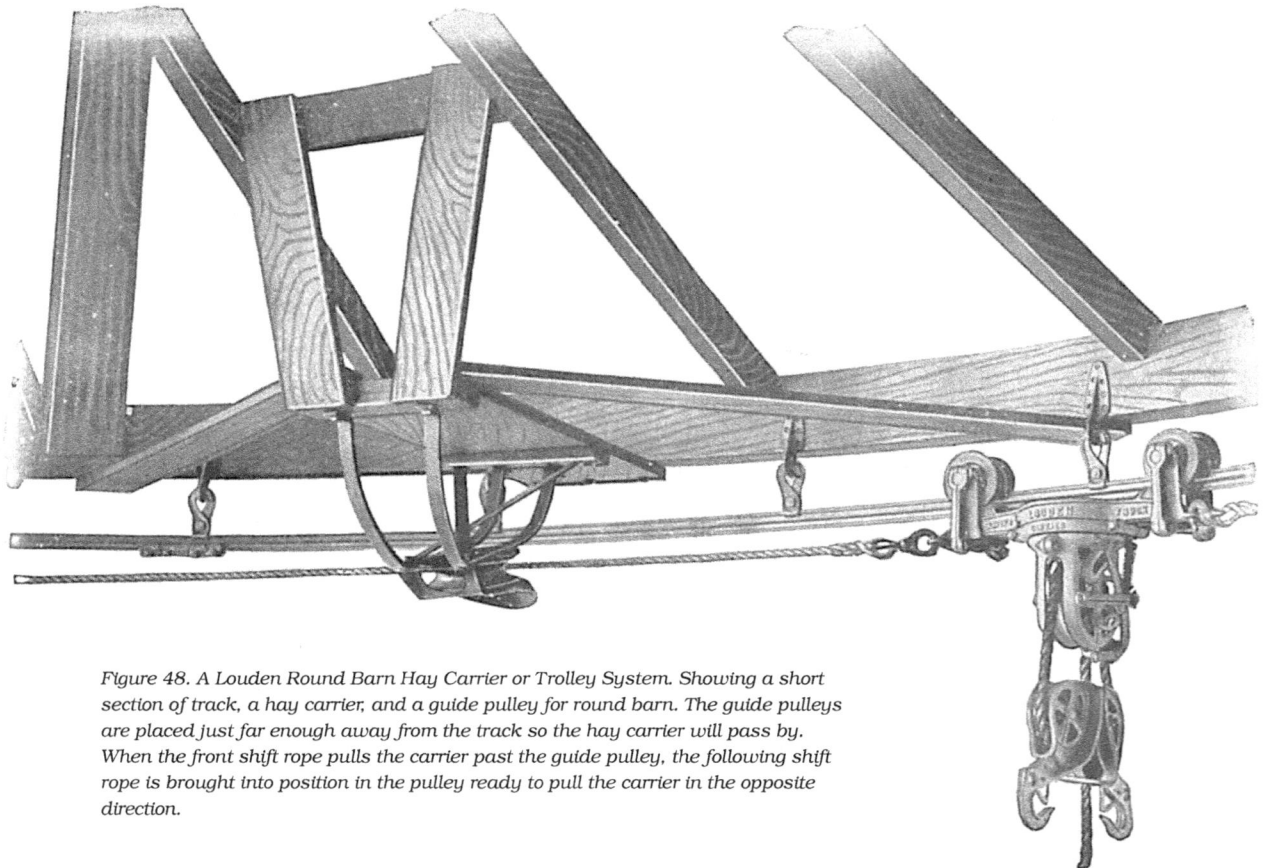

Figure 48. A Louden Round Barn Hay Carrier or Trolley System. Showing a short section of track, a hay carrier, and a guide pulley for round barn. The guide pulleys are placed just far enough away from the track so the hay carrier will pass by. When the front shift rope pulls the carrier past the guide pulley, the following shift rope is brought into position in the pulley ready to pull the carrier in the opposite direction.

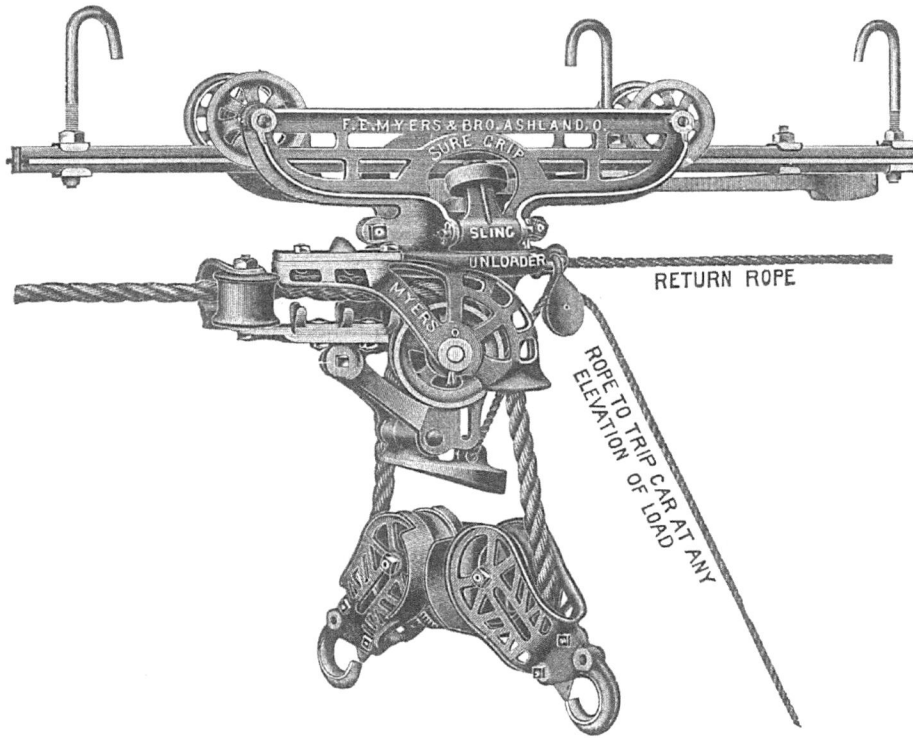

Figure 49. Another steel track trolley with sling carrier.

RETURN ROPE

ROPE TO TRIP CAR AT ANY ELEVATION OF LOAD

Figure 50. A homemade framework to position into barn stored hay when added circulation is thought to be needed. Sometimes green hay goes in to the barn mow. To avoid spontaneous combustion one farmer thought to build these frames and insert them as the hay went up to allow greater circulation.

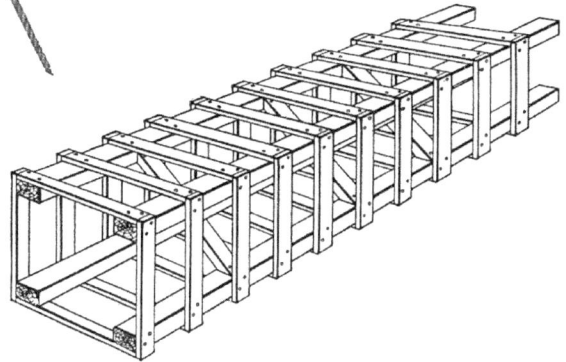

Figure 51. homemade ventilators in place in a hay mow.

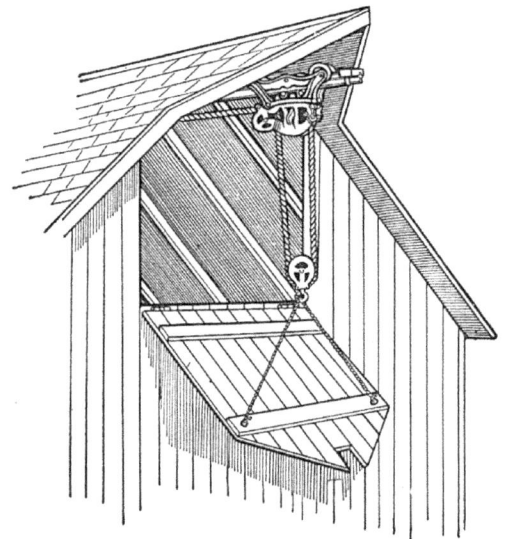

Figure 52. Using the trolley to raise and close the mow door.

279

Chapter Seventeen

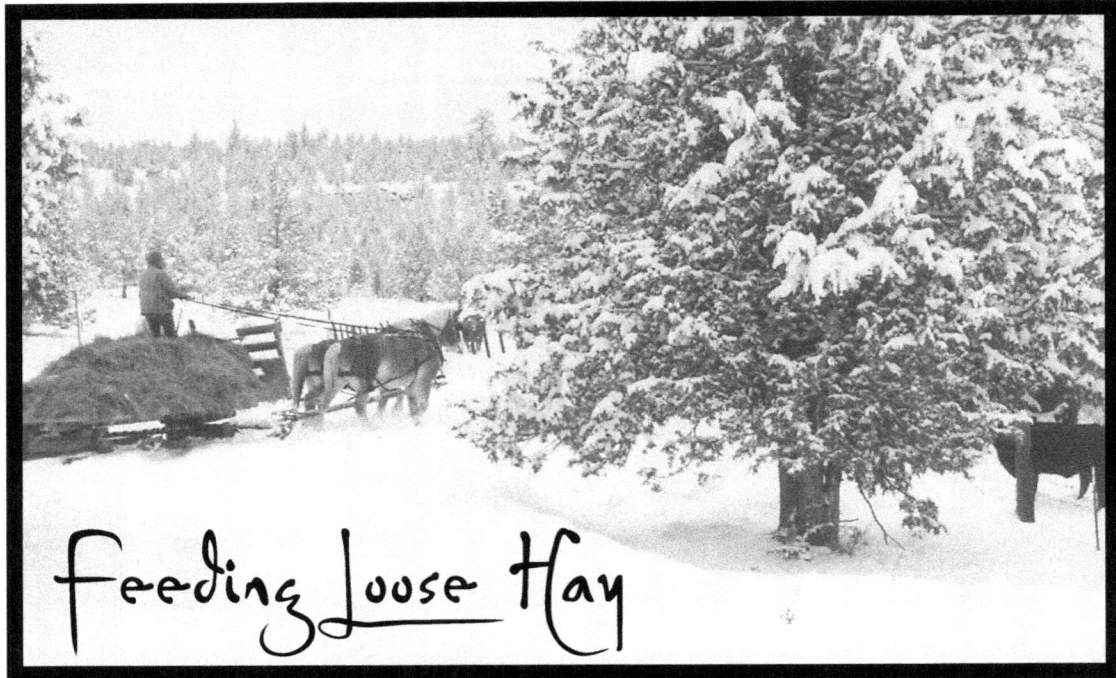

Feeding Loose Hay

Figures 1 & 2. Two views of winter feeding chores at Singing Horse Ranch with Belgian mares and bobsled.

This chapter is specifically about loose hay feeding with only a passing mention of bales. The small conventional bales are simple to feed. If your livestock are not located in or near the hay storage or barn, you stack them on the wagon or bobsled and take them to wherever the feeding occurs. With large bales either mechanical assistance or cleverness needs to come into play to load the bales. The end of this chapter shows a commercially available device for moving round bales with a team.

With loose hay feeding we divide the information into four categories; fed in barns, fed by hand out on the land, fed at the stack, and fed with mechanical assistance. These categories are far too simplistic but the ideas we are about to share should lead the readers to their own best solutions.

Figure 3. Feeding bales on Jack Walthers ranch in Nevada

Figure 4. Doug Hammill's Clydes with a Montana load of bales.

Fed in the barn.

If the livestock are housed in winter (in the barn where the hay is stored) feeding may be a simple matter of forking hay in to waiting mangers. If the hay is stored upstairs, above the animals, it will need to be dropped down through some opening. The smarter approach is to be certain that the "hay hole" in the loft floor is a large one (say at least 4 foot by 4 foot) and located in such a place (for example over the center alley) so that the hay when it falls does not land on livestock or manure. Ideally it should be possible to park some sort of conveyance (i.e. wagon or cart) directly under the hay hole to facilitate taking some hay away from the barn storage and to distant livestock.

An aside: a large well-built barn with an upstairs loft full of hay may be, in effect, too well insulated. If there is not "proper" and adequate ventilation and animals are housed below, this barn may become a very unhealthy environment in winter. Make sure that there is lots of air circulation or the animals may get quite ill.

Fed by hand outside.

Spreading loose hay out on a snowy landscape or dry pasture or filling outside feed bunks necessitates the use of a conveyance (wagon, truck, sled, slip etc.) and some way to get the hay loaded. All through this book we've referred to our own experiences and experiments at Singing Horse Ranch. For some of you it is important to know the scale of our operation to determine if the specifics might fit for you. We are a small ranch in the mountains and usually have winter snow and freezing. For these last ten years we've averaged a hundred animal units to feed. Usually we separate the horse herd (anywhere from 25 to 40 head) from the cow herd (from 40 to 75). With some animals in training, hospital pens, lockup, etc., we need to move around some when feeding. Most of the horses and cattle are fed out in the winter pastures. Our feed conveyances have been trucks, wagons, and a big bobsled with feedrack. We lay out our stackyards so

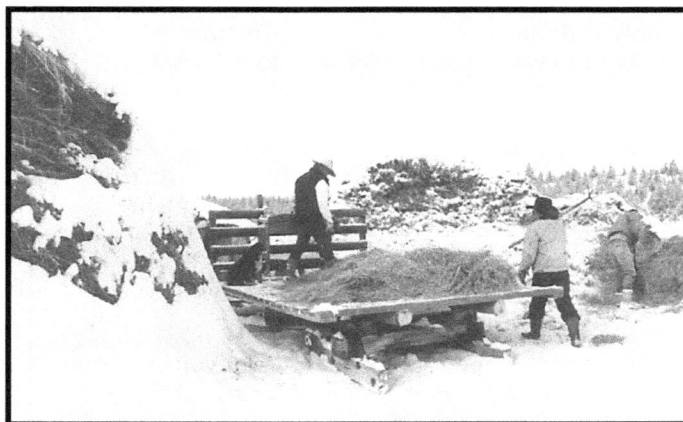

Figure 5. Lynn, Tony and Justin Miller loading hay on to the Singing Horse bobsled.

Figure 6. Feeding with team and sled in new snow can be pure joy.

that a bobsled can be driven into the yard and in between the stacks and out the other side.

With large stacks such as ours we will always be taking a portion of the stack to feed, anywhere from one third ton to a ton. Whether it is from a barn loft or outside stack one thing is constant; it is always easier to pull hay off the top than off the side. We use a heavy duty potato fork (bent tines) to pull the hay off the top in big pancake flakes of considerable weight. After the "cap" rind has been removed, it is possible for one man to "pull" off (and onto the parked wagon or sled) a ton of hay in a short time. The load is then taken to the animals and fed out.

Two men, working with pulling forks from the top of the stack, can load up to 3 tons of hay on wagon or sled in a short time.

Fed at the stack

Though we have no experience with this approach, some outfits set up feeders or feedlots around, up against, or adjacent to stacks and stackyards. Tony always wanted to try such an approach but we couldn't figure out how to prevent the ground from becoming sloppy from the livestock trampling and manure and thereby ruining easy access to the stacks with trucks, wagons or sleds. Certainly with smaller stacks, where the time will be somewhat limited, allowing livestock to self-feed would be the laziest approach. But having seen how much damage a herd of cattle can do to a hay stack in one short day after the bull broke the fence, we are certain that self-feeding will not be the most efficient approach - unless some sort of moving feeder fence were employed.

Fed with mechanical assistance

In Montana, Wyoming, Colorado and other mountain states some ranch operations involved with loose hay use a hydraulically operated arm with clasping forks - or hydra fork - mounted on sleds, wagons or trucks. This apparatus is a smaller cousin to a loader on a self-loading log truck. The operator

sits at its base and works the controls, grabbing and lifting jags of hay off the stack and then loading it on the feeding conveyance. These are expensive units and require someone on the controls who is accustomed to working around heavy equipment (such as backhoes or cranes). They do allow that large amounts of hay can be loaded quickly. But it does need to be powered by a motor, which at zero degrees may be difficult to get started. (Healthy horses always start.)

Some outfits use tractors which are equipped with front end loaders that feature forks (i.e. stackhands) but these can be difficult with larger stacks.

Some cable stackers can have their operation reversed thereby using a set of grapple forks and the trolley to lift hay.

Not quite in the category of mechanical assistance is the practise of dragging great chunks of hay off the stacks in the loop of cables or chains as is demonstrated in the Pete Lorenzo sidebar which follows. This same method has been used to move entire stacks.

In these practises it is helpful if the sled or wagon bed can be as low to the ground as possible.

One last mention in this category is the use of a combo mechanical loader with a feed wagon set up with a conveyor bed to mechanically spread the hay on the ground or into feed bunks. The wagon bed can be ground drive (ala manure spreaders) or motorized.

Figure 7. Jean Christophe Grosettete of France drives Polly and Anna while the author feeds the cows. The muddy conditions had them switch from bobsled to wagon.

To Keep Your Horses and Your Hands Happy When Its Cold

When the temperature goes below freezing and stays there day after day think about taking the feed team bridles and the lines into the house or a warm room after you're through feeding. The horses will appreciate the bits not sticking to their tongues and your hands will appreciate pliable leather lines.

Nebraska Stack Feeding
by Pete Lorenzo

With stacks ... can you leave them where they are made or do you have to yard them? In Nebraska on most ranches they are yarded and the cattle are fed on feed ground outside the meadows. I know in Wyoming some stacks are made long, maybe 60'-100' long and fenced out. ... I usually had to feed at least a stack a day. Sometimes more depending on number of cattle. And we would cable the hay on a low "sled". We moved our stacks with a stack mover machine pulled by a Ford 7000 tractor. I'm glad I never had to run it. "It's boring." Back when it was all horse power, they used to use a big timber, I think about a 12' by 20"-24" beam, long as needed to get behind a stack. And a big hitch of at least 6 head on each end, and would skid the stacks to yard them. They probably had more smaller yards to fence if needed, though and more gates to get to the yards in the meadow. Another thing with more yards is less chance of lightning burning up winter feed before lightning season ends. We usually tried to wait for lightning season to end.

Figure 10. On feed ground, Lightning Valley.

Figure 11. You can see some of my sled here, how long it is.

Figure 8. The author's feed team, (L-R) Homer-colt, Jin-swing, Rum & Silver-tongue, Duke-swing, Jethrow-colt. End of calving, warm weather at the quarter camp, Aurther, Nebraska.

Figure 12. About 2/3 of a stack loaded on sled by cabling.

Figure 9. You can see 5th wheel front end of sled. I liked 5th wheel best, short turns. Note forks, pitch & drag jack.

Figure 13. Calving time, snow is gone pre-pasture no grass yet. Before turn out to summer range and branding.

Feeding Loose Hay

Figure 14. Feeding my cow bunch over 250 cows in Lighting Valley. Over 4 mile ride from camp I live at. Hooked up in open all winter.

Tools needed
Some of my feed ground was over 4 miles horse back from camp. No barns.
2 Pitch forks
2 Drag Forks (to pull off hay) 6' heavy handled with bent over tines heavier then pitch fork
1 handy man jack
1 hammer
couple chain link repairs
1 cabling chain 5/16" - 3/8" - 40' high test with grab hook on sled side only
1 ring and big slip hook. Keep hook on cart.

Just a Note

I hooked up all winter, before calving time at my camp, in the open. I kept my harness on my sled corned to a stack yard. I jingled in my horses everyday to the corner and a rope for a corral. Sun, snow, 60 below, cows had to be fed.

One time we were getting a ground blizzard. Foreman called and said not to feed early. It broke some around 10 a.m. so I took off on one of my older horses, to be safer on the ride. Mist onset of a blow out and fell in. I had to get off. I flounder my horse out. When I got to the feed ground I could just see some of the cows. Horses close to sled humped up. Bad part the six horses had to face wind to harness. Me and horses got cows fed fast. Never had any drift off feed ground. Blizzard started again about time I got harness off the team. Broke some ice, and started for home heading south which helped. Only thing you couldn't hardly see. I was heading for a gate north of my camp. Came to a fence, it didn't look quite like it should. I thought a minute to get my bearings then gave my horse his head and went home. I did that on another ranch I worked too.

This is the type sled I used to feed hay. 5th wheel type. Others were auto steer. But mine could turn tighter.

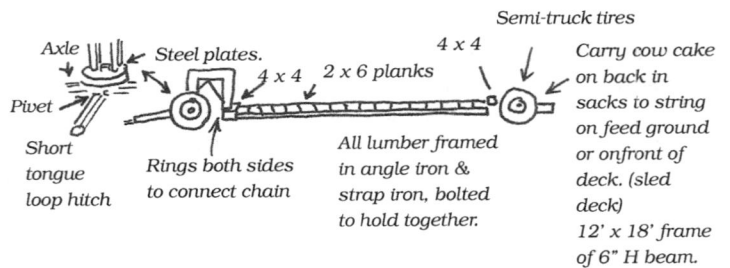

Axle
Pivet
Short tongue loop hitch
Steel plates.
Rings both sides to connect chain
4 x 4
2 x 6 planks
All lumber framed in angle iron & strap iron, bolted to hold together.
4 x 4
Semi-truck tires
Carry cow cake on back in sacks to string on feed ground or onfront of deck. (sled deck)
12' x 18' frame of 6" H beam.

Rig front for forks, jack, etc.

Welded chain rings each side.

Note: These are 7-8 ton stacks we're feeding.

2 x 6 planks to hold deck on, it really sets loose on the H beams.

1. To cable top
Close to stack

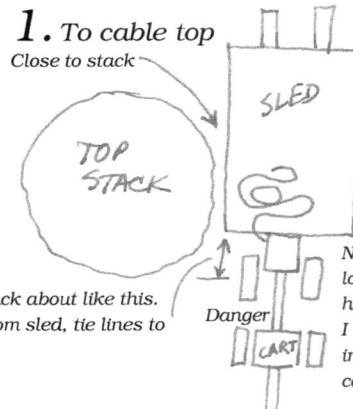

TOP STACK
SLED
Come up to stack about like this. Unhook cart from sled, tie lines to sled
Danger
CART

Note: Our lines are long enough to drive horses from sled. I will mark (Danger) in places to be extra careful.

2. Set chain first pull

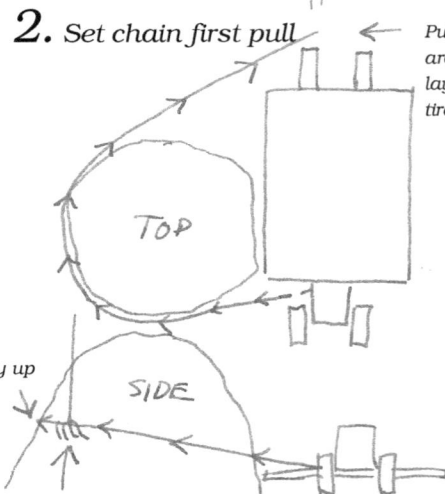

Pull chain around tight, lay behind tires.

TOP

About 2/3 down
If chain won't stay up use pitch forks

SIDE

3. With chain around, drive and back team close to chain and sled. Put on ring and hook. Hook on chain.

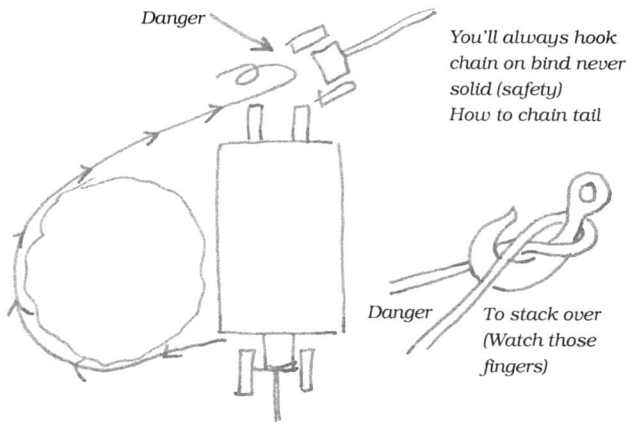

Danger

You'll always hook chain on bind never solid (safety)
How to chain tail

Danger

To stack over (Watch those fingers)

To pull any cabling
When you hook chain pull snug to hold bind. I always talked my horses (up a step) to take out slack, before the pull. I always worked the chain slow so horses wouldn't start to out guess you. Don't let them want to hit the chain.

1st pull should end like this, stop team. Unhook chain.

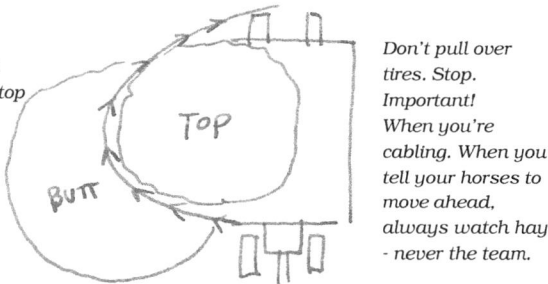

TOP

BUTT

Don't pull over tires. Stop. Important! When you're cabling. When you tell your horses to move ahead, always watch hay - never the team.

Note: Most of the time I was loaded and on the feed ground within 20 minutes after catching horses. Loading doesn't take long once you get the pattern, and your motion.

One year we were just getting into calving good and got an Easter blizzard. Snowed in deep for a week all I could feed was tops.

In a way it's mostly good work. You'll always have trouble sometimes. The coldest times was always the ride to feed and then home.

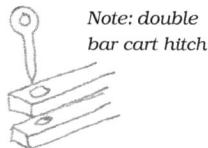

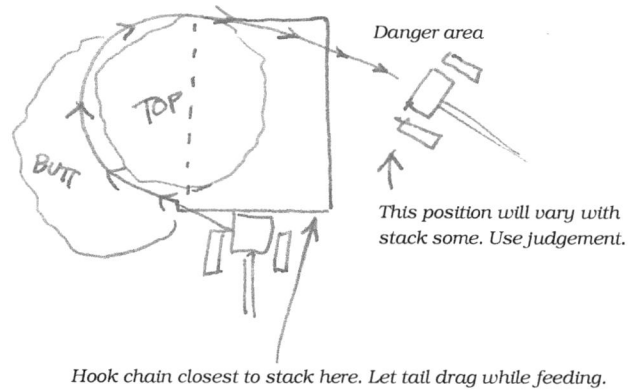

Note: double bar cart hitch

Danger area

TOP

BUTT

This position will vary with stack some. Use judgement.

Hook chain closest to stack here. Let tail drag while feeding.

4. Second Pull to Finish Load.
Put chain in this position, about 2-3 feet from sled. Move team to chain. Hook chain the same as before. Pull chain up, let tail of chain drag, and finish putting on load. Remember watch the hay. Sled should be almost, level. Don't forget your pitch fork if used for chain.

5. Loading Butt.

BUTT

Make sure you don't ride over hay. Leave some clearance.

The loading is basically the same other than setting chain, which is a kind of weave over and under.

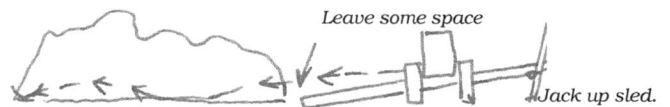

Leave some space

Jack up sled.

You might have to use your pitch fork some. This is important. You need to have a good size butt to load or it will be harder to load. Chain will either cut under or slide over. This just takes practice. They say when you load a butt, if you only have about a square bail of hay to pitch on, you made a good load.

Figure 15. A good old homemade feed wagon from 1925.

Figure 16. In this old photo the farmer is pushing a 1/2" steel rod through the hay stack like a needle. This will be pulled through with a cable or chain fastened on the end.The horses are hooked to the two ends and pull the portion of the stack on to a waiting slip, low wagon or sled.

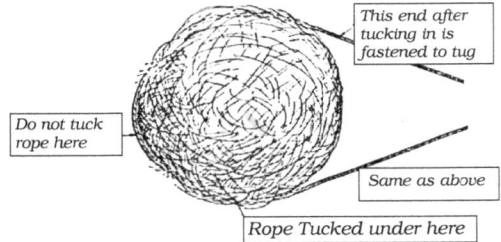

This end after tucking in is fastened to tug

Do not tuck rope here

Same as above

Rope Tucked under here

Below, a long pointed pole is laid up against a small stack. A rope or chain with a ring on one end and a hook in the other goes from pole end, around the stack, around the pole end and to the single tree.

Figure 17. There is no end to what ingenuity and need will form when it comes moving mounds of hay.

One Round Bale Mover

To load, back up to bale until roller hits bale. Apply brakes and back up for folding action. The hook drops down and sets into bale. Apply brakes and pull forward.

The hook releases and drops forward. To unload, apply brakes and back up until bale rolls off.

A bunch of farm shop wizards have invented hundreds of devices and systems for moving big bales with horses or mules. Some of those ideas have even made it to the general marketplace. This is the Grabilt offered by Graber Welding of Bloomfield, Iowa.

Equipment Roundup

Chapter Eighteen

Walter A. Wood's 1857 New Enclosed Gear Mower for one horse

This final chapter offers illustrations of various makes and models of the equipment which has been covered in this book. It is by no means a complete listing. Most of the pre 1910 equipment we have left out because it is so difficult to find any parts for. In some cases, as with the above, their curiosity value won out.

This chapter also features some attachments and other side notes pertinent to the equipment for haying with horses.

Pitchforks

Yes, it does make a difference. A large well curved four tine fork is this author's choice. With the many helpers we've had over these last two decades we've noticed that the three tine fork seems most popular. The two tine is primarily a grain bundle fork (or a lazy hay farmer's fork.)

Oval steel tines. Straight handle. Bronze finish.

Oval steel tines. Bent handle. Bronze finish.

Oval steel tines. Straight handle. Bronze finish.

THREE TINE

FOUR TINE

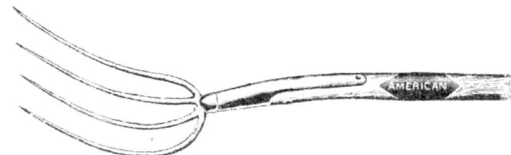

Oval steel tines. Bent handle. Bronze finish.

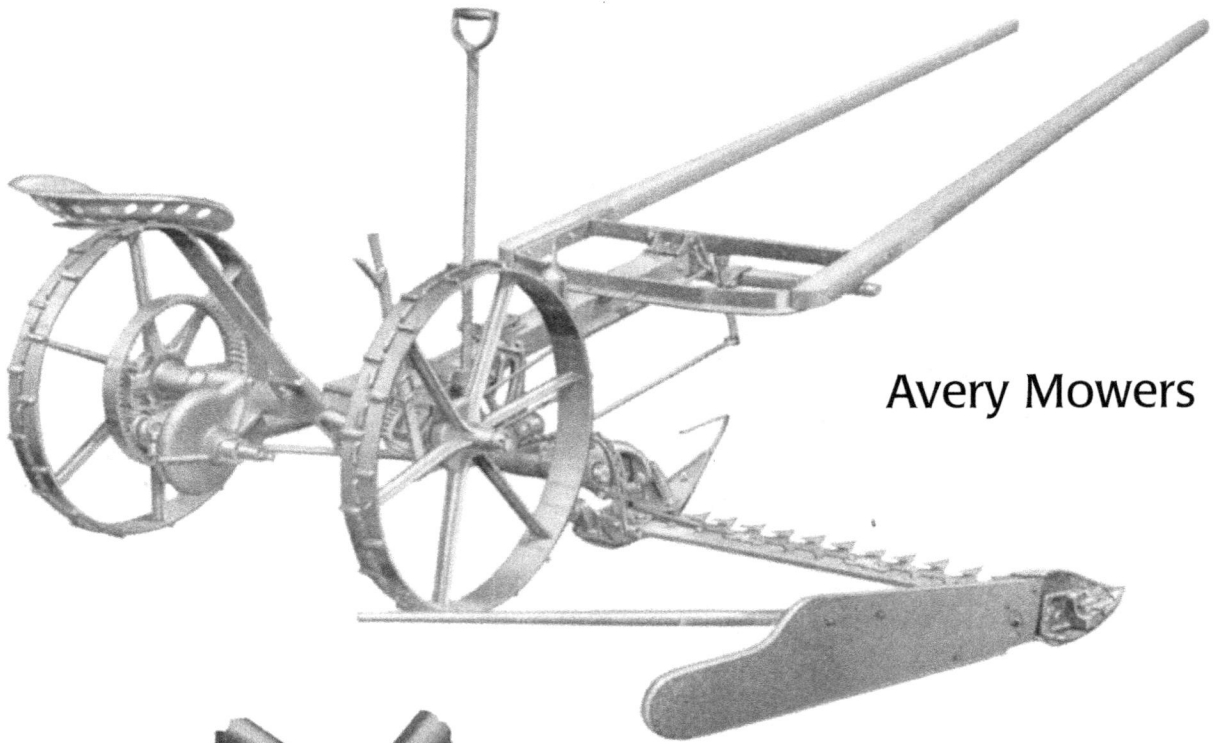

Avery Mowers

Avery Pitman
Flywheel

Avery Mowers

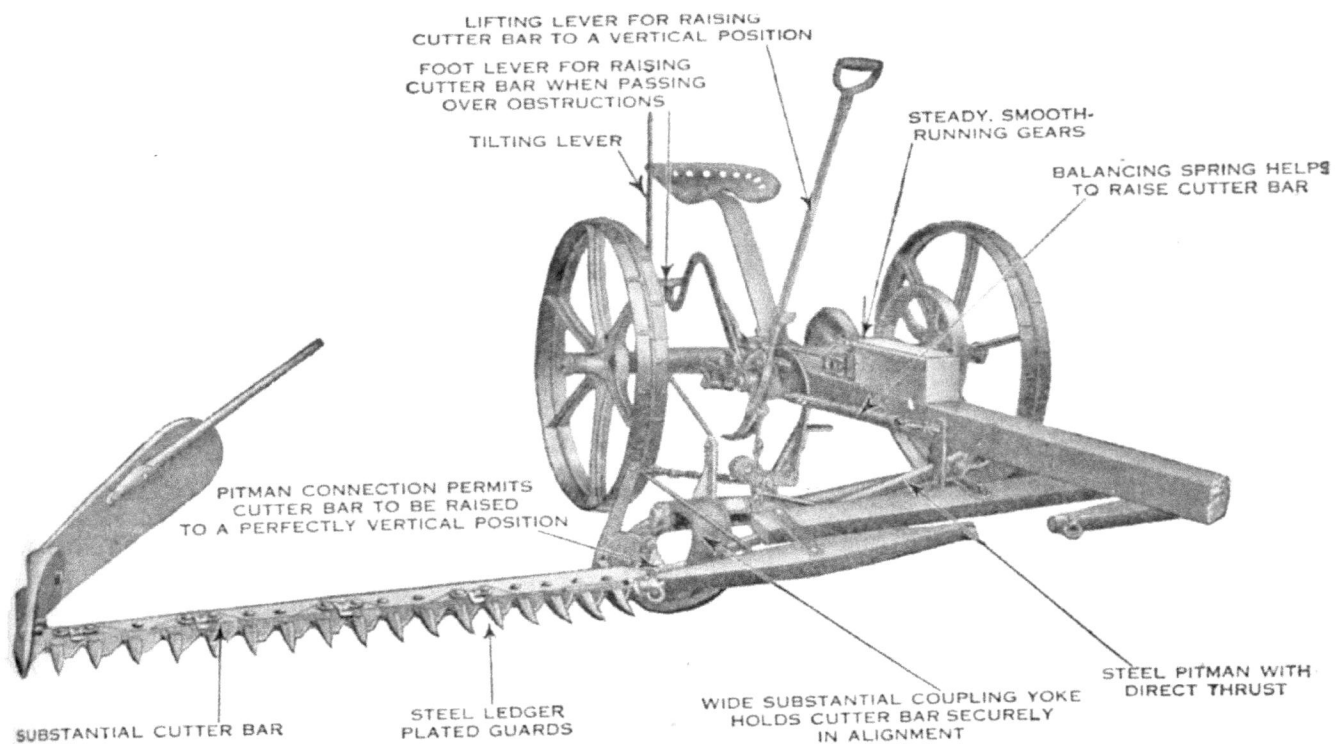

LIFTING LEVER FOR RAISING
CUTTER BAR TO A VERTICAL POSITION

FOOT LEVER FOR RAISING
CUTTER BAR WHEN PASSING
OVER OBSTRUCTIONS

TILTING LEVER

STEADY, SMOOTH-
RUNNING GEARS

BALANCING SPRING HELPS
TO RAISE CUTTER BAR

PITMAN CONNECTION PERMITS
CUTTER BAR TO BE RAISED
TO A PERFECTLY VERTICAL POSITION

STEEL PITMAN WITH
DIRECT THRUST

SUBSTANTIAL CUTTER BAR

STEEL LEDGER
PLATED GUARDS

WIDE SUBSTANTIAL COUPLING YOKE
HOLDS CUTTER BAR SECURELY
IN ALIGNMENT

CHAMPION POPULARITY HAS ALWAYS BEEN DUE TO SUPERIOR CUTTING ABILITY AND LONG LIFE

PITMAN AND CONNECTION

Knife Head Showing Automatic Adjusting Device for
Taking up Wear

Every user knows the importance of having these parts right. The Champion knife head connection has a broad bearing surface, extra long, extending over four sections, and the parts are made of case hardened steel. Note how the two parts fit together, no lost motion.

Method of Attaching Pitman
to Crank Pin Bushing

The crank pin connection here illustrated is found on the Champion only. The bushing slips over the crank pin, and is held on by a steel plate bolted to the wheel. It is closed at the end, thereby retaining the oil and making perfect lubrication possible. The bar is raised by means of convenient levers assisted by a powerful lifting spring

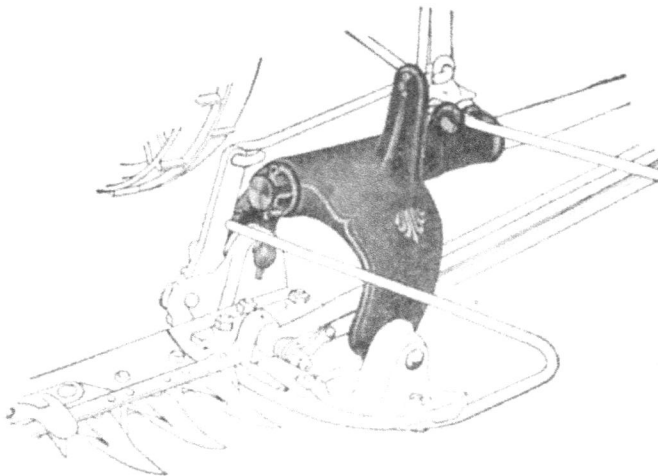

Champion
Yoke Pins are
Exceptionally
Large.
More Strength
at this point.

Note the extra wide and substantially made coupling yoke. The width of this yoke gives it great leverage over the cutter bar and holds the bar securely in alignment.

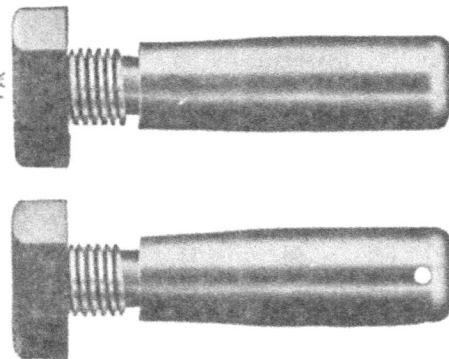

Parallel Draft Link permits cutter bar to conform to all conditions of ground, regardless of tongue. Used only on the Champion. Wearing plates are case hardened and extra long.

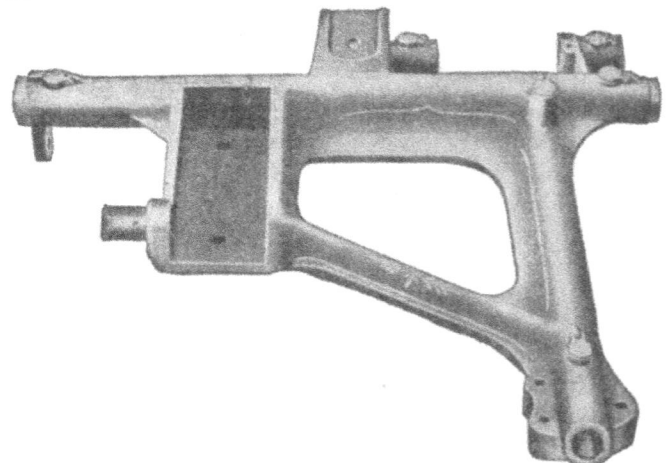

Substantial Champion Main Frame

Champion Mower & Swamp Wheel

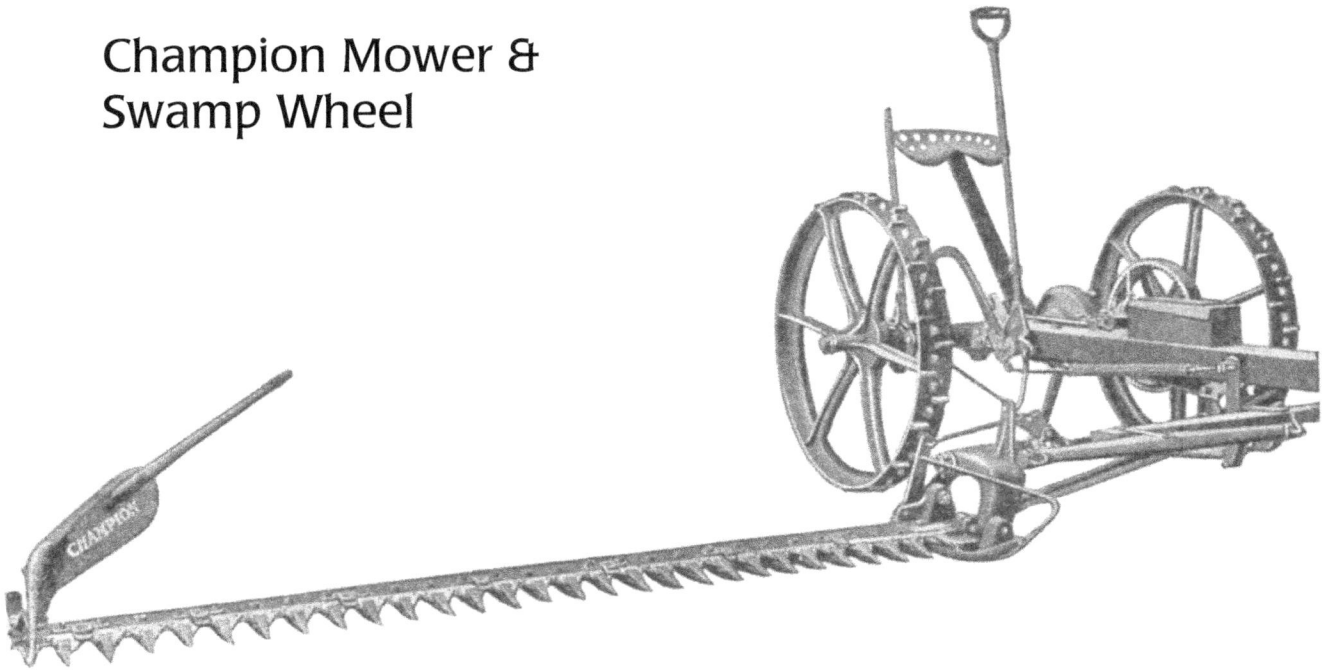

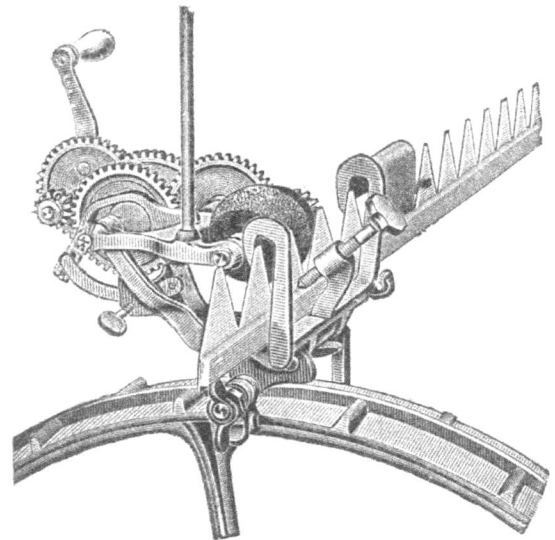

Champion Wheel-mount sickle sharpener

Walter A. Wood's Mower

Walter A. Wood's 1878 New Iron Frame Mowing Attachment for Sweep Rake Reapers

Champion Mower

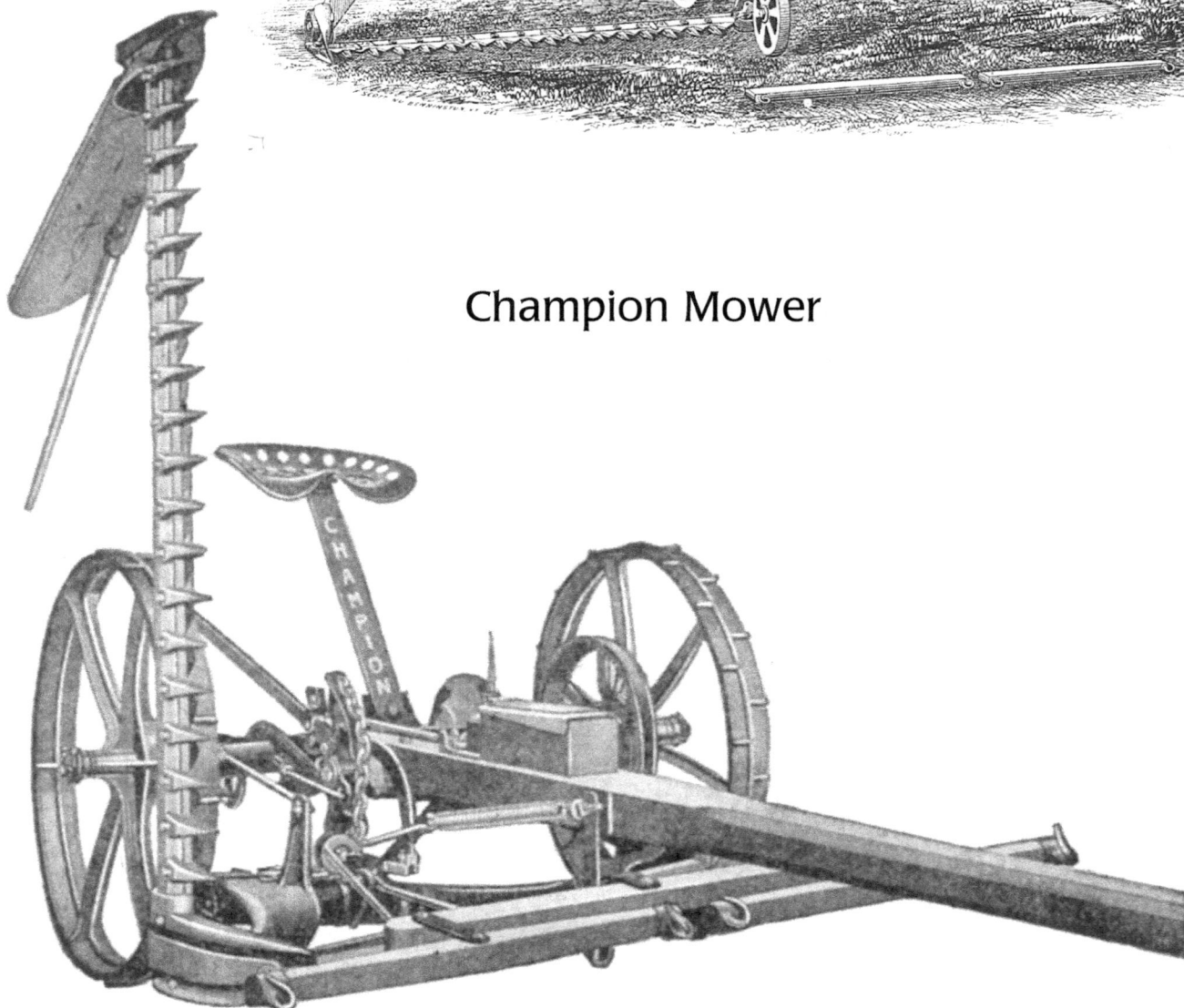

Bud Dimick (age 80) of Madras, Oregon mowing on Singing Horse Ranch

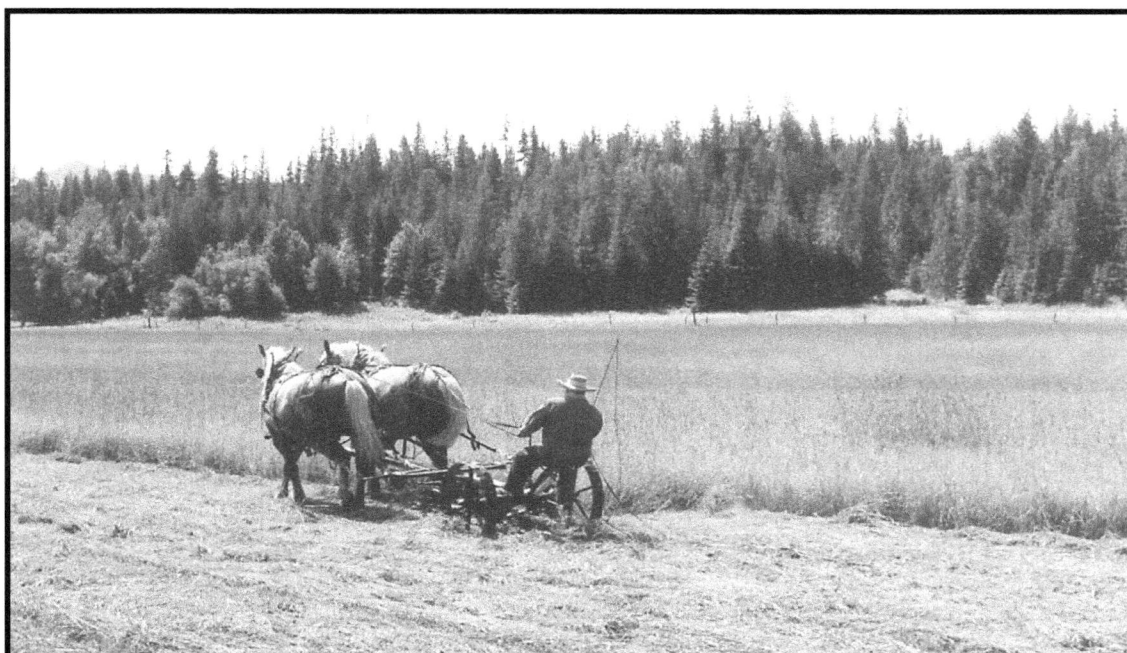

Bulldog Fraser of Northwestern Montana mowing with young team in training.

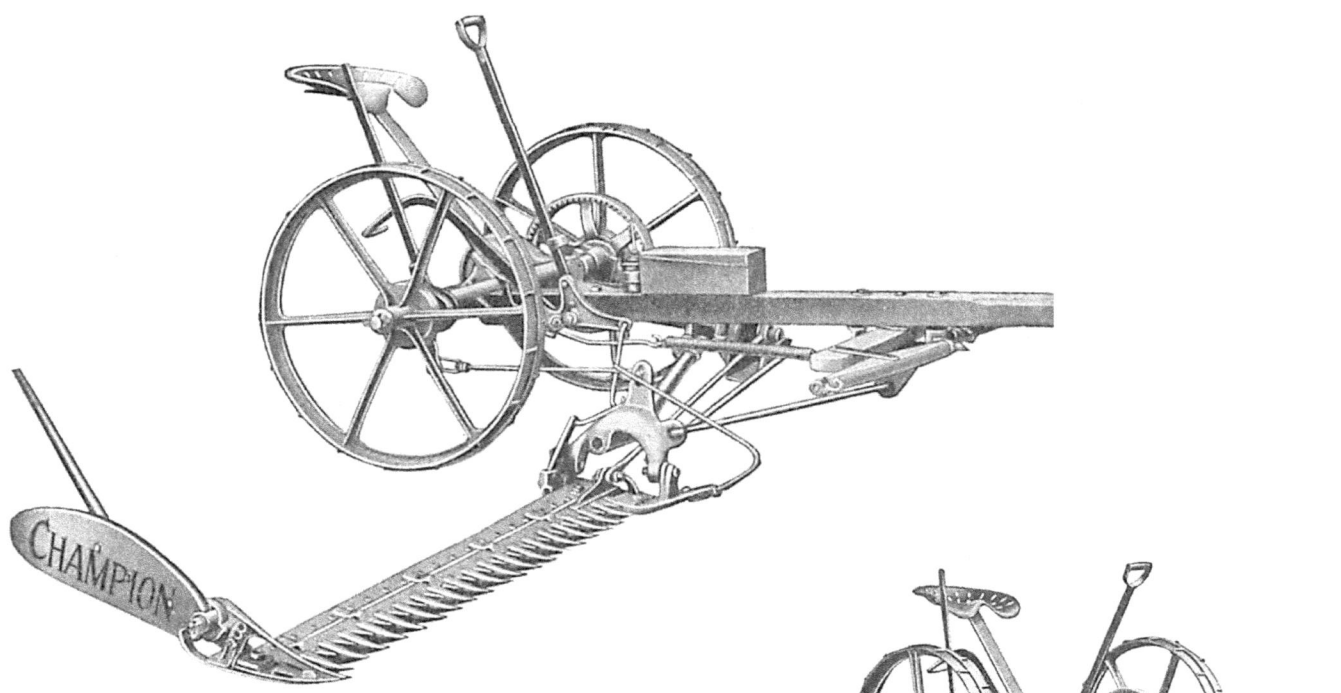

Champion Mowers

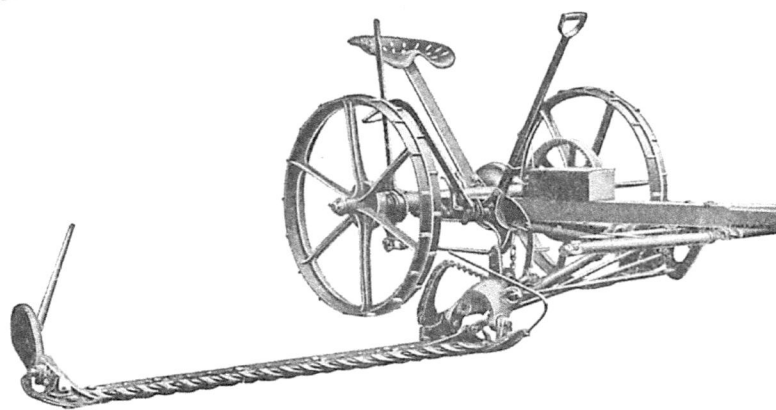

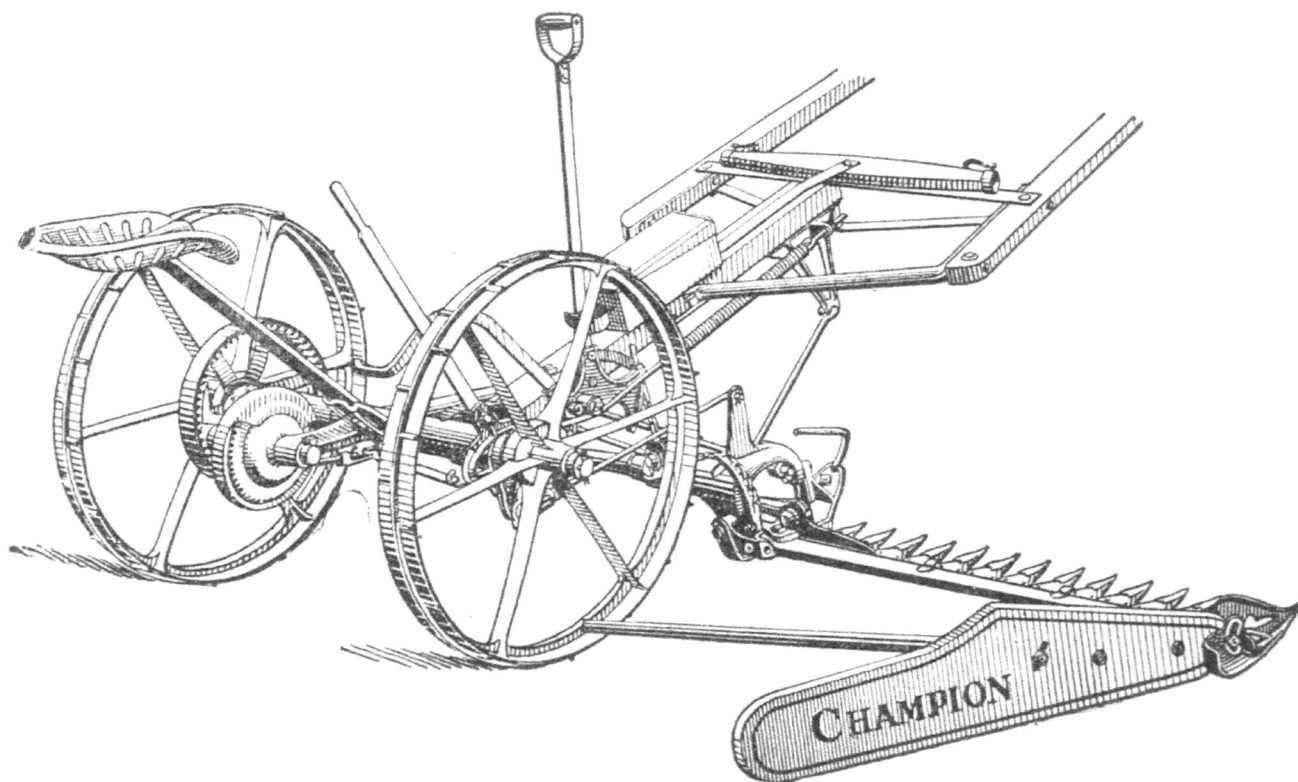

Emerson Brantingham / Osborne Mowers

Emerson Brantingham / Osborne Mower

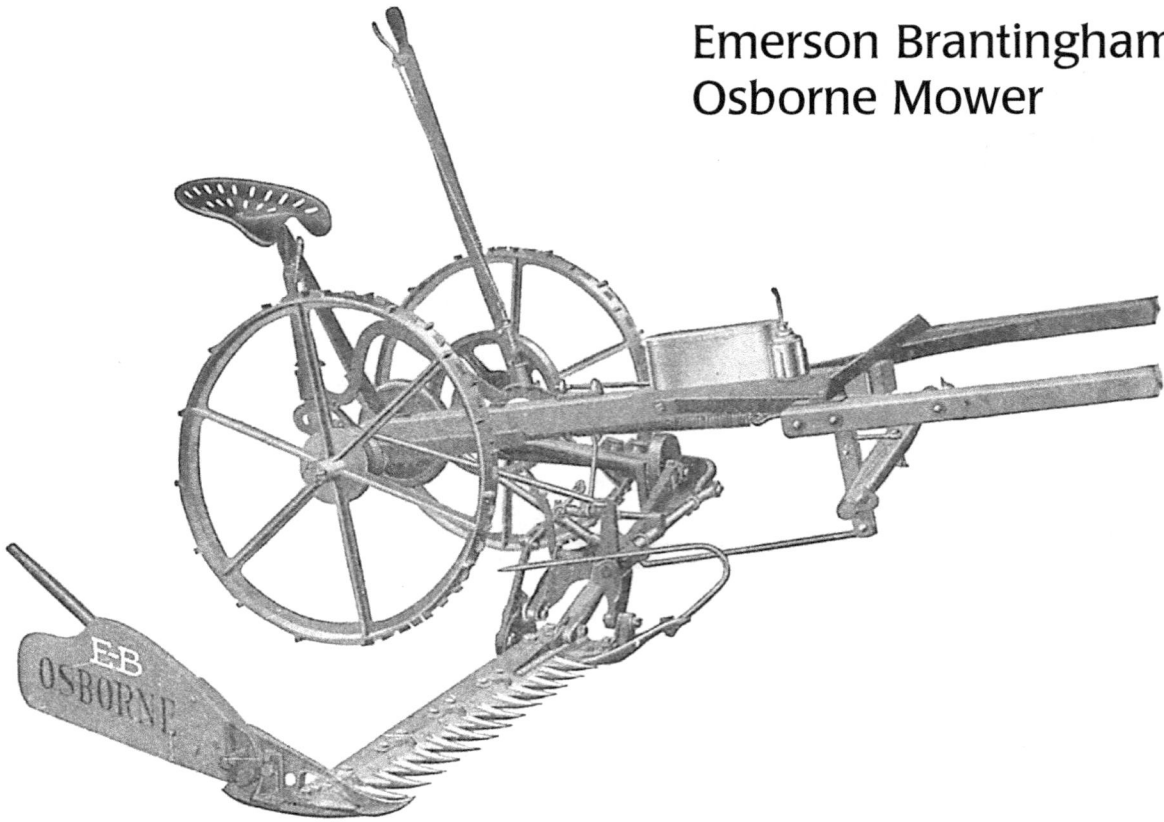

Amy Evers and Lynn Miller with two Singing Horse Ranch teams and McCormick Deering / International #9 mowers.

The late Jack Bissel with his team on a Case mower.

An old photo of a John Deere Big 4 mower.

Another predepression photo of a John Deere Big 4 mower.

John Deere Mower

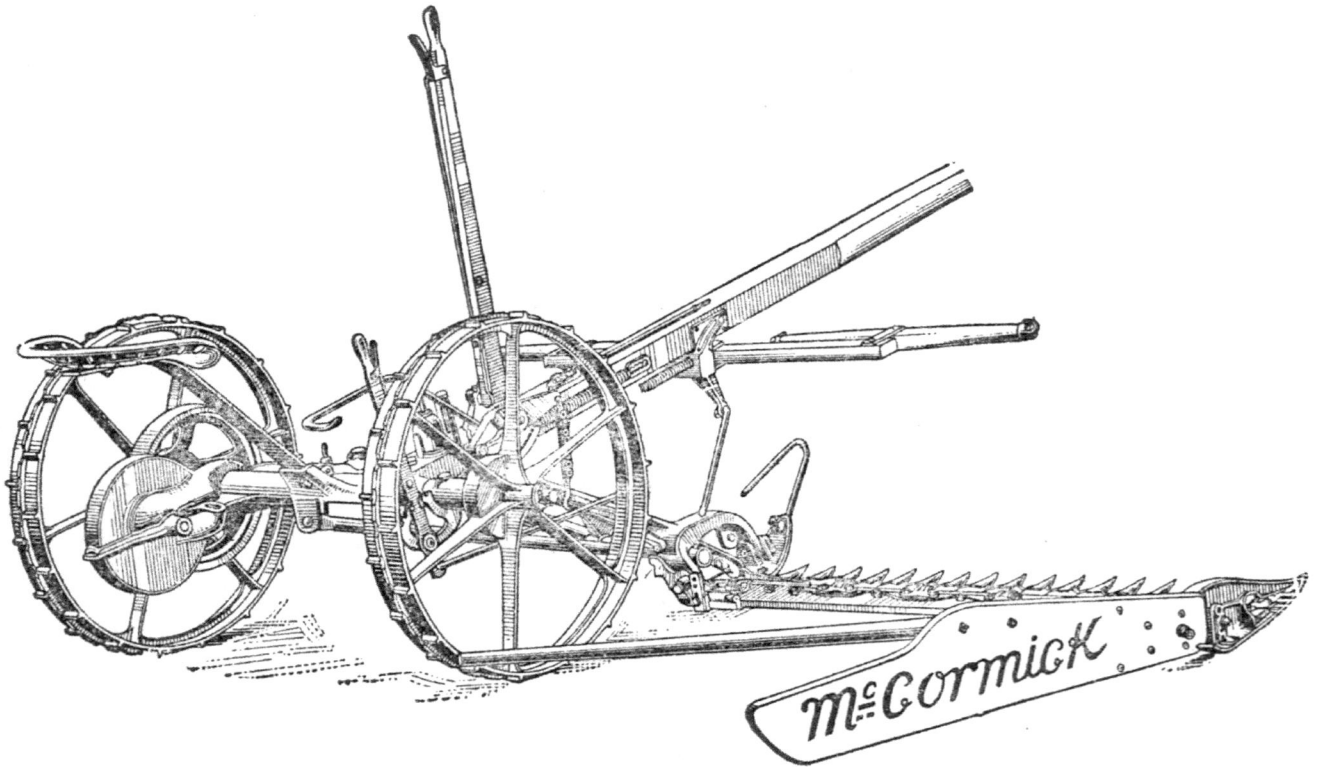

McCormick Mower

M783
M489
M643

M626
M539
D 2033
M782
M753B

M765

MB488

M1146
HA71

MA 669

L363½   M675

M 801
MA 656
MB 658

P171
M294
M782A

M807
M808
M 674

M799
M800
M803½
MA803

MB 743
M666
MA754

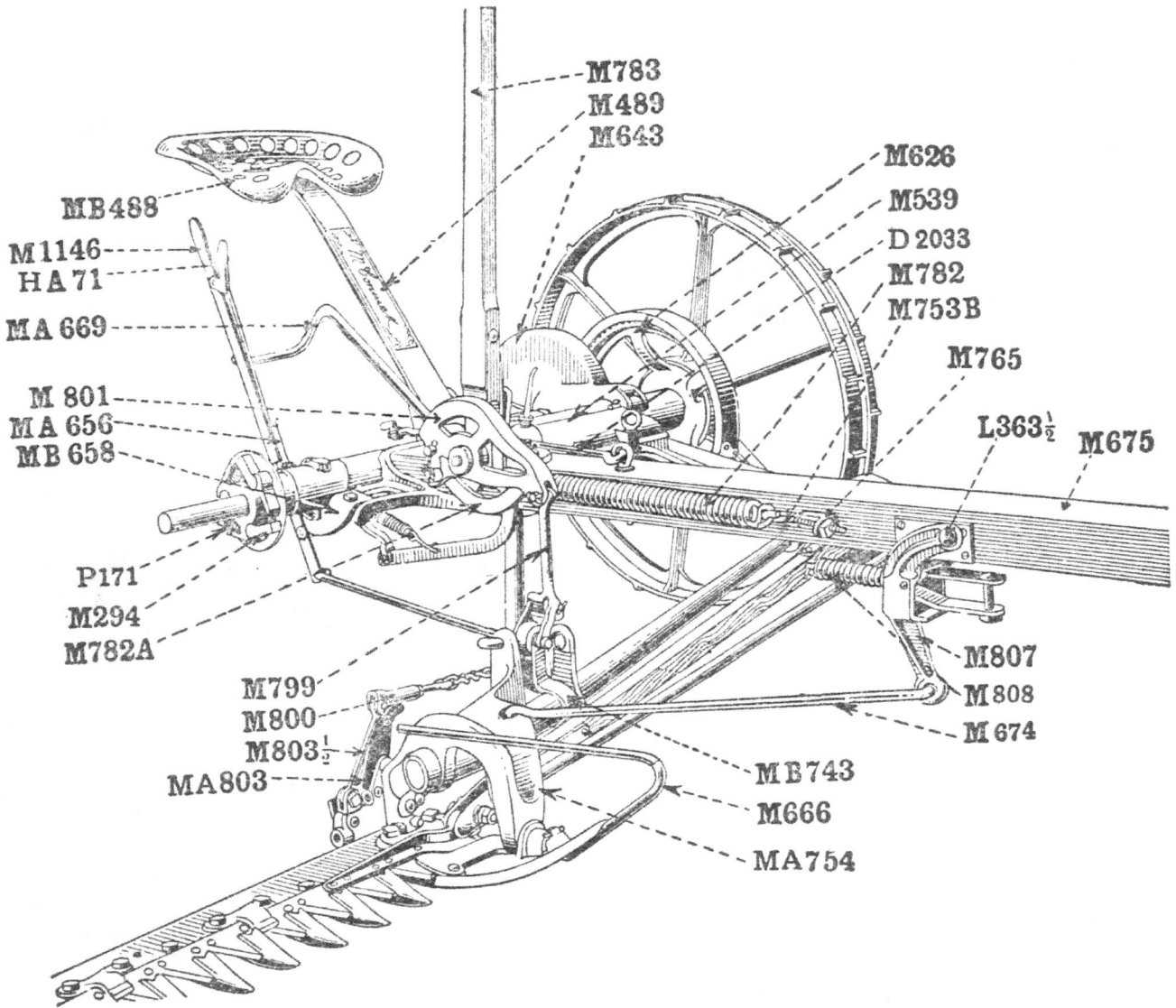

McCormick Mower

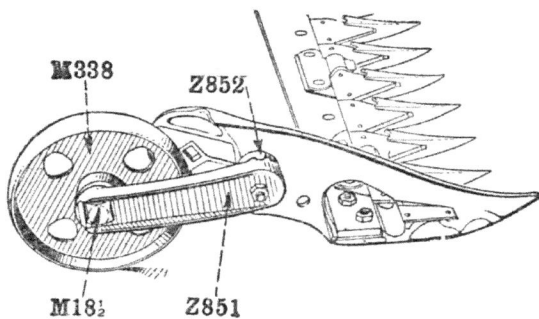

M338   Z852

Model Big 6

M18½   Z851

Weed clipping wheel.

McCormick Mowers

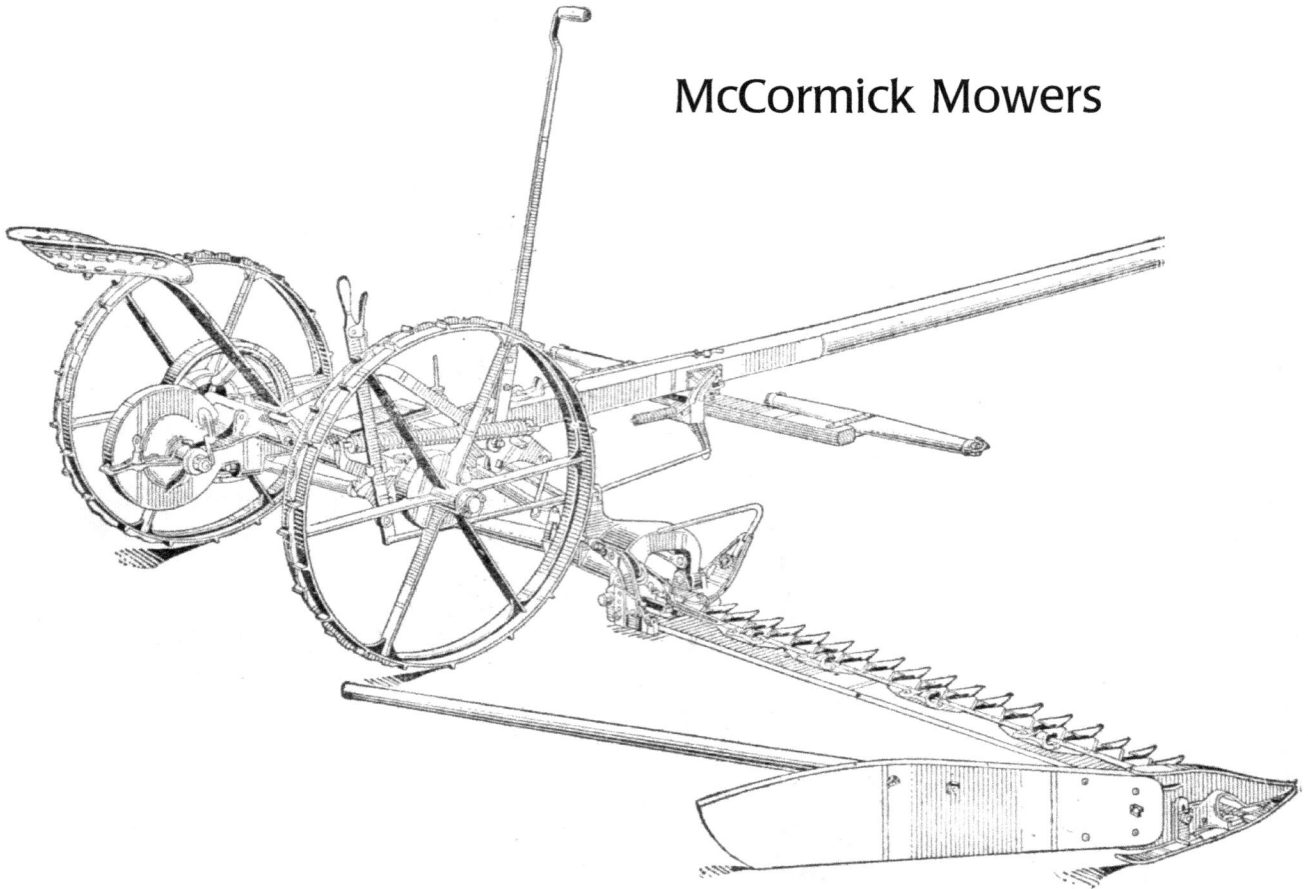

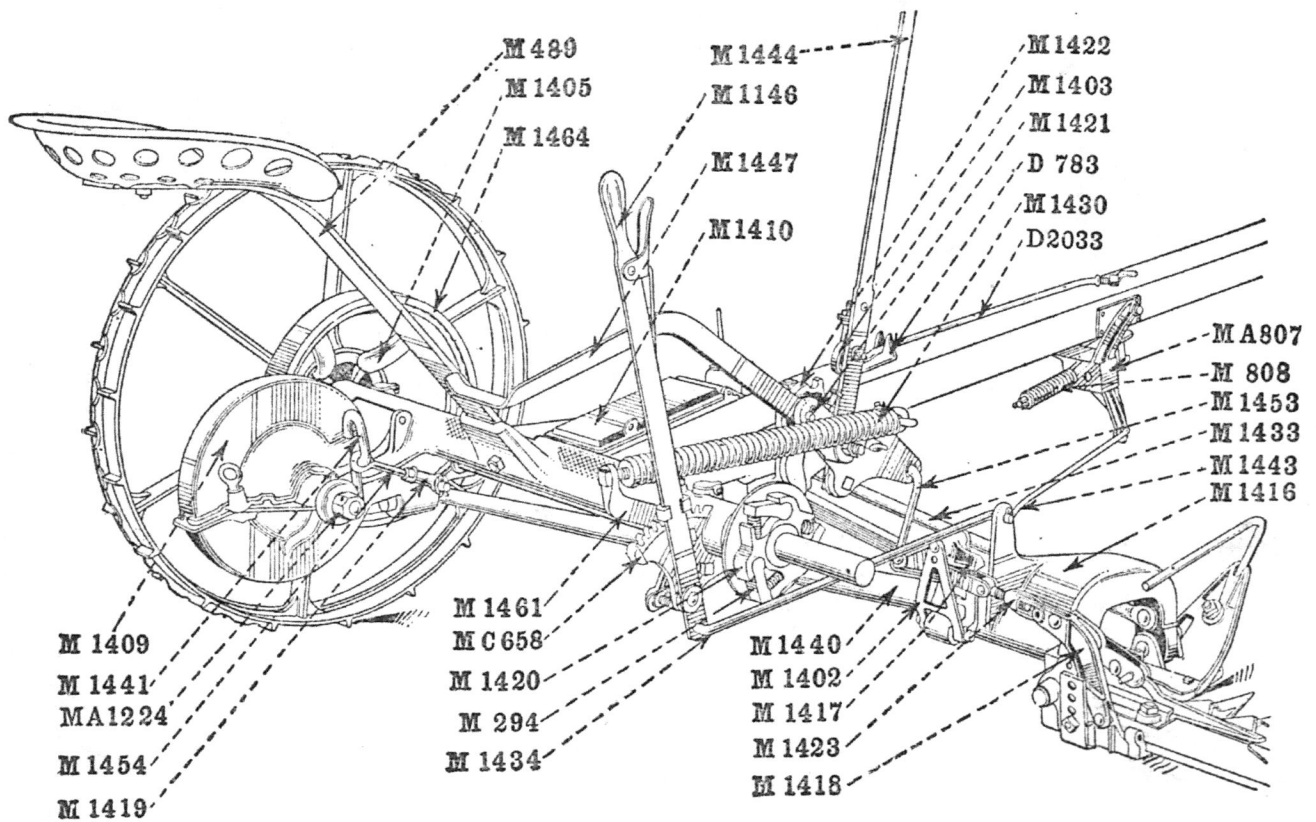

M 480
M 1405
M 1464

M 1444
M 1146
M 1447
M 1410

M 1422
M 1403
M 1421
D 783
M 1430
D 2033

M A807
M 808
M 1453
M 1433
M 1443
M 1416

M 1409
M 1441
MA1224
M 1454
M 1419

M 1461
M C 658
M 1420
M 294
M 1434

M 1440
M 1402
M 1417
M 1423
M 1418

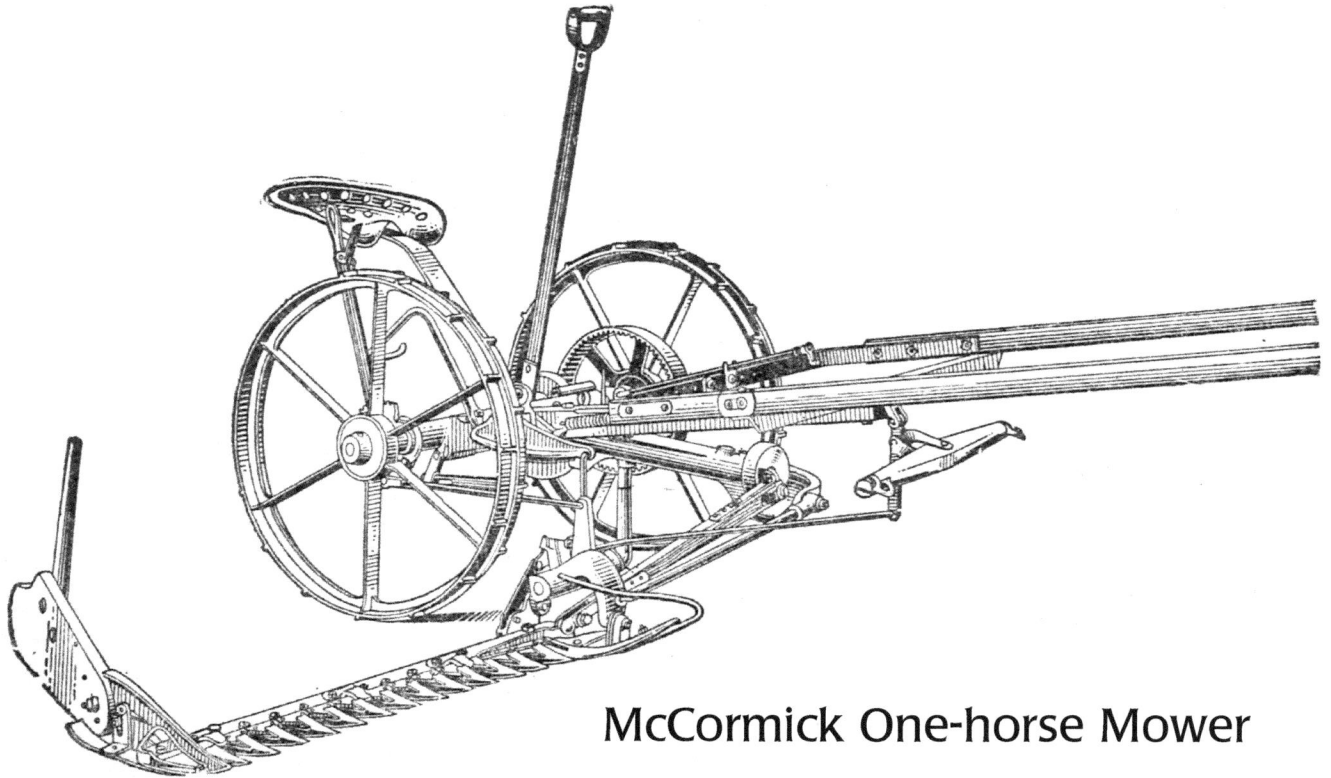

McCormick One-horse Mower

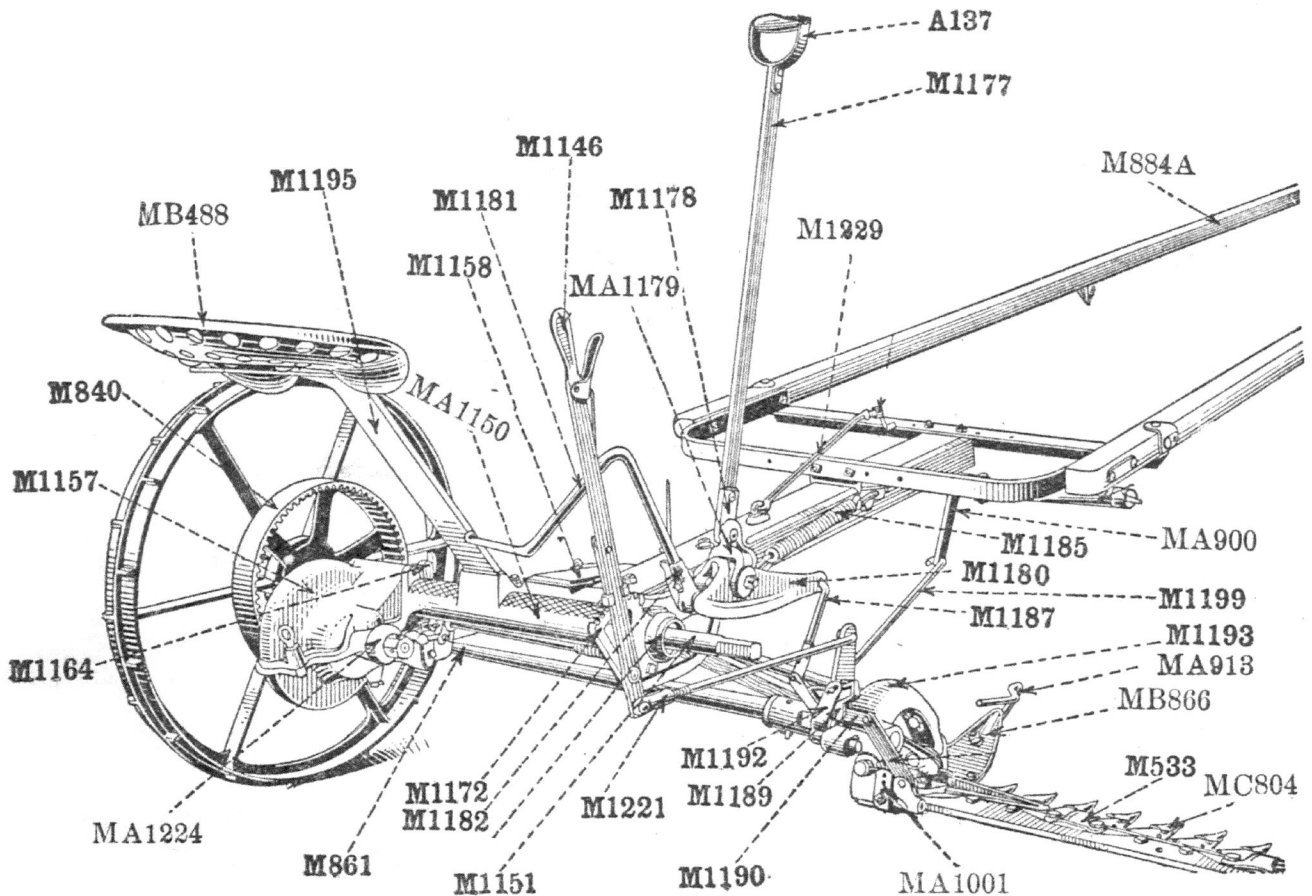

A137
M1177
M884A
M1146
M1195
M1181
M1178
MB488
M1158
M1229
MA1179
M840
MA1150
M1157
M1185 — MA900
M1180
M1199
M1187
M1193
MA913
M1164
MB866
M533
MC804
MA1224
M1172
M1182
M1221
M1192
M1189
M861
M1151
M1190
MA1001

McCormick Mower

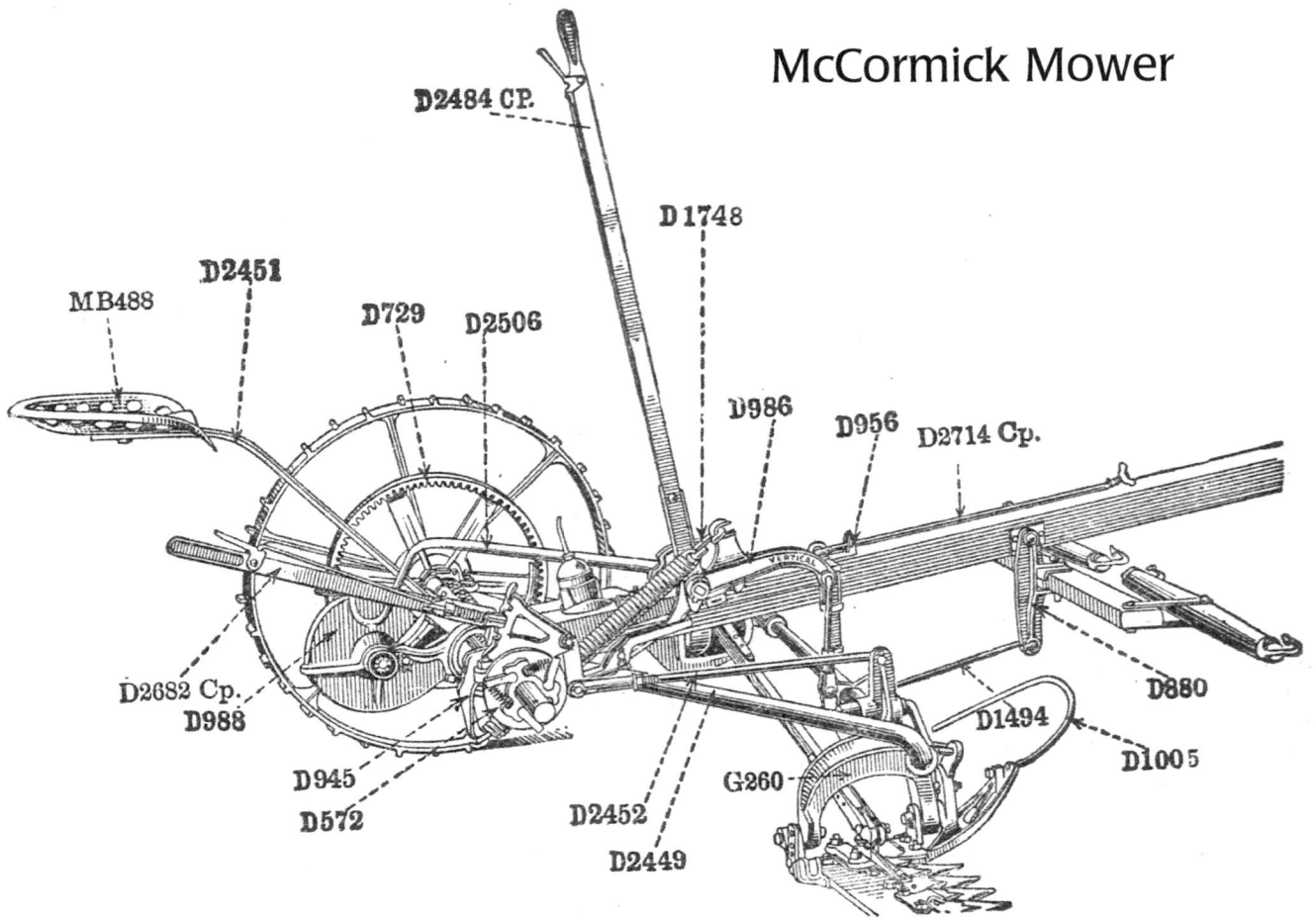

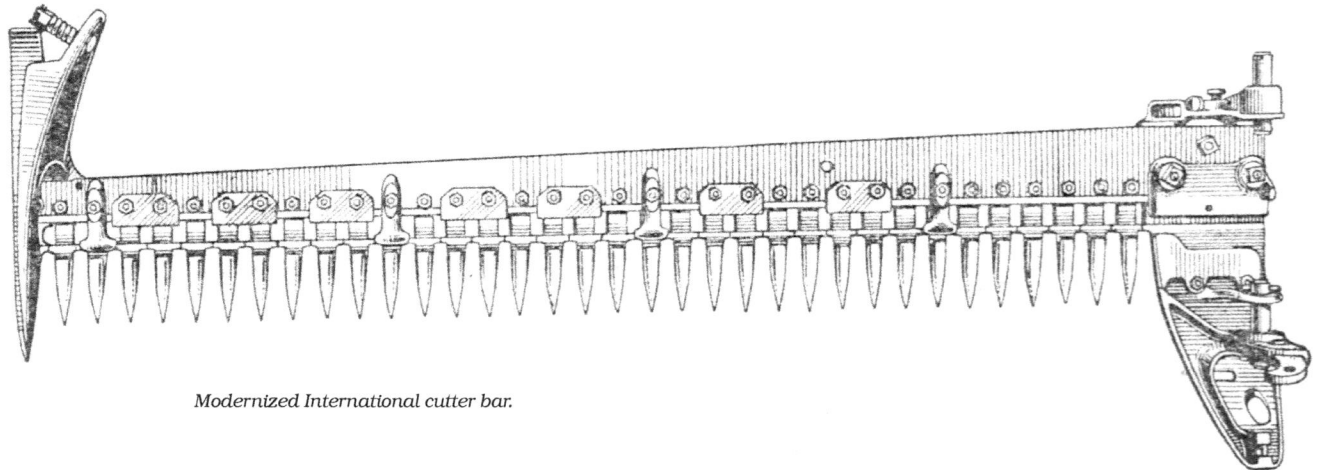

D2484 CP.

D1748

D2451

MB488

D729

D2506

D986

D956

D2714 Cp.

D2682 Cp.

D988

D880

D1494

D1005

D945

G260

D572

D2452

D2449

Modernized International cutter bar.

D 652

D 687

G198

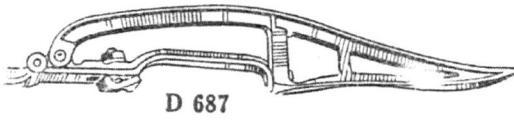

Various pea-lifter attachments for cutterbars.

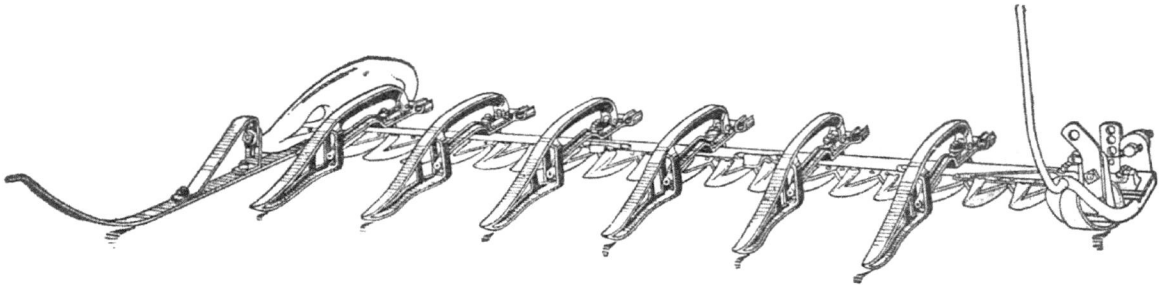

McCormick Pedal
grinder/ sharpener

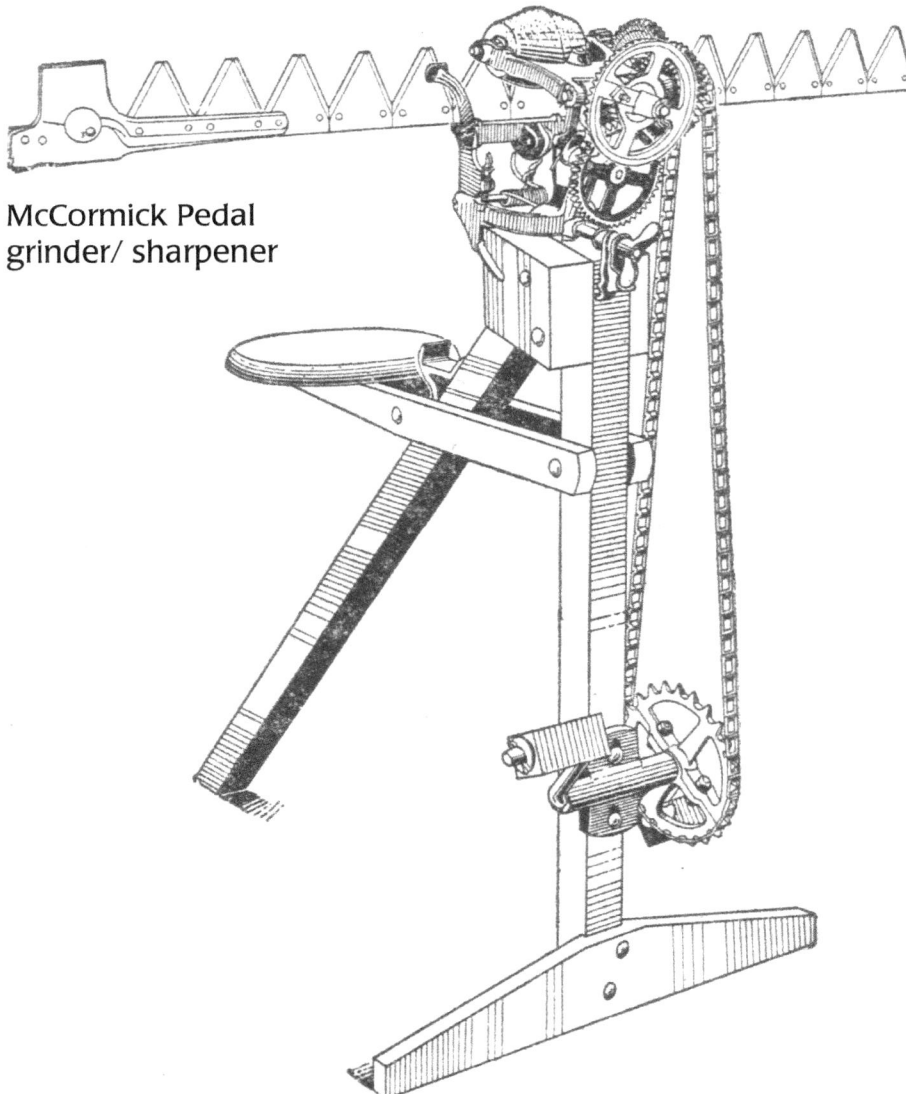

Deering Mower

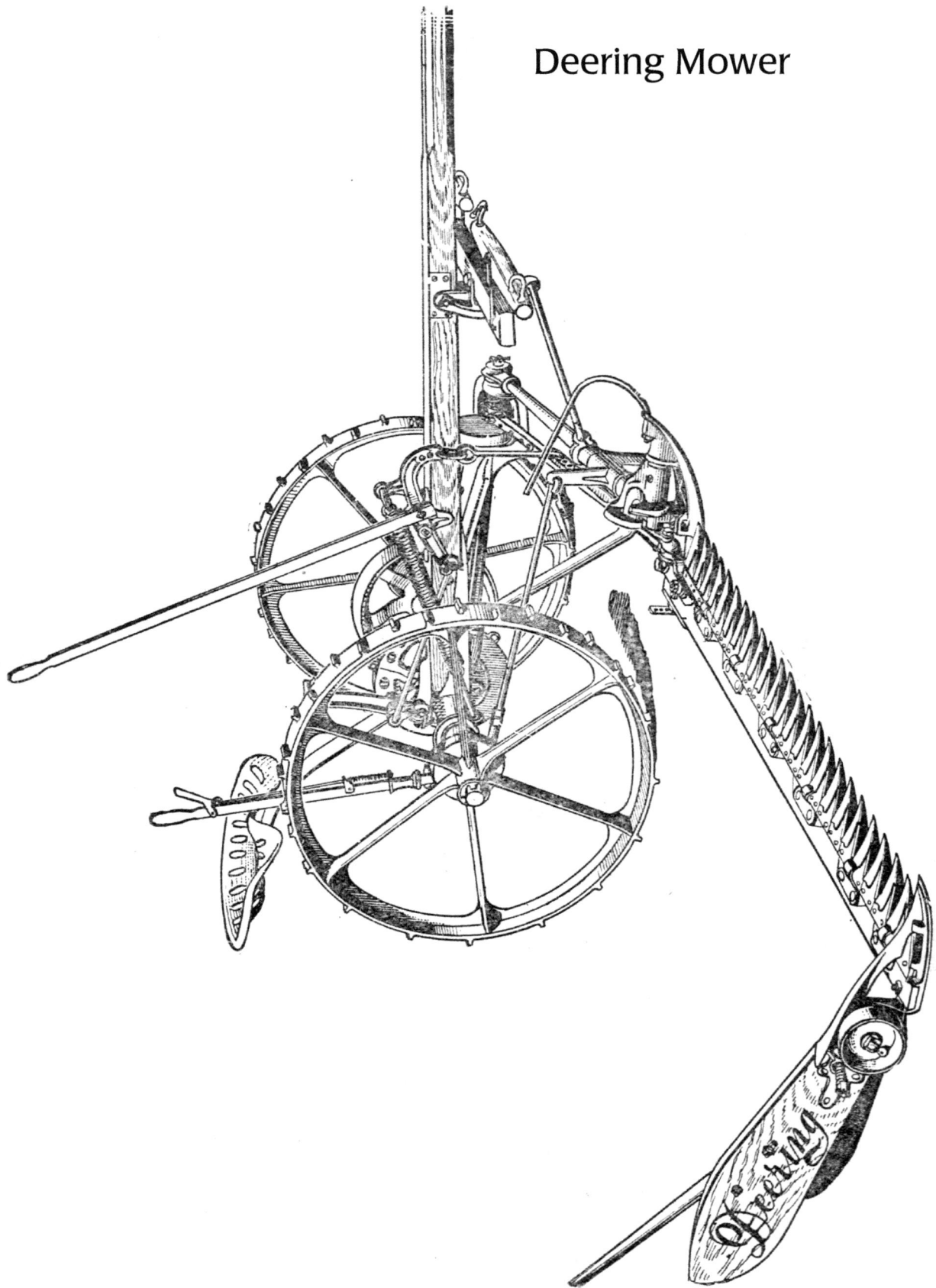

Two of Bulldog Fraser's students mowing in Montana.

Bulldog putting the finishing touches on a team of Percheron trainees.

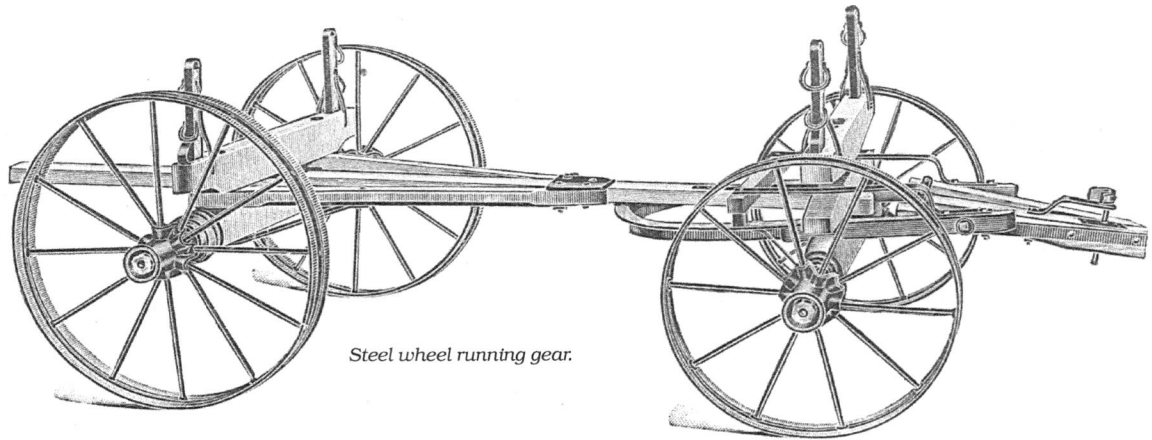

Steel wheel running gear.

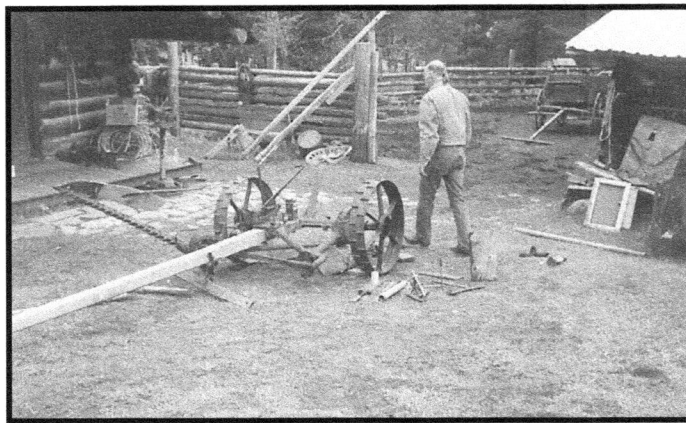

Doug Hammill rebuilding a No. 9 mower.

Abram Ellwood
Mower

Very rare.

Wood Brothers Mower

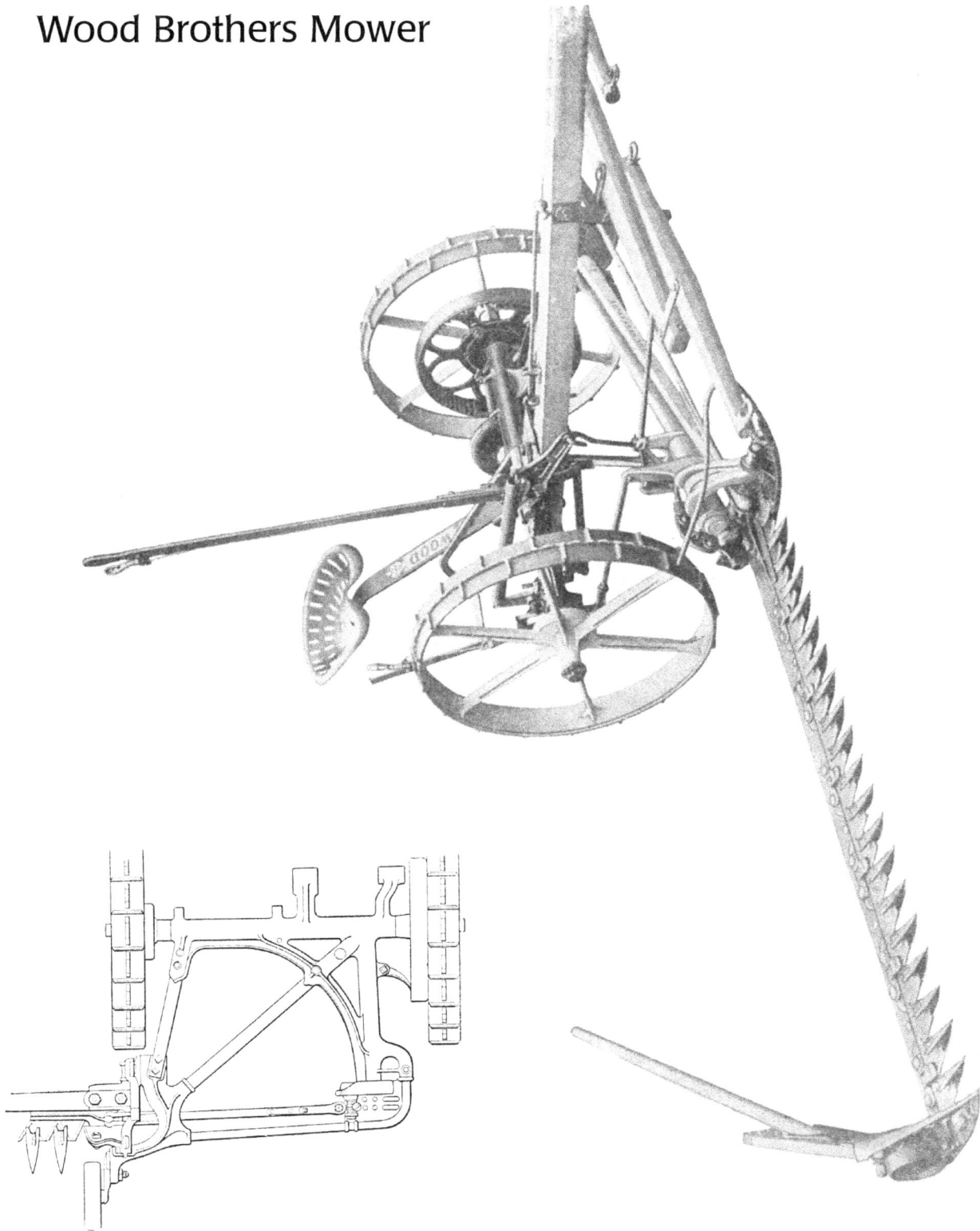

Walter A. Wood's New Enclosed Gear Mower
for two Horses

1878

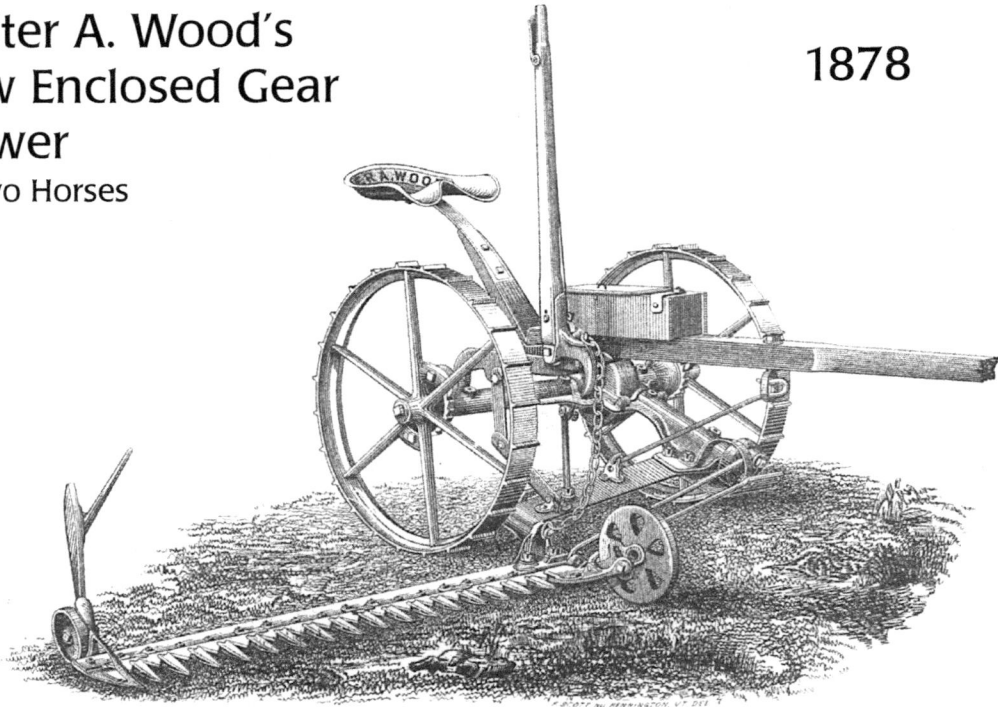

Showing driving gears enclosed and exposed.

A

B

C

D

From 1916 sales literature

"Diagram to show why the lower balance wheel of the Admiral Mower gives it greater power than any other mower can develop. The lower position delivers 50% more power."

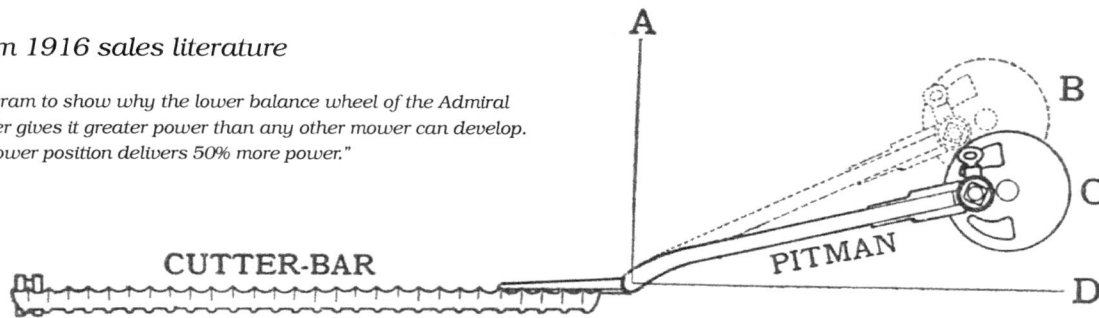

CUTTER-BAR

PITMAN

"Illustrating the backward or forward swing (as much as 6 inches on some makes) to other cutter bars when tilted and the rigid alignment of the Admiral's cutter bar no matter how it is tilted."

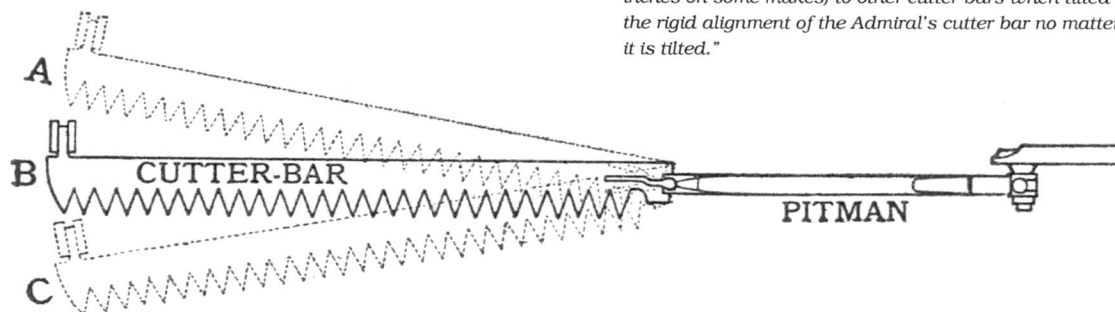

A

B

C

CUTTER-BAR

PITMAN

Walter A. Wood Admiral Line of Mowers

In 1916 the Wood Company was proud of their new underdraft feature and footlift dynamics.

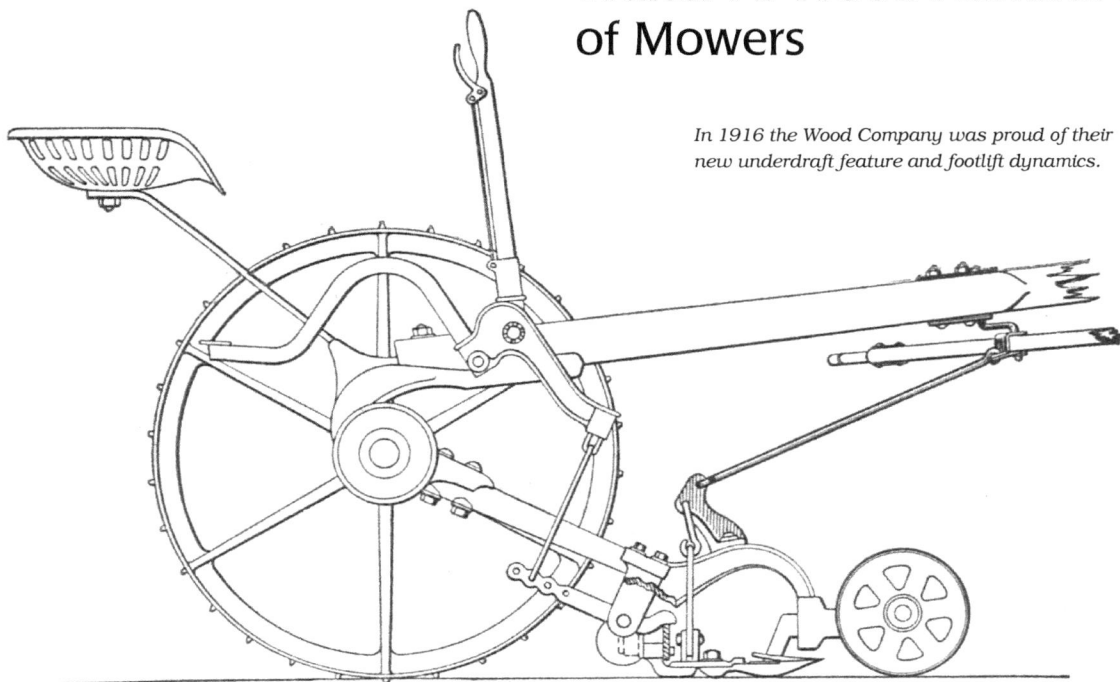

Walter A. Wood Admiral Line of Mowers

Weed cutting attachment.

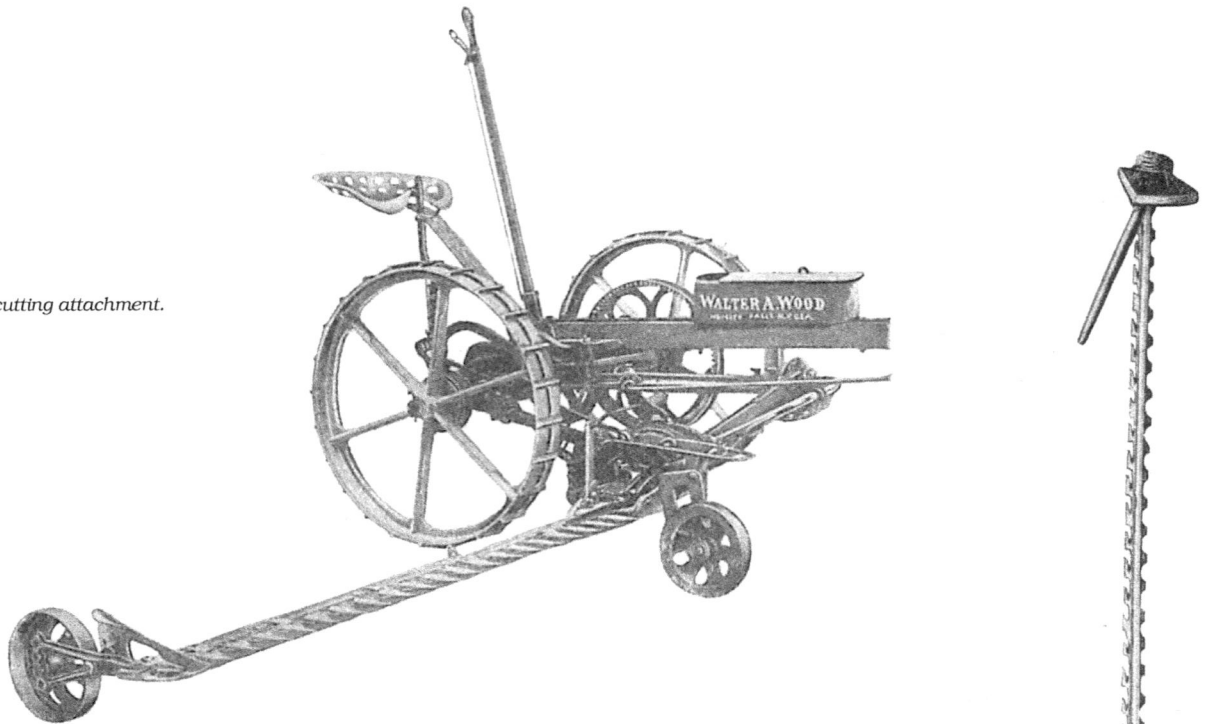

Giant Admiral mower with reaper wheels.

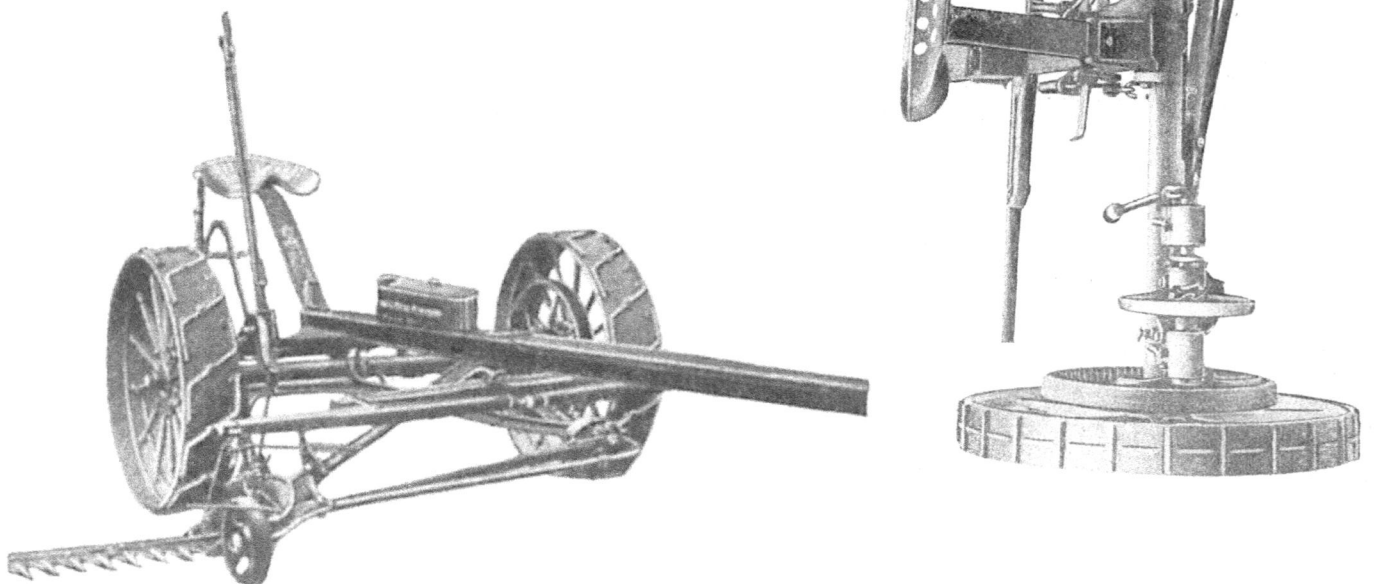

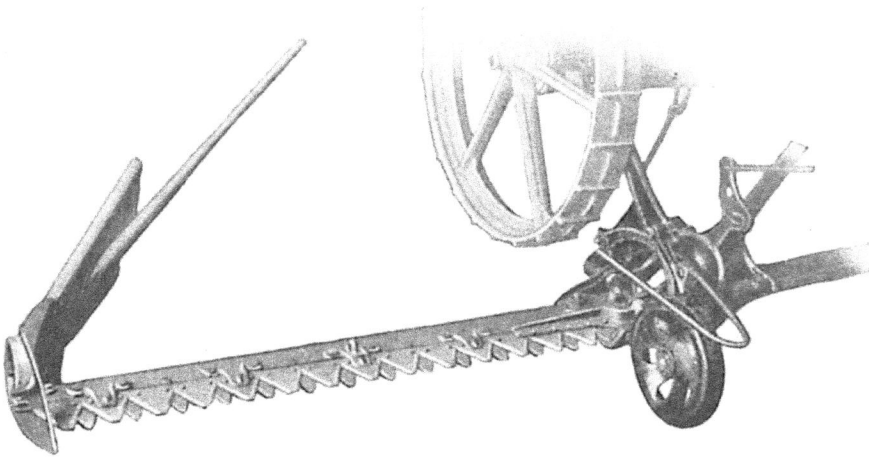

Specially designed brush cutting bar.

Walter A. Wood Admiral Line of Mowers

Reaping attachment for Wood mowers.

Consists of a slatted platform which is bolted to the rear of the cutter bar. It has a guard for the mower wheel which keeps the grain away from the wheel, an outside divider, a hand rake for the raker who sits in the extra seat which is easily attached to the mower.

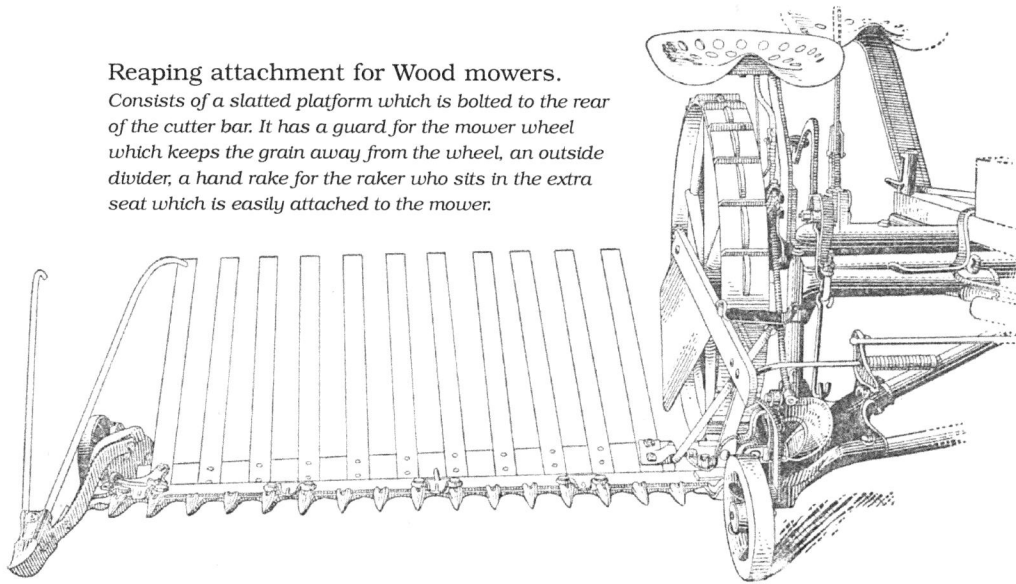

The all-purpose mower guard.

A pea-lifter which hooks over guard and lifts the tangled vines to help with

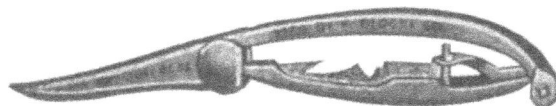

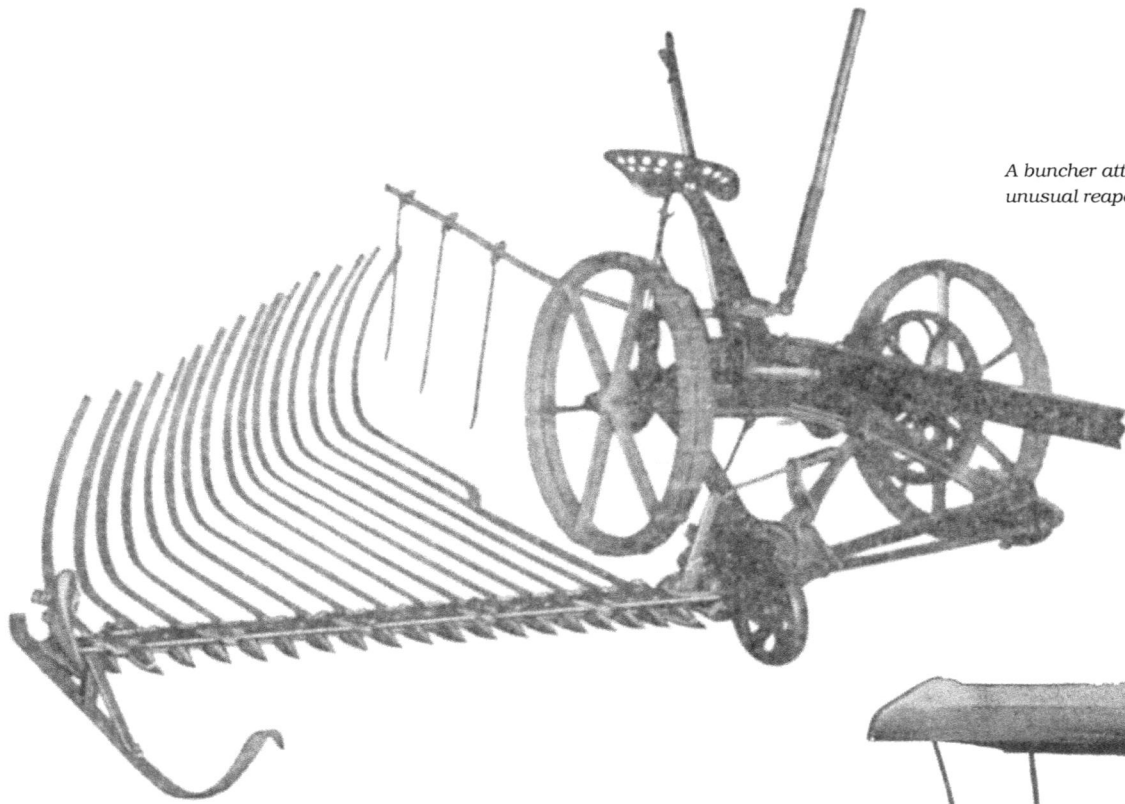

A buncher attachment with an unusual reaper style sweep.

RIDE AND REST
ON THE E-Z
IN THE SHADE

In 1920 this seat was sold to replace the regular mower seat. We don't know why it didn't catch on.

Two different sickle sharpeners designed to bolt on the mower wheel.

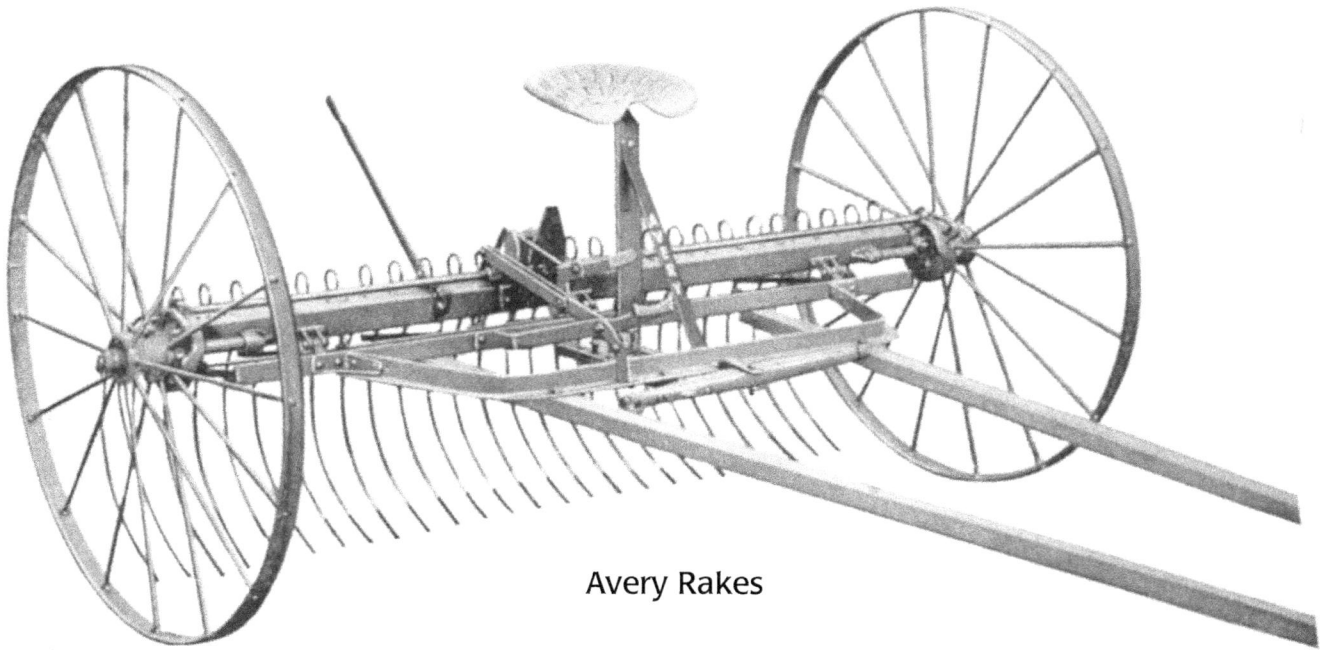

Avery Rakes

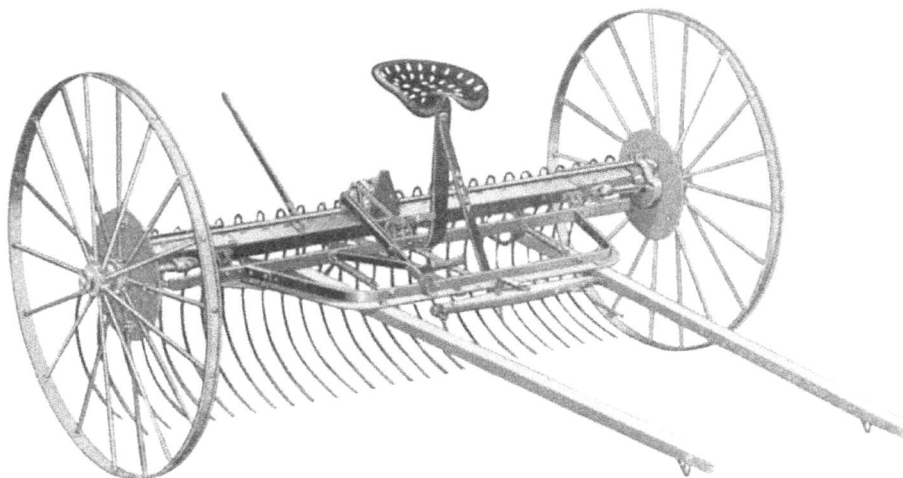

Champion Rakes

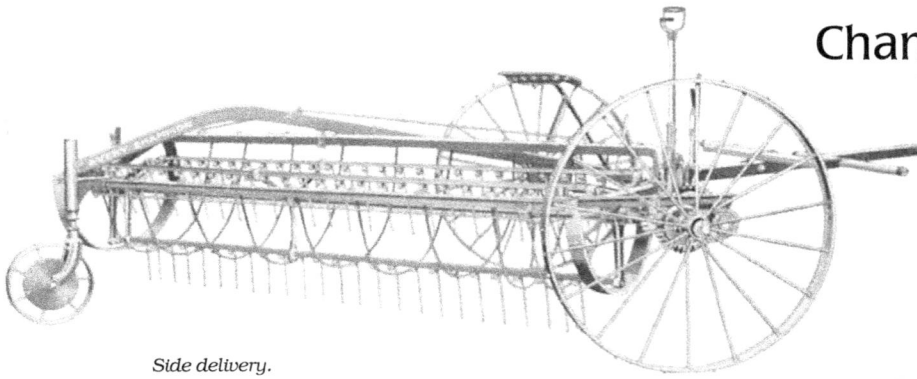

Side delivery.

Dump rake catches.

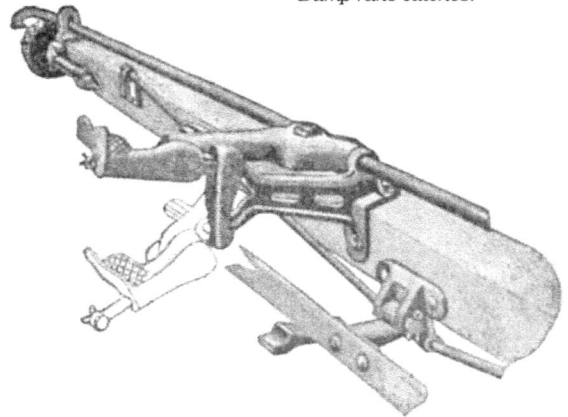

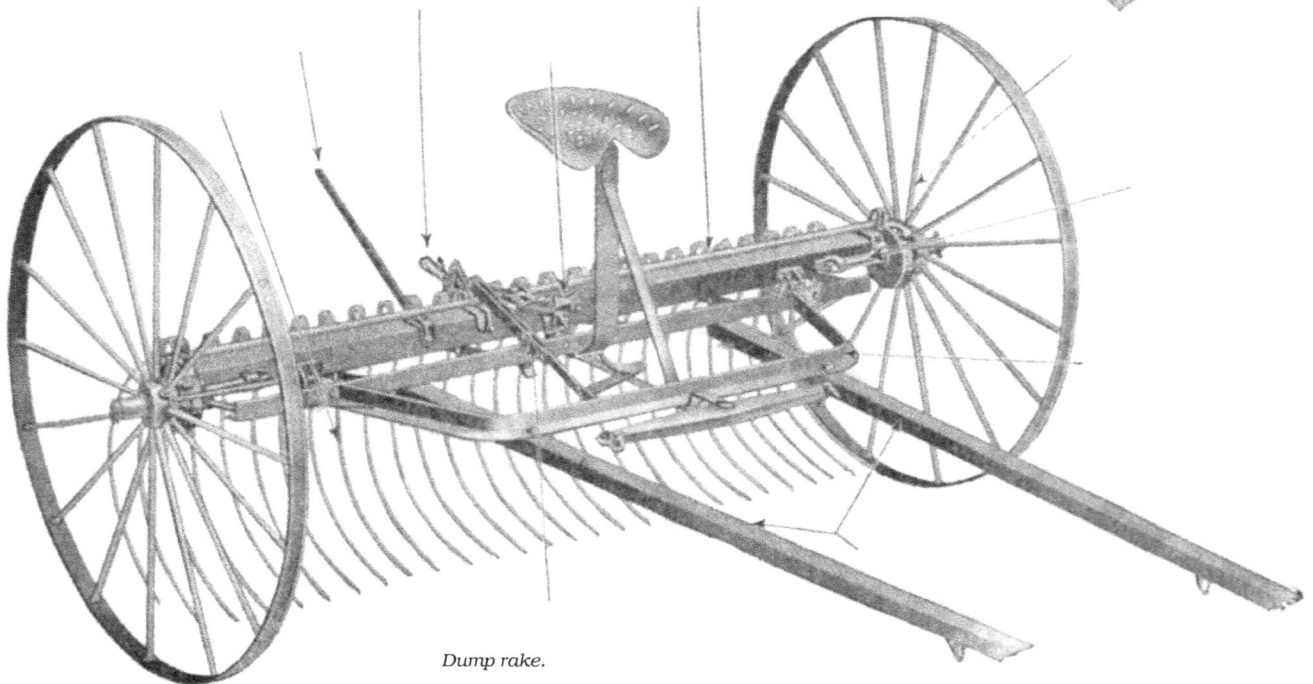

Dump rake.

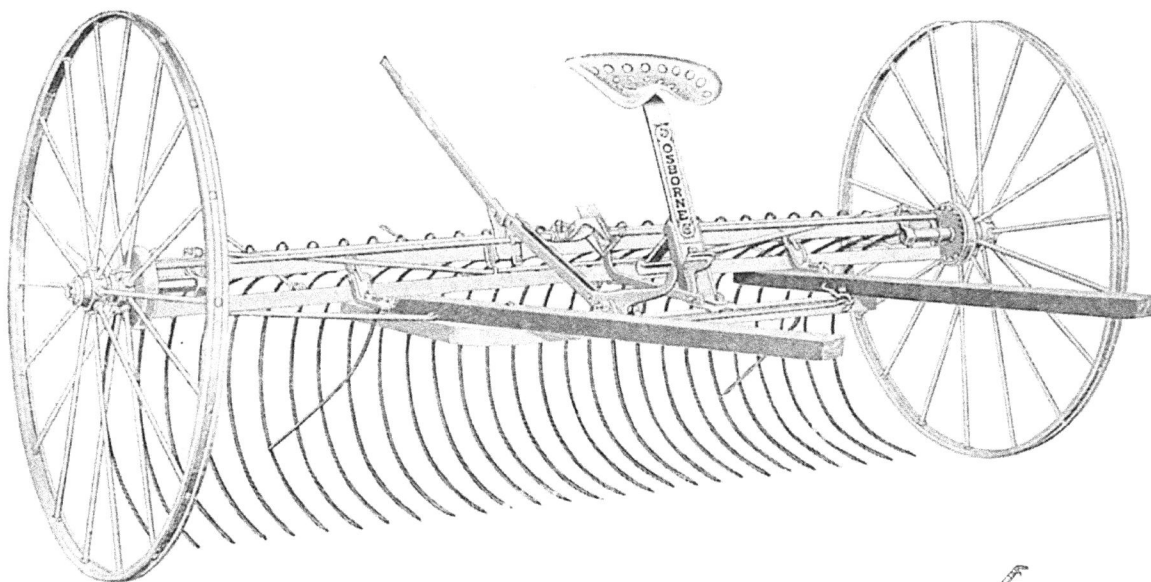

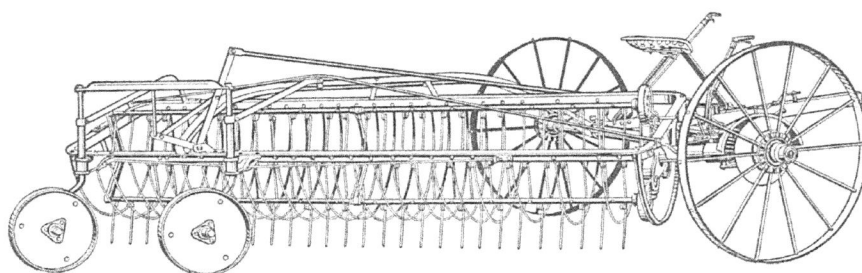

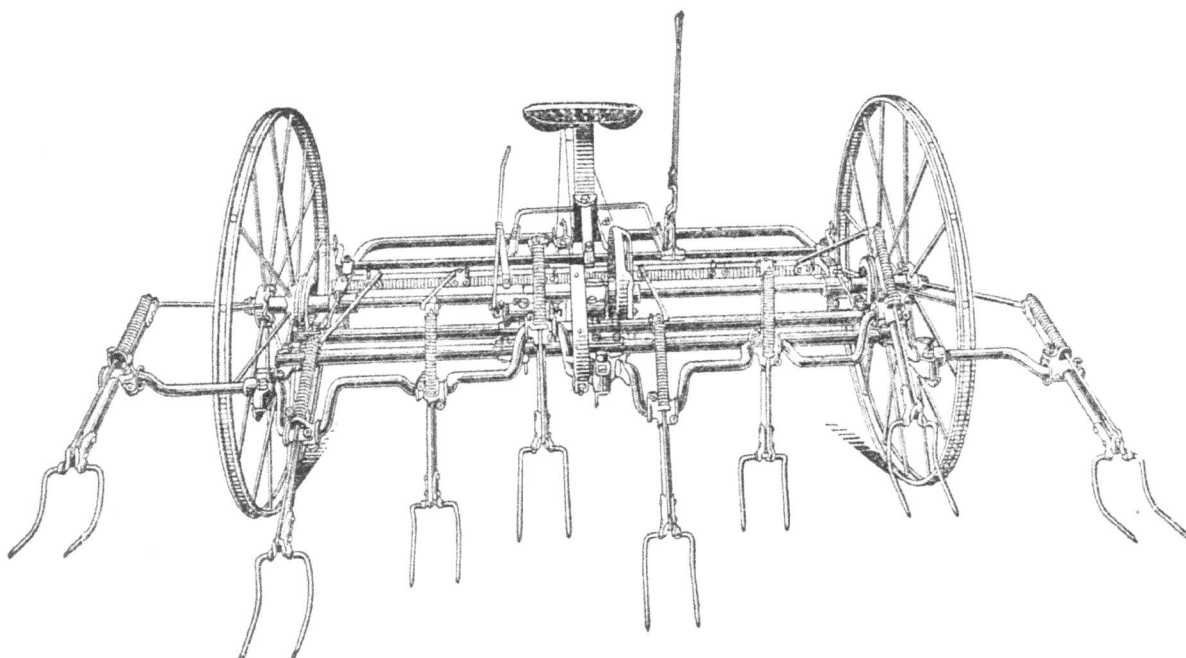

Emerson Brantingham / Osborne Rakes & Tedder

John Deere Rakes

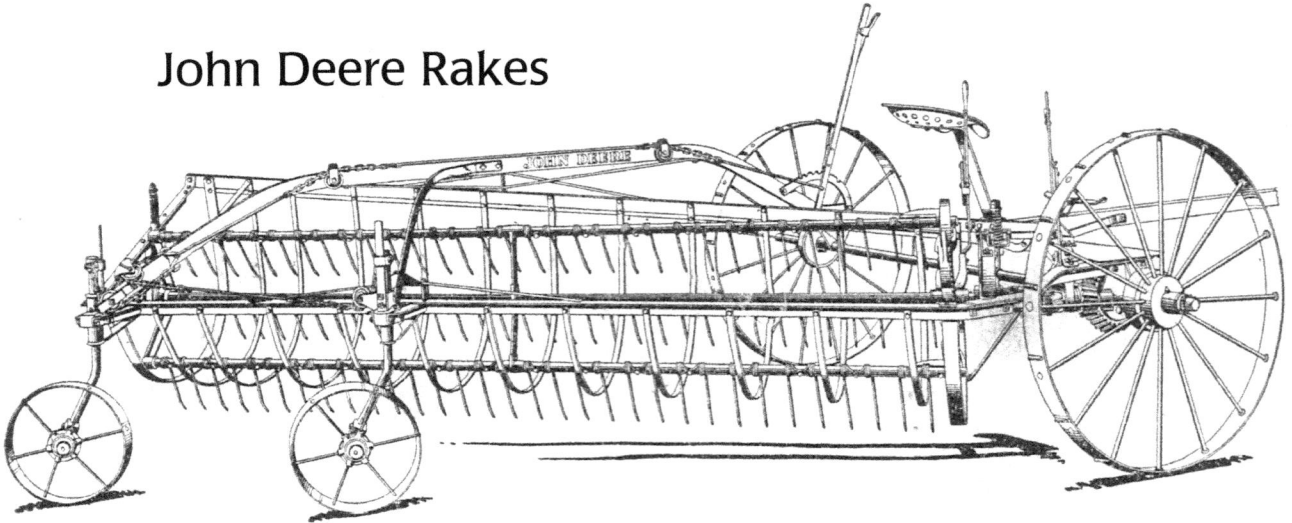

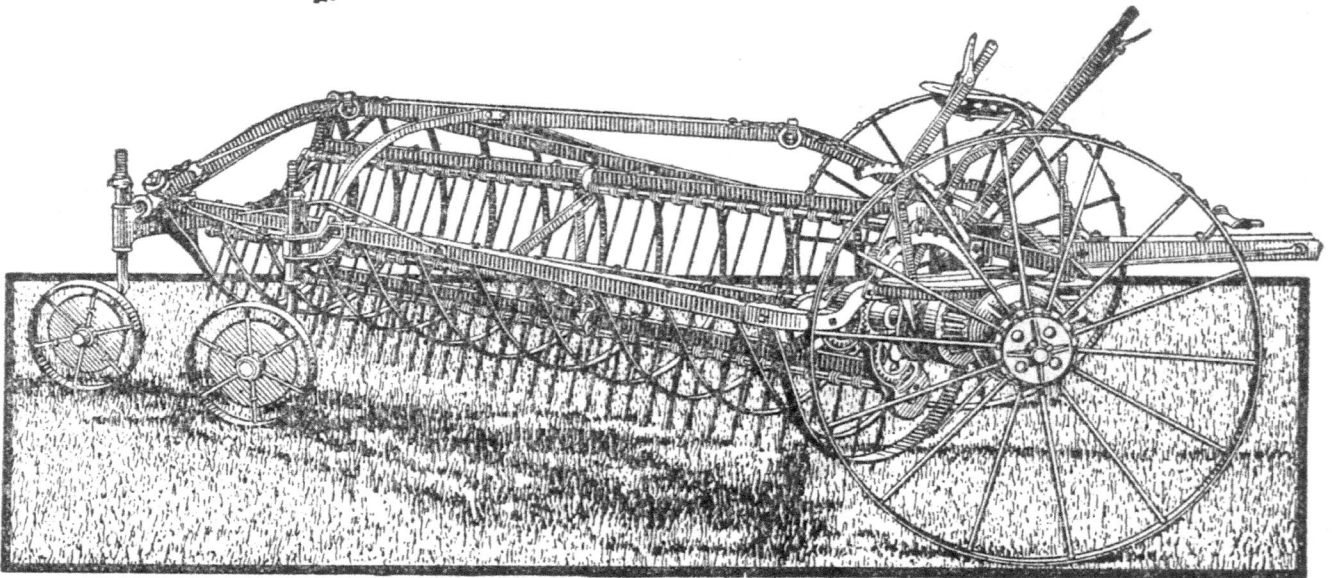

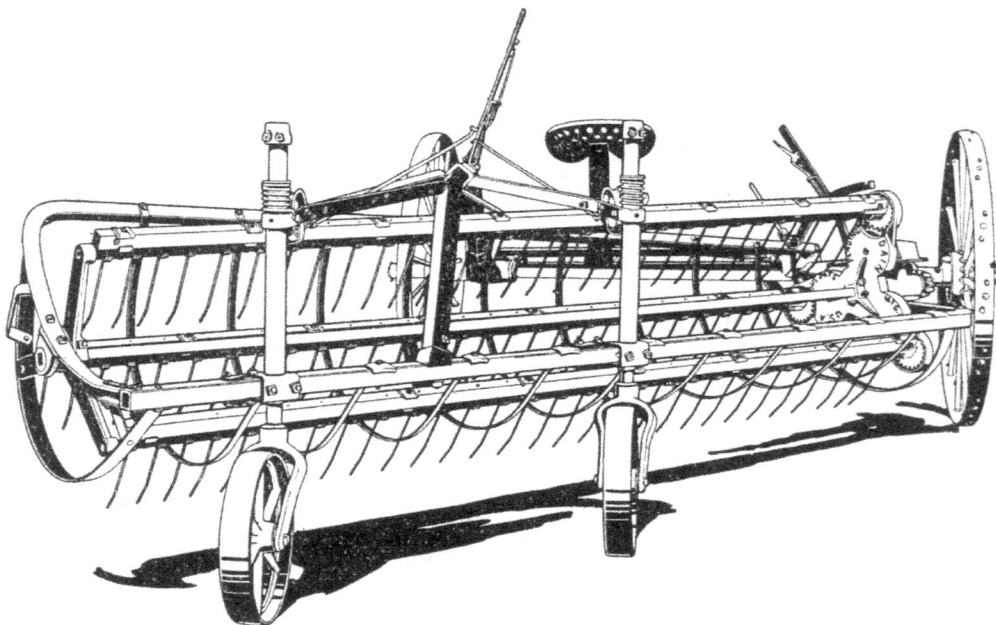

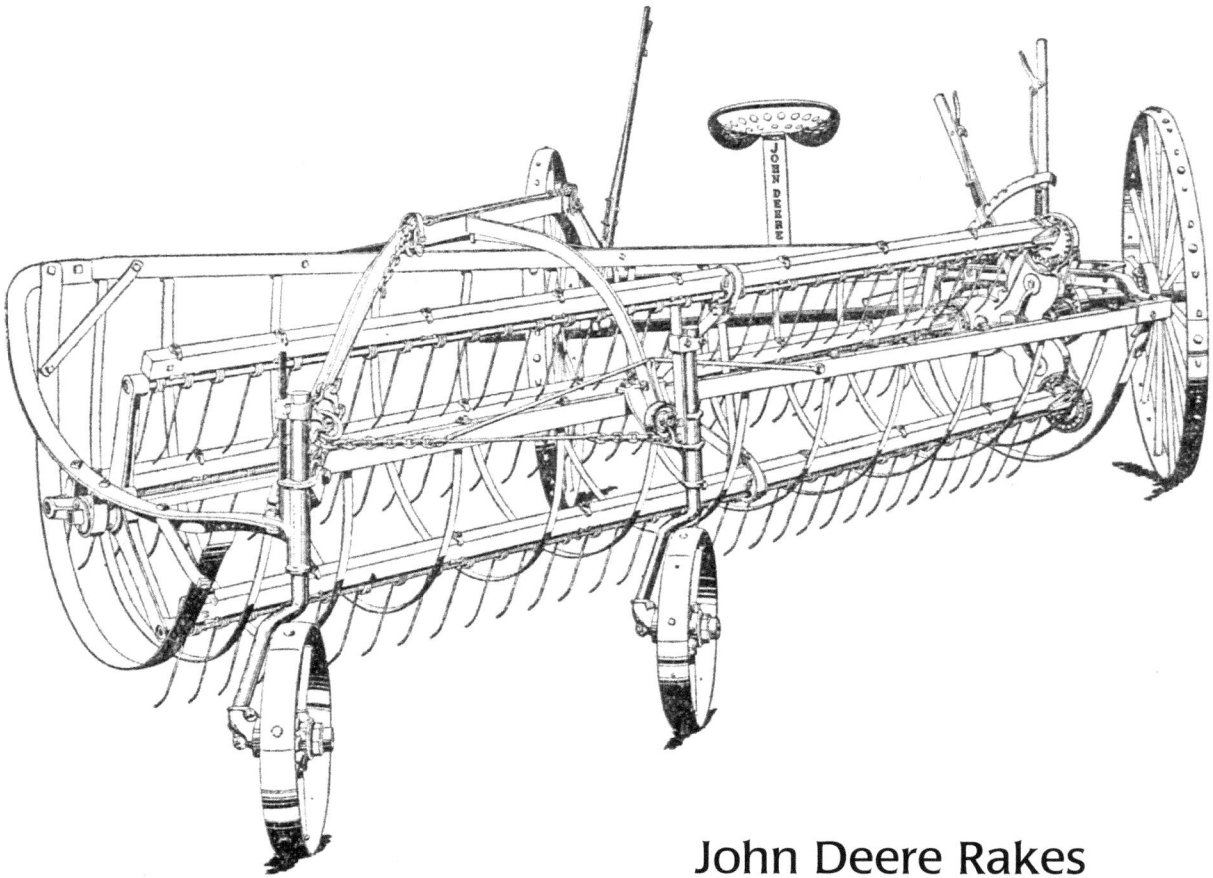

John Deere Rakes

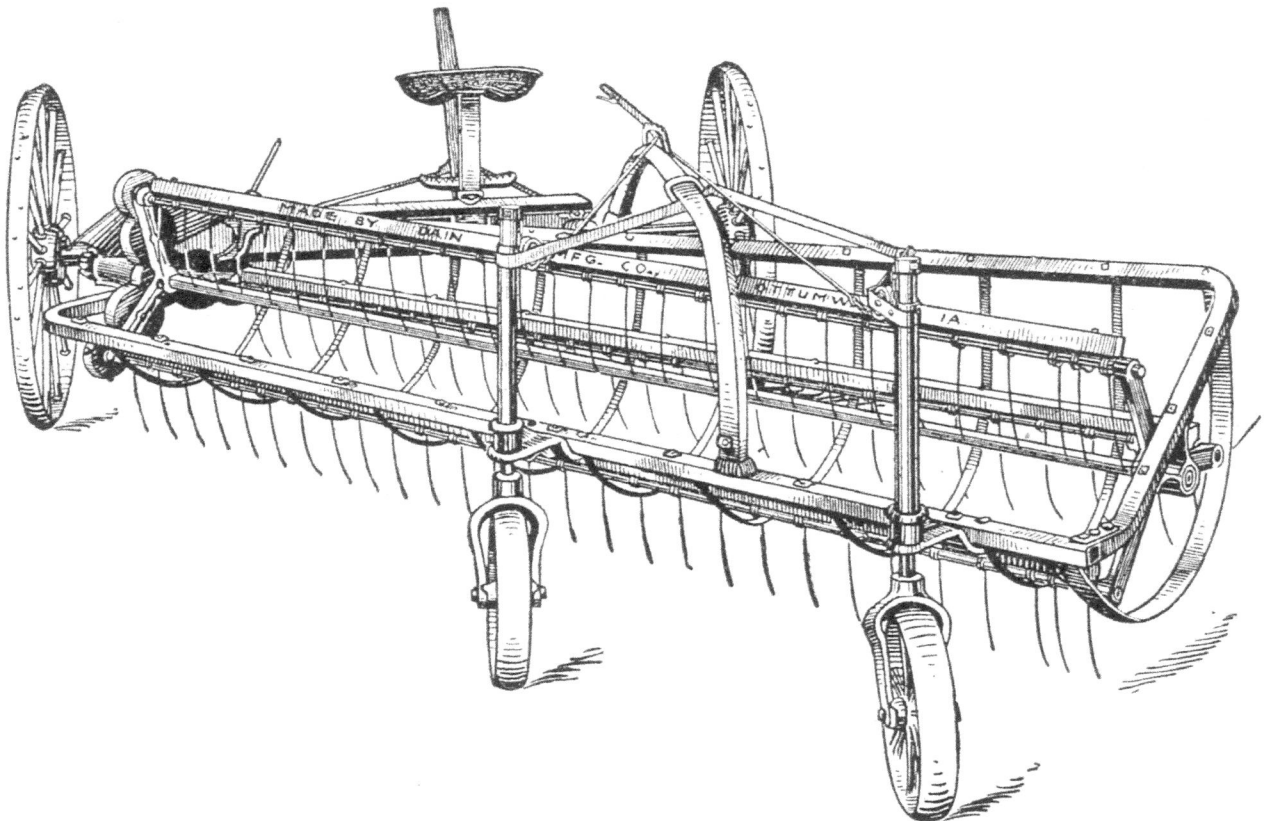

RB 108
M 731

R 491

S 221

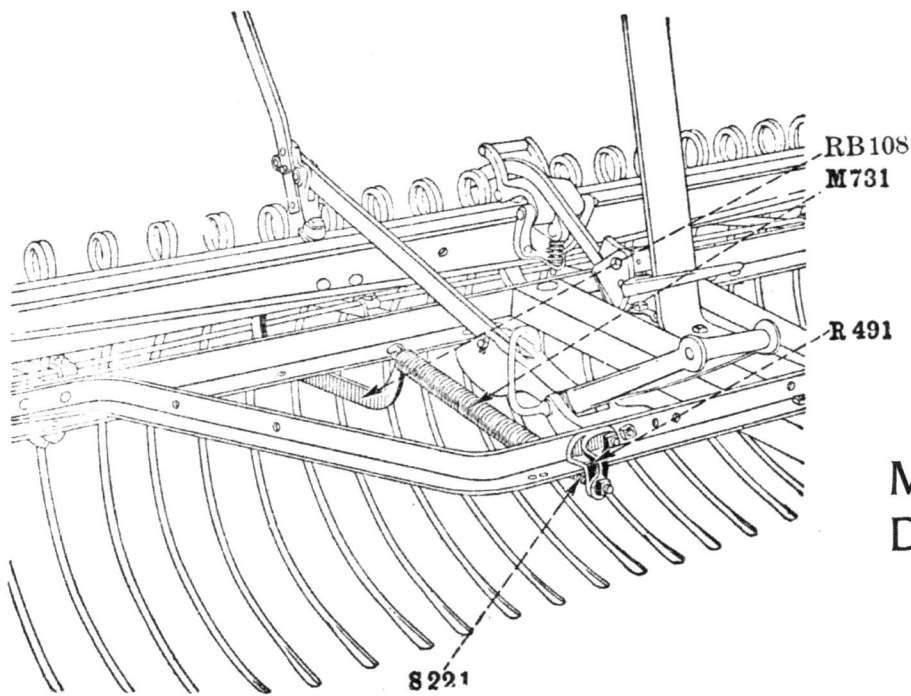

McCormick Deering Rakes

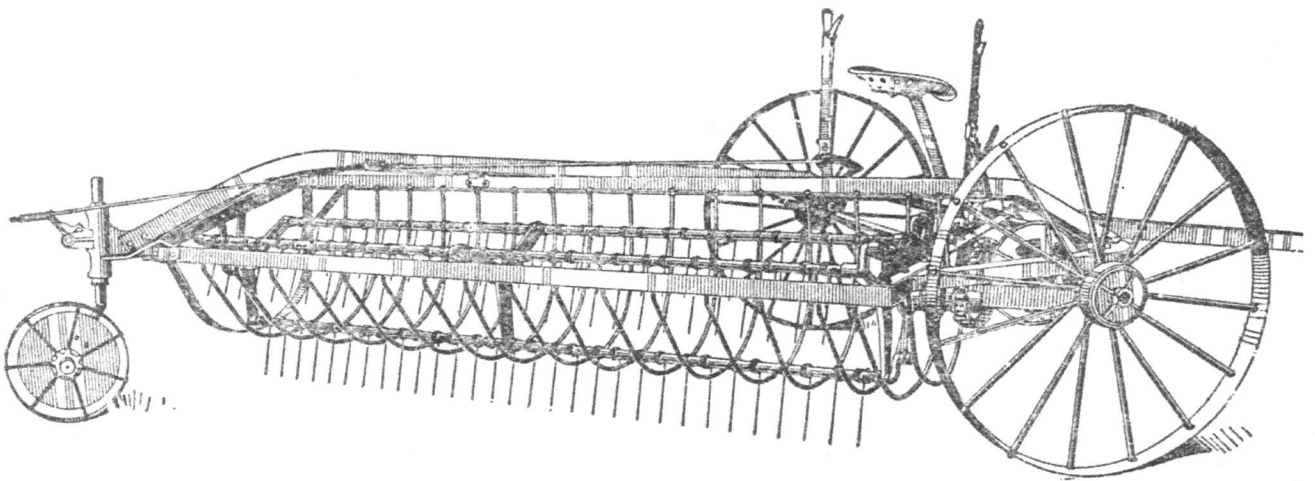

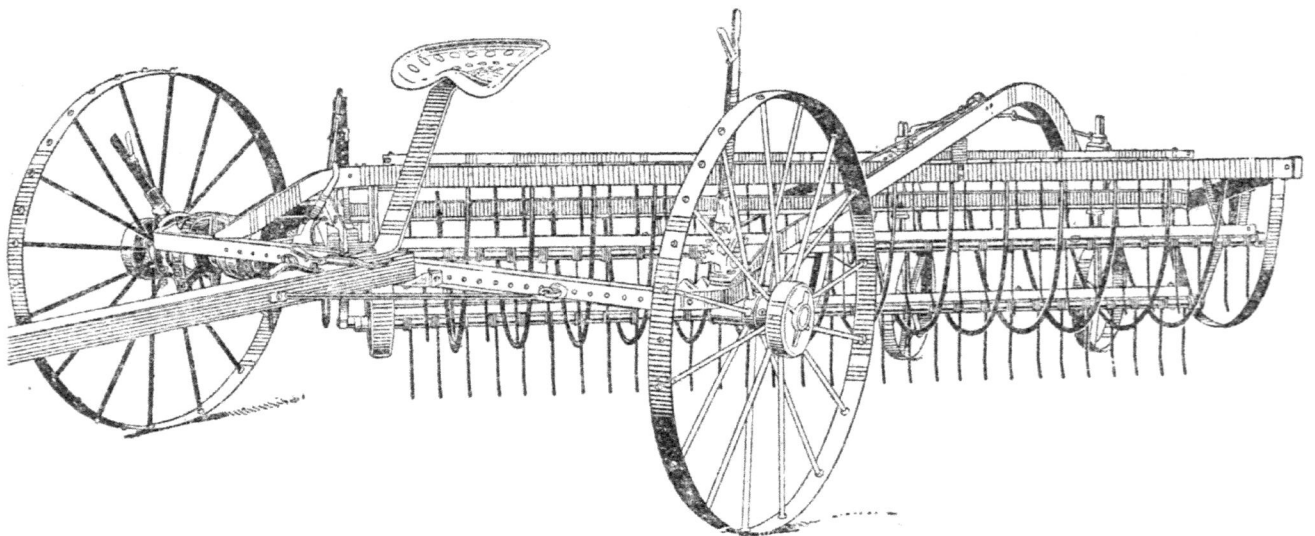

McCormick Deering Rakes

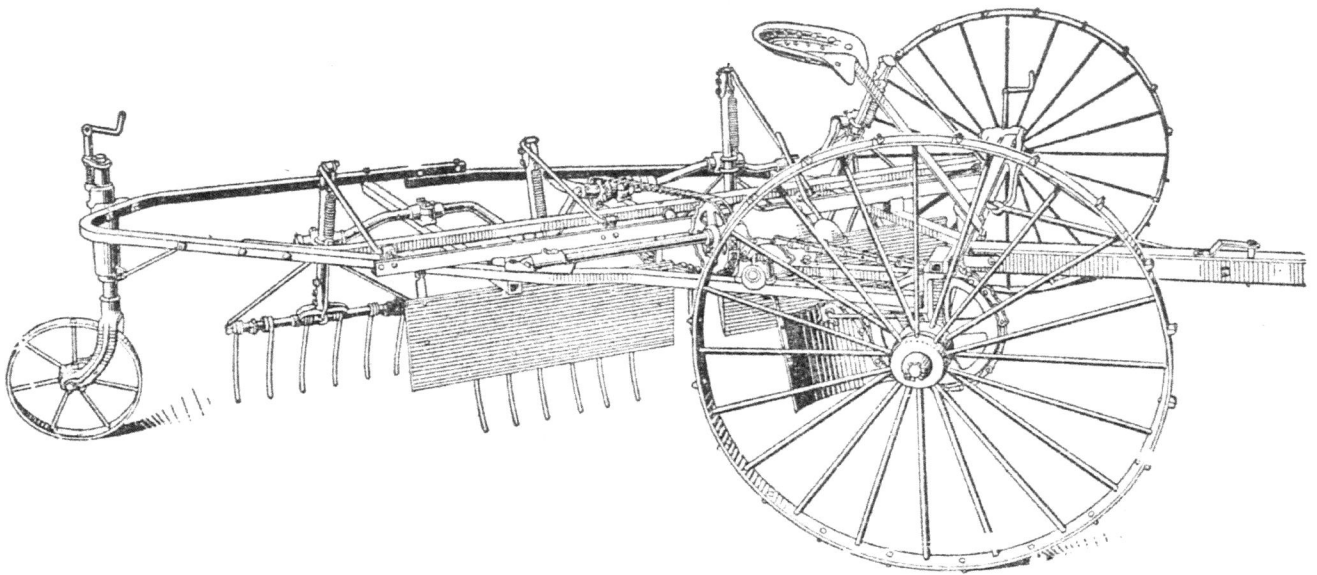

*Jess Ross up front, the author behind
waiting with two buckrake jags.*

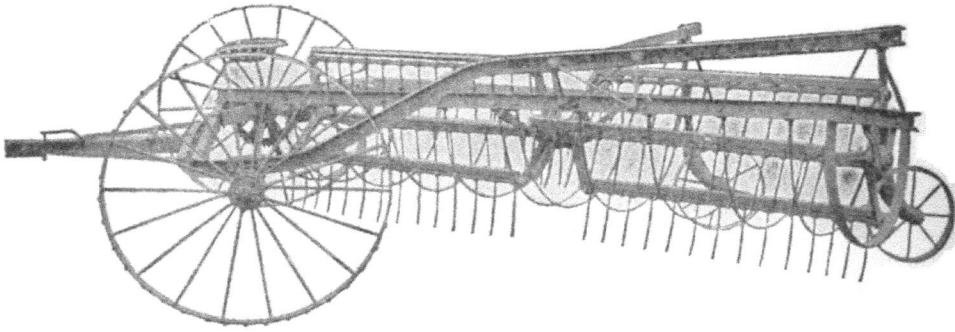

Rock Island Side Delivery Rakes

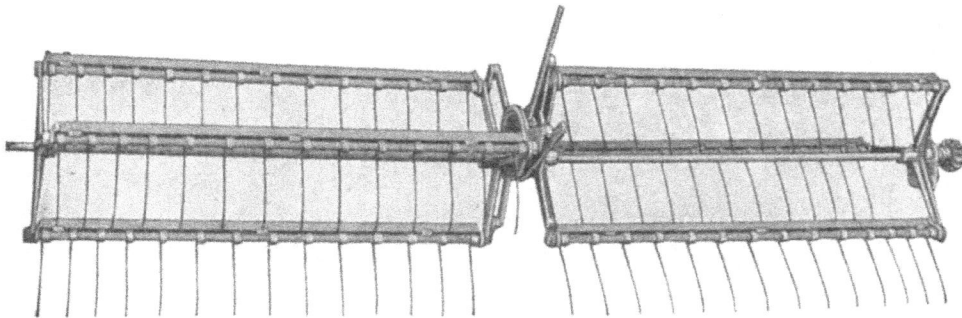

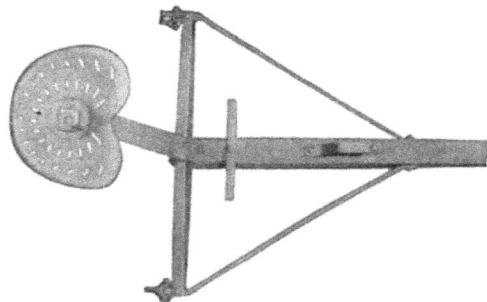

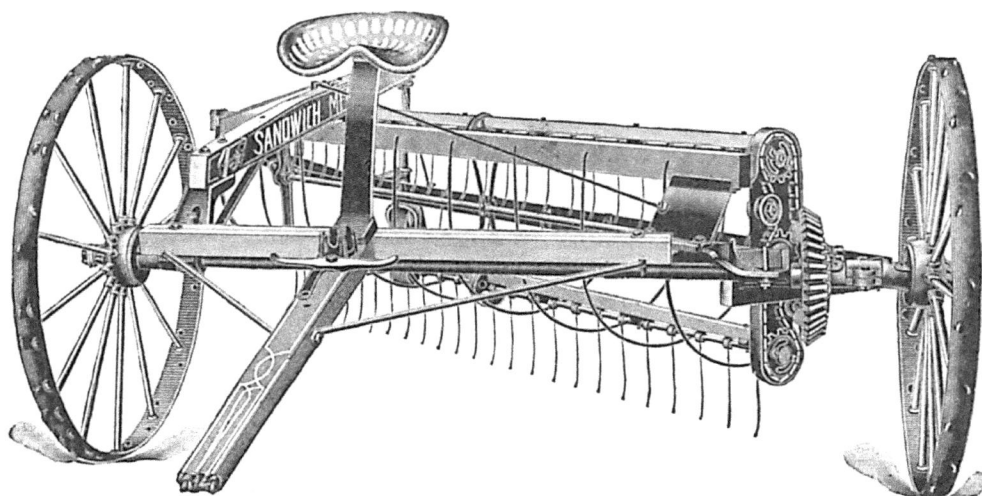

Sandwich brand Side Delivery Rake

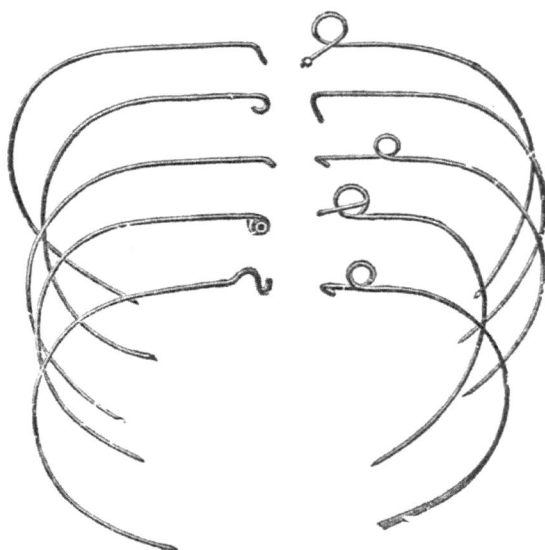

Dump rake teeth.

Custom built double wide Scatter Rake in use in Montana.

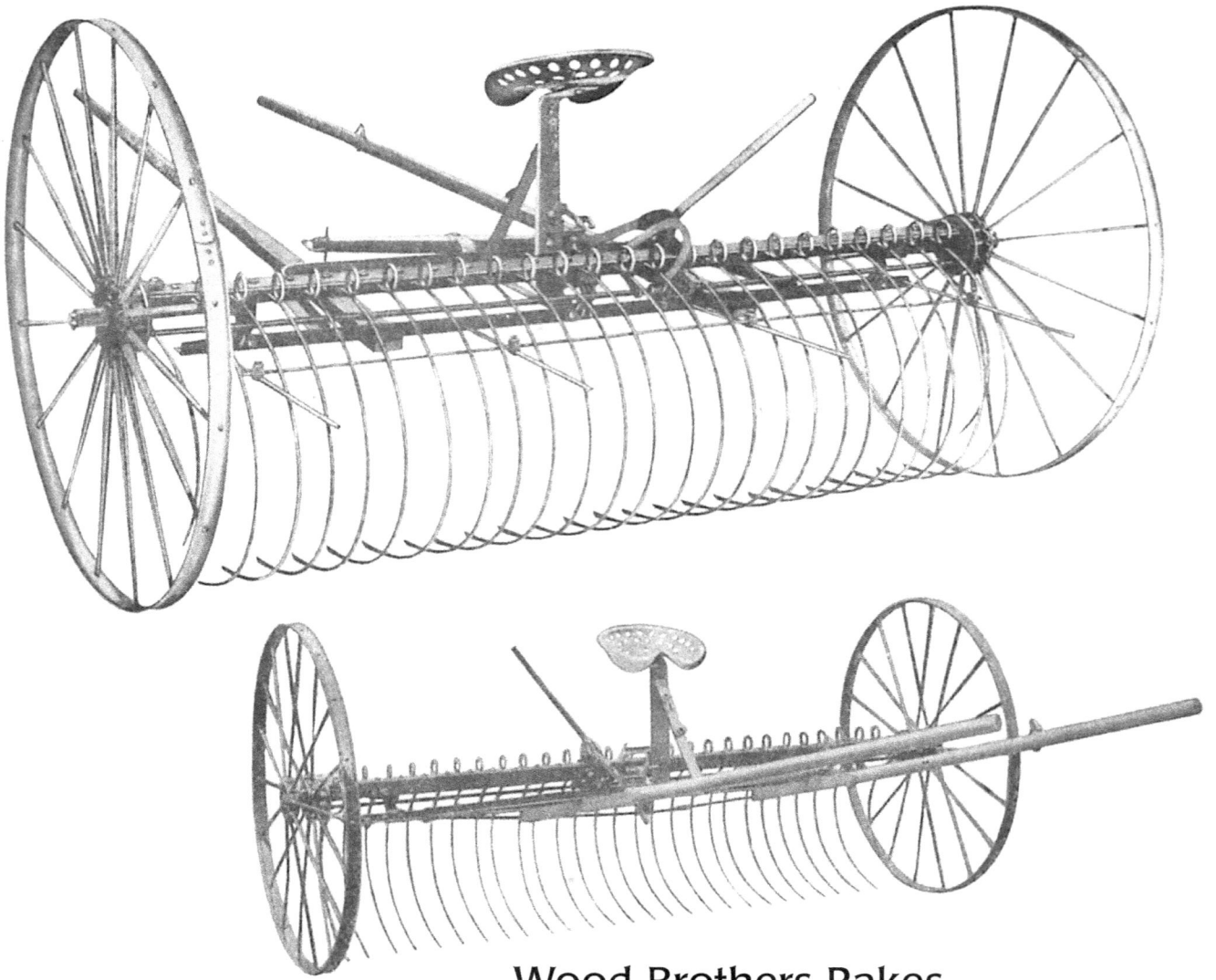

Wood Brothers Rakes

*Bud Evers mowing clover at
Singing Horse Ranch.*

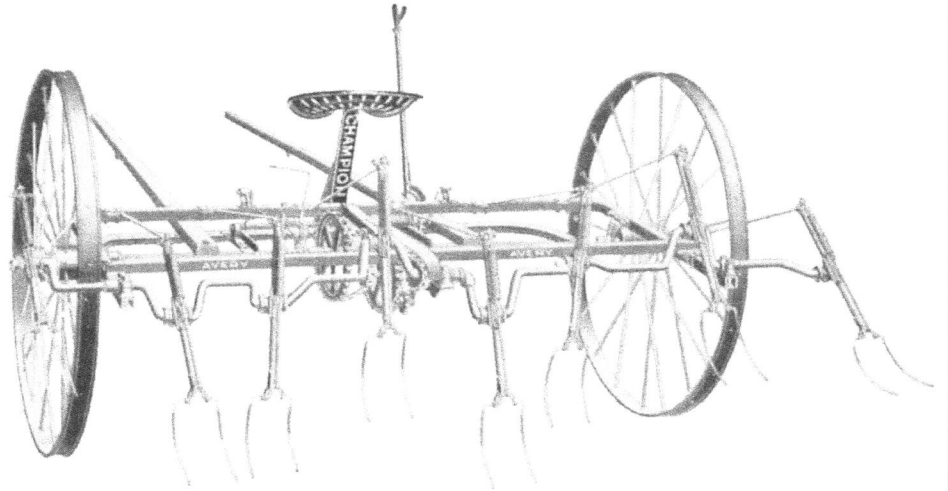

Champion Tedders

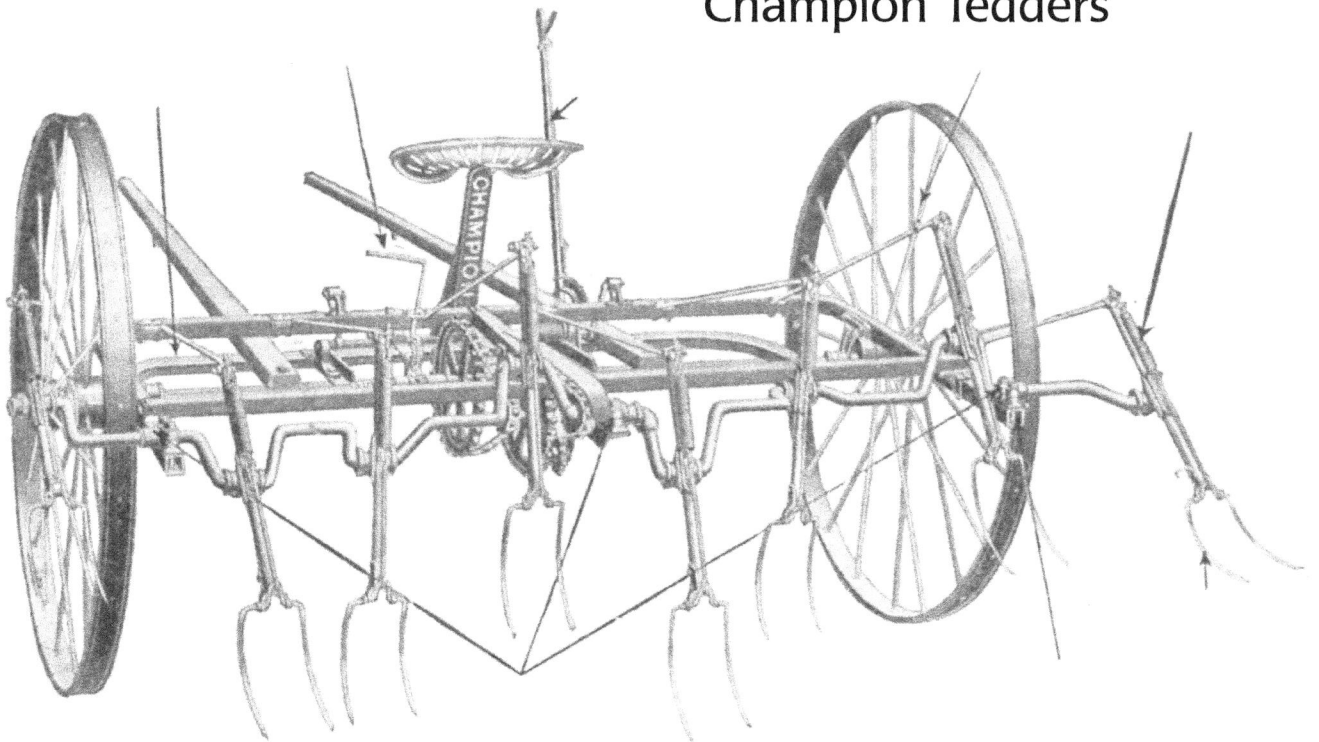

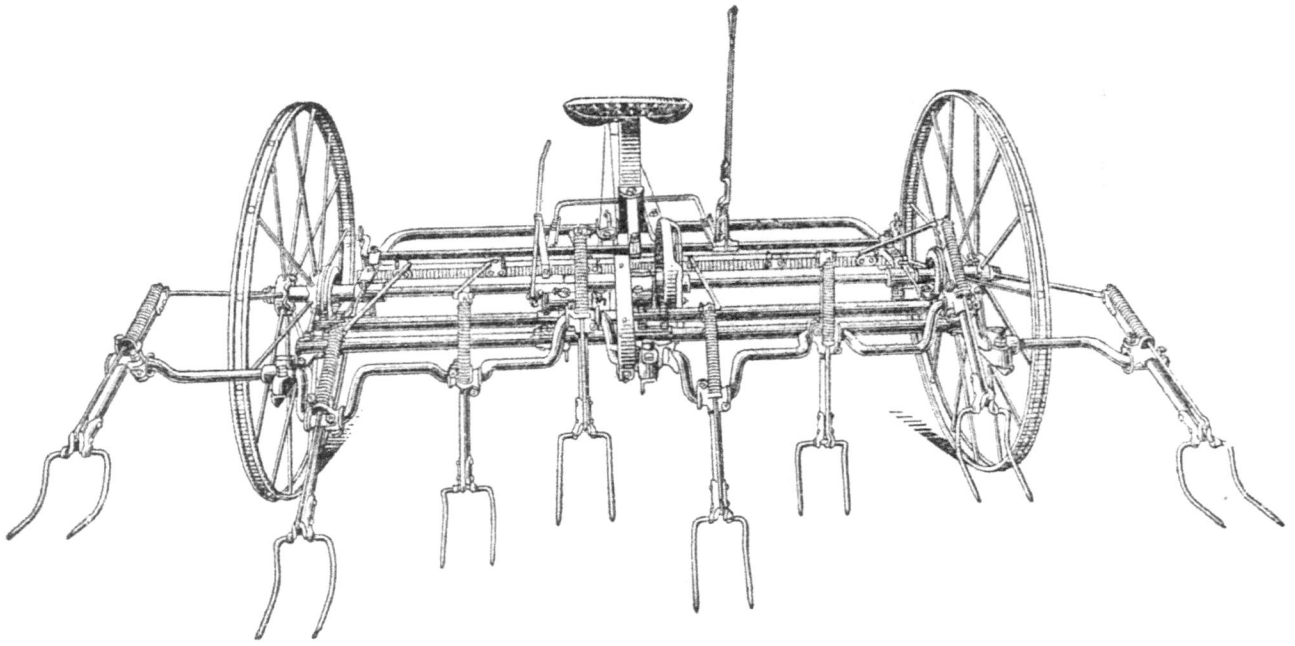

McCormick Tedders

Rock Island Rakes

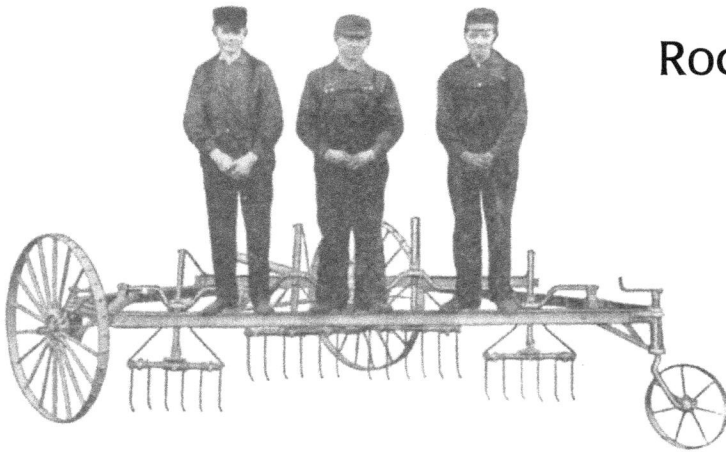

*Cuts demonstrating the strength
of Rock Island rake frames and
the reversible gearing.*

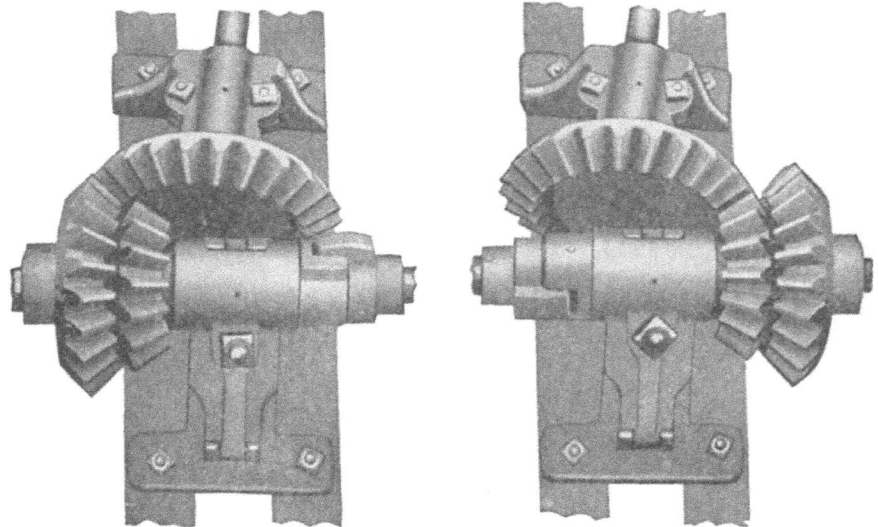

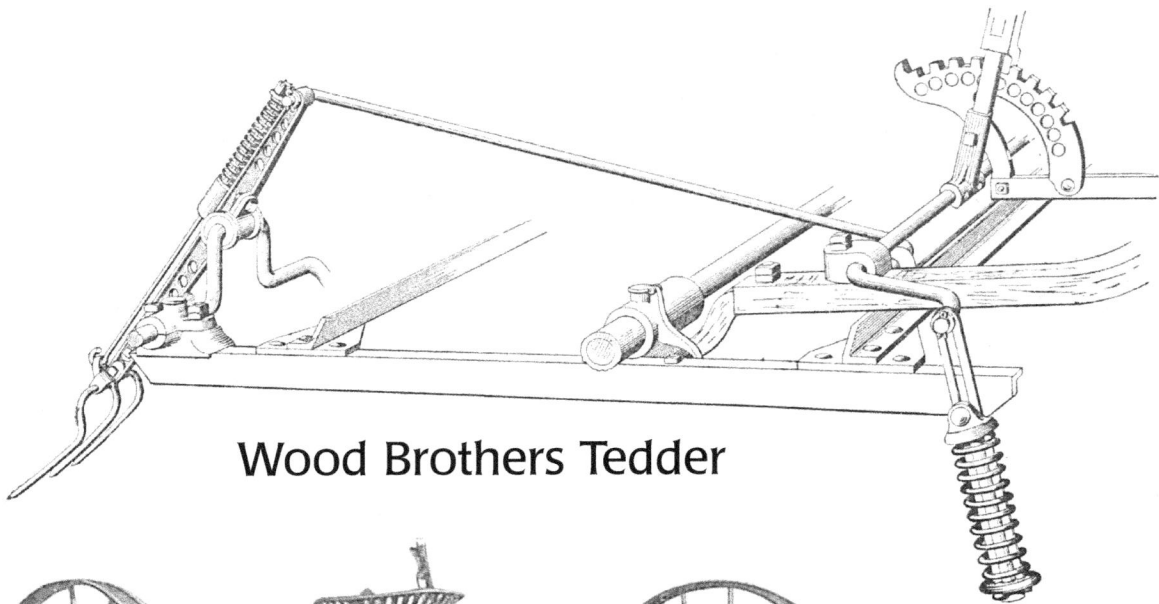

Wood Brothers Tedder

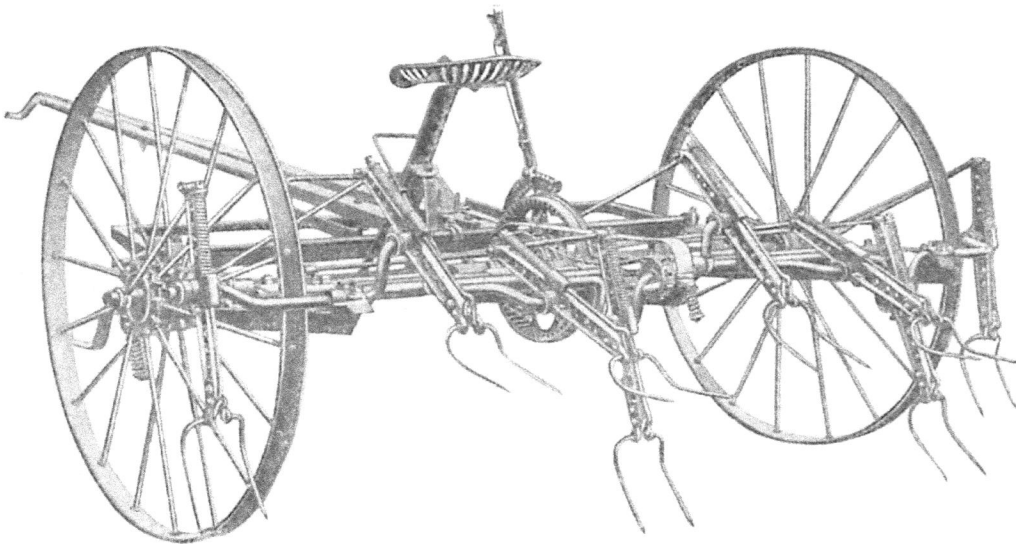

"Quail" Tedder

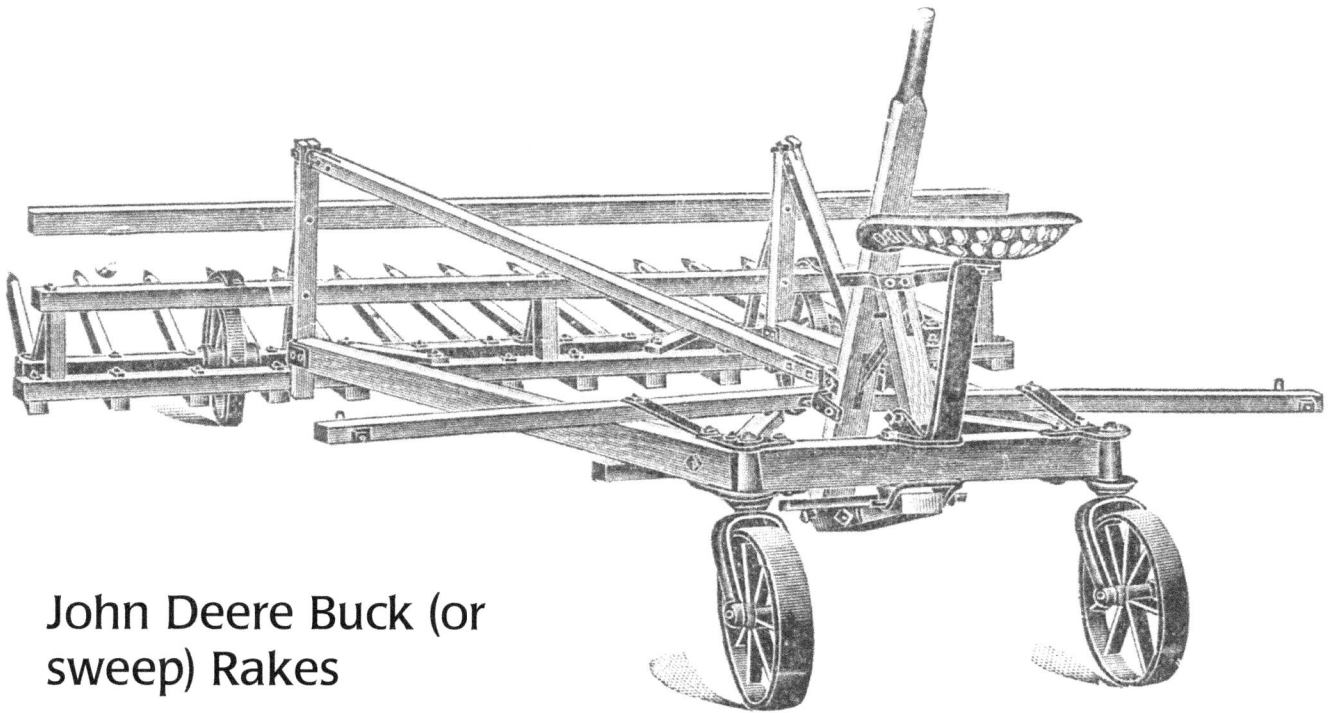

John Deere Buck (or sweep) Rakes

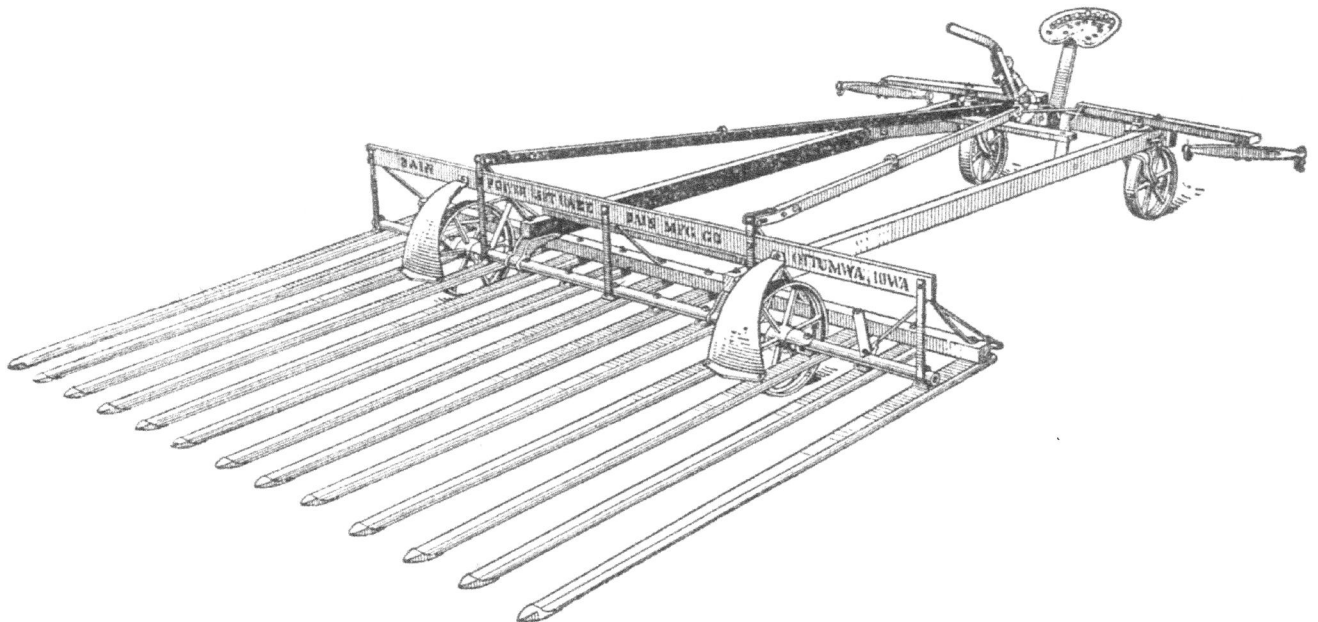

John Deere Buck (or sweep) Rakes

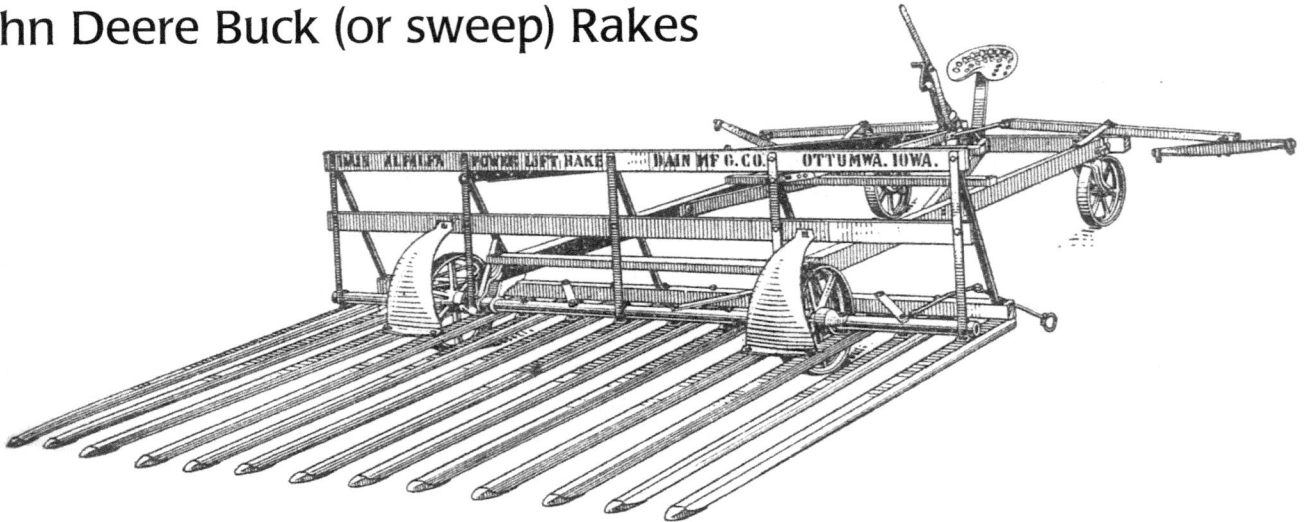

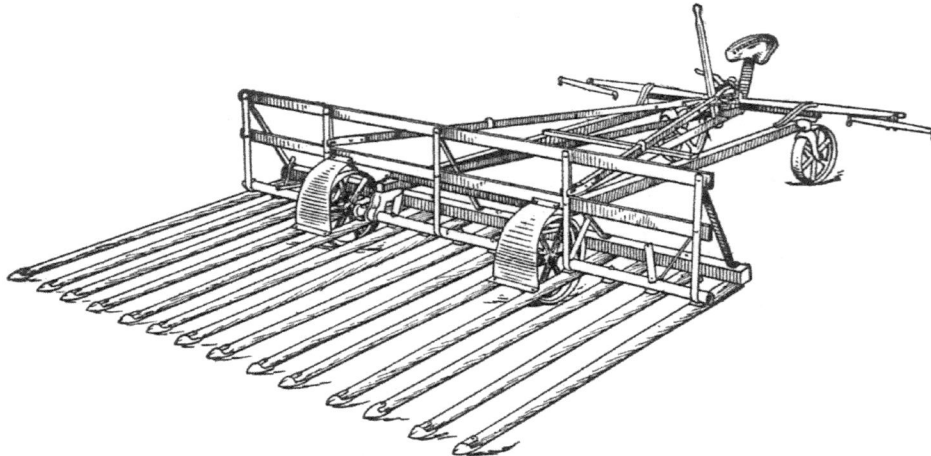

A "push-off board" for John Deere
Buckrakes.

W5575
WA5527
WA5170
W5564
WA5527

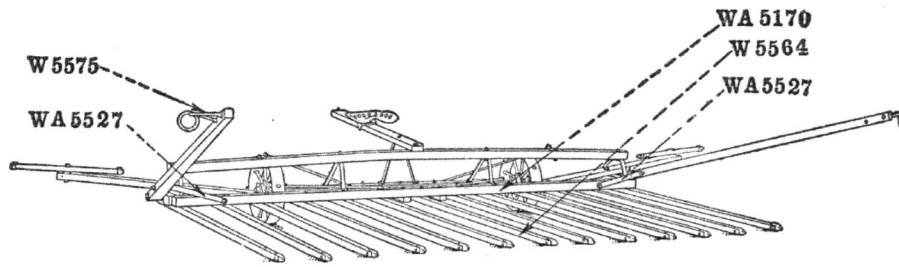

No. 1A Two-Wheel Sweep Rake (Side Hitch) (front view)

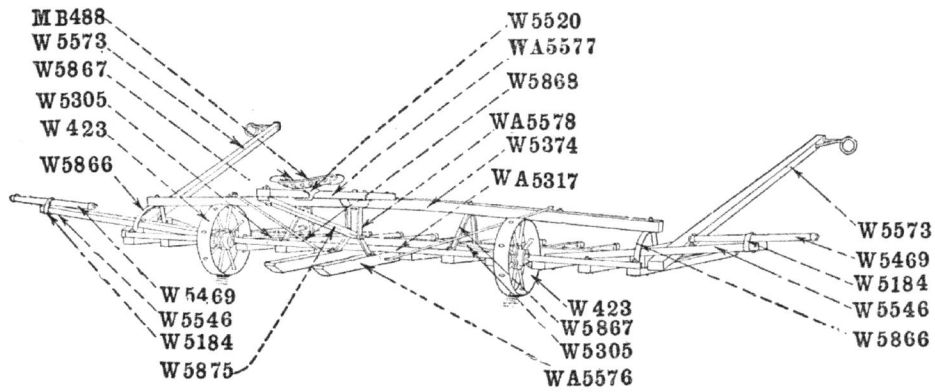

MB488
W5573
W5867
W5305
W423
W5866

W5520
WA5577
W5868
WA5578
W5374
WA5317

W5469
W5546
W5184
W5875

W423
W5867
W5305
WA5576

W5573
W5469
W5184
W5546
W5866

No. 1A Two-Wheel Sweep Rake (Side Hitch) (rear view)

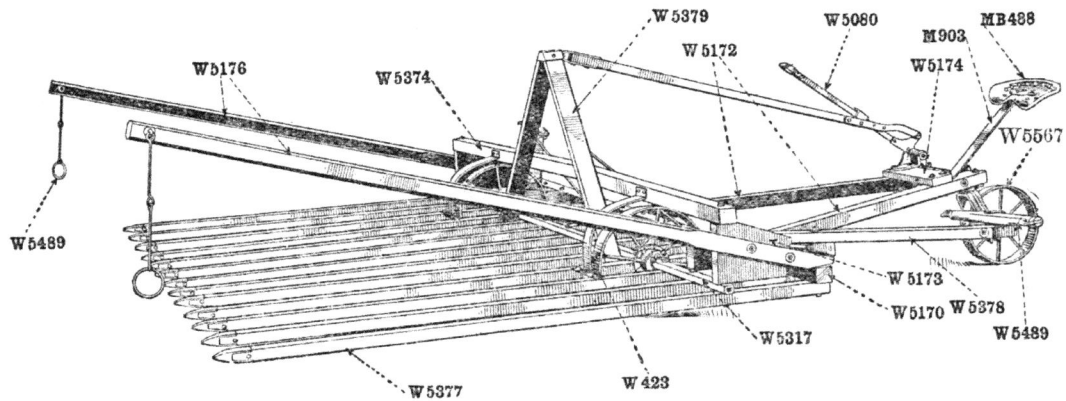

W5379
W5080
MB488
M903
W5172
W5174
W5176
W5374
W5567
W5489
W5173
W5170
W5378
W5489
W5317
W5377
W423

McCormick Deering Buckrakes

W5309 W5295 W5290 W5372 W5299 W5297 MB488
W5314 M903
W250 MA589
W5312 W5371 W5567
W5565 W5564 W5317 W5170

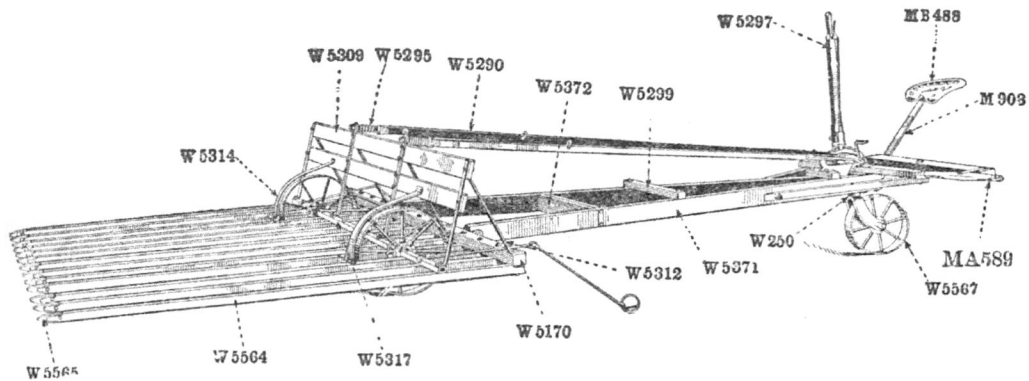

No. 3 Three-Wheel Sweep Rake (Rear Hitch)
No. 4 is like No. 3, but has Two Castor Wheels

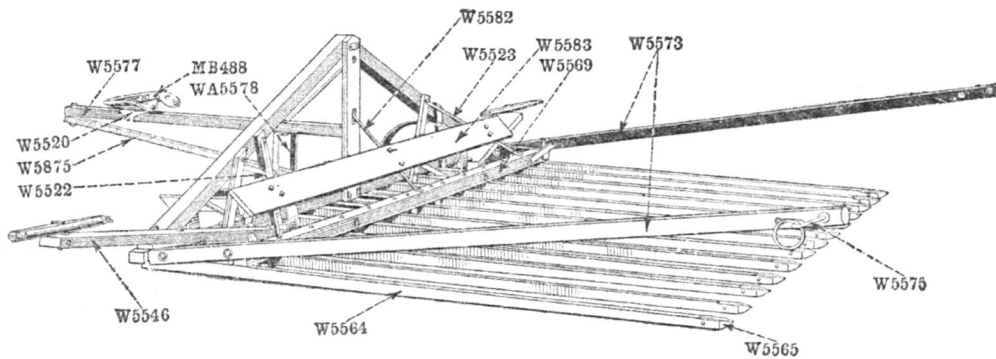

W5582
W5523 W5583 W5573
W5577 W5569
MB488
WA5578
W5520
W5875
W5522
W5575
W5546 W5564 W5565

No. 5 Two-Wheel Side Hitch Sweep Rake

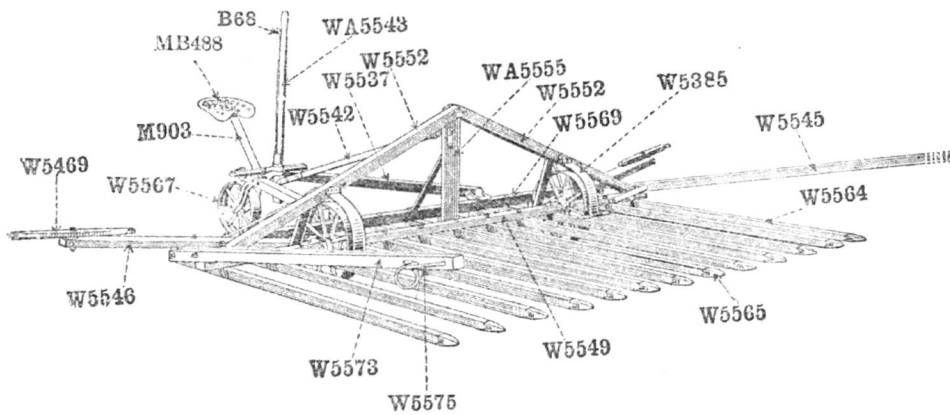

B68 WA5543
MB488 W5552
W5537 WA5555 W5385
W5542 W5552 W5545
M903 W5569
W5469 W5564
W5567
W5546 W5565
W5573 W5549
W5575

McCormick Deering Buckrakes

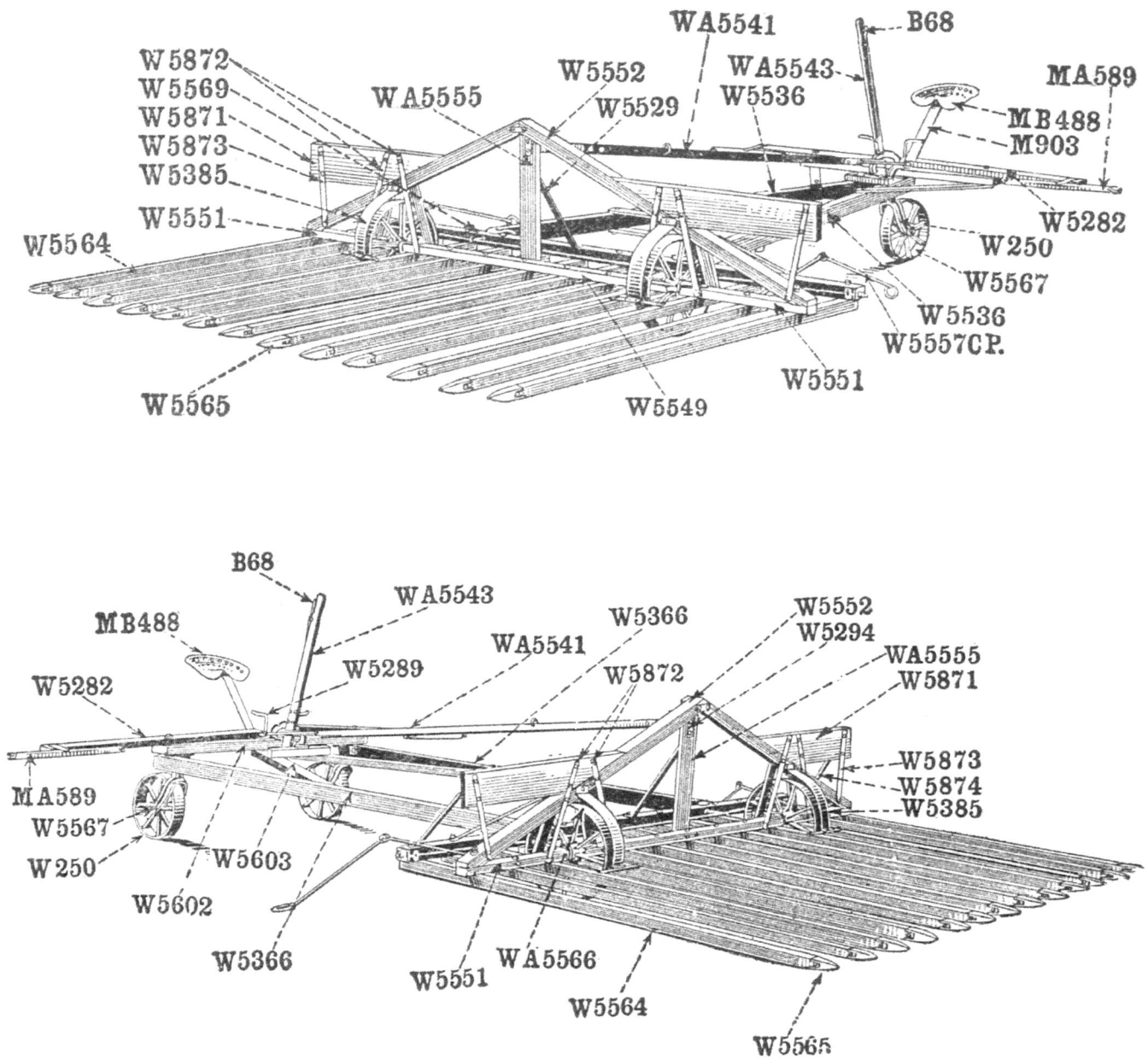

W5872
W5569
W5871
W5873
W5385
W5551
W5564
WA5555
W5552
W5529
WA5541
WA5543
W5536
B68
MA589
MB488
M903
W5282
W250
W5567
W5536
W5557CP.
W5551
W5565
W5549

B68
MB488
W5282
MA589
W5567
W250
W5602
W5366
WA5543
W5289
WA5541
W5366
W5872
W5552
W5294
WA5555
W5871
W5873
W5874
W5385
W5603
W5551
WA5566
W5564
W5565

McCormick Deering Buckrakes

John Deere Hay Loaders

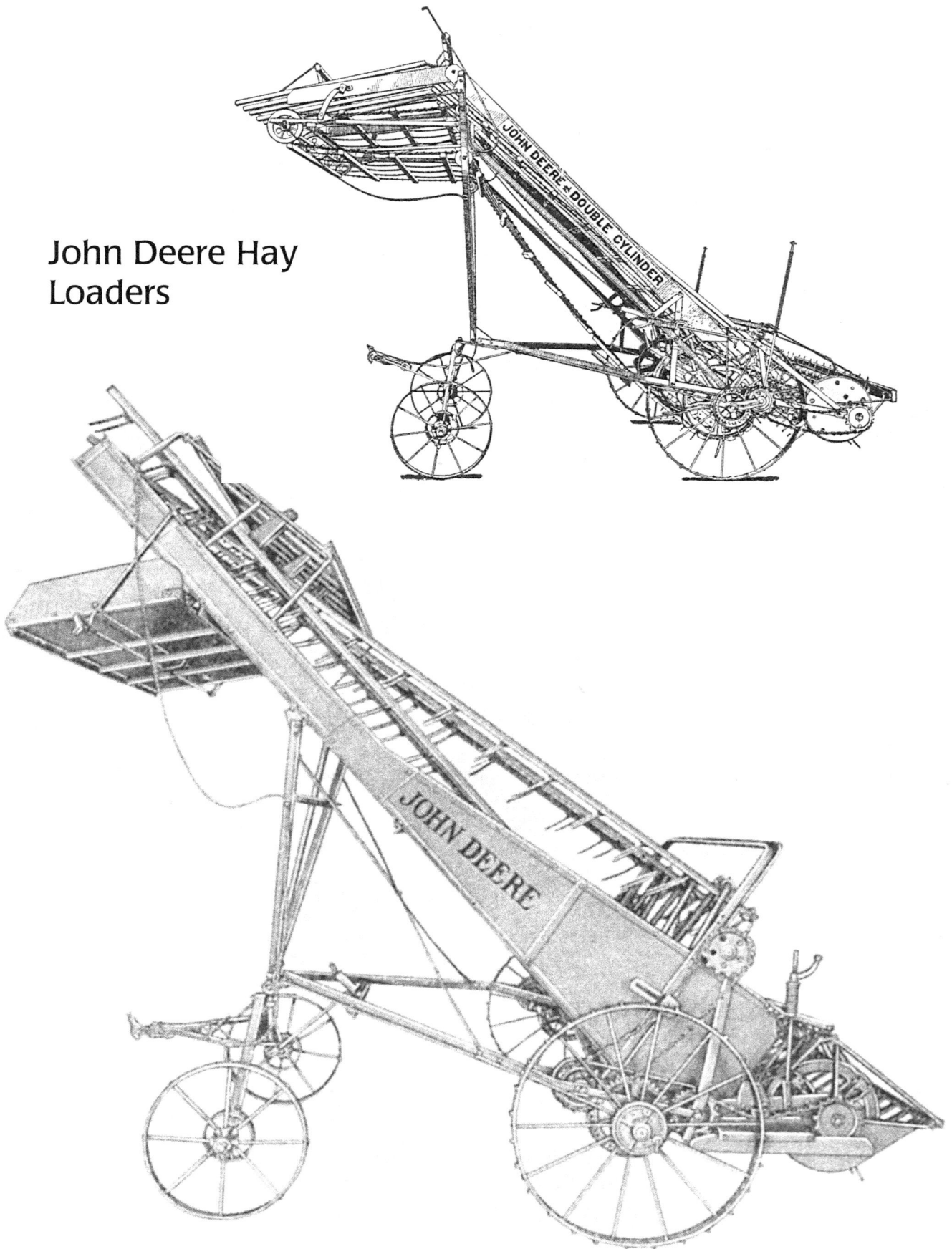

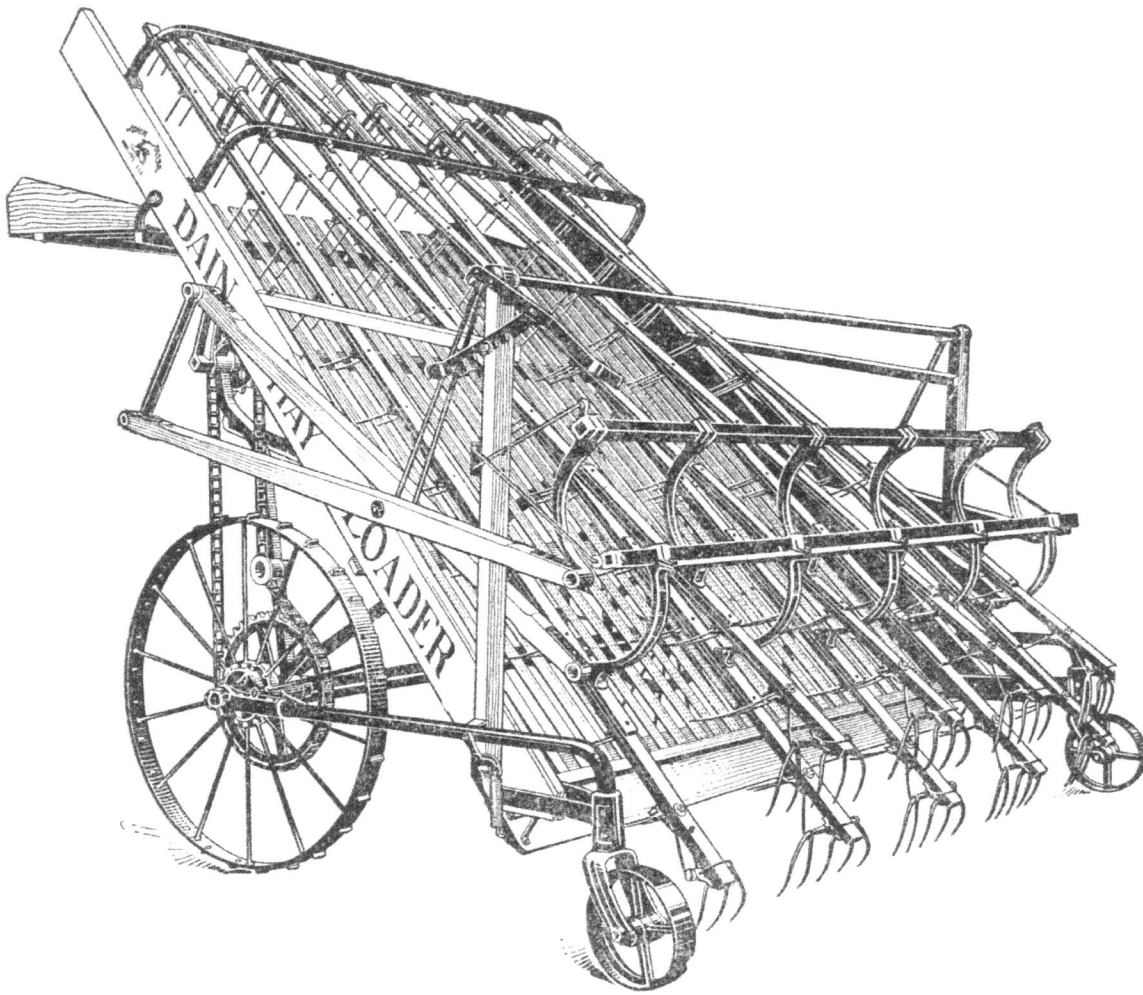

John Deere Hay Loaders

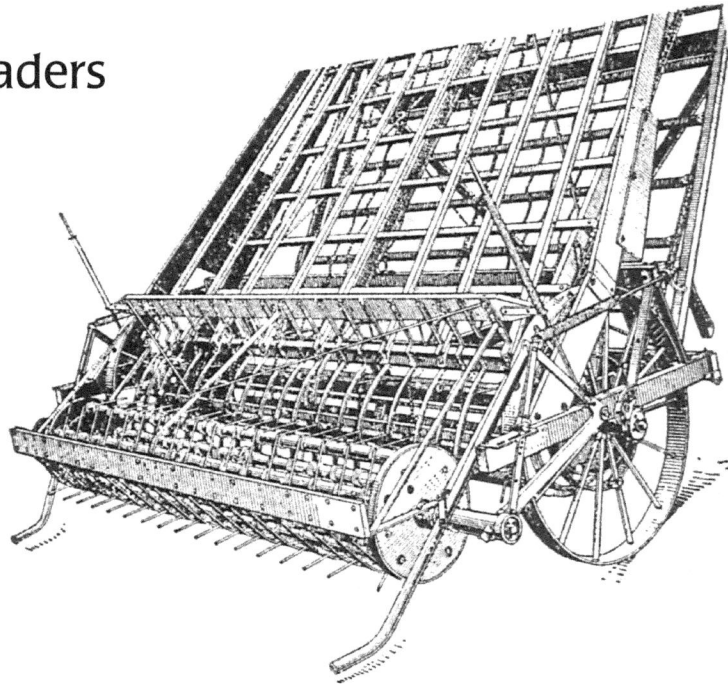

John Deere Hay Loader

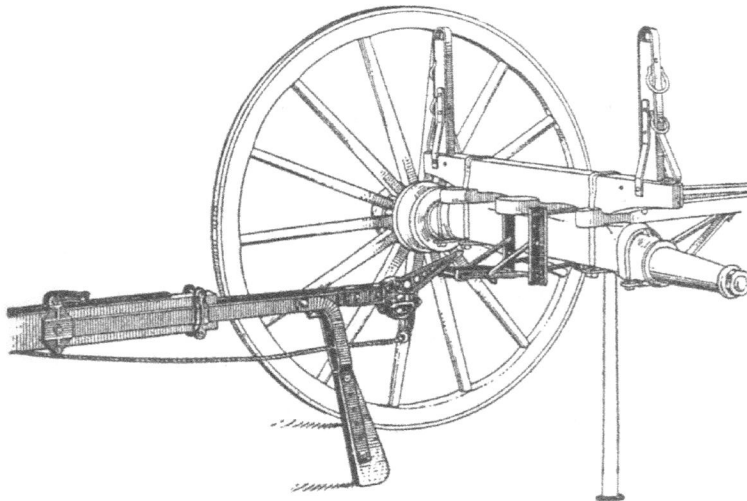

A trigger leg was available for the John Deere Hay loaders to use rather than truck wheels. When moving forwrd the trigger folded back and drug free. If the wagon was backed a foot or more the trigger came forward and held up the loader tongue.

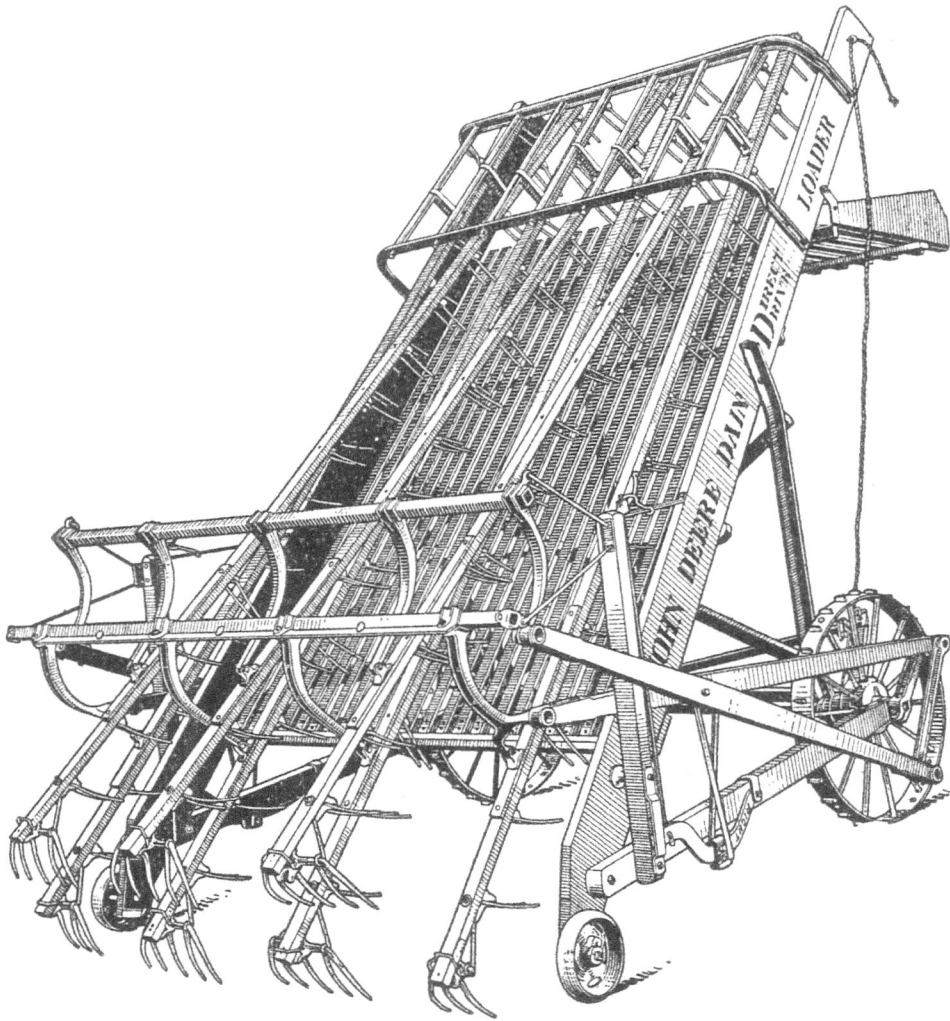

John Deere Hay Loaders

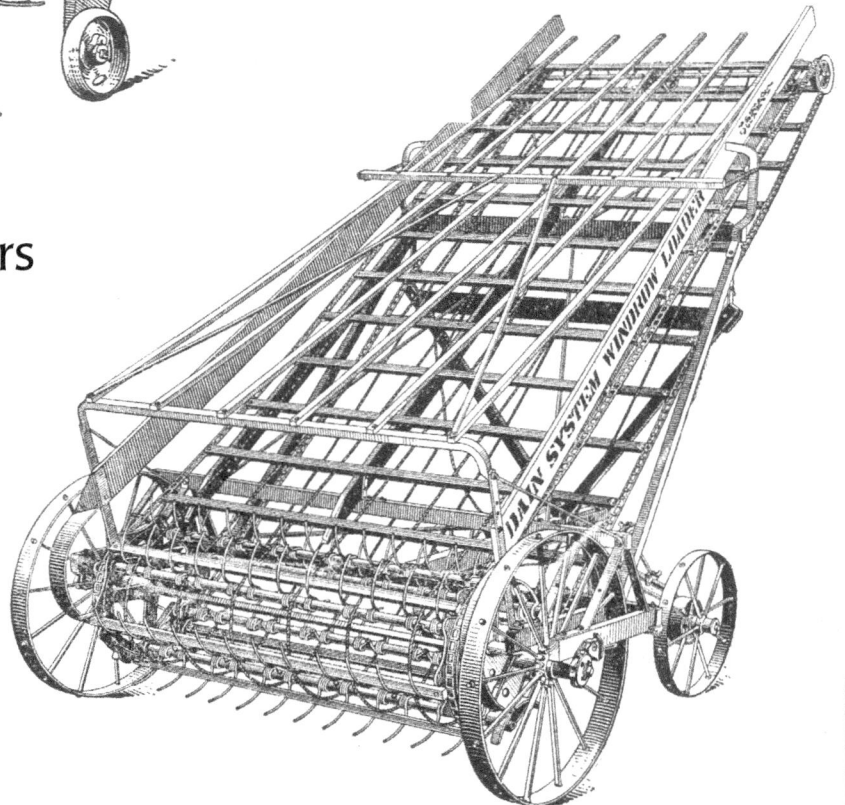

McCormick Deering Hay Loaders

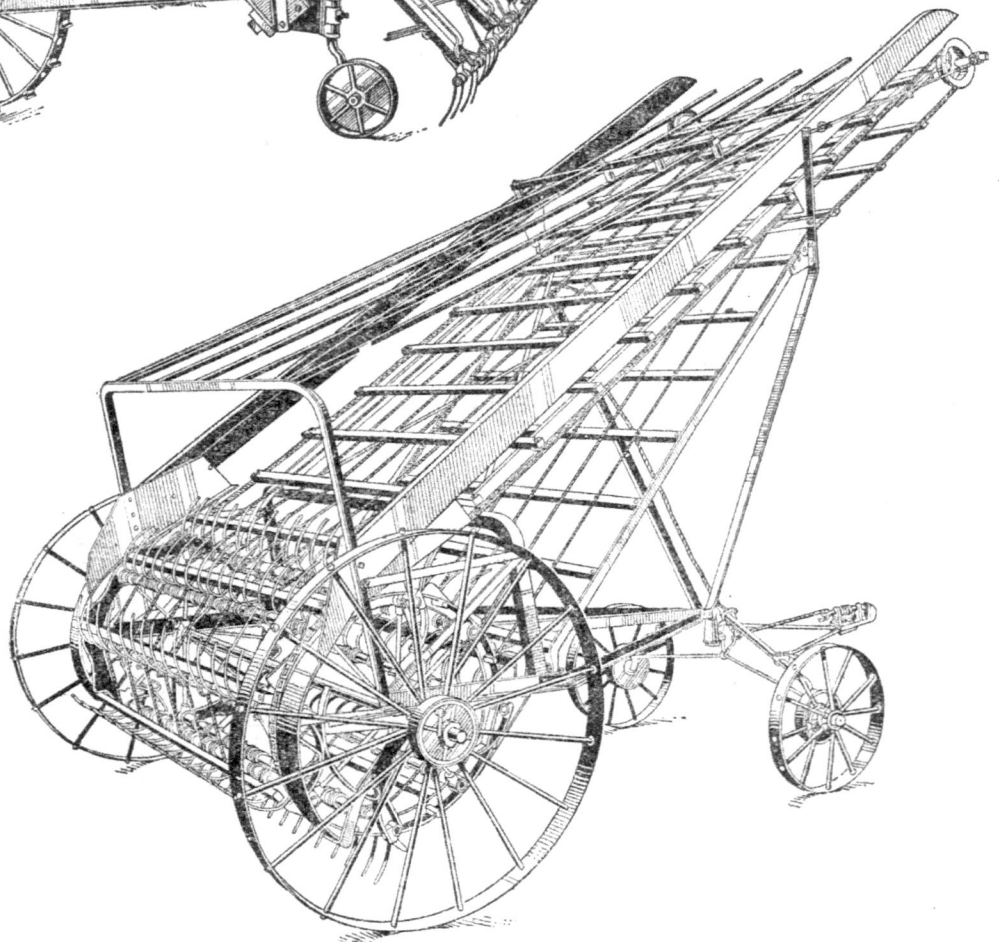

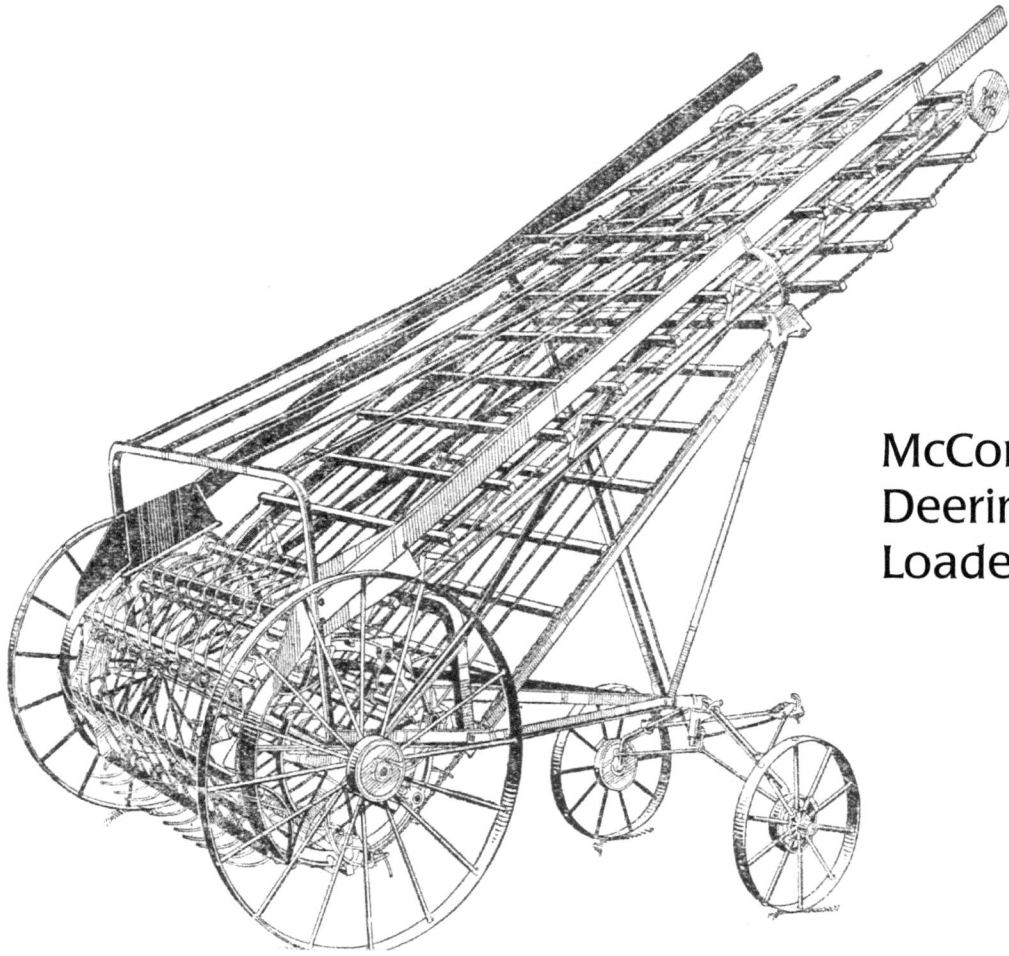

McCormick Deering Hay Loaders

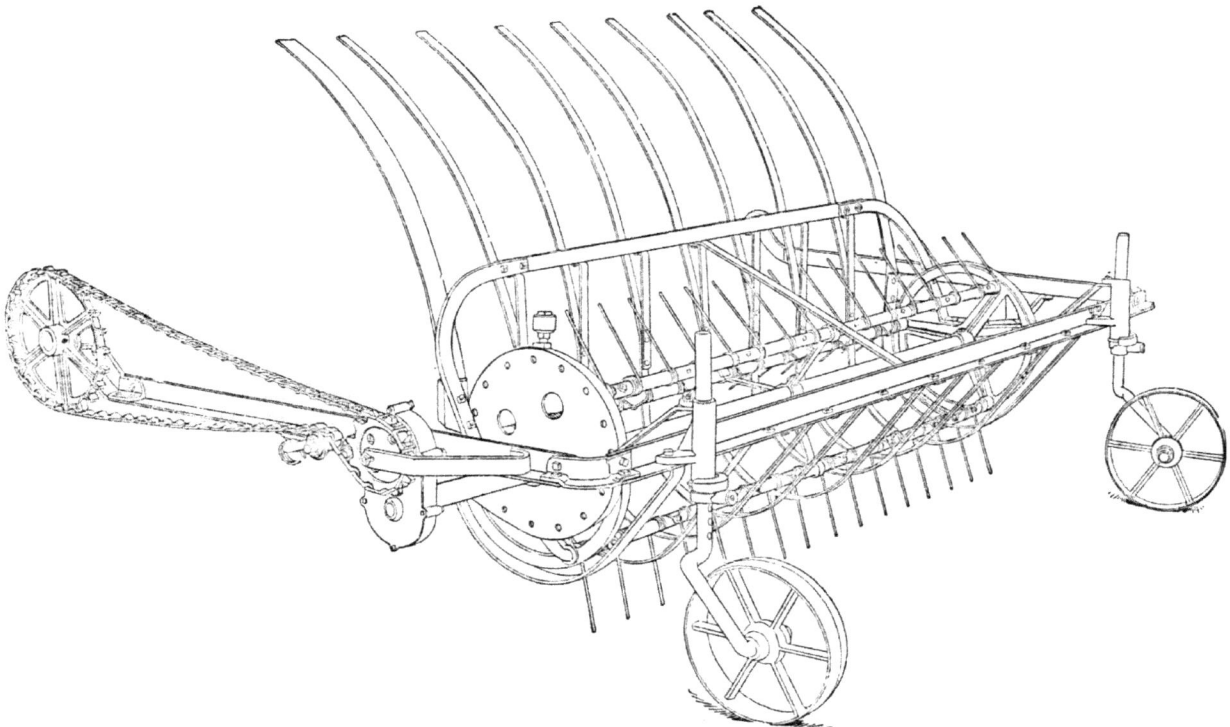

Keystone Hay Loaders

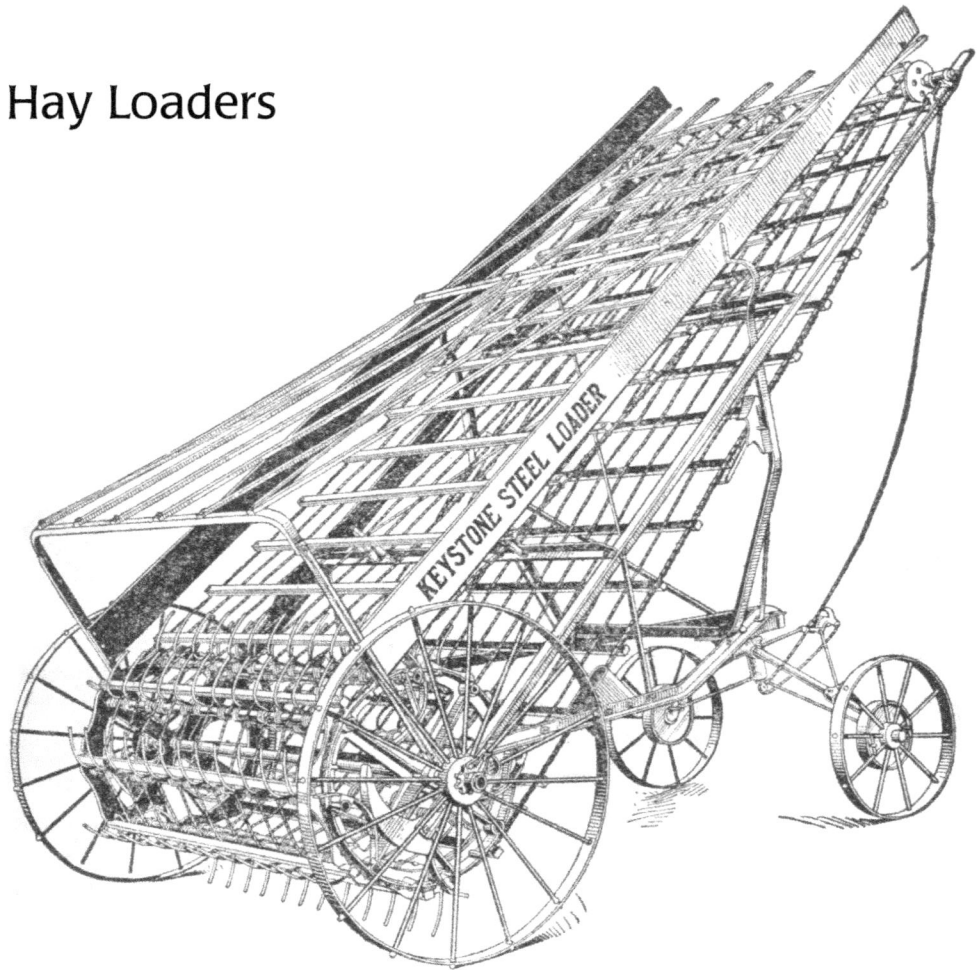

Rock Island Hay Loaders

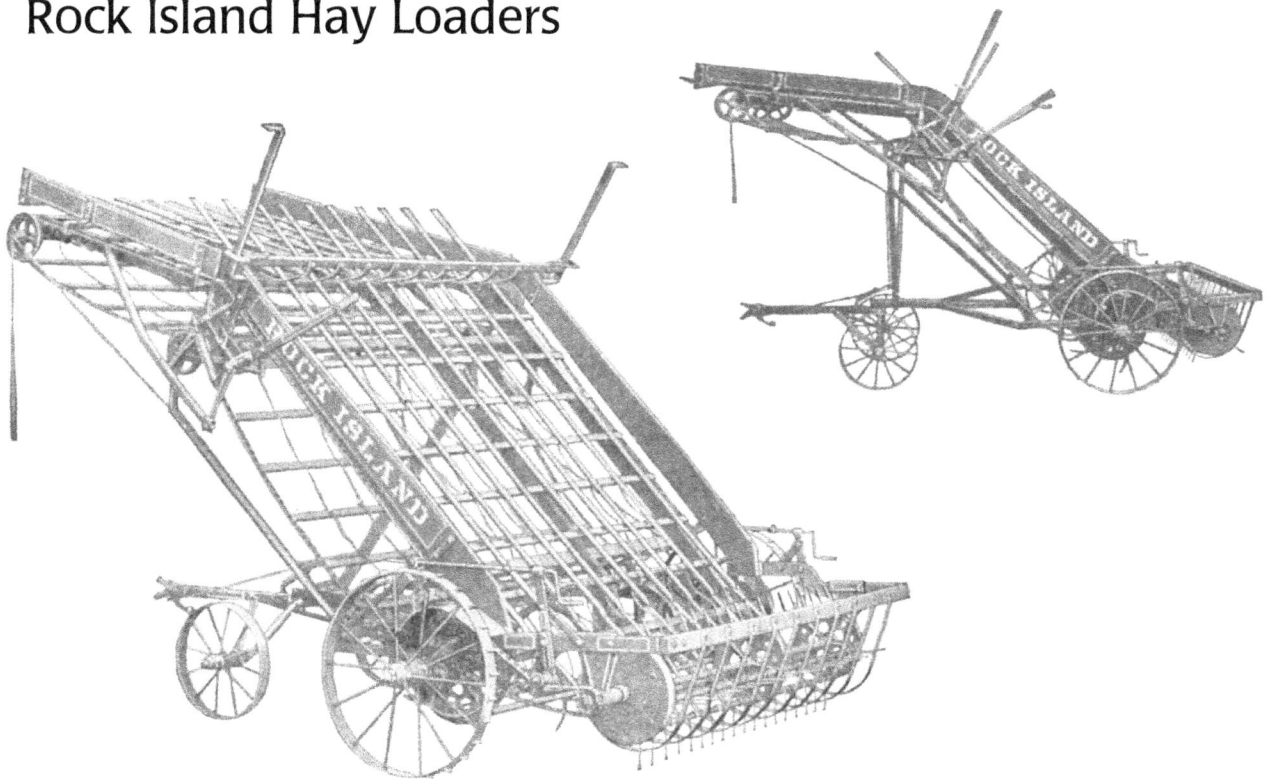

Wood Brothers
Hay Loader

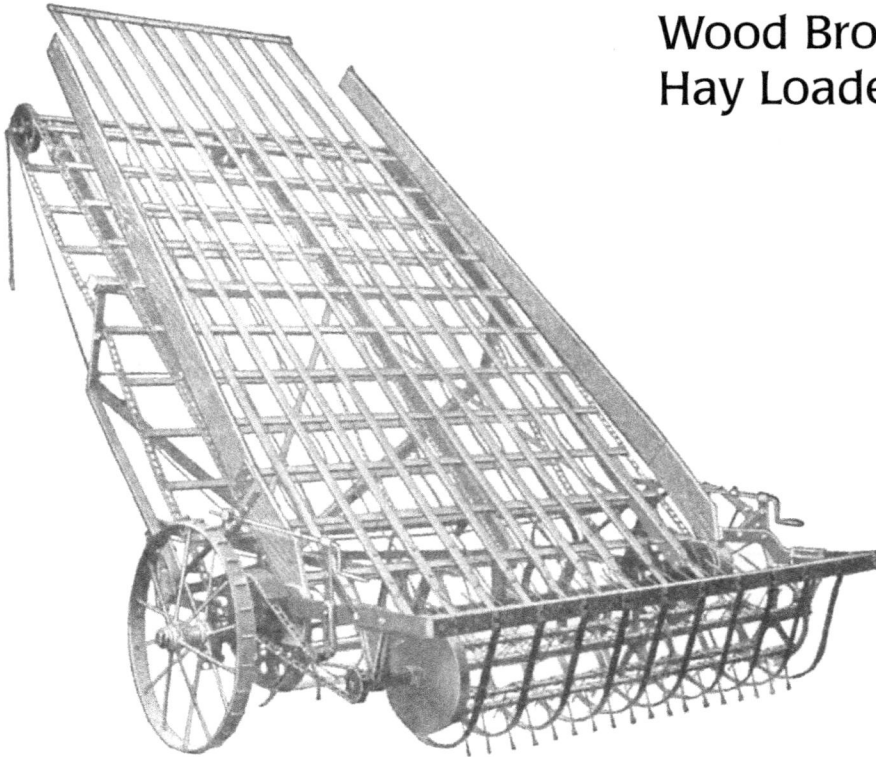

Unusual Rock Island hay loader at work.

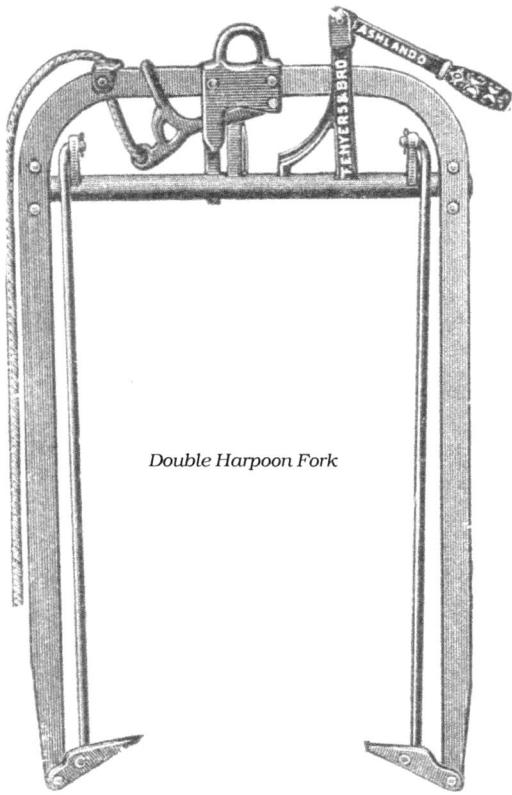

Double Harpoon Fork

Harpoon fork

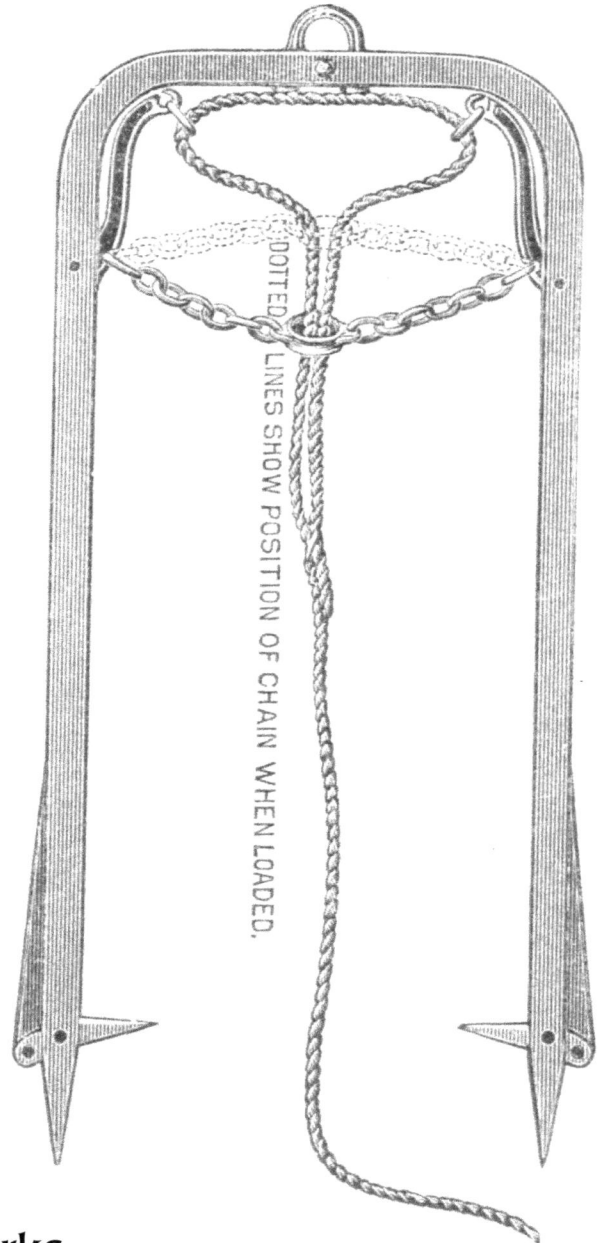

DOTTED LINES SHOW POSITION OF CHAIN WHEN LOADED.

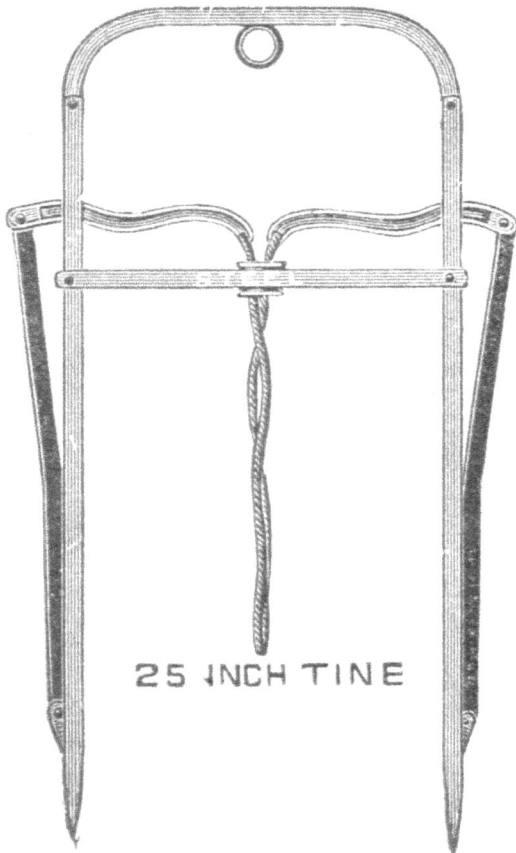

25 INCH TINE

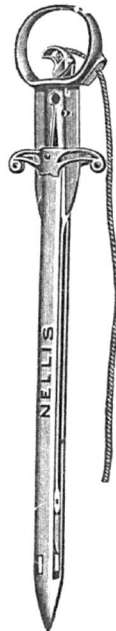

Hay Forks

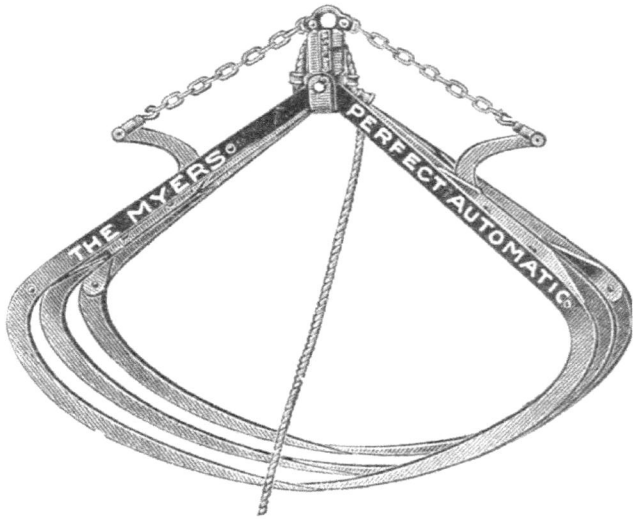

Grapple Fork

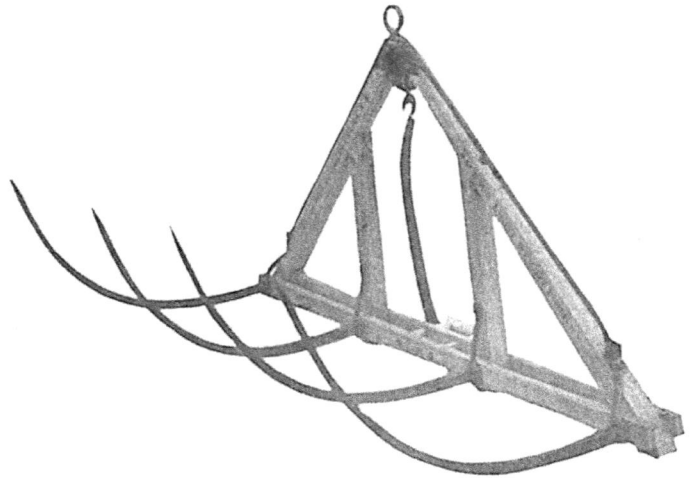

Jackson Derrick Fork

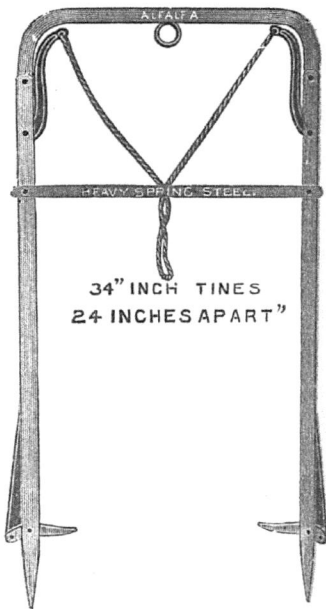

34" INCH TINES
24 INCHES APART"

FORK READY
TO ENTER HAY

31 INCH TINE

Double Harpoon

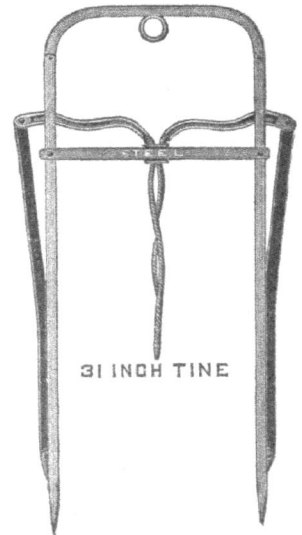

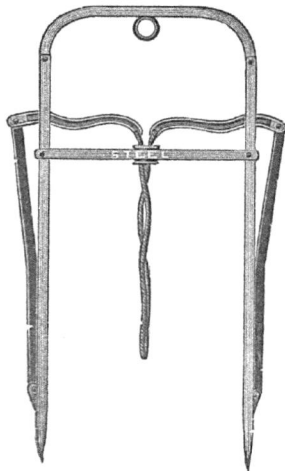

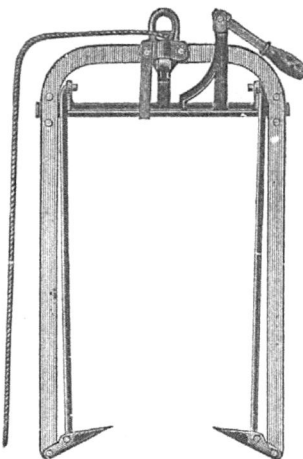

Jackson Fork latched and tripped

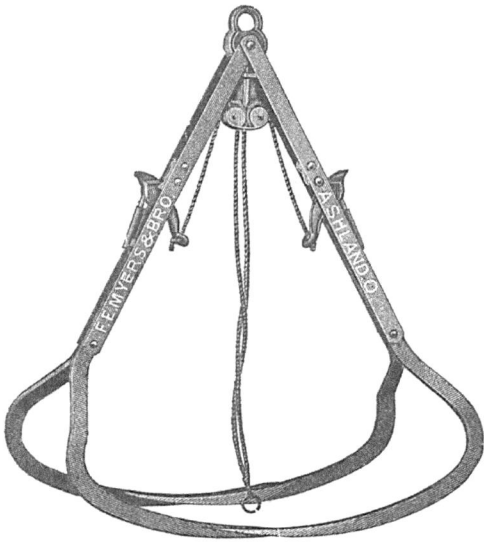

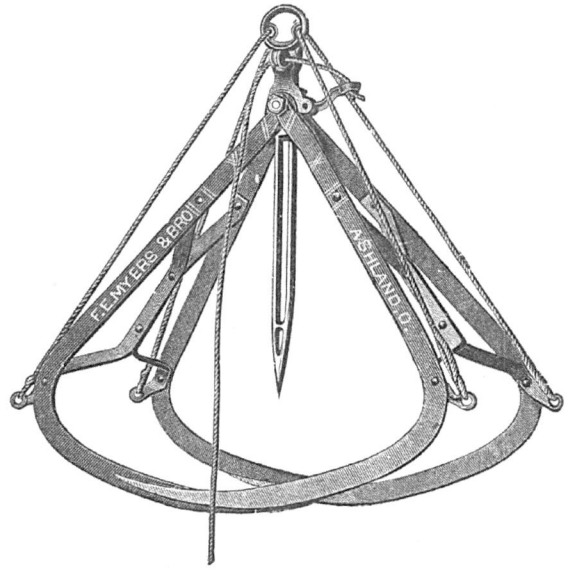

Grapple Forks

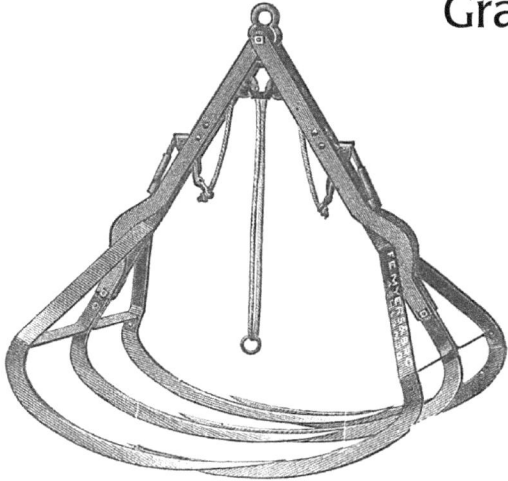

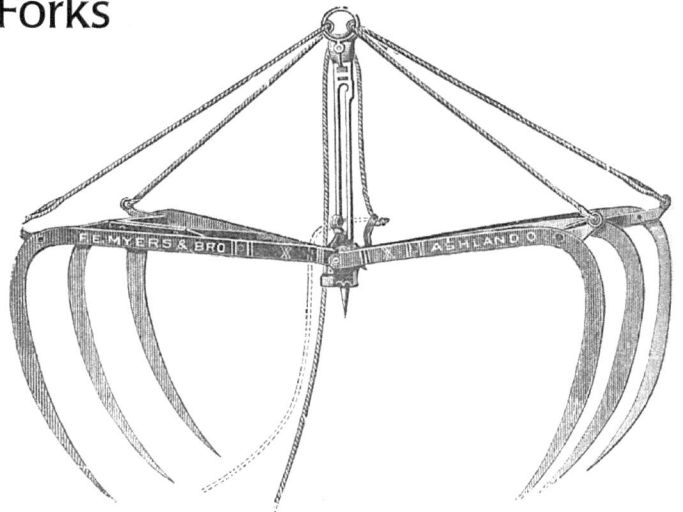

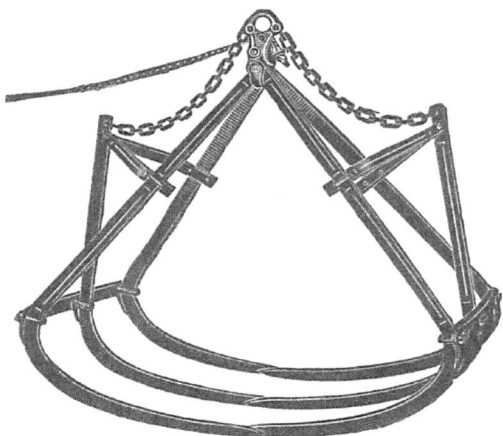

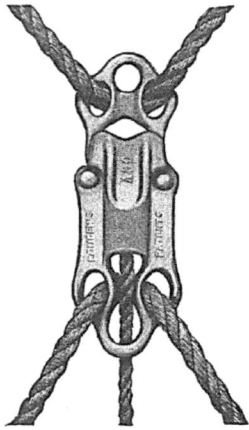

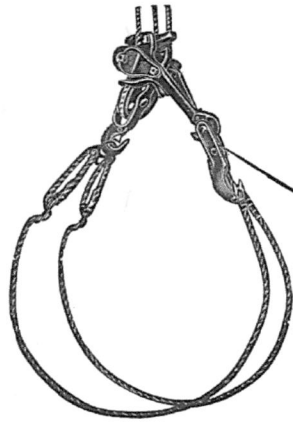

Sling Hardware

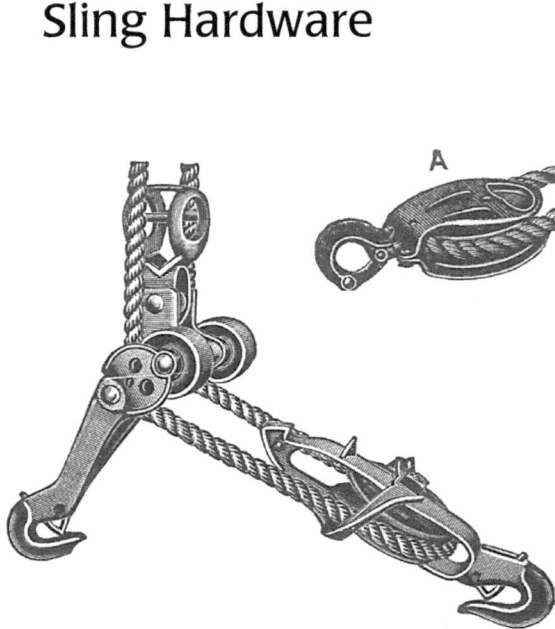

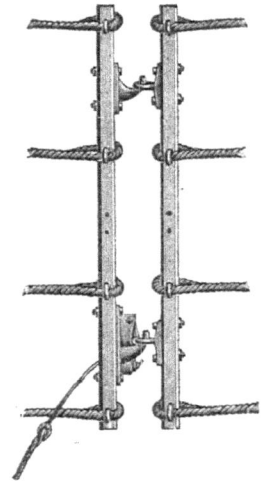

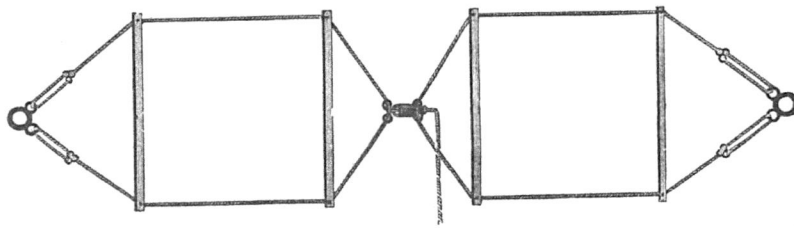

Sling Trolley

Hay Slings

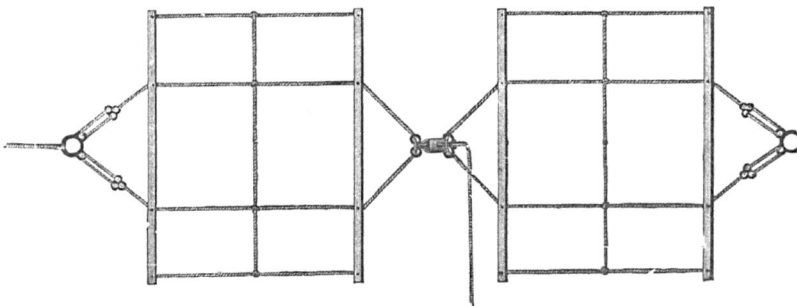

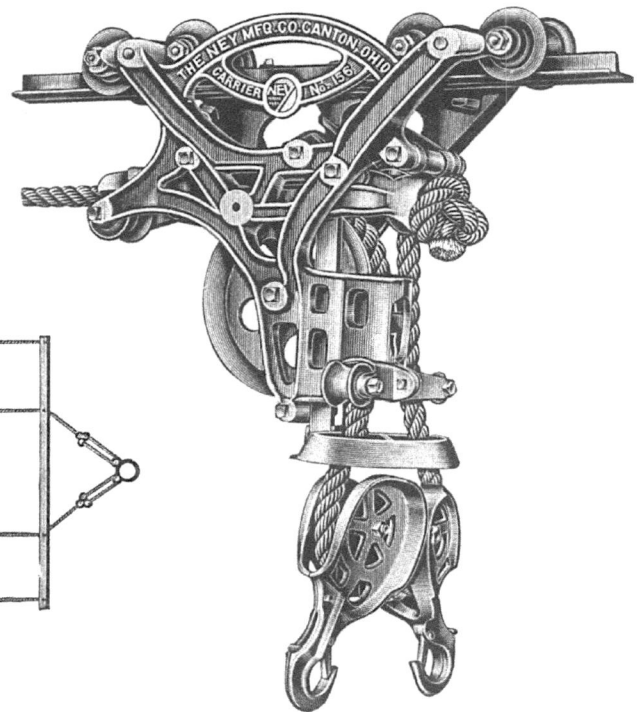

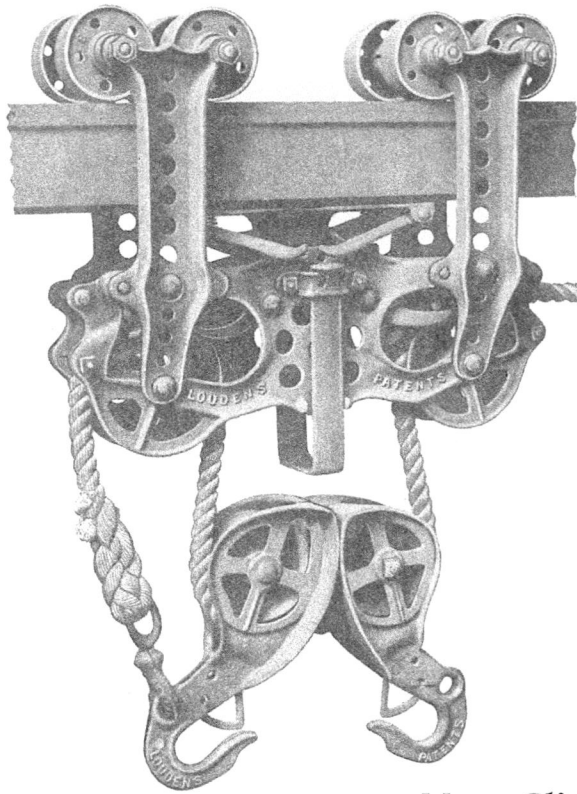

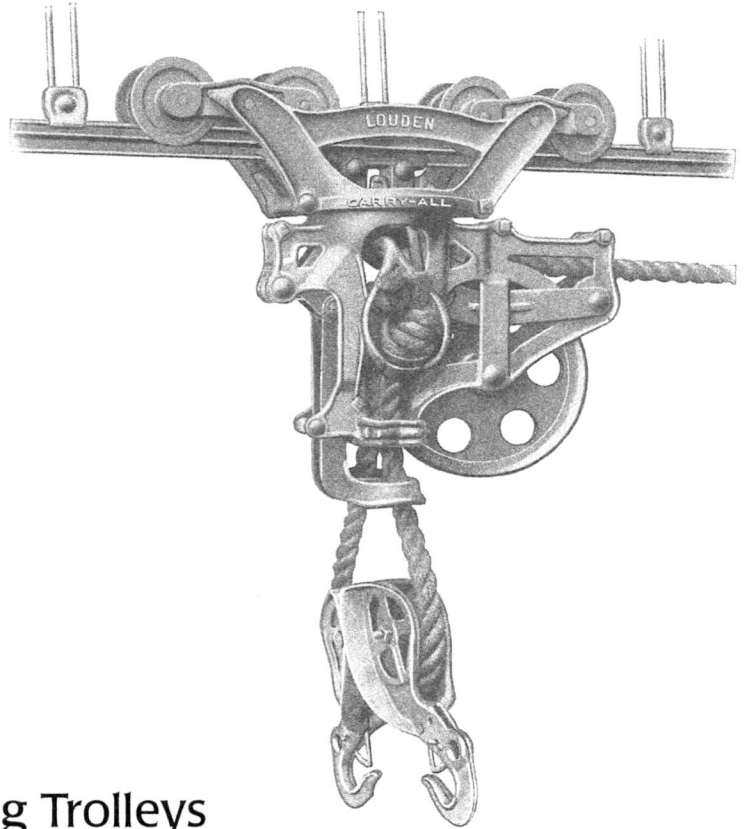

Hay Sling Trolleys

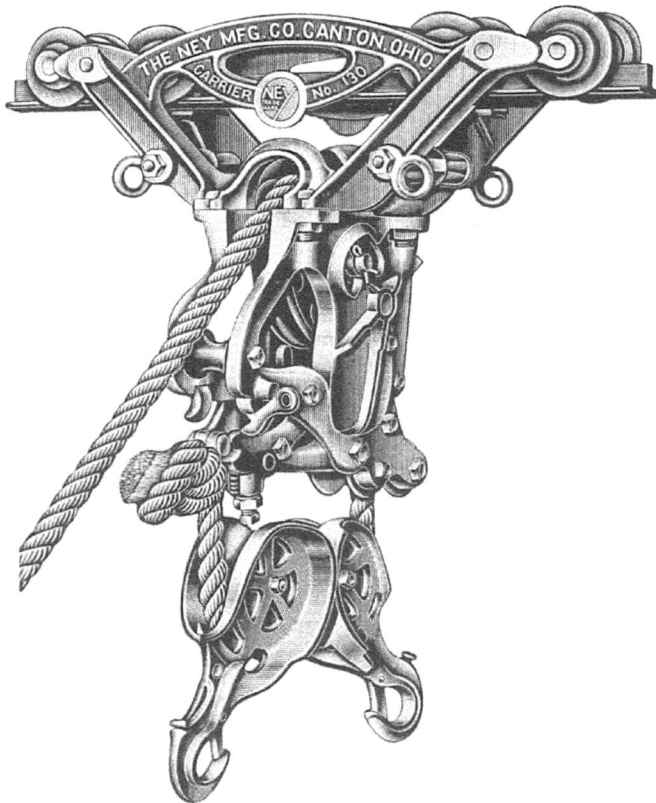

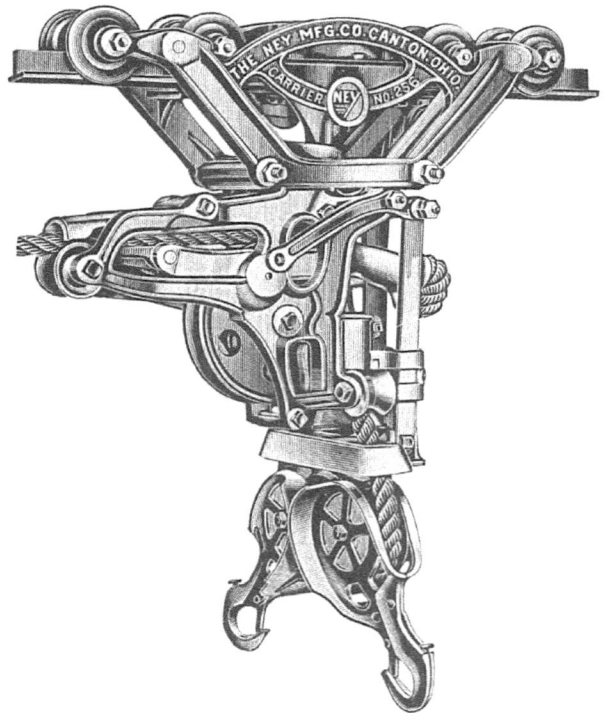

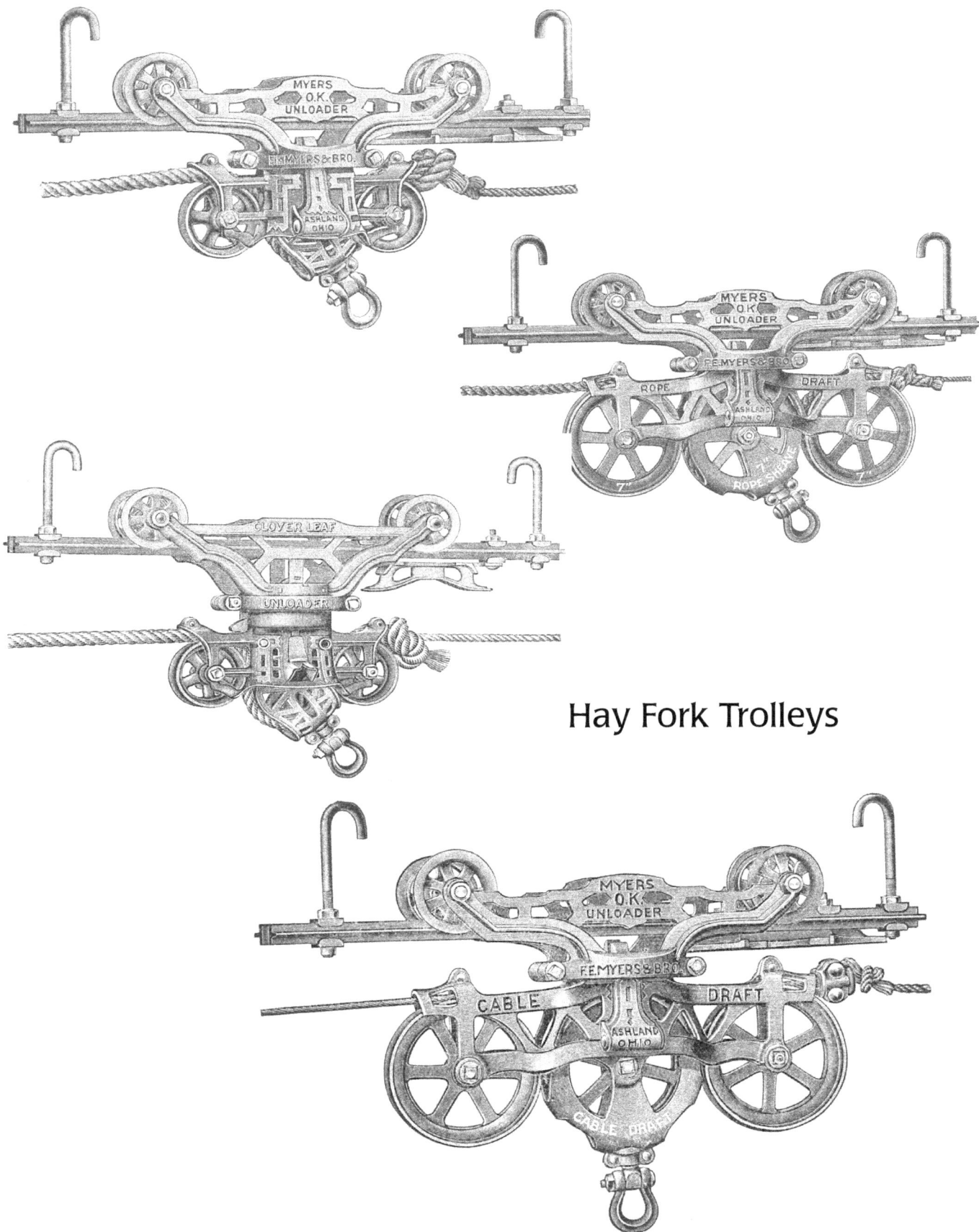

Hay Fork Trolleys

Barn Trolleys

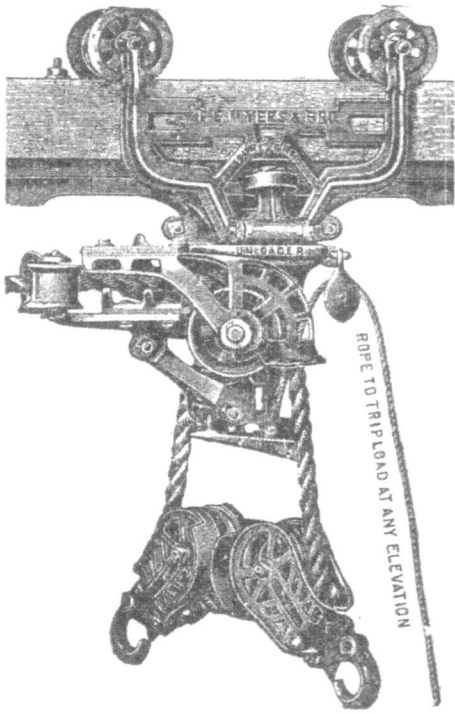

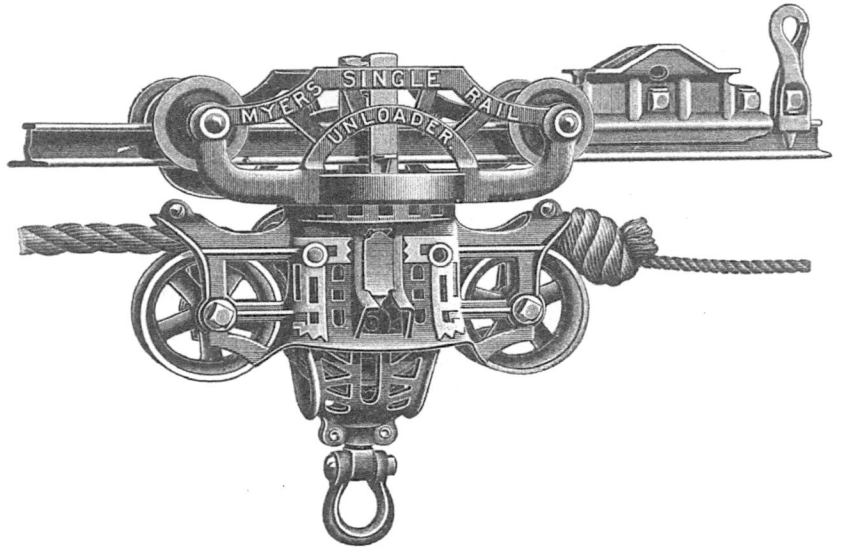

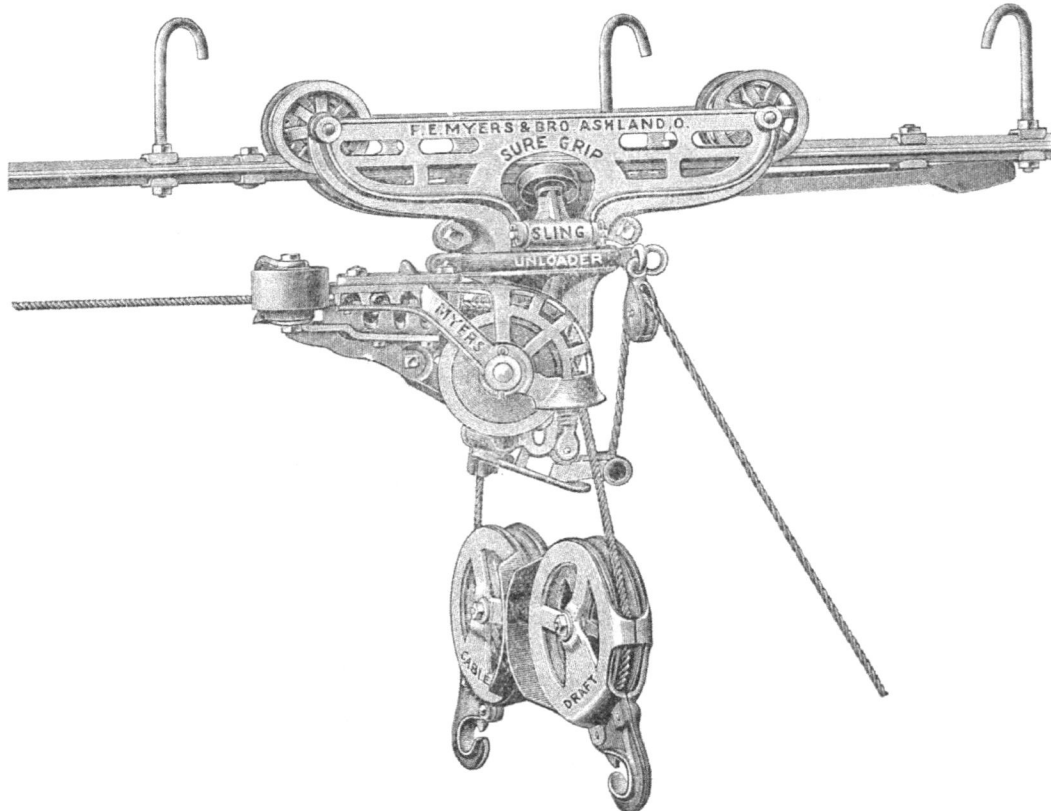

A Store Display for a Fork Trolley

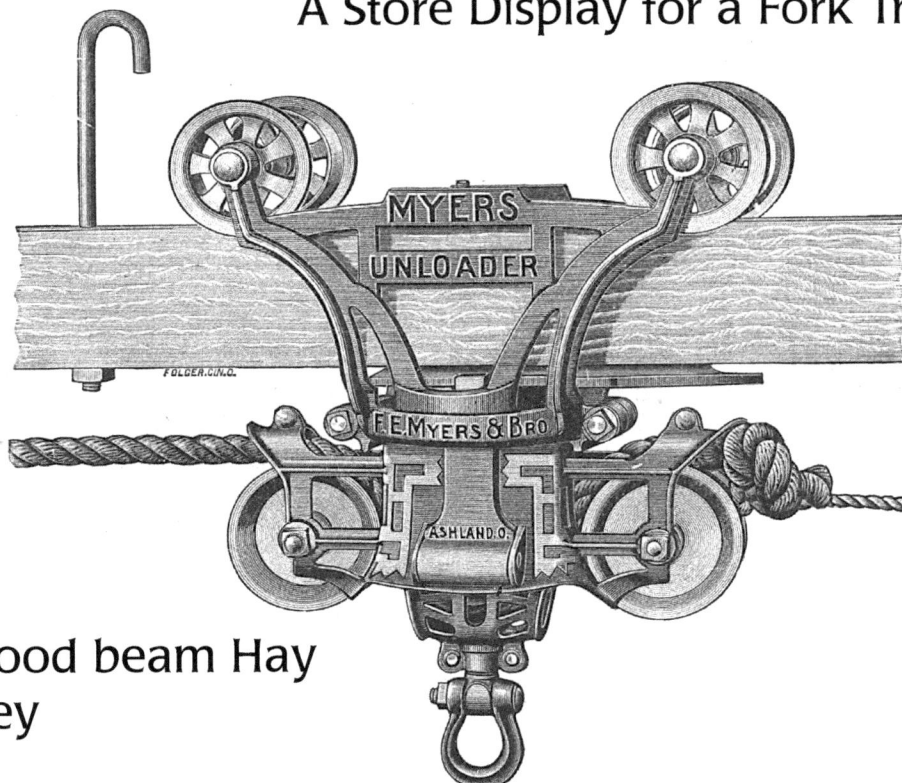

A Basic wood beam Hay Fork Trolley

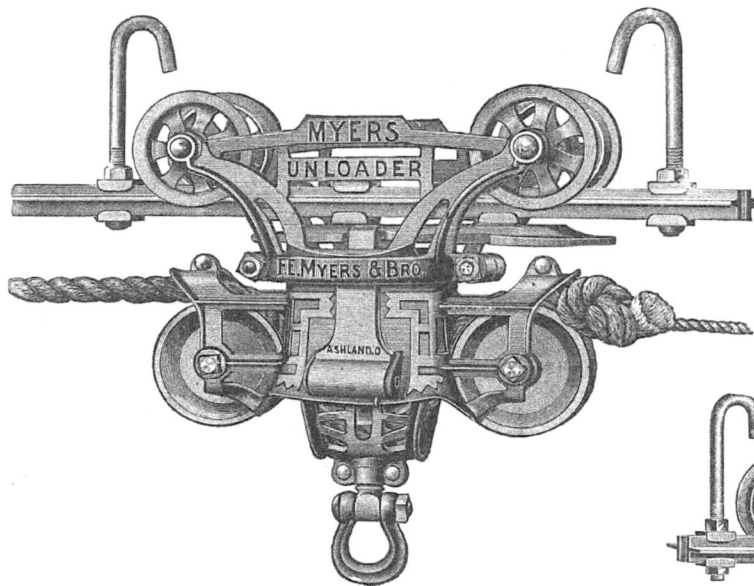

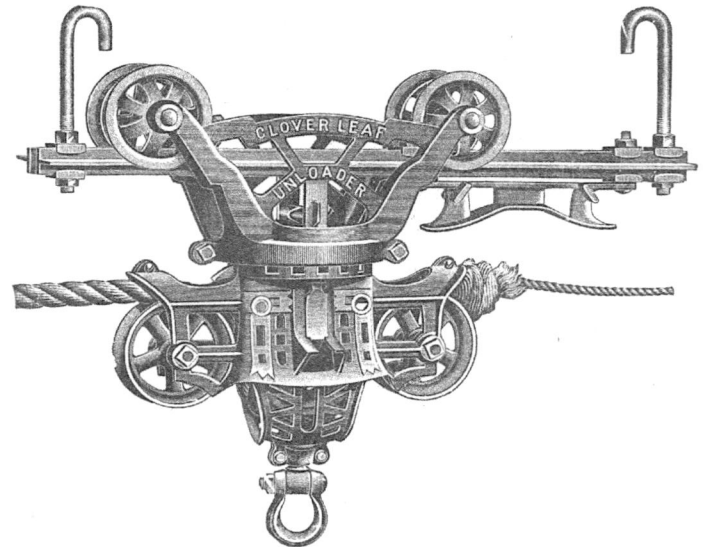

Hay Fork Trolleys

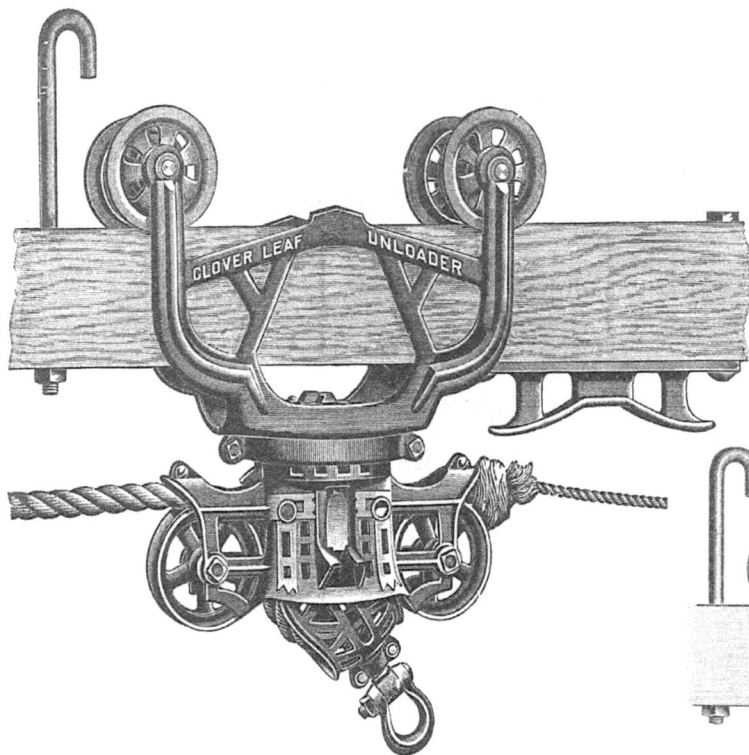

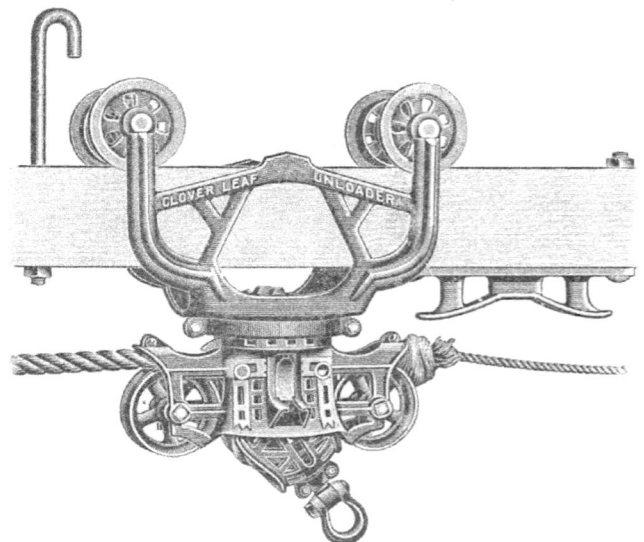

Miscellaneous Hay Rope Pulleys

Miscellaneous Hay Rope Pulleys

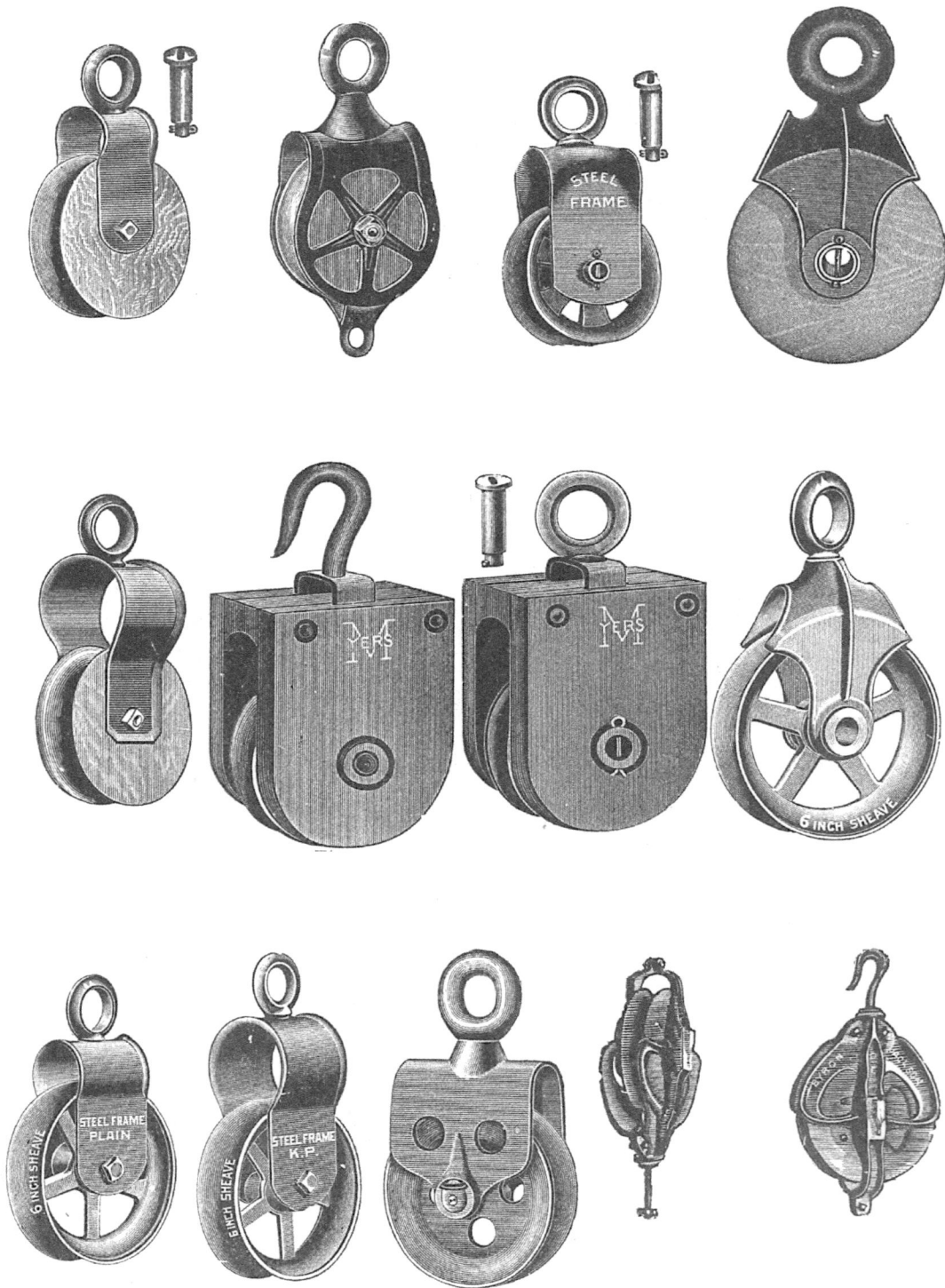

Miscellaneous Hay Rope Pulleys

Fork Pulley Sling Adapters

In Closing

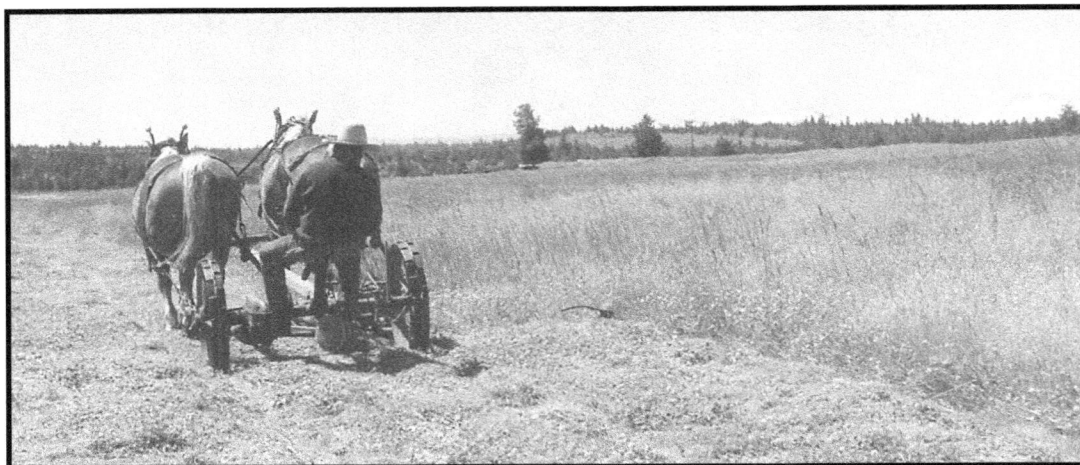

The author in his church.

This book has been, like each one before it, a bigger project than was anticipated. It's a big project if it keeps me from doing other things for anymore than two weeks. A good friend of mine and a brilliant writer, David Giampetroni, wondered at how it was I could sit down and spit out a book like this so quickly. I tried to explain to him that a book like this is not a job of authorship so much as it is a review of process and a sharing of diagrams. I tried to explain that - because this is a book about something (haying with horses) which I do with eyes wide open, and which I do with love - telling its story is close to singing in the shower. I tried to explain to him that spitting a book like this out (if you can call a month and a half's crash compilation after thirty years of research a spitting out) is easy because I am not a writer in any classical sense. I have zero respect or anxiety for the craft of writing. To me this writing business is like a lump of clay I might push around into some funny little creature shape just so I might get a smile or giggle from my baby daughter. As a painter I have a crippling and profoundly loving respect for the deliberate creative process and so also for those who, like David, truly are writers. It may seem strange to close this book with a request that you forgive this old cowboy/farmer/storyteller for making such a mockery of the book building process. But, I must send this lump to the printer in a couple of days and my eyes burned when I finished my last review. Although it is exciting to finally see all this information in one place, it is disheartening to discover, too late, that it is poorly organized and perhaps more than a little daunting for someone new to the subject. Yet, doubtless no writer would have wanted to do this book. So I did, we did, and I ask your consideration. I'm not apologizing for the book. It's a good one. It needed to be done. I'm glad we were able to get it in print.

Carl Leonhardy (in Scout's swing) & Lynn Miller enjoying a shade break from haying. It's about people working together.

Now that it's done we can move on.

A last word on this business of haying with horses.

If you should choose to try any aspect of the systems outlined in this book because they are attractive in an aesthetic or romantic sense - good for you. Be prepared to discover a world of efficiency and practicality that will doubtless surprise you.

If you should try because you think it might be a better way to work - good for you. Be prepared to discover a beauty and a romance to the work which will wrap you up in goosebumps.

At the front of this volume is a dedication to my children. Looking at the photo below, of my baby daughter Scout Gabrielle and I, it's hard not to think about my long standing conscious wish that this way of farming, from which I have derived so much pleasure and satisfaction, could be passed along. For that to happen work needs to be done to preserve its key ingredients in a form that can be easily drawn from. We trust and hope that these books, in some small yet important way, fill the bill. So, for the kids, the planet, the horses and mules and the landscape we wish you safety and tremendous good fortune because it is your working example that guarantees the torch be passed.

L.R. Miller

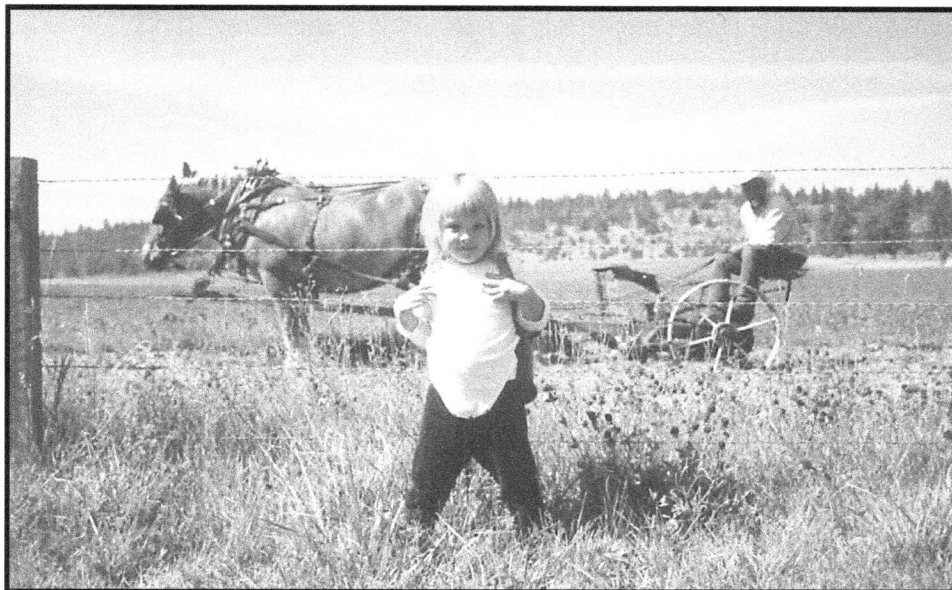

Scout Gabrielle Miller and her dad with Cali and Lana on the mower.

Bibliography

Avery, B. F. & Sons, General Catalogue No. 104, Manufacturers of Plows, Tillage Implements, Harvesting Machinery, Louisville, KY

Bressman, E. N., Growing Alfalfa in Montana, University of Montana, Agricultural Experiment Station, Bozeman, MT, August 1923

Burke, J. D., The Efficient Use of Horses on New York Farms, Cornell Extension Bulletin No. 530, Published by the New York State College of Agriculture at Cornell University, Ithaca, New York, September 1942

Country Guide, The, Farm Workshop Guide, published by the Extension Department, Winnipeg, Manitoba, Canada

Courtney, Wilshire S., The Farmer's and Mechanic's Manual, Revised and enlarged by George E. Waring, Jr., New York, E. B. Treat & Co. 1869

Crozier, William and Peter Henderson, How the Farm Pays, The Experiences of Forty Years of Successful Farming and Gardening, New York: Peter Henderson & Co., 1902

Dain Manufacturing Co., Directions for Setting Up and Operating John Deere Side Delivery Rakes Nos. 553, 554, 554P, 563, 564, 573, 574, 574P 583, 584, 593, 594, 594B, 594M, 594P, and 594PM, Ottumwa, IA

Dain Manufacturing Co., Repair Catalog No. 101-E Effective May 11, 1931 for John Deere-Dain Hay Machinery, Dain Manufacturing Co., Ottumwa, IA, 1931

Dain Manufacturing Co., Repair Catalog No. 104-E for John Deere-Dain Hay Machinery, Dain Manufacturing Co., Ottumwa, IA, 1944

Davis, Kary Cadmus, Ph. D. (Cornell University), Productive Farming, J. B. Lippincott Company, Philadelphia, PA

Engbretson, Albert E. and George R. Hyslop, Forage Crops for Oregon Coast Counties, Oregon Agricultural College, Corvallis, OR, 1924

Farm Implement News, The, A Monthly Illustrated Newspaper Devoted to the Manufacture, Sale and Use of Agricultural Implements and Their Kindred Interests, Vol. X, No. 1, Chicago, IL, January, 1889

Harrison, E. S., Savage, E. S., Hartwig, H. B., The Production of Quality Hay, Cornell Extension Bulletin No. 461, The New York State College of Agriculture at Cornell University, Ithaca, New York, May 1941

Herschel Manufacturing Co., R., Catalog No. 79, Peoria, IL

International Correspondence Schools, Scranton, PA, Hay and Pasture Crops, International Textbook Company, London, 1911

International Harvester Company of New Jersey, Directions for Setting Up and Operating the Deering New Ideal and New Ideal Giant Plain Lift Mowers (Type B), International Print Shop, Chicago, IL

International Harvester Company, No. 1-DM Repairs Catalog for Deering Ideal Mowers Built 1893 to 1911, Chicago, IL

International Harvester Company of America, Deering Division, The Golden Era, Deering, Chicago, IL, 1903

John Deere Harvester Works of Deere & Co., Directions for Setting Up and Operating John Deere No. 4 Big Enclosed Gear Mower (High Lift), John Deere, East Moline, IL

John Deere, Operator's Manual for Side Delivery Rakes No. 594, John Deere, East Moline, IL

John Deere, Pasture - A Paying Crop, John Deere, Moline, IL

John Deere Sulky Rake, A General-Purpose Rake for all hay crops, John Deere, East Moline, IL

John Deere, Supplementary Notes for Farm Machinery Charts in the form of Questions and Answers for use by Agricultural Instructors, John Deere, Moline, IL

John Deere Sweep Rakes and Stackers, Handle Hay at a Big Saving, John Deere, Moline, IL

Johnson, Paul C., Farm Inventions in the Making of America, Wallace-Homestead Book Co., Des Moines, IL, 1976

Jones, Mack M., M.S., Shopwork on the Farm, McGraw-Hill Book Co., Inc., New York, 1945

Lipscomb & Company, H. G., Wholesale Hardware and Associate Lines, Nashville, TN, 1913

Massey-Harris, No. 8 Hay Loader, Massey-Harris Co., Ltd., Toronto, Canada, 1950

Massey-Harris, Setting Up and Operating Instructions Massey-Harris No. 11 Side Rake, The Massey-Harris Co., Inc., Racine, WI

McClure, H. B., Hay Stackers, Contribution from the Office of Farm Management, Washington, D.C., 1919

McCormick-Deering Blue Ribbon Service Training Course, Hay Machines, International Harvester Co., Chicago, IL

McCormick-Deering, Instructions for Setting Up and Operating the McCormick-Deering No. 7 Vertical Lift Two-Horse Mowers, International Harvester Company of Canada, Hamilton, Ontario, Canada

McCormick-Deering, Owner's Manual Setting Up Instructions, Parts List, McCormick-Deering Self-Dump Hay Rake (Type M), International Harvester Co., Chicago, IL

McCormick-Deering, Repairs Catalog for McCormick-Deering Hay Machines No. 24-HM, International Harvester Co., Chicago, IL

McCormick-Deering, Repairs Catalog for McCormick-Deering Hay Machines No. 26-HM, International Harvester Co., Chicago, IL

Mitchell, Lewis & Staver Co., Catalog No. 16, Agricultural Implements, Wagons, Pumps, Engines, Pipe, Wind Mills, etc., etc., Portland, OR, 1916

Moline Flying Dutchman Farm Implements, General Catalog No. 60, Tillage, Seeding and Haying Machinery, Moline Implement Co., Moline, IL

Moline Plows and Other Flying Dutchman Farm Tools, Moline Plow Co., Moline, IL

Montgomery Ward & Co., 1897 Special Catalogue B, Agricultural Implements and General Farm Machinery

Morrison, Frank B., Feeds and Feeding, A Handbook for the Student and Stockman, The Morrison Publishing Co., Ithaca, NY, 1948

National Hay Assoc., The, Grades of Hay and Straw, Inspection and Weighing Rules, 1917

New Idea, Inc., Instructions and Repair List No. L-136 for Setting Up and Operating The New Idea Easy-Way Hay Loader, New Idea, Inc., Coldwater, OH

New Idea, Inc., Instructions and Repair List No. L-137 for Setting Up and Operating The New Idea Easy-Way Hay Loader, New Idea, Inc., Coldwater, OH

New Idea, Inc., Instruction and Repair List No. R-137 for Setting Up and Operating the New Idea Side Rake and Tedder, New Idea, Inc., Coldwater, OH

New Idea, Inc., Repair List No. L-43 for Clean Sweep Hay Loader, New Idea, Inc., Coldwater, OH, 1942

New Idea Farm Equipment Co., Repair Parts List No. L-47 for New Idea Easyway Hay Loader, New Idea, Coldwater, OH, 1947

New Idea Farm Equipment Co., Repair List No. M-32 for New Idea Mowers, New Idea, Coldwater, OH, 1949

New Idea, Inc., Repair List No. HP28 for Sandwich Hay Presses, New Idea, Inc., Coldwater, OH 1942

New Idea Farm Equipment Co., Repair Parts List No. L-48 for Sandwich-New Idea Easyway No. 2 and No. 3 Hay Loader, New Idea, Coldwater, OH, 1947

New Idea, Inc., Repair List No. R-42 for Sandwich Steel Frame Two and Three Bar Side Delivery Hay Rake, New Idea, Inc., Coldwater, OH, 1942

Ney Manufacturing Co., The, General Catalog No. 61 Haying Tools, The Ney Manufacturing Co., Canton, OH

Oregon Agricultural College, Experiment Station Bulletin 203, Forage Crops for Oregon Coast Counties, Oregon Agricultural College, Corvallis, OR 1924

Oregon State Agricultural College, Agricultural Experiment Station Bulletin 241, Cost and Efficiency in Producing Alfalfa Hay in Oregon, Oregon State University, Corvallis, OR 1928

Ramsower, Equipment for the Farmstead

Rock Island Implements, Catalog No. 41, The Rock Island Plow Company, Manufacturers of Farm Machinery, Rock Island Plow Co., Rock Island, IL

Rock Island Implements, Making Farm Life Easier, Rock Island Plow Co., Rock Island, IL

Roehl, L. M., Hay Racks, Cornell Extension Bulletin No. 244, New York State College of Agriculture at Cornell University, Ithaca, NY, 1932

Sampson, H. O., B.Sc., B.S.A., Effective Farming, The MacMillan Co., New York, 1918

Serviss, George H. And H. B. Hartwig, Growing Hay Crops and Making Quality Hay, Cornell Extension Bulletin No. 568, New York State College of Agriculture at Cornell University, Ithaca, New York, 1943

Shearer, Herbert A., Farm Mechanics, Machinery and Its Use to Save Hand Labor on the Farm, Frederick J. Drake & Co., Chicago, IL, 1918

Smith, Harris Pearson, A.E., Farm Machinery and Equipment, McGraw-Hill Book Co., Inc., New York, 1937

Stippler, H. H., M. T. Buchanan, and A. G. Law, Methods of Harvesting Hay Fields and Pastures in Northwestern Washington (North Coast Area), 1945, Bulletin No. 502, The State College of Washington Institute of Agricultural Sciences, Agricultural Experiment Stations, Pullman, WA, 1948

Taylor, Dr. W. E., Soil Culture and Modern Farm Methods, Deere & Webber Company, Minneapolis, MN

Thomas, John J., Farm Implements and Farm Machinery and the Principles of their Construction and Use, Orange Judd Co., New York, 1883

United States Department of Agriculture, Leaflet No. 72, Measuring Hay in Stacks, Washington, D.C., 1931

United States Department of Agriculture, Leaflet No. 119, White Clover, Washington, D.C., 1936

United States Department of Agriculture Leaflet No. 160, Crimson Clover, Washington, D.C., 1938

United States Department of Agriculture, Farmers' Bulletin No. 797, Sweet Clover, Growing the Crop, Washington, D.C., 1917

United States Department of Agriculture, Farmers' Bulletin 836, Sweet Clover, Harvesting and Thrashing the Seed Crop, Washington, D. C., 1917

United States Department of Agriculture, Farmers' Bulletin 947, Care and Repair of Farm Implements, No. 4 Mowers, Reapers, and Binders, Washington, D.C., 1918

United States Department of Agriculture, Farmers' Bulletin 956, Curing Hay on Trucks, Washington, D.C., 1918

United States Department of Agriculture, Farmers' Bulletin 977, Hay Caps, Washington, D.C., 1918

United States Department of Agriculture, Farmers' Bulletin 987, Labor Saving Practices in haymaking, Washington, D.C., 1918

United States Department of Agriculture, Farmers' Bulletin 990, Timothy, Washington, D.C., 1918

United States Department of Agriculture, Farmers' Bulletin No. 1148, Cowpeas, Culture and Varieties, Washington, D.C., 1920, rev. 1924

United States Department of Agriculture, Farmers' Bulletin No. 1151, Alsike Clover, Washington, D.C., 1920, rev. 1924

United States Department of Agriculture, Farmers' Bulletin No. 1153, Cowpeas Utilization, Washington, D.C., 1920

United States Department of Agriculture, Farmers' Bulletin No. 1254, Important Cultivated Grasses, Washington, D.C., 1922, rev. 1931, rev. 1934

United States Department of Agriculture Farmers' Bulletin No. 1339, Red Clover Culture, Washington, D.C., 1924

United States Department of Agriculture Farmers' Bulletin No. 1341, Mule Production, Washington, D.C., 1923, rev. 1938, rev. 1948

United States Department of Agriculture Farmers' Bulletin No. 1525, Effective Haying Equipment and Practices, Washington, D.C., 1927, rev. 1931

United States Department of Agriculture Farmers' Bulletin No. 1539, High-Grade Alfalfa Hay, Methods of Producing, Baling, and Loading for Market, Washington, D.C., 1929, rev. 1952

United States Department of Agriculture Farmers' Bulletin No. 1573, Legume Hays for Milk Production, Washington, D.C., 1928

United States Department of Agriculture Farmers' Bulletin No. 1574, Preparing Johnson Hay for Market in the Black Prairie Belt of Alabama and Mississippi, Washington, D.C., 1928

United States Department of Agriculture Farmers' Bulletin No. 1583, Spring-Sown Red Oats, Washington, D.C., 1929

United States Department of Agriculture Farmers' Bulletin No. 1597, The Production of John Grass for Hay and Pasturage, Washington, D.C., 1929

United States Department of Agriculture Farmers' Bulletin No. 1605, Soybean Hay and Seed Production, Washington, D.C., 1929

United States Department of Agriculture Farmers' Bulletin No. 1722, Growing Alfalfa, Washington, D.C., 1954

United States Department of Agriculture Farmers' Bulletin No. 1754, Care and Repair of Mowers and Binders, Washington, D.C., 1936, rev. 1942

United States Department of Agriculture Farmers' Bulletin No. 1770, High-Grade Timothy and Clover Hay, Methods of Producing, Baling and Loading for Market, Washington, D.C., 1937, rev. 1952

United States Department of Agriculture Farmers' Bulletin No. 1784, Nitrogen-Fixing Bacteria and Legumes, Washington, D.C., 1937

United States Department of Agriculture Farmers' Bulletin No. 1839, The Uses of Alfalfa, Washington, D.C., 1940

United States Department of Agriculture Farmers' Bulletin No. 1929, Persian Clover, Washington, D.C., 1943

United States Department of Agriculture Farmers' Bulletin No. 2024, Soybean Production for Hay and Beans, Washington, D.C., 1950

United States Department of Agriculture Farmers' Bulletin No. 2113, Annual Lespedezas, Culture and Use, Washington, D.C., 1958, rev. 1961

United States Department of Agriculture Farmers' Bulletin No. 2191, Trefoil Production for Pasture and Hay, Washington, D.C., 1963

Wallace, Corcoran & Co., Agricultural Implements Catalog No. 6, Portland, OR

Wood, Walter A., Enclosed Gear Mower for Two Horses

Wood, Walter A., Mowing & Reaping Machine Co., Hoosick Falls, NY

Wood, Walter A., World Renowned Harvesting Machines, Hoosick Falls, NY

Index   Haying with Horses

Photo Credits

Kristi Gilman-Miller   pages 14, 16, 17, 18, 19, 20, 21, 24, 40, 41, 51, 52, 53, 56, 57, 60, 61, 66, 69, 84, 85, 86, 87, 88, 144, 145, 149, 151, 167, 177, 205, 211, 212, 221, 222, 223, 229, 237, 239, 240, 241, 263, 264, 265, 271, 280, 281, 282, 296, 297, 306, 319, 321, 323

Doug Hammill pages 14, 23, 76, 77, 147

Laurie Hammill pages 76, 90, 110, 151, 281

Judith Hoffman pages 58, 63, 74, 75, 88, 89, 96, 97, 98, 150, 151

Pete Lorenzo pages 213, 214, 283, 284

Lynn Miller   pages 54, 61, 62, 63, 64, 77, 94, 95, 96, 149, 150, 219, 220, 223, 224

Glenn Weber pages 84, 85, 92, 94, 140, 152, 293, 305

For Parts & More Information.....

For MOWER PARTS
B.W. Macknair & Son
3055 US Highway 522 North
Lewistown PA   17044
(717) 543-5136

BEAUTIFUL CALENDARS & BOOKS
Mischka Press
P.O. Box 2067
Cedar Rapids, IA 52406-2067
(319) 362- 3027
www.mischka.com

Assorted Implements
I & J Manufacturing
5302 Amish Rd.
Gap, PA 17527
(717) 442-9451

Reel Mowers
Mascot Sharpening
434-B Newport Rd
Ronks, PA 17572
(717) 656-6486 ext. 1

Assorted Implements
Mullet's Machinery
6870 S State Route 5
Topeka, IN 46571
(260) 593-2960

NEW FORECARTS
Pioneer Equipment Inc.
16875 Jericho Rd
Dalton OH 44618
(330) 857-6340

Assorted Implements
Shipshe Farm Supply
2380 N 925 W
Shipsewana IN 46565
(260) 768-7271

Assorted implements
White Horse Machine
5566 Old Philadelphia Pike
Gap PA   17527
(717) 768-8313

DEMONSTRATIONS
Horse Progress Days
Around the 4th of July each year
in PA or OH or IN or IL or MI
1-800-465-4156
www.horseprogressdays.com

Small Farmer's Journal

Farming as a way of life
versus
Farming as industrial process

Dear Folks at SFJ,
 You folks put out <u>a great publication, a real gem</u> in this world of blitz media and chemical/corporate farming. We look forward eagerly to its arrival even though we are not horse farmers. Instead we grow about 5 acres of vegetables and 5 of cover crops on our 80-acre farm using old tractors. We haven't quite made it into the 1950's yet with our machinery. At age 38, when we bought this place I had decided that I didn't have time to learn about horse farming. Then a friend told me about SFJ. Oh boy! Just maybe I still have a chance! Or at least our three children do.
 ... thanks again for a great magazine.
  Sincerely, Ted Fisher
  Stockholm, WI

Small Farmer's Journal
featuring Practical Horsefarming
Published Quarterly

please contaact us for pricing
Small Farmer's Journal
P.O. Box 1627
Sisters, OR 97759

Phone 800-876-2893
email: agrarian@smallfarmersjournal.com
w w w . s m a l l f a r m e r s j o u r n a l . c o m

Also Available from the Author & Small Farmer's Journal

THE WORKHORSE HANDBOOK 2ND EDITION.

This book has become a classic and the standard reference. This popular, highly regarded text is filled with current information and hundreds of photographs and drawings. It is a sensitive and intelligent examination of the craft of the teamster.

From care and feeding through hitching and driving; every aspect is covered. Find out for yourself why this book is considered by thousands of people to be THE volume on working horses in harness. $45 soft cover, plus shipping.

TRAINING WORKHORSES / TRAINING TEAMSTERS,

is a text combining two books in one and including 482 photographs and hundreds of drawings on 352 pages. This text covers the subjects of: training horses to work in harness on the farm, in the woods and on the road, correcting behavior problems with work horses and training people to drive and work horses.

WHO MIGHT BENEFIT FROM THIS BOOK?

*Those who want to learn how to train their own work horses. *People who want to correct problems with their horses. *Anyone wishing to learn how to drive work horses. *People needing to improve their skills with handling work horses. *Trainers and teachers. *Folks who think they might want to farm or log with horses. $45 soft cover, plus shipping.

HORSEDRAWN PLOWS AND PLOWING

368 pages with 2,000 drawings and photos covering how to plow with horses using older equipment and new implements. Here you will find simple diagrams explaining tricky adjustments for both riding and walking plows. Detailed engineer's drawings of John Deere, Oliver, McCormick Deering, Parlin Orendorff, Avery, and many other older manufacturers will be immensely helpful to folks restoring equipment. Also includes closeup photos and information on new makes of animal-drawn plows including Pioneer and White Horse. $45 soft cover plus shipping.

HORSEDRAWN TILLAGE TOOLS

368 pages with 2,000 drawings and photos covering row cultivators, field cultivators, harrows, ridge busters, listers, and rollers. Illustrated information on shovel and point types. Detailed engineer's drawings of John Deere, Oliver, McCormick Deering, Parlin Orendorff, Avery, and many other older manufacturers will be immensely helpful to folks restoring equipment. $45 soft cover, plus shipping.

THE HORSEDRAWN MOWER BOOK

With hundreds of photos and drawings, the profusely illustrated text covers restoration, rebuilding, repair, and tuneup with a focus on the very popular McCormick Deering (International) No. 9. It also includes references to other makes and models as well as resource information for updating cutter bar assemblies to new materials and functions. Mr. Miller, along with being a long time horsefarmer, has restored mowers for 25 years and taught several workshops on the subject. $45 soft cover, plus shipping.

Small Farmer's Journal
P.O. Box 1627, Sisters, OR 97759

Phone 800-876-2893

email: agrarian@smallfarmersjournal.com

w w w . s m a l l f a r m e r s j o u r n a l . c o m